# A HISTORY OF SOVIET RUSSIA

# A HISTORY OF SOVIET RUSSIA

by E. H. Carr

*in fourteen volumes*

*with R. W. Davies

# SOCIALISM
# IN ONE COUNTRY
# 1924-1926

BY

E. H. CARR

*Fellow of Trinity College, Cambridge*

VOLUME ONE

*First published 1958*
*Reprinted 1964, 1972, 1978*

*Published by*
THE MACMILLAN PRESS LTD
*London and Basingstoke*
*Associated companies in Delhi*
*Dublin Hong Kong Johannesburg Lagos*
*Melbourne New York Singapore Tokyo*

*Printed in Hong Kong by*
CHINA TRANSLATION AND PRINTING SERVICES

**British Library Cataloguing in Publication Data**

Carr, Edward Hallett
  Socialism in one country, 1924–1926
  Vol. 1. — (Carr, Edward Hallett. History of Soviet
  Russia; 5)
  1.  Russia — Social conditions — 1917–
  I.  Title
  309.1′47′0842                    HN523

  ISBN 0-333-03442-2
  ISBN 0-333-24216-5   Boxed set

b 5917222

# PREFACE

THE present volume, the first of three under the title *Socialism in One Country*, *1924–1926*, brings me to the heart of my subject. As I said in the preface to the first volume of *The Bolshevik Revolution*, *1917–1923*, my ambition was "to write the history, not of the revolution . . . but of the political, social and economic order which emerged from it". The volumes hitherto published have been, in a certain sense, preliminary to this main purpose. While history knows no hard-and-fast frontiers between periods, it is fair to say that the new order resulting from the revolution of 1917 began to take firm shape only in the middle nineteen-twenties. The years 1924–1926 were a critical turning-point, and gave to the revolutionary régime, for good and for evil, its decisive direction.

By way of introduction to this central section, four chapters have been grouped together under the general title "The Background". In the first, I have attempted to define the relation of the revolution to Russian history, which first became clearly apparent in this period (part of this chapter appeared in the volume of *Essays Presented to Sir Lewis Namier* in 1956) ; in the second, to illustrate the moral and intellectual climate of the period by drawing on peripheral fields neglected in the earlier volumes ; in the third, to investigate the obscure and crucial issue of the motive forces of the new society ; in the fourth, to portray the personal characteristics of some of the principal actors and to indicate the place which they occupy in the story. The remainder of the volume is devoted to the economic history of the period from the spring of 1924 to the spring of 1926. In the second volume, the sixth of the whole series, I shall describe the party struggle leading to the break-up of the triumvirate and to the first defeat of Zinoviev, and the political and constitutional developments of the period. The following volume will deal with external relations.

As always, the most difficult problem of presentation has been that of arrangement. Precedence has been given to the narrative of economic developments ; for, though the rivalry between party leaders was the most conspicuous, and superficially the most dramatic, feature of these years, the forms which it took were dependent on basic economic issues. This arrangement, though necessary, has the disadvantage that I have been obliged to touch in this volume on certain aspects of the party struggle and of relations between the party leaders, the main treatment of which is reserved for the next volume. Even within the economic chapters some overlapping could not be avoided. In order to make the material manageable, different sectors of the economy had to be treated separately ; yet it was obvious that current problems and current decisions of policy, even if they ostensibly related to one sector, had repercussions on the other sectors. If the chapter on agriculture in this volume is by far the longest, this is no doubt in part a just tribute to the predominance of agriculture in the Soviet economy and in the preoccupations of Soviet politicians. But it is also due in part to the fact that, since this is the first of the economic chapters, issues that cut across all sectors of the economy arise here for the first time, and call for general treatment here rather than later. I must ask indulgence for some repetitions and for a perhaps tedious abundance of cross-references.

The progress of the work has produced, as generally happens, a growing sense of the complexity of the issues with which I am dealing. What I take to be the conventional view of Soviet history in the years after the revolution, i.e. that it was the work of determined men — enlightened pioneers on one view, hardened villains on another — who knew exactly what they wanted and where they were going, seems to me almost wholly misleading. The view commonly expressed that the Bolshevik leaders, or Stalin in particular, were inspired primarily by the desire to perpetuate their rule, is equally inadequate. No doubt every government seeks to retain its authority as long as possible. But the policies pursued were not by any means always those apparently most conducive to the undisturbed exercise of power by those in possession. The situation was so complex, and varied so much from place to place and from group to group of the population, that the task of unravelling the decisive factors in the process has been unusually baffling. This is a field where material is abundant, but often vague and sometimes contradictory, and where I have had few predecessors and few signposts to follow : few specialist studies have yet been written on particular points or aspects of the story. This must be my excuse for cumbering some parts of my narrative with, perhaps, an unnecessary profusion of detail. I have preferred to run the risk of including the superfluous rather than of omitting features which may prove significant when a more complete picture finally emerges.

A lengthy visit to the United States in the winter of 1956–1957 delayed the completion of this volume, but enabled me to obtain much additional material both for it and for its successor. The Russian Research Center at Harvard offered me generous hospitality and assistance; and it gives me particular pleasure to record my warm appreciation of the help and kindness which I received from Professor William Langer, the director, Mr. Marshall Shulman, the deputy director, as well as from other members of the Center. The Widener Library and the Law Library at Harvard are both rich in Soviet material of the period, and I was privileged to work on the Trotsky archives preserved in the Houghton Library: Professor George Fischer is at present preparing a catalogue of the Trotsky archives, which will make them more readily accessible and facilitate systematic reference to them. In addition to the Harvard libraries I visited the unrivalled collections of the New York Public Library and the Hoover Library at Stanford. I was also able to borrow from the Library of Congress and from Columbia University Library; the Library of Brandeis University (where I lectured during the first semester of my stay) gave me invaluable help in locating books for me and borrowing them on my behalf. I should like to express my warm thanks to the librarians of all these institutions and their staffs. I am particularly indebted to Professor Herbert Marcuse of Brandeis University for stimulating discussion of theoretical problems; to Mrs. Olga Gankin of the Hoover Library for much detailed help and advice in the pursuit of rare sources; to Dr. S. Heitman for the loan of his unpublished bibliography of Bukharin's writings; and to many other American friends who have given me in many different forms valued assistance and encouragement.

While, however, the final stages of research for this volume were carried out in the United States, the foundations were laid in this country, and it is here that most of the work has been done. Mr. J. C. W. Horne and the staff of the Reading Room of the British Museum have once more been unfailingly helpful; and the resources of the Museum have been supplemented by those of the libraries of the London School of Economics, of the School of Slavonic Studies and of the Department of Soviet Institutions in the University of Glasgow. Coming nearer home, Cambridge University Library has a most useful collection, recently supplemented by fresh acquisitions, of microfilms of Soviet documents and periodicals; and the Marshall Library of Economics possesses the copy presented to the late Lord Keynes in Moscow in September 1925 of the extremely rare first *Control Figures* of Gosplan — the volume described on p. 501 below. The Librarian and Sub-Librarian of Trinity College have earned my special gratitude by the kindness and patience with which they have met my extensive requests for borrowings from other libraries.

It would prolong this preface intolerably if I were to name all those friends who have in one way or another, by lending me pamphlets or books, by drawing my attention to sources which I had overlooked, or by discussing the problems of the period, provided me with fresh material or fresh stimulus. I hope they will forgive me for acknowledging their generous help in this global and anonymous expression of thanks, which is none the less sincere. I should, however, particularly mention Mr. R. W. Davies, author of a recently published book on *The Development of the Soviet Budgetary System*, who has given me help in the financial chapter. Mrs. Degras has once more put me in her debt by undertaking the laborious task of proof-reading; Dr. Ilya Neustadt has again rendered indispensable assistance to the reader and to myself by compiling the index; and Miss J. E. Morris bore a major part of the burden of typing this and earlier volumes.

Since I have worked on this volume and its successor more or less simultaneously, the latter is now nearing completion, and should be published next year. The third volume, dealing with external relations, will, if my present hopes and intentions are fulfilled, be substantially briefer than the other two, and should not long be delayed. A bibliography will appear at the end of the third volume.

E. H. CARR

*May 28, 1958*

# CONTENTS

## PART I
## THE BACKGROUND

## PART II
## THE ECONOMIC REVIVAL

*PART I*

# THE BACKGROUND

# CHAPTER I

# THE LEGACY OF HISTORY

THE tension between the opposed principles of continuity and change is the groundwork of history. Nothing in history that seems continuous is exempt from the subtle erosion of inner change ; no change, however violent and abrupt in appearance, wholly breaks the continuity between past and present. Great revolutions — the conversion of the Roman Empire to Christianity, the English revolution of the seventeenth century, the French revolution, the Bolshevik revolution — represent this tension in its most acute form. Dramatic turning-points in history, they reflect, and set in motion, new social forces which alter the destinies and the outlook of mankind. Tocqueville, in his classic study of the French revolution, drew attention to the two essential characteristics of revolutionary change — the sudden shock of its impact and its quasi-universal significance :

> In the French revolution . . . the mind of man entirely lost its anchorage ; it no longer knew what to hold on to or where to stop ; revolutionaries of an unknown kind appeared who carried boldness to the point of insanity, whom no novelty could surprise, no scruple restrain, and who never flinched before the execution of any purpose. Nor must it be thought that these new beings were the isolated and ephemeral creations of a moment, destined to pass away with it ; they have since formed a race which has reproduced and spread itself in all the civilized parts of the world, and which has everywhere retained the same physiognomy, the same passions, the same character.[1]

The Bolshevik revolution in no way fell behind its prototype in these respects. Never had the heritage of the past been more sharply, more sweepingly or more provocatively rejected ; never had the claim to universality been more uncompromisingly

---

[1] A. de Tocqueville, *L'Ancien Régime et la Révolution Française*, Book III, ch. ii.

3

asserted ; never in any previous revolution had the break in continuity seemed so absolute.

Revolutions do not, however, resolve the tension between change and continuity, but rather heighten it, since the dynamic of revolution stimulates all the forces in play. In the heat of the moment, the desire for change appears to triumph unreservedly over the inclination to conserve. But presently tradition begins to unfold its power as the antidote to change : indeed, tradition is something which remains dormant in uneventful times, and of which we become conscious mainly as a force of resistance to change, through contact with some other " tradition " which challenges our own. Thus, in the development of the revolution, the elements of change and continuity fight side by side, now conflicting and now coalescing, until a new and stable synthesis is established. The process may be a matter of a few years or a few generations. But, broadly speaking, the greater the distance in time from the initial impact of the revolution, the more decisively does the principle of continuity reassert itself against the principle of change. This appears to happen in three ways.

In the first place, revolutions, however universal their pretensions and their significance, are made in a specific material environment and by men reared in a specific national tradition. The programme of the revolution must be empirically adapted to the facts of the environment and is limited by those facts ; the ideas of the revolution are unconsciously seen and interpreted through the prism of preconceptions moulded both by that environment and by a historical past. The main theme of Tocqueville's study was to show how processes already at work, and measures already taken, under the French monarchy had paved the way for the French revolution, which thus not only interrupted, but continued, the orderly course of French historical development. The Bolshevik revolution of October 1917 was also a Russian revolution, and was made by Marxists who were also Russians. To say that it was a revolution inspired by Marxist doctrine, but realized in a country with a predominantly peasant population and still largely pre-capitalist economy, is merely to indicate the simplest and crudest of the antinomies that had to be resolved in the amalgam of " socialism in one country ".

Secondly, the character of the revolution is altered, and

altered to the advantage of the principle of continuity, by the very victory which transforms it from a movement of insurrection into an established government.  In certain technical aspects all governments are alike, and stand at the opposite pole of thought and action to revolution : once the revolution has attained its goal and enthroned itself in the seats of authority, a halt has to be called to further revolutionary change, and the principle of continuity automatically reappears.  It is, however, a common experience of revolutions that hatred of a particular government tends, in the heat of destructive enthusiasm, to breed hatred of government in general, so that when the victorious revolutionaries face the necessary task of constructing their own government and of making it strong, they incur not only the enmity of the man in the street and the peasant on his farm, to whom all governments look alike, but the criticism of the more hot-headed or more consistent of their supporters, who accuse them of betraying their own ideals and principles and attribute the change of attitude to a process of degeneration or decay.  This diagnosis has frequently been applied to the change which came over Christianity when it emerged from its primitive underground stage to assume a position of authority :

> Every contact with the secular . . . reacts strongly on the religious.  An inward decay is inevitably associated with the rise of its secular power, if only because quite other men come to the fore than at the time of the *ecclesia pressa*.[1]

In the French revolution, " the last vices of the monarchy corrupted democracy at its birth " ;[2] the absolutism of kings was succeeded by the absolutism of the Jacobins and, later, of an emperor.  The victorious leaders of the Russian revolution quickly incurred, from the Russian " Left communists " of March 1918, from Rosa Luxemburg in her German prison, and then from every opposition leader down to and including Trotsky, the charge of establishing a dictatorship in the likeness of the defunct autocracy of the Tsars.  The mere act of transforming revolutionary theory and practice into the theory and practice of government involves a compromise which inevitably breaks old links with the revolutionary past and creates new links with a national

---

[1] J. Burckhardt, *Reflections on History* (Engl. transl. 1943), p. 120.
[2] A. Sorel, *L'Europe et la Révolution Française* (1885), i, 222-223.

tradition of governmental authority. The paradoxical phrase " revolutionary legality "[1] aptly expressed this dilemma.

Thirdly, the victory of a revolutionary movement, by transforming it into the government of a state, places on it the practical obligation to conduct relations of some kind, whether friendly or hostile, with other states. In other words, it is forced to have a foreign policy ; and, since every foreign policy is governed, in part by immutable geographical factors and in part by economic conditions which cannot be changed overnight, it is in this field that continuity with the policy of previous governments is most rapidly and conspicuously asserted. *Raison d'état* is tough enough to emerge unscathed from the revolutionary turmoil. One of the first tasks of the victorious revolution is to effect a working compromise between its professedly universal ideals and the empirically determined national interests of the territory over which it has established its authority. The way in which the French revolution achieved this end has been described by a French diplomatic historian in a famous passage :

> The French republicans believe themselves cosmopolitan, but are cosmopolitan only in their speeches ; they feel, think, act and interpret their universal ideas and abstract principles in conformity with the traditions of a conquering monarchy which for 800 years has been fashioning France in its image. They identify humanity with their fatherland, their national cause with the cause of all nations. Consequently and quite naturally, they confuse the propagation of the new doctrines with the extension of French power, the emancipation of humanity with the grandeur of the republic, the reign of reason with that of France, the liberation of the peoples with the conquest of states, the European revolution with the domination of the French revolution in Europe. In reality they follow the impulses of the whole of French history. . . . Humanity takes over the title-deeds of the monarchy and claims its rights.[2]

The parallel of the Russian revolution is extraordinarily close. While Marxist doctrine pointed to the view that national interests

---

[1] See pp. 74-75 below.

[2] A. Sorel, *L'Europe et la Révolution Française* (1885), i, 541-542. It is significant that Tocqueville, who became Minister for Foreign Affairs in 1848, and Sorel, the diplomatic historian, are the two outstanding writers who have emphasized most strongly the continuity of the French revolution with previous régimes.

are no more than a cloak for class interests, and while the Bolshevik leaders, absorbed in the vision of a progressively expanding revolution, expected to have no need of a foreign policy, the Brest-Litovsk crisis led to the rapid evolution of a working compromise between the revolutionary programme and the interests of the Soviet state. In defiance of its intentions, the Soviet Government became the wielder and defender of Russian state power, the organizer of what was in all but name a national army, the spokesman of a national foreign policy. Both in the French and in the Russian revolutions, the stimulus of foreign intervention sufficed to revive popular nationalism. In France, masses of ordinary Frenchmen " identified love of France with love of the revolution as they had formerly identified it with love of the king ".[1] In Soviet Russia the uncovenanted beginnings of a " national " foreign policy, and the equally unforeseen strength of the appeal to a tradition of " Russian " patriotism,[2] were the first and most potent factors which paved the way for a reconciliation with survivors of the old régime and laid the psychological foundations of " socialism in one country ".

But, though the analogy of the Russian revolution with the French revolution holds thus far, the tension between the elements of change and continuity in the aftermath of the Russian revolution presented peculiar features. In the French revolution, as in the English revolution of the seventeenth century, the forces in play on either side had worn the same national colour. Though the French revolution quickly assumed an international rôle, the initial impetus, the dominant ideas of the revolution, had come from within the nation itself. The genesis of the Bolshevik revolution was infinitely more complex. While in one aspect it could be said to stem from a native revolutionary tradition which went back to Pugachev, and had been an obsessing theme in Russian politics, thought and literature throughout the nineteenth century, the irruption of Marxism into Russia, like the irruption of Christianity into the Roman empire, meant the acceptance of a creed, claiming indeed universal validity, but carrying the stigmata of an alien origin. The direct inspiration of the Bolshevik

---

[1] *Ibid.* i, 540.
[2] For these see *The Bolshevik Revolution, 1917–1923*, Vol. 3, ch. 2 *passim*, and pp. 272-274.

revolution and the basis of its ideology came from western Europe ; its principal leaders had spent long years there ; their training and outlook were predominantly western.   The revolution which they made in Russia was conceived by them not primarily as a Russian revolution, but as the first step in a European or world-wide revolution ; as an exclusively Russian phenomenon, it had for them no meaning, no validity and no chance of survival.   Hence the re-emergence of the features of the old order, after the revolutionary flood had receded, took the form not merely of the restoration of an earlier ideological and institutional framework, but of a national restoration.   The defeated social forces which now re-emerged to make their compromise with the new revolutionary order, and insensibly to modify its course, were also national forces reasserting the validity of a native tradition against the influx of foreign influences.   What happened in the aftermath of the revolution, and especially after Lenin's death, had a dual character. Seen in the perspective of the revolution, it represented the familiar reaction of the principle of continuity against the onset of revolutionary change.   Seen in the perspective of Russian history, it represented an attempt of the Russian national tradition to reassert itself against the encroachments of the west.

The Bolshevik revolution followed in this respect a pattern firmly set in the process of Russian national development.   The problem of Russia's backwardness, which haunted the Bolshevik leaders and was discussed by them in the Marxist terminology of successive bourgeois and socialist revolutions, had long overshadowed Russian policy and Russian thought.   The episodes which marked the earlier stages of Russian history — the rift between eastern and western Christendom, the fall of Constantinople, the Mongol invasion — were probably less important influences than certain basic geographical and economic factors which maintained and widened the divergences between east and west, and caused Russia's material progress to lag behind that of western Europe.   The vast expanse of territory, unbroken by any well-defined geographical features or ethnographical divisions, which went to make the Russian state, the inclement climatic conditions prevailing over the greater part of it, and the unfavourable distribution of its mineral resources [1] were the real foundation of Russia's backwardness in comparison with the material development of western Europe. The great distances over which authority had to be organized made state-building in Russia an unusually slow and cumbrous process ; and, in the unpropitious environment of the Russian steppe, forms of production and the social relations arising from them lagged far behind those of the more favoured west. And this time-lag, continuing throughout Russian history, created disparities which coloured and determined all Russian relations with the west. The first contacts of the rising Russian state with western Europe, which began on an extensive scale under Ivan the Terrible in the latter part of the sixteenth century, revealed all the disadvantages of Russia's backwardness in face of the west ; and these disadvantages were still more conspicuously shown up in the ensuing " period of troubles " and of the Polish invasions. Henceforth the development of state power in Russia proceeded at a forced pace under the watchword of military necessity. The outstanding place occupied by Peter the Great in Russian history is due to his success in building in Russia a power capable of confronting western European countries on comparable, if not equal, terms.

This historical pattern of the development of the Russian state had three important consequences. In the first place, it produced that chronically ambivalent attitude to western Europe which ran through all subsequent Russian thought and policy. It was indispensable to imitate and " catch up with " the west as a means of self-defence against the west : the west was admired and envied as a model, as well as feared and hated as the potential enemy. Secondly, the pattern of development rested on the conception of " revolution from above ". [2] Reform came, not through

[1] This point is persuasively argued by A. Baykov in *Economic History Review*, vii, No. 2 (December 1954), pp. 137-149.

[2] The phrase appears to have been first used by the French liberal journalist Girardin, who, in *La Presse* of June 6, 1848, distinguished between two types of revolution : " *from above (par en haut)*, which is revolution by initiative, by intelligence, by progress, by ideas : *from below (par en bas)*, which is revolution by insurrection, by force, by despair, by the streets ". Proudhon, quoting this passage in *Confessions d'un Révolutionnaire*, attacked as " revolutionaries from above " not only Louis XIV, Robespierre, Napoleon and Charles X, but also Saint-Simon, Fourier, Owen, Cabet and Louis Blanc, who favoured the organization of labour " by the state, by capital or by what authority soever " (*Œuvres complètes de P.-J. Proudhon* (1876), ix, 26-27).

pressure from below, from an under-privileged class or from
oppressed masses, expressing itself in demands for social justice
or equality, but through pressure of external crisis, resulting in a
belated demand within the ruling group for an efficient authority
and for a strong leader to exercise it.  Hence reform, which in the
west normally led to a curbing and dispersal of state power, meant
in Russia a strengthening and concentration of that power.
Thirdly, the pattern imposed by these conditions was one, not of
orderly progress, but of spasmodic advance by fits and starts —
a pattern not of evolution but of intermittent revolution.  The
function of Peter the Great, succeeding to the unfinished work of
Ivan the Terrible, was, within the space of a single lifetime, to
transform a mediaeval into a modern society, and, using European
models, to drive his backward and reluctant subjects by forced
marches to new tasks in a new world.  Progress in Russia thus
acquired a spasmodic and episodic character.

> In Europe, in most civilized countries [wrote Nicholas
> Turgenev], institutions have developed by stages ; everything
> that exists there has its source and root in the past ; the Middle
> Ages still serve, more or less, as the basis for everything that
> constitutes the social, civic and political life of the European
> States.  Russia has had no Middle Ages ; everything that is
> to prosper there must be borrowed from Europe ; Russia can-
> not graft it on her own ancient institutions.[1]

And the same point was made by a western traveller :

> Russia alone, belatedly civilized, has been deprived by the
> impatience of her leaders of the profound fermentation and the
> benefit of slow natural development. . . . Adolescence, that
> laborious age when the spirit of man assumes entire responsi-
> bility for his independence, has been lost to her.  Her princes,
> especially Peter the Great, counting time for nothing, made
> her pass violently from childhood to manhood.[2]

Nor was such progress wholly maintained.  Peter's death in 1725
was followed by a period of nearly forty years in which weak
successors went as far as they dared to nullify his work by transforming it on traditional Russian lines.  The alternation of violent

[1] N. Turgenev, *La Russie et les Russes* (1847), iii, 5.
[2] De Custine, *La Russie en 1839* (Brussels, 1843), iv, 153-154.

advance and no less violent reaction continued to mark the uneven course of Russian history.

The consequence of this development was to leave in simultaneous existence, within the loose and ample structure of the Russian state, social, economic, political and cultural forms which in western Europe seemed to belong to different stages of civilization and were regarded as incompatible with one another. In Russia elements of servile, feudal and capitalist society continued to exist side by side ; and this anomaly could not fail to create new divisions and set up new tensions. In the eighteenth century the complex of traditions and beliefs known in the west by the vague name of " humanism " at length reached Russia. But it came in the form of a foreign extravagance imported from the west, and scarcely penetrated beneath the surface of Russian society and Russian consciousness. Its effect was to deepen and perpetuate the wide cleavage that separated rulers from ruled : Russia was now more sharply than ever divided between a " society " which solaced itself for the backwardness of Russian life in the contemplation of western ideas and the enjoyment of the trappings of civilization, and the " dark " mass of the Russian people plunged in the immemorial Russian tradition of poverty and ignorance. Russia became the land of extremes — of the extremes of luxury and indigence, of the most advanced thought and the most primitive superstition, of uninhibited freedom and untempered oppression. The gulf between west and east in Europe was doubled by a gulf within Russia itself, between a superficially westernized society and an authentic Russian people. The rift between east and west was no longer purely external. It had inserted itself into the composite fabric of the Russian state.

These complexities reached their peak in Russian nineteenth-century history — a fruitful period which revealed all the contradictions and all the potentialities of Russian development in exuberant profusion. Throughout the nineteenth century Russian political activity and political thought was polarized on the vital question : for or against the west. Was Russia to overcome her backwardness by following the well-marked western path of development or by striking a new and unique trail of her own ?

And this carried the further question whether the west was to be revered as a mentor and forerunner, or looked on askance as a stranger whose achievements were alien and hostile to the Russian spirit.[1] Behind these time-honoured questions, which divided westerners and Slavophils, there began to emerge, as the century wore on, the increasingly intrusive issue of the rift in Russian society. But this merely reopened the older questions in a new setting. The same fateful ambiguities divided those who sought to change as well as those who desired to conserve, the radicals and revolutionaries as well as the champions of order and autocracy. Both groups contained imitators, as well as enemies, of the west.

Before the end of the century the whole issue of the attitude to Europe had come to a head in the movement for the industrialization of Russia on western lines and with the support of western finance. Incongruously, it ranged the Marxists with Witte and the industrialists, and the *narodniks* with the land-owning gentry and the court. But while, at first, industrialization appeared to have conquered all along the line, the response was once more ambivalent. The west could not be rejected, nor yet wholeheartedly accepted. What was taken, was taken and reshaped in a unique and traditional way. The process of industrialization in Russia at the end of the nineteenth century exhibited many of the characteristic features of Russian development in the period after Peter the Great. First of all, Russian heavy industry, almost from the moment of its birth, was geared to the production of " war potential ", including railway construction, rather than to the needs of a consumer market ; in a population consisting largely of peasants, who were self-sufficient at a low subsistence level, a large-scale consumer market could not come into existence. Industry was " planned " in the sense that it depended primarily on government orders, not on spontaneous market demand, and was financed by loans accorded for political reasons rather than for the traditional " capitalist " motive of earning commercial profits ; in these respects it anticipated much that was to happen under the Five-Year Plans thirty years later. Secondly, the tardy

---

[1] For the nineteenth-century setting of this argument see "'Russia and Europe' as a Theme of Russian History," in *Essays Presented to Sir Lewis Namier* (1956), pp. 363-385.

arrival of industrialization in Russia meant that it skipped over
many of the earlier stages through which the much slower growth
of industrialization had passed in western Europe — the gradual
transformation from the single-handed craftsman to the small
workshop, and from the first primitive factory to the giant
agglomeration employing hundreds and thousands of workmen.
Russian industry, the youngest in Europe and in other respects
the most backward, was the most advanced in respect of the concentration of production in large-scale units.

Thus, the hot-house development of Russian industry, in its
haste to catch up the time-lag by an intensive borrowing from
western models, once more skipped the gradual, formative stage
of adolescence, and carried it at one step from infancy to adult
stature.  In so doing, it created a social structure sharply
differentiated from that of the older industrial communities of
western Europe, so that western influence, and even conscious
imitation of western models, failed to reproduce in Russia the
characteristic western pattern.  The rapidity and belatedness of
Russian industrial development shaped the human factor on both
sides of industry on distinctive lines of its own.  In the west,
something of the spirit of the earlier entrepreneur, attentive to
the changing conditions of the market and in close personal
contact with his workers, survived even in the manager of modern
industry; in Russia, the industrial manager was, from the first, the
administrator, the organizer, the bureaucrat.  In the west, the
industrial worker contrived to retain, even in the age of mass
production, something of the personal skills and independent
spirit of the artisan.  In Russia, the vast majority of the new
generation of industrial workers were still peasants in factory
clothes.  A " grey mass " of peasants was transformed overnight
into a " grey mass " of factory workers.  But to drive the peasant
into the factories and force on him the rigours of factory routine
required — before, as after, the revolution of 1917 — a harsh and
relentless discipline, which shaped relations between industrial
management and the industrial worker on lines of a sharply
defined class hostility.  Weak and backward as it was, the Russian
proletariat provided a more fertile soil than the advanced proletariats of the west for the proletarian revolution.  What had
begun in the traditional Russian fashion as a " revolution from

above " was for the first time creating some of the conditions for
a " revolution from below ".  Once again, a process set in motion
under western influence and in imitation of the west had developed
a peculiar national character of its own.

The political history of Russia in the latter half of the nineteenth
century reflected its economic foundations.  Just as the emancipation of the serfs was a belated attempt to modernize the Russian
economy on western lines, so the political reforms which accompanied it were an attempt to bring an obsolete system of government up to date by borrowing and adapting western liberal and
democratic institutions.  The courts were reformed, rudimentary
social services established, and an enlightened — though scarcely
democratic — machinery of local self-government grafted on to
the rigid, age-old trunk of autocratic power.  But, just as the
Russian economy developed in a forcing-house at a temperature
maintained by pressures from without, so the political reforms
grew not from the strength of their own indigenous roots, but
under alien impulses from western Europe ;  and the product was
something which, though ostensibly imitated from the west, had
a national character all its own.  The long-standing failure to
develop an active bourgeoisie and independent urban communities
could not be repaired in a moment and had far-reaching consequences.  The constitution of 1906 was a pinchbeck imitation
of western constitutional monarchy, and lacked all reality.  Like
German liberalism in 1848, Russian liberalism lacked the solid
social basis which western liberalism found in an energetic and
prosperous class of manufacturers and merchants.[1]  The Russian
liberal was an isolated intellectual, the conscious imitator of a
western model.  Personally sincere, he was without political
weight ;  in time of crisis he could not play the rôle of his western
counterpart.  From the Russian political equation, as from the
economic equation, the middle term was absent.  The Russian
intelligentsia was no substitute for the western middle class.

[1] Trotsky had written in 1901 :  " Pure liberalism with all its Manchester
symbols of faith faded in our country before it blossomed : it did not find any
social soil in which to grow.  Manchester ideas could be imported . . . but
the social environment which produced those ideas could not be imported "
(L. Trotsky, *Sochineniya*, xx, 85-86) ;  ten years later he wrote of the " bour-
geoisification " and " Europeanization " of the Russian intelligentsia, meaning
by this that it had lost its intellectual independence and become the tool of the
ruling class (*ibid*. xx, 351-352).

Institutions and social groups, deriving directly from imitation
of western models, were quickly transformed in Russian conditions into something alien to the west and distinctively national.

The history of the Bolshevik revolution fitted perfectly into
this complicated national pattern. No previous innovator in
Russian history had drawn so frankly and unreservedly as Lenin
on the experience and example of the west, or had spoken in terms
of such open contempt of Russia's native backwardness. The
doctrine that the Russian revolution was merely the forerunner
of the much more important German, European and eventually
world-wide proletarian revolution, and was indeed dependent on
such a revolution for its own survival, was an extreme expression
of the traditional belief of Russian reformers in the backwardness
of Russia and in the need to imitate, and learn from, the west.
The Russian national tradition was weighed and found wanting in
almost every field. The Russian past was condemned root and
branch. The very name of Russia disappeared from the official
title of the new authority, which, with presumptuous universality,
described itself simply as a " workers' and peasants' government ".
If the temporary headquarters of the proletarian world revolution
had been set up in Russia, this was no more than an unexpected
and rather disconcerting accident. Yet within a few years, innovation undertaken in time of emergency under the inspiration of
the west was reabsorbed into a national setting, and took on a
specifically national colour. In this sense, " socialism in one
country " was a repetition of what had happened countless times
before in Russian history.

Premonitory symptoms of this development might have been
detected, even before the revolution, in the revolutionary movement itself. Marxism came to Russia, not merely as a western
doctrine, but as a doctrine requiring the development of Russia
on capitalist lines in direct and conscious imitation of the west ; [1]
only when Russia had followed the west on the path of industrialization could she fulfil her Marxist destiny. " Let us recognize our

---

[1] For the not very successful efforts of some early Russian Marxists, and of
Marx himself, to evade this requirement, see *The Bolshevik Revolution, 1917–
1923*, Vol. 2, pp. 388-393.

uncultured condition, and go to school to the capitalists ", was
the conclusion of a famous article by Struve, the founder of
" legal Marxism ".[1]  In the eighteen-nineties Russian Marxists
stood in the anomalous position of sharing and applauding the
aims of Witte, the arch-capitalist and protagonist of the policy
of industrialization.  The first Russian Marxist group was
founded in the eighteen-eighties by Russian émigrés in western
Europe.  The Russian Social-Democratic Workers' Party, which
was created at the turn of the century, borrowed, in token of its
creed and ambitions, the name of the German Social-Democratic
Workers' Party, which it did not cease to regard as its model and
mentor.  Nothing in Russian history seemed so unimpeachably
and unreservedly western, so free of any national taint, as the
Russian Marxist movement.

Yet contrary symptoms were not slow to develop.  Lenin was
early alive to the impracticability of simply reproducing western
models on Russian soil.

> A movement beginning in a young country [he wrote in
> 1902 in *What is to be Done ?*] can only be successful if it trans-
> forms the experience of other countries.  And for such trans-
> formation it is not enough merely to be acquainted with this
> experience and to copy out the latest resolutions : one must
> know how to adopt a critical attitude to this experience and
> test it independently.[2]

Scarcely had the Russian Social-Democratic Workers' Party begun
to organize itself when, at the congress of 1903, the split occurred
between Bolsheviks and Mensheviks.  The apparently trivial
differences proved significant, the split deep and lasting.  Henceforth Russian Marxists were divided on the issue whether their
party should stick to its western model or adapt itself to specifically
Russian conditions, whether it should organize itself as a broad
party of opinion or equip itself for the conspiratorial activities
which were the only means of action open to the Left in Russia.
Unconsciously, but from the very first moment, the Mensheviks
were the westerners in the party, the Bolsheviks the easterners.
And the issue quickly broadened out into fundamental questions
of Marxist doctrine.  The Bolsheviks, as practical revolutionaries,

---

[1] See *The Bolshevik Revolution, 1917–1923*, Vol. I, p. 9.
[2] Lenin, *Sochineniya*, iv, 380.

were brought face to face with the dilemma of the Russian peasant, who constituted more than 80 per cent of the population of Russia. Lenin understood that no Russian revolution could be made except in a broad-based alliance with the Russian peasantry, whose revolutionary potentialities were amply attested in Russian history ; and, while he firmly rejected the *narodnik* hypothesis with which Marx had toyed in his last years, he postulated as the culminating point of the first phase of the Russian revolution a " democratic dictatorship of workers and peasants ". Finally, in 1917, by ostentatiously borrowing the *narodnik* agrarian pro- gramme of the Social-Revolutionaries and embodying it in the land decree,[1] Lenin firmly anchored the Bolshevik revolution to the Russian national tradition of peasant land-hunger and peasant revolt. Already in 1917 Bolshevism was Marxism applied to Russian conditions and interpreted in the light of them.

The incorporation of this " eastern " element in the amalgam of Bolshevism had not escaped the attention of critics. As early as 1904 the keen-eyed Trotsky, then in his Menshevik period, had noted that the main Bolshevik strongholds in Russia, outside the two capitals, were the factories in the Urals, and taunted the Bolsheviks with striving to " preserve their social-democratic Asia ".[2] A Menshevik journal which appeared spasmodically in Petersburg after the 1905 revolution dubbed the Bolsheviks " Slavophilizing Marxists ".[3] Plekhanov, as well as the Men- sheviks, denounced Lenin's attitude towards the peasantry as non-Marxist and a revival of *narodnik* heresies.[4] In 1912 the Menshevik Axelrod[5] was preaching the need " to Europeanize, *i.e.* radically to change, the character of Russian social-democracy, . . . and to organize it on the same principles on which the party structure of European social-democracy rests " ; and Lenin angrily retorted that " the notorious ' Europeanization ' about which Dan and Martov and Trotsky and Levitsky and all the liquidators talk in season and out of season " was " one of the

[1] See *The Bolshevik Revolution, 1917-1923*, Vol. 2, p. 35.
[2] N. Trotsky, *Nashi Politicheskie Zadachi* (Geneva, 1904), p. 69.
[3] *Sotsial-Demokrat* (Petersburg), No. 2, October 6, 1906, p. 5.
[4] *Chetvertyi (Ob"edinitel'nyi) S"ezd RSDRP* (1934), pp. 133-134.
[5] In 1896 Plekhanov had written to Axelrod : " You are first and foremost a European, and that is someone whom it is important to have in any Russian party " (*Perepiska G. V. Plekhanova i P. B. Aksel'roda* (1925), i, 138) — a remark equally revealing for both.

chief points of their opportunism ". How was the character of *any* social-democracy, how were " *radical* changes " in it, to be determined ? Clearly, argued Lenin, in terms of " the general economic and political conditions of the country in question ". Axelrod was like " a naked savage who puts on a top-hat and imagines himself for that reason a European ".[1] Trotsky retaliated in similar style in 1916 when, in reviewing the collection of articles by Lenin and Zinoviev on *Socialism and the War*, he dubbed the authors " *narodniks* from Chelyabinsk ".[2] When Lenin proclaimed the ambition of the Bolsheviks to seize power from the hands of the Provisional Government, it was a common charge that he was acting as a disciple of Bakunin, not of Marx ; [3] and no less an opponent than Milyukov compared him with the Slavophils : " Gentleman Lenin merely repeats gentleman Kireevsky or Khomyakov when he asserts that from Russia will come the new word which will resuscitate the aged west ".[4]

Such criticisms left Lenin unmoved. He felt himself heart and soul a westerner : in his conception of the party, he could appeal to an older western tradition — the tradition of the Jacobins : he had proudly claimed the name when it was first hurled at him by Trotsky as a term of abuse.[5] In his reliance on Marxism, he appealed more often to the Marx of the period before 1848, to Marx the active propagandist of revolution, than to the later Marx, the student of the contradictions and inevitable downfall of capitalism. It was the earlier Marx who had lived and worked in conditions most nearly comparable to those now confronting Lenin ; and the episode of the Paris Commune showed that, even much later, he had abated nothing of his enthusiasm for the practice of revolution. That the Marxism of the Bolsheviks was

---

[1] Lenin, *Sochineniya*, xvi, 41-42.

[2] Quoted in G. Zinoviev, *Litsom k Derevne* (1925), p. 24 : I have not been able to trace the original. Zinoviev, as late as 1925 (*ibid.* p. 26), retorted that the party would " not concede an inch to ' European ' pseudo-Marxism dressed up in a ' Left ' Trotskyist guise ".

[3] See, for example, an incident cited in *The Bolshevik Revolution, 1917–1923*, Vol. 1, p. 79

[4] Quoted in Bunyan and Fisher, *The Bolshevik Revolution, 1917–1918* (Stanford, 1934), p. 42.

[5] See *The Bolshevik Revolution, 1917–1923*, Vol. 1, pp. 33, 35 ; Plekhanov also accepted the imputation of " Jacobinism ", contrasting this favourably with Axelrod's humanitarian liberalism (*Perepiska G. V. Plekhanova i P. B. Aksel'roda* (1925), i, 44, 192 ; ii, 118).

as authentic, and therefore as " western ", as the Marxism of the Mensheviks was a perfectly tenable view. But the discrepancies between them were patent. Of the two strands which went to make the composite fabric of Marxist teaching, the Bolsheviks represented primarily the revolutionary, voluntarist element,[1] the Mensheviks the evolutionary, determinist element. The Bolsheviks spoke of the need to act in order to change the world, the Mensheviks of the need to study the forces which were changing it and to conform their action to these forces. Finally the Bolsheviks put their faith in a conscious minority which would lead the masses and galvanize them into action ; the Mensheviks more cautiously awaited the moment when the hidden forces of change would ripen and penetrate the consciousness of the masses, this last divergence being directly reflected in their views of party organization. On all these issues the views of the Mensheviks coincided far more closely than those of the Bolsheviks with the prevailing attitude of western Marxists ; and this alone sufficed to give Bolshevism, whatever the sources of its inspiration, a certain Russian, or non-western, colour. The belief in the need for a group of highly conscious and highly organized professional revolutionaries to direct the unconscious and " spontaneous " action of the mass of the workers was a far more accurate response to Russian than to western conditions. On a longer view, it might also be said to have equipped the Bolsheviks to cope, far more effectively than the Mensheviks, with the irrational tendencies permeating modern mass society.

The traditions of the Russian past created a soil in which Bolshevism could easily develop the latent anti-western elements in its composition, and merge its Marxist messianism in an older Russian messianism. " History ", said Sokolnikov a few weeks after the revolution, " clearly shows that the salt of the earth is gradually moving eastwards. In the eighteenth century France was the salt of the earth, in the nineteenth century Germany ; now it is Russia." [2] But the shift entailed the introduction of

---

[1] Plekhanov in 1905 accused the Bolsheviks of introducing into Marxism the voluntarism of Mach and the idealists (Lenin, *Sochineniya*, vii, 267, note 121)

[2] *Protokoly Tsentral'nogo Komiteta RSDRP* (1929), p. 206.

specifically Russian elements. The time-honoured Russian pattern of spasmodic advance, hastening to catch up with the west, and, in the process, skipping over intermediate stages through which western progress had passed, was repeated in the preparations for the Russian revolution. Trotsky's theory of " permanent revolution " was devised to meet the dilemma arising from specifically Russian conditions — the absence in Russia of a powerful bourgeoisie capable of realizing the bourgeois revolution which was a necessary stage in western conceptions of Marxist development ; and Lenin, while formally rejecting the theory, adopted in 1917 what was virtually the same expedient of making the Bolshevik seizure of power do simultaneous duty as the last act of the bourgeois revolution and the first of the socialist revolution. Russian history had experienced one more violent and abrupt transition from " childhood " to " manhood ".[1] Even the initial appeal of the " workers' and peasants' government " to the world for peace and brotherhood among the nations might have seemed to reflect the long-standing claim of the Russian people to fulfil a universal, and not a purely national, rôle. As the new régime found itself isolated and driven to the wall by its enemies, domestic and foreign, and exposed to the hazards of civil war, the old pattern of revolution from above began, imperceptibly at first, to substitute itself for the revolution from below which had carried the Bolsheviks to victory in October 1917 ; and the dictatorship of the proletariat fell into the mould of reforming autocracy. Finally, when peasant discontent forced the " retreat " into NEP, another jarring, but irresistible, Russian force had imposed itself on the original Marxist conception of the revolution. The question which the Bolshevik leaders had to ask themselves in 1921 was essentially the question which had divided the westerners and the Slavophils. Would the triumph of socialism in Russia be achieved by following the western path, or by following a specifically Russian line of development ? If the first answer were accepted, reliance must be placed on the development of industry and of the proletariat, if necessary, at the expense of the peasant.

---

[1] A writer in the *émigré* symposium *Smena Vekh*, published in 1921, declared that " Russia, in the few months of the Provisional Government, had run through all those illusions of the democratic order which it had taken Europe more than a hundred years to outlive " (*Smena Vekh* (2nd ed. Prague, 1922), p. 109).

If the second answer were accepted, reliance must be placed on conciliating the peasant and winning his support for increased agricultural production as the prerequisite of an advance to socialism. As always in Russian history, a clear-cut choice between the two answers was impossible. Russia could neither unconditionally pursue nor totally reject the western path. In NEP Lenin found the compromise between the two answers — the " link " between proletariat and peasantry which would for a time make it possible to travel the two roads simultaneously. But the compromise, which was also a " retreat ", had ideological implications; and these implications also carried reflections of the Russian past. The resistance of the Russian peasant to Marxism was the resistance of the traditional Russian way of life to western innovation.

Thus, during the first years of the régime, while the revolutionary impetus continued to predominate, familiar features of the Russian landscape and the Russian outlook slowly emerged from beneath the revolutionary flood. As the Soviet Government became more and more openly the heir of Russian state power and attracted to itself traditional feelings of Russian patriotism, it proclaimed its mission in terms which conveyed to sensitive ears unmistakable echoes of the Russian past. Moscow, the third Rome and now the centre of the Third International, was once again conscious of its mission to renew, out of the fullness of its uncorrupted youth and vigour, the decrepit and decadent west, was once again courting a hostility from the west which it attributed to the envy and malice inspired by its achievements, and was once again covering its material backwardness by boastful assertions of its superior spiritual essence. The fulfilment of the eschatological promises of Marxism was delayed, like the Second Advent, far beyond the original expectations of the faithful; and, when this delay bred the inevitable current compromises with power and expediency, the process of degeneration from the pure ideal took on specifically Russian forms in a Russian context. Primitive Christianity decked itself in the trappings of imperial Rome, communism in those of the Russian national state. Though it soon transpired that the compromise was not all on one side, the transformation was incongruous, and scandalized some believers. But, as the cause of Russia and the cause of Bolshevism began to coalesce into a single undifferentiated whole, the resulting amalgam showed clear traces of both the original components out of which it had been formed ; the idiom was a blend of both elements. This process, subtle and undeclared, was well advanced when Stalin first propounded the hybrid doctrine of " socialism in one country "

CHAPTER 2

# THE CHANGING OUTLOOK

T HE general change of outlook which set in with the introduction of NEP was in part psychological, and resulted from
the lowering of tension after the years of revolutionary ex-
citement and the stresses of the civil war.   It was materially impossible to go on living in the conditions of unspeakable hardship
and privation to which a large part of the population had been
subjected for four or five years.   It was psychologically impossible
to maintain the exalted mood of faith and enthusiasm in which
present turmoil and horror could be welcomed as the birth-pangs
of the new world of the future.   The development which, between
the years 1921 and 1924, shifted the balance of emphasis from
political programmes to the routine of everyday life, from icono-
clastic theory to traditional practice, from revolution to organization,
from visionary utopianism to hard-headed realism, from an internationalism that knew no frontiers to an astute calculation of the
national interests of the USSR, affected almost every aspect of
Soviet life and thought.   In public affairs it brought with it a
shift in emphasis from adventure to administration, from sweeping
revolutionary design to the meticulous execution of day-to-day
decisions.   Lenin devoted to this theme a long and repetitive
passage in a speech of December 1921 to the ninth All-Russian
Congress of Soviets :

> Political problems and military problems could be solved
> in an access of enthusiasm. . . . We look back and imagine
> that economic problems can be solved in the same way.   There
> is the mistake. . . . Learn to work at a different tempo,
> reckoning your work by decades not by months, and gearing
> yourself to the mass of mankind who have suffered torments
> and who cannot keep up a revolutionary-heroic tempo in every-
> day work.

And later in the same speech :

> Here is work for whole decades . . . it cannot be carried
> on at the tempo, with the speed and in the conditions in which
> we carried on our military work.[1]

It was a mood of patience, caution and compromise. The key to
the situation was no longer " in policy, in the sense of a change of
direction ", but in finding the right man for the right job :

> This is a prosaic task, a small task. These are petty affairs,
> but we live in the aftermath of the greatest of political upheavals,
> in conditions where we must continue to exist for a certain time
> in the midst of a capitalist setting. . . . Choose the people who
> are necessary, and verify the practical execution of decisions :
> this the people will appreciate.[2]

Immediately after Lenin's death Kamenev echoed the same
theme :

> We have come out of the period of landslides, of sudden
> earthquakes, of catastrophes, we have entered on a period of slow
> economic processes which we must know how to watch.[3]

It was no longer the bold revolutionary, but the law-abiding, hard-
working citizen, who was held in honour.

The fading of the revolutionary vision, and the cult of common
sense in administration and attention to everyday affairs, bred a
conservative frame of mind. Every successful revolution pre-
cipitates a division — at first, perhaps, only a difference of
emphasis, later, a more radical split — between those who are
still fired by the ambition for further revolutionary achievement
and those who are mainly concerned to stabilize what has been
achieved. The former now easily incur the charge of utopianism.
The division first appeared after the Bolshevik revolution in the
debates about Brest-Litovsk ; and in the ensuing years Lenin
twice took the field against the " left-wing infantilism " of the
revolutionaries à outrance. The tenth party congress which
adopted NEP in March 1921 also condemned the programme
and activities of the " workers' opposition ", which attacked the
party leaders for betraying the principles of revolution. After

---

[1] Lenin, *Sochineniya*, xxvii, 137-139.                    [2] *Ibid.* xxvii, 256.
[3] *Trinadtsatyi S"ezd Rossiiskoi Kommunisticheskoi Partii (Bol'shevikov)* (1924),
P. 393.

the ending of the civil war and the establishment of NEP, it was reasonable to treat the revolution as a *fait accompli* — and this in a double sense.  On the one hand, nobody except a few fanatics any longer expected or desired to undo the work of the revolution or to return to the past.  On the other hand, only party extremists and doctrinaires now seriously thought in terms of further revolutionary action ;  the completion of the revolution through " socialist construction " would consist of the consolidation and expansion of existing positions by orderly and peaceful means.  The radicalism of revolutionary doctrine was succeeded by the conservatism of administrative empiricism.

In such an atmosphere a falling off in the revolutionary idealism of the first years was unavoidable, especially in the younger generation.  It was to a Komsomol congress in 1922 that Bukharin spoke of " a sort of demoralization, a crisis of ideas among communist youth, and among youth in general ", resulting from NEP.[1]  As Trotsky afterwards wrote : " the ascetic tendencies of the civil war gave way in the period of NEP to a more epicurean, not to say gay, mood ".[2]  Even Komsomol journals of the period were preoccupied with such questions as what kind of trousers a komsomol should wear " with or without a crease ", how many bottles of beer he might drink, whether he should give up his seat to a woman in the tram.[3]  For the party stalwarts a sense of flatness and disappointment supervened :  the contrast between the heroic, glorious days when the revolution had to be made and fought for, and the dull, monotonous days of economic reconstruction — what were called in the catchword of the time " Soviet week-days " — was a constant theme of the period.  A party report of 1924 spoke with concern of the number of recent suicides in the party " for ideological reasons, for the reason that they could not adapt themselves to the new stage, an extremely difficult stage, but lived in the mood of the period of the offensive, the period of war communism ".[4]

[1] *Pyatyi Vserossiiskii S"ezd RKSM* (1927), p. 113.
[2] L. Trotsky, *La Révolution Trahie* (n.d. [1936]), p. 187.
[3] *Molodaya Gvardiya*, No. 1, January 1926, p. 235.
[4] Report by Yaroslavsky to the party central control commission in *Pravda*, October 9, 1924.  I. Bobryshev, *Melkoburzhuaznye Vliyaniya sredi Molodezhi* (1928), p. 97, records a defence of suicide heard at a Komsomol meeting : " They said that formerly, in the days of the civil war, suicides were unheard

The same atmosphere made possible the guarded and qualified reconciliation between the Soviet régime and the survivors of the former régime which was a striking feature of the early NEP period.. It was a meeting on unequal terms. The victors were able to dictate the terms of the cooperation which many of the defeated were now ready to offer. But, in the framing of the policies to be pursued, and of the ideas by which these policies were inspired or supported, the inequality was less marked, and the balance was tilted rather in the opposite direction. The lack of " culture " and administrative experience among communists, on which Lenin constantly dwelt in his last years, had the result of placing the business of administration and management largely in the hands of survivors of the former régime, who established in their person a continuity between the old and the new.

> Our state apparatus, with the exception of Narkomindel [wrote Lenin in his last article], represents for the largest part a survival of the old, which has only in a very small part been subjected to any kind of serious changes.[1]

Not all the problems of the new Russia differed fundamentally from those of the old. It sometimes happened that the old official, confronted by the old questions, returned the same answers as he had returned in the past and took the same decisions. Those former pillars of a bourgeois society and administration who rallied to the Soviet cause and now transacted much of the necessary business of the Soviet Government, did so in the conviction that the government had come to represent Russia and to act in the name of Russia ; and it was natural that, consciously or unconsciously, they should strive to uphold a Russian national tradition. Nor did this any longer imply hostility to the revolution as such. Nobody was thinking any longer in terms of restoration, or of the overthrow of Soviet power. The achievements of the revolution were accepted, stabilized and added to the national record.

of among party or Komsomol members, that then heroic deeds could be accomplished. Now we have to do very prosaic things, things which cannot arouse enthusiasm or kindle the revolutionary flame. To support this argument they mentioned the alleged fact that men from the old Bolshevik underground could not bear this everyday life, and departed for ' the other world '."

[1] Lenin, *Sochineniya*, xxvii, 402 ; for further discussion of this theme see pp. 114-119 below.

But, above all, the universal feeling generated by NEP was one of relaxation and immense relief. Even those who most fervently insisted that NEP was only " a breathing space " and a prelude to fresh effort, admitted that the breathing space was indispensable. People were once more able to occupy themselves with their ordinary personal affairs. Life resumed its once familiar routine. And this return to what was thought of as normal was necessarily a return to former ways, a re-establishment of continuity with the past, an acceptance of half-forgotten tradition. Trotsky, in an article of 1923, noted' this phenomenon with some apparent surprise : " Politics are flexible, but life is immovable and stubborn. . . . It is much more difficult for life than for the state to free itself from ritual." [1] In the early years of NEP every field of Soviet life and thought was affected by this almost instinctive reaction from a mood of innovation to a mood of conformity. But the changing outlook was most conspicuous in fields that lay on the periphery of politics and were traditionally recalcitrant to political interference. It may be illuminating, as a study of the background of opinion in this period, to trace the landmarks of change in four such fields, in the current attitude to the family, to the Orthodox Church, to literature and to law.

## (a) The Family

Radical theories of sex relations and of the family, originally drawn from the literature of western romanticism, had been familiar for more than half a century in Russian revolutionary writings. The secret *Young Russia* proclamation of 1862, often quoted as the first manifesto of the modern revolutionary movement, demanded the abolition of marriage as a " highly immoral phenomenon and one incompatible with the full equality of the sexes ", and argued that, in order to give freedom to women, the care and education of children should become a function of society.[2] Official party doctrine, shared by the Bolsheviks with other Marxist parties, derived from the dictum in Engels's major work on *The Origins of the Family, of Private Property and of the State* that " the liberation of women presupposes as its first

---

[1] L. Trotsky, *Sochineniya*, xxi, 18, 39.
[2] *Za Sto Let*, ed. V. Burtsev (London, 1897), p. 43.

preliminary condition the return of the whole female sex to social labour ", that women must be relieved of domestic cares through the institution of communal dining-rooms and communal nurseries, and that the individual family would then cease to be " the economic unit of society ".[1]  Neither Marx nor Engels drew any practical conclusion from this theoretical analysis of the economic conditions of equality between the sexes.  But some Marxist thinkers were prepared to deduce from it the hypothesis that the family, like the state, was a feudal or bourgeois institution destined to die away in a communist society.  The assertion of the full equality of women with men appeared to require that both domestic services and the rearing of children should become a communal responsibility instead of weighing as an individual burden on the wife and mother.  It also implied a rejection of the so-called dual standard of morality of nineteenth-century bourgeois society, and a corresponding change of outlook on sexual relations.  The woman was to enjoy the same freedom as the man. " The satisfaction of the sexual impulse ", wrote Bebel in his authoritative work on *Woman and Socialism*, " is everyone's private affair just like the satisfaction of any other natural impulse " :[2] it was an act of no more moral significance than, in a much favoured comparison, the drinking of a glass of water. But, while such speculations were common, they did not occupy any important place in social-democratic theory, and did not influence the conduct of social-democratic leaders, whose private lives were, in general, irreproachable by any current bourgeois standard. This was as true of the Russian Social-Democratic Workers' Party as of any other social-democratic party.  The question was ignored in the party programme of 1903, and did not figure in any of the subsequent party discussions.  When a woman Bolshevik, Inessa Armand, drafted a pamphlet in 1915 on women's demands which included the " demand for free love ", Lenin vigorously protested that this was a bourgeois, not a proletarian, conception.[3]

The enactments of the first period of the Soviet régime on

[1] Marx i Engels, *Sochineniya*, xvi, i, 56.
[2] A. Bebel, *Die Frau und der Sozialismus* (10th ed. Stuttgart, 1891), p. 338 ; a Russian translation of this work appeared in Petrograd in 1918 with an introduction by Kollontai.        [3] Lenin, *Sochineniya* (4th ed.), xxxv, 137-138.

marriage and the family, like its first economic enactments, were not specifically socialist in character, and would have been endorsed by bourgeois radical opinion in many western countries. The first of them made civil registration obligatory for all marriages, thus abolishing the legally binding ecclesiastical marriage of the past.[1] This was followed by a decree authorizing the automatic dissolution of marriage on the demand of either or both of the partners.[2] In the autumn of 1918, these principles were embodied in a detailed marriage code, which also made provision for the complete equality of the sexes in all matrimonial relations, and accorded to illegitimate children the same rights as to legitimate children, thus taking the first step towards the legal recognition of what later came to be called " de factò marriage ".[3] Finally, in November 1920, a decree was issued making abortion legal when performed by a qualified doctor in a public hospital, " for so long as the moral survivals of the past and economic conditions of the present compel some women to resort to this operation ".[4]

While, however, legislation on marriage and the family was confined within these comparatively modest limits, the implications of socialism for relations between the sexes were widely canvassed, and, for the first time, began to acquire practical significance in the light of current policy and behaviour. The employment of women in productive work, and the enjoyment by them of full equality of rights and responsibilities with men, were no longer items in a theoretical programme, but necessities of a period of economic breakdown and civil war. Acute food shortages rather than the exigencies of socialist theory led to a large extension of communal feeding. The vast problem of homeless children imposed on reluctant and overburdened authorities the establishment of children's homes and settlements. In this aspect of war communism, as in others, doctrine was invoked to prove that what was done in the emergency of war was identical with what had long been included in the cherished precepts of socialist programmes. Lenin in 1919 demanded the creation of " model institutions, dining-rooms and crèches which would free women from domestic

[1] *Sobranie Uzakonenii, 1917–1918*, No. 11, art. 160.
[2] *Ibid*. No. 10, art. 152.
[3] *Ibid*. No. 76-77, art. 818 ; for the recognition of *de facto* marriage see p. 37 below.          [4] *Sobranie Uzakonenii, 1920*, No. 90, art. 471.

labour ", and described such labour as " petty and containing nothing that can in any way further the development of women ".[1] He seems to have shared the common opinion of the time that, for this and for other reasons, the bringing up of children in communal institutions was a goal to be aimed at.  " Only by these means ", he told Clara Zetkin in 1920, " can woman be delivered from the old house slavery and from all dependence on the man " ; and he added that, when the performance of these functions is transferred to society, " the children enjoy more favourable conditions than at home ".[2]  Such utterances must be read in part against the background of current Russian life.  The traditional peasant's or worker's family, with its subjection and maltreatment of women and exploitation of child labour, was too familiar a consequence of Russian poverty, and symbol of Russian backwardness, to be anything but a bugbear to progressive Russian thinkers, while in Asiatic Russia the polygamous and patriarchal family structure formed the main bulwark of resistance to the modern world.  Even in more advanced regions, the family seemed the enemy of everything that the revolution sought to achieve ;  the programme of the Komsomol adopted in 1920 mentioned " the conservatism of parents " side by side with " the influence of priests and *kulaks* " among the adverse conditions of the environment of peasant youth.[3]  As late as 1924 Bukharin called the family " the most conservative stronghold of all the squalors of the old régime ", and thought it a matter for congratulation that the young pioneer movement was conducting " a gradual mining operation " against the traditional pattern of family relations.[4] The revolutionary attitude to the family can be understood only as a reaction to pre-revolutionary conditions ;  and the achievement of the revolution in inculcating acceptance of the equality of the sexes and in promoting a higher regard for women was real and indubitable.

Apart, however, from these conscious strivings to remove abuses of the old order, the sequence of war, revolution and civil war had produced many of the same unpremeditated and

---

[1] Lenin, *Sochineniya*, xxiv, 470.
[2] C. Zetkin, *Erinnerungen an Lenin* (Vienna, 1929), p. 75.
[3] *Tretii Vserossiiskii S"ezd RKSM* (1926), p. 306.
[4] *Trinadtsatyi S"ezd Rossiiskoi Kommunisticheskc ' Partii(Bol'shevikov)*(1924), p. 545.

disintegrating effects on family and sex relations as on other aspects
of social life.  Here, too, " war communism " marked a specific
period ;  and here, too, what in other conditions would have been
treated as the unwelcome result of chaos, confusion and licence
was now retrospectively justified in terms of socialist doctrine.
Alexandra Kollontai was the only leading Bolshevik who carried
this theory to its extreme conclusion, arguing that stable marriage
was a function of bourgeois society rendered necessary only by
the importance attached to property relations, and that " in the
working classes greater ' fluidity ' and less fixity in the relations
of the sexes completely coincide with, and directly result from,
the fundamental tasks of those classes ".[1]  In a widely circulated
pamphlet of the civil-war period Kollontai sounded the death-
knell of the family :

> *The family ceases to be necessary.*  It is not necessary to the
> state because domestic economy is no longer advantageous to
> the state, it needlessly distracts women workers from more
> useful productive labour.  It is not necessary to members of
> the family themselves because the other task of the family —
> the bringing up of children — is gradually taken over by
> society.

In the future " the socially conscious worker-mother will rise to
a point where she no longer differentiates yours and mine, and
remembers that there are henceforth only *our* children, the children
of communist workers' Russia ".[2]  A number of popular novels
and stories from Kollontai's pen cast ridicule on the bourgeois
prejudices of the past, and preached the uninhibited satisfaction
of the sexual impulse, supported by the assumption that it was the
business of the state to take care of the consequences.  Bukharin
later recalled the time when " it was thought very revolutionary
to spit on all and every sense of shame in sex relations " by way
of protest against " the blind prejudices of society ", " so-called
' family law ' ", and " the debasement of women ".[3]  These
views never received official party endorsement.  Lenin especially
disliked them.  In conversation with Clara Zetkin in 1920 he
inveighed against " the famous theory that in communist society

[1] A. Kollontai, *Novaya Moral' i Rabochii Klass* (1919), p. 59.
[2] A. Kollontai, *Sem'ya i Kommunisticheskoe Gosudarstvo* (1920), pp. 20, 33.
[3] *Byt i Molodezh'*, ed. A. Slepkov (1926), p. 8.

the satisfaction of sexual desire, of love, is as simple and unimportant as drinking a glass of water ". This theory, which was
" completely un-Marxist and unsocial into the bargain ", has
" driven our young people mad, quite mad ".[1] But, so long as
civil war conditions prevailed, cover would be sought in party
doctrine for a relaxation of standards of sexual behaviour, and
the theories of Kollontai remained widely popular in party
circles.[2]

It was the changed outlook associated with the ending of the
civil war and the introduction of NEP which brought the first
reaction against these views. The new legislation on marriage
and divorce was not challenged : this indeed, belonged to the
bourgeois rather than to the socialist stage of the revolution. But
Kollontai's prestige declined sharply owing to her association
with the " workers' opposition ", which was condemned by the
tenth party congress in March 1921 ;[3] and the theories of the
family and of sexual relations of which she had been the protagonist gradually gave way to more conventional attitudes. The
fifth Komsomol congress in October 1922 heard Bukharin attack
the prevailing "anarchy in the realm of rules of conduct " with
specific reference to lax sex morals as well as to excessive indulgence in alcohol and tobacco ; and the congress passed a resolution
condemning all these evils.[4] Trotsky in 1923 conducted a symposium of party workers which revealed a marked return to traditional views of the rôle of the family. " The theses of comrade
Kollontai " were criticized as ignoring " the responsibility of
father and mother to their child " and leading to the abandonment of children — a growing evil in Moscow. Because " we
wrongly emphasized the conception of ' free love ' ", party
members in the civil war had begotten children without caring
what became of them. Workers had been encouraged by party

[1] C. Zetkin, *Erinnerungen an Lenin* (Vienna, 1929), pp. 62-63.
[2] P. Romanov, in a once famous short story, *Bez Cheremukhi*, originally
published in *Molodaya Gvardiya*, No. 6, June 1926, pp. 13-21, put into the
mouth of his heroine the complaint that " those who seek in love something
more than physiology are looked on with contempt as if they were mentally
deficient or sick " ; the " heated discussions " of this story in Komsomol
meetings were later recalled by one of the participants (*Yunii Kommunist*,
No. 12, 1931, p. 54).
[3] See *The Bolshevik Revolution, 1917–1923*, Vol. 1, pp. 197-200, 210.
[4] *Pyatyi Vserossiiskii S"ezd RKSM* (1927), pp. 114, 124-125, 315-317.

teaching to divorce their wives. Women communists neglected their duties as wives and mothers for party work ; on the other hand, cases were quoted of women communists who left the party on the insistent demand of their husbands.[1] The doctrine of Engels on the liberation of women from domestic labour and the obsolescence of the " individual family " continued to be preached. But it was confined to formal expositions, and both practice and opinion diverged more and more widely from it.[2] Even the increased employment of women in the later nineteen-twenties did not restore it to favour ; and the Soviet family continued to follow traditional patterns.

Other symptoms of a return to conventional attitudes quickly declared themselves. By 1924 another achievement of the revolution — the legalization of abortion — had begun to incur criticism. In a report to the central control commission of the party, Yaroslavsky, while insisting that the party was not a " monastic sect " and had no desire to preach " purely and simply a parsonical morality ", referred to the figures of abortion in Moscow and Leningrad as " horrifying ", though he claimed that they were lower than in bourgeois countries.[3] An article published in 1925 by the People's Commissar for Health of the RSFSR was a curious attempt to reconcile the conventional attitudes of the past with formal recognition of communist theories. " Of course ", wrote Semashko, " the ideal would be if the state took on itself to regulate all the consequences of the sexual act (rearing of children, etc.)." But since this was impracticable, he could only recommend " sublimation " (the word appeared in inverted commas with a coy reference to the dubious authority of Freud) of sexual instincts in social work. Semashko denounced the " old wives' tale " that restraint was harmful and sexual indulgence necessary to health.

[1] L. Trotsky, *Voprosy Byta* (2nd ed. 1923), pp. 121-125. The difficulty of reconciling party and conjugal duties for the wife is one of the themes of Gladkov's well-known novel *Cement*, published in 1924 ; no solution appears to be offered.

[2] In the Trotsky symposium one speaker complained that a party lecturer on " family and marriage " had confined himself to repeating the substance of Engels's essay, whereas " some conclusion had to be drawn from this work of Engels for the present day, and this is exactly what we are unable to do " (L. Trotsky, *Voprosy Byta* (2nd ed. 1923), p. 125).

[3] *Pravda*, October 9, 1924.

Drown your sexual energy [he concluded] in public work. ....
If you want to solve the sexual problem, be a public worker, a
comrade, not a stallion or a brood-mare.[1]

Bukharin at the fourteenth party congress in December 1925
denounced the prevalence among the young of " decadent and
semi-hooligan groups with such names as 'Down with innocence',
' Down with shame ' " ; [2] and the Komsomol journal followed
this up with another broadside attacking the heresies of Kollontai.[3]
A crying evil which played its part in modifying the initial
attitude to the family was that of " homeless " children.  The
revolution and the civil war had left behind them immense numbers
of children, orphans or separated without trace from their parents,
who, being without homes or protectors and without normal means
of subsistence, roamed in gangs through cities and countryside,
living by their wits and engaging in every form of crime and
violence.  At the time of Lenin's death VTsIK announced the
establishment of a " Lenin fund " in his memory for aid to
" homeless " children, " especially victims of the civil war and the
famine ".[4]  An extensive press campaign followed ; and six
months later, in July 1924, a sum of 50 million rubles was voted
from the budget to the fund, the expectation being to raise another
50 millions from voluntary contributions and local levies.[5]
Hitherto the official remedy for this evil had been to put the
children in publicly run children's homes, where they would be
trained for suitable occupations.  But the homes had begun to
acquire an unenviable reputation.

If you were to read [said Bukharin at this time] about the
present condition of the " educational institutions " in which the
homeless children are maintained, your hair would stand on end.[6]

As Lunacharsky, who, as People's Commissar for Education of
the RSFSR, was in charge of these homes, confessed, they were

[1] *Izvestiya*, May 15, 1925.
[2] *XIV S"ezd Vsesoyuznoi Kommunisticheskoi Partii (B)* (1926), p. 815.
[3] *Molodaya Gvardiya*, No. 3, March 1926, pp. 136-148.
[4] *2ⁱ S"ezd Sovetov Soyuza Sovetskikh Sotsialisticheskikh Respublik :
Postanovleniya* (1924), p. 8.
[5] *Sobranie Zakonov, 1924*, No. 3, art. 33 ; a decree of the RSFSR on the
raising of local funds is in *Sobranie Uzakonenii, 1925*, No. 8, art. 53.
[6] *Trinadtsatyi S"ezd Rossiiskoi Kommunisticheskoi Partii (Bol'shevikov)* (1924),
pp. 545-546.

hopelessly overcrowded and inadequate, and lacked both money for clothing the children and facilities for training them : some provincial authorities complained that the children's homes already swallowed up half their budget. Nor was an end of the problem in sight. Those dealing with it were in the position of a " squirrel going round and round in a cage ".[1] In August 1924, when the prospects of a partial harvest failure inspired fears of a further increase in the number of abandoned children, Rykov at the party central committee launched an attack on the whole policy :

> In the children's homes we are bringing up idlers, who do not know how to work and will in future be a burden to the state. In order to prevent this we must take measures to stop the divorce of these children from all productive work, and to prevent an increase in the number of homeless children : we have given a directive in the regions where the harvest is bad to avoid increasing the population of the children's homes by bringing in children who have a family. In cases where the family is not in a position to feed the child, it is better to help the family than to take the child and feed it in a children's home.[2]

Article 183 of the original family code of 1918 explicitly prohibited the adoption of children — a surprising provision for which three different explanations were commonly given : unwillingness to open the door to artificial increases in the membership of peasant households, leading to claims for larger shares in the redistribution of land ; fear that adoption would serve as a cover for the exploitation of juvenile labour ; and belief that orphan children would be more satisfactorily looked after in public institutions.[3] Two months after Rykov's speech, Lunacharsky announced an official policy of " putting out the children to the population ".[4] By the autumn of 1925 many homeless children from the towns

---

[1] *Vserossiiskii Tsentral'nyi Ispolnitel'nyi Komitet XI Sozyva : Vtoraya Sessiya* (1924), pp. 116-118.

[2] A. I. Rykov, *Sochineniya*, iii (1929), 194.

[3] D. Kursky, *Izbrannye Stat'i i Rechi* (1948), pp. 147-148.

[4] *Vserossiiskii Tsentral'nyi Ispolnitel'nyi Komitet XI Sozyva : Vtoraya Sessiya* (1924), pp. 117-118. A detailed decree of the RSFSR of March 8, 1926, provided for the placing of homeless children in " families of toilers with the consent of the latter ", allowances being granted for the purpose from public funds (*Sobranie Uzakonenii, 1926*, No. 19, art. 143) ; a further decree of April 5, 1926 (*ibid.* No. 21, art. 168), laid down that a peasant household adopting a homeless child was entitled to an allocation of land in respect of it, which was to be free of tax for three years.

had been settled in peasant families ; [1] according to the modest
claim made in official statistics, 55,000 children were handled by
the commission set up for the purpose in 1924, 75,000 in 1925
and 85,000 in 1926.[2] As the legacy of the civil war was left
behind and life became more orderly and regular, the problem of
the homeless children gradually assumed the more normal form
of a problem of juvenile unemployment, though in some parts
of the country it proved extraordinarily stubborn and persistent.[3]
What was now clear was that the idea of creating a vast network
of children's homes for the rearing of children was " pure utopia
in our economic conditions ".[4] The care of children was once
more being considered in the traditional framework of a restora-.
tion of family life. The state could not disinterest itself in the
institution of marriage, declared a speaker at the TsIK of the
RSFSR in November 1925, " because on the stability of marriages
depend a number of consequences of undoubted importance for
society ", and went on to attribute the problem of homeless
children to " the disintegration of the family ".[5]

One particular aspect of the return to more conventional views
of family and marriage deserves notice. The changed outlook
was in part a change in the attitude of individuals, marking a
retreat from the fervour of revolutionary doctrine. But it was
also a change in the relative weight attached to the opinions of
town and country. The " advanced " views current in the early
days of the revolution, and the practices corresponding to them,
were representative of the towns rather than of the country as a
whole, and of party circles rather than of the population at large.
Precise information is both difficult to obtain and difficult to
assess. Divorce statistics showed that in the RSFSR in the last

---

[1] *Izvestiya*, January 2, 1926.

[2] *Statisticheskii Spravochnik SSSR za 1928 g.* (1929), pp. 896-897.

[3] In April 1926 a resolution of TsIK still spoke of the need for " measures
directed to the liquidation of the phenomenon of homeless children " in the
Ukraine (*SSSR : Tsentral'nyi Ispolnitel'nyi Komitet 3 Sozyva : 2 Sessiya :
Postanovleniya* (1926), p. 23).

[4] *Izvestiya*, February 20, 1926. A pamphlet by A. Sabsovich published in
1929, quoted in R. Schlesinger, *The Family in the U.S.S.R.* (1947), pp. 169-171,
still treated " the bringing up of children from their earliest days in special
state institutions at the expense of the government " as the ultimate ideal ;
later this became heretical, even as a remote prospect.

[5] *Vserossiiskii Tsentral'nyi Ispolnitel'nyi Komitet XII Sozyva : Vtoraya
Sessiya* (1925), pp. 254-255.

three months of 1924 there had been seven divorces per 10,000
of population in provincial capitals, three in smaller towns and
two in the villages.[1]  A hotly contested debate in the TsIK of
the RSFSR in 1925 and 1926 on a proposal to make the legal
consequences of " de facto marriage " identical with those of
registered marriage [2] revealed a strong prejudice among the
peasants, which was almost entirely absent in the towns, in favour
of maintaining the exclusive rights and obligations of conventional
marriage, and even of limiting automatic freedom of divorce.  A
woman delegate put the case with pungency for the peasant view :

> The villages do not wish to bring to the rural areas the
> instability of town marriages.  Who is responsible for the home-
> less orphans ?  The villages ?  No, by your leave, the towns.
> What will happen if the 85 per cent of the population of our
> country formed by the peasantry do as the towns do ?  We
> should flounder in disintegration.  Registration of marriage is a
> useful check in this respect. . . .  Marriage should be annulled
> only by a court.[3]

It was many years before limitations were imposed on the right of
divorce.  But here too NEP, representing the reaction of the
peasant against the towns, brought with it a certain reaction
against revolutionary dogmatism and in favour of traditional ways
of life in a national setting.[4]

[1] Ibid. pp. 304-305.
[2] Under article 133 of the marriage code of 1918 (see p. 29 above) the rights
of illegitimate children were in no way different from those of legitimate children,
but claims to alimony and division of property in case of divorce were valid
only if the marriage had been registered.
[3] III Sessiya Vserossiiskogo Tsentral'nogo Ispolnitel'nogo Komiteta XII
Sozyva (1926), pp. 689-690.  Izvestiya, January 9, 1926, reported on the
discussion which had been taking place throughout the provincial press.  In
the towns the women were said to favour the proposal to equate de facto with
registered marriage, the men to oppose it (the practical effect would be to
strengthen the financial claim of the mother against the father of her child) ;
in the country, opinion was unanimous against it.  Party members were said to
approve it in principle, but many of them regarded it as impracticable in view
of " the ignorance of the masses and especially of the peasant population ".
According to Izvestiya, January 31, 1926, " reports which come in from the
different regions and republics are almost unanimous for rejection ".
[4] In the RSFSR, where the weight and prestige of the cities in party counsels
turned the scale, the proposal to recognize de facto marriage as conferring the
same legal rights and obligations as registered marriage won a short-lived victory
and was inscribed in the marriage code of November 1926 (Sobranie Uzakonenii,
1926, No. 82, art. 612).  In the other republics, where peasant influence was
dominant, no such recognition was ever accorded.

## (b) The Orthodox Church

The Bolshevik revolution had overtaken the Orthodox Church at a moment of internal crisis. The collapse of the monarchy stimulated a movement for the re-establishment of the patriarchate abolished by Peter the Great. This was advocated by some as a necessary condition of efficiency in the church, and condemned by others as incompatible with the spirit of Orthodoxy, which rejected any kind of Papacy and held that the custody of the true faith was vested in the whole body of believers. A holy synod, which met in August 1917, decided by a narrow majority, at the very moment of the revolution, to restore the patriarchate, and on November 5/18, 1917, before the power of the new government had been actually established in Moscow, chose a patriarch by lot (from three candidates nominated by voting) in the person of Tikhon, the metropolitan of Moscow.[1] A clash between the church and the Bolsheviks was inevitable. After Tikhon had pronounced an anathema against the usurpers, a decree was issued pronouncing the separation of church and state and the nationalization of church property.[2] The church was not formally banned. The constitution of the RSFSR adopted in July 1918 recognized " freedom of religious and anti-religious propaganda ". The party programme adopted in 1919 proposed to counter religion by education and propaganda rather than by state action, and even recommended a measure of caution in dealing with it :

> The RKP is guided by the conviction that the realization of planned order and consciousness in the whole social-economic activity of the masses can alone bring with it a complete dying out of religious prejudices. The party aims at a complete destruction of the link between the exploiting classes and the organization of religious propaganda by assisting the effective liberation of the toiling masses from religious prejudices and by organizing the broadest propaganda in favour of scientific enlightenment and against religion. At the same time it is necessary carefully to avoid any insult to the feelings of

---

[1] *Orientalia Christiana Analecta*, No. 129 (Rome, 1941), contains the best informed and most dispassionate available account with a bibliography ; Metropolit Evlogii, *Put' Moei Zhizni* (Paris, 1947), is the autobiography of a participant in the synod of 1917–1918.
[2] See *The Bolshevik Revolution, 1917–1923*, Vol. I, p. 153, note I.

believers which can lead only to the strengthening of religious fanaticism.[1]

The principle, however, remained and was strengthened by the experience of the civil war. Religion, wrote Trotsky at this time, was the " principal moral arm of the bourgeoisie ".[2] Persecution was widespread. Killings of priests occurred, and many churches were taken for secular uses. Yet the intensity of the struggle varied from place to place, and depended in part on the character and attitude of the local Soviet authority and the local priest. A case was quoted from the year 1919 of a group of village Soviets which met with members of the local party cell to elect a church council for the parish church, and petitioned for the exemption of the precentor from military service on the ground of his indispensability ; such examples of toleration were said to have been not rare.[3] Measures of repression adopted by the Soviet authorities in the first years of the régime were spontaneous and spasmodic rather than uniform or calculated.

The ending of the civil war and the coming of NEP did not at first affect the Soviet attitude towards the church. At the end of 1921 the Soviet Government took cognizance of the vessels and ornaments in the possession of the churches, ordering that these should be classified in three categories — articles of historical or artistic value, articles of material, but no historical or artistic, value, and articles in ordinary use — and that nothing should be removed without the consent of the museum administration.[4] Then, at the height of the famine which raged throughout this winter, the Soviet Government issued on February 16, 1922, a decree ordering that articles containing gold, silver and precious stones in the possession of the church, " the removal of which cannot essentially affect the cult ", should be handed over to Narkomfin and sold abroad for the benefit of the hunger-stricken population : an instruction following the decree made it clear that gold and silver vessels used in church services were not exempt from requisition.[5] Tikhon gave orders to the faithful to

[1] VKP(B) v Rezolyutsiyakh (1941), i, 289.
[2] Trotsky, Sochineniya, xii, 141.
[3] Sovetskoe Stroitel'stvo : Sbornik, iv-v (1926), 138.
[4] Sobranie Uzakonenii, 1922, No. 19, art. 215.
[5] Ibid. No. 19, arts. 217, 218 ; a previous decree of February 9, 1922, related to the sale of treasures in museums for famine relief (ibid. No. 19, art. 216).

resist. The orders were carried out in many places. Riots occurred with numerous casualties and arrests, and were extensively reported in the Soviet press.[1] Numbers of priests were put on trial, and several sentenced to death. Tikhon himself was finally arrested. These proceedings were accompanied and justified by a propaganda campaign in which the church was accused of being in league with counter-revolutionary forces abroad and of counting on the weapon of hunger to bring about the downfall of the Soviet régime. Anti-religious themes became prominent in the party press. In the spring of 1922 a publishing house was set up to publish a monthly journal, *Bezbozhnik*, which engaged in a popular campaign to discredit religion ;[2] and the Orthodox Christmas of 1922 was made the occasion for a much-publicized anti-religious festival.[3] Throughout the winter trials of priests for resistance to the orders of the government, or sometimes more specifically for counter-revolutionary activities, continued intermittently ; sentences of death were frequently pronounced, and more rarely carried out. In March 1923 the trial of a group of Catholic bishops and priests, and the execution of one of them, led to world-wide protests, and was one of the items which figured in the Curzon ultimatum.[4]

Simultaneously with this campaign, however, another and more significant development occurred. A group of priests, who rejected the institution of the patriarchate, were personally opposed to Tikhon, and claimed to represent reforming and modernizing tendencies in the church, issued a letter denouncing Tikhon for his refusal to surrender church treasures. This letter was published in the Soviet press [5] and formed the starting-point of a new movement, which evidently enjoyed the qualified approval of the Soviet authorities. Early in May 1922 a new journal, *The Living Church*, was created to further the aims of the movement ; and a few days later a manifesto of the group was published in *Izvestiya* [6] accusing the existing church leaders of a conspiracy against the secular power and appealing to the Soviet Government to sanction the holding of a synod to put the affairs of the church in order and condemn the offending bishops. The leaders

[1] See, for example, *Izvestiya*, March 28, 1922 ; *Pravda*, May 19, 1922.
[2] *Izvestiya*, August 5, 1922.                    [3] *Ibid.* January 10, 1923.
[4] See *The Interregnum, 1923–1924*, p. 168.
[5] *Izvestiya*, March 29, 1922.                    [6] *Ibid.* May 14, 1922.

of the movement, who claimed — a point which was later contested — to have received some kind of provisional powers from
Tikhon in prison, convened an ecclesiastical assembly which met
in Moscow at the end of May 1922, reconstituted the church
under the name of " the living church ", and replaced the patriarch
by a " supreme church administration ".[1]   In August 1922 a
conference of the Living Church met in Moscow to consolidate
its position and to organize an attack on parishes and priests
remaining faithful to the patriarchate and to Tikhon.   A deputa-
tion from the conference was received by Kalinin.[2]   On the eve
of the conference a decree had been issued under which all
" associations not serving purposes of material gain " were obliged
to seek registration with the state authorities ; those which failed
to secure registration were to be closed down.[3]   This provided
an opportunity for a vigorous campaign to deprive Tikhon's
adherents of legal status and to hand over churches and buildings
occupied by them to nominees of the Living Church.   Bitterness
on both sides was extreme.   The Living Church was loudly
denounced by Tikhon's supporters as a tool of the Soviet Government, and its leaders accused of instigating and supporting the
persecution of the faithful.

These developments were significant as constituting the first
formal recognition of religious bodies by the Soviet state.   Trotsky
called the new policy " an ecclesiastical NEP ".   The rather far-
fetched comparison rested on the argument that, while socialism
could ultimately have no truck with religion, concessions analogous
to those made to capitalists under NEP could be temporarily

[1] The rise of the Living Church is described in *Orientalia Christiana*
(Rome), No. 46 (June 1928), pp. 8-15 ; in M. Spinka, *The Church and the
Russian Revolution* (N.Y., 1927), pp. 190-224 ; and in W. C. Emhardt, *Religion
in Soviet Russia* (1929), pp. 304-332 (this section of the book was written by
an émigré Orthodox theologian).   Much was made in Orthodox accounts of
the allegation that Vvedensky, one of the leaders of the Living Church movement, was " a baptized Jew " (*ibid.* p. 312).   The Living Church was referred
to as " the Jewish church ", and its existence attributed to " some Jewish
agitators ", in a document quoted at length in *Orientalia Christiana* (Rome),
No. 4, July-September 1923, pp. 214-217 ; the head of the Living Church
in the Ukraine (where anti-Semitic propaganda was particularly effective) was
denounced as " the vicar of the circumcised Jew Bronstein " (*ibid.* No. 4,
pp. 132-133).   The frankest discussion of anti-Semitism in this period is in
L. Trotsky, *Voprosy Byta* (2nd ed. 1923), pp. 143-145.
[2] *Pravda*, August 23, 1922.
[3] *Sobranie Uzakonenii, 1922*, No. 49, arts. 622, 623.

extended to a group which, like Protestantism in the west, stood
for a bourgeois, capitalist and quasi-rationalist revolt against the
extreme superstitions of the old feudal religion — " a bourgeois
graft on a feudal trunk ".[1]  The leaders of the Living Church
were thus in the position of nepmen or *kulaks*, recognized as
essentially bourgeois and discredited in principle, but tolerated
for the temporary contribution which they could make to the
survival of the régime.  The comparison between the new ecclesiastical policy and NEP was valid in one respect.  Both denoted a
certain reaction against the exaggerated optimism of the first
years of the revolution, when it had seemed possible to overthrow
the power of capitalism and the power of the church by direct
assault.  Just as it had proved necessary to make concessions to
buyers and sellers of commodities, so it was necessary to conciliate in some measure those who still clung to the practices of
the church.  Religion had not been eliminated at a single stroke
by the revolution.  Even among the workers old habit died hard,
and all sorts of compromises were practised.  The worker, in the
words of one witness, " does not buy new ikons, but does not
throw the old ones away ".  According to another, he " does not
go to church — and reads *Bezbozhnik*, but sends for a priest to
christen his child — just in case ; he does not go to confession,
but when he is dying sends for a priest ".[2]  Another thought
that the Russian was basically irreligious, but that religion had
hitherto been the only form of distraction open to him :

> Today when some non-party people go to church, they go
> only perhaps because they have nothing to fill the emptiness of
> their lives. . . . He denies god, but at the same time goes to
> church.  Why does he go ?  Because we have broken up what
> existed, and created nothing on the ruins.  We, communists,
> must create something new.[3]

In the country, and especially among the women, the hold of
religion had scarcely been shaken at all.  Attacks on religion in
the countryside tended to provoke unfavourable reactions ; and

---

[1] L. Trotsky, *Literatura i Revolyutsiya* (1923), p. 29.  An apter comparison
was suggested by a Soviet writer who wrote of the Living Church movement
under the title *Smena Vekh v Tserkvi* (quoted in W. C. Emhardt, *Religion in
Soviet Russia* (1929), p. 80) ; for the *smenovekh* movement see pp. 56-59
below.          [2] L. Trotsky, *Voprosy Byta* (2nd ed. 1923), pp. 143, 145.
[3] *Ibid.* p. 146.

party workers had more than once to be warned of the danger of indulging in them.

The Living Church, having originated in a split from the parent church, itself proved fissiparous, giving birth within the first year of its existence to two sects calling themselves respectively " the renovators " and " the primitive apostolic church ". Common hostility to the patriarchate and common reliance on the support of the Soviet Government sufficed, however, to hold the three groups together, and all were represented at a holy synod convened in May 1923. This gathering began by defining its attitude to the Soviet Government :

It recognizes the justice of the social revolution : it sees in the Soviet power the force that is leading the world to fraternity, equality and peace among nations ; it condemns the counter-revolution, and treats the anathema of Patriarch Tikhon as invalid.

It then denounced the patriarchal church in no uncertain terms, declared Tikhon deposed and the patriarchate abolished, established a Supreme Council of the Russian Orthodox Church as the highest ecclesiastical authority, dissolved the monasteries and adopted various reforms including the marriage of bishops and the Gregorian calendar.[1] " The jargon of our days on the lips of the ' Red fathers ' ", wrote a correspondent of *Izvestiya* signing himself " Unbeliever ", " sounds the death-knell of the Tikhonite church." [2]

For the moment the triumph of the Living Church and its associate groups seemed complete. But this victory proved a turning-point. The Soviet authorities had no intention of committing themselves unconditionally to their new protégés. What motives weighed most strongly in the new shift of policy is uncertain. At home the Living Church had failed to appeal to the peasant who was traditionally attached to ancient religious forms : it was no accident that the change of course came at a moment when the party was particularly conscious of the need to

[1] Documented accounts of the synod are given in *Orientalia Christiana* (Rome), No. 11 (September-November 1924), pp. 22-26, No. 46 (June 1928), pp. 32-40, and in M. Spinka, *The Church and the Russian Revolution* (N.Y., 1927), pp. 232-249 ; brief reports of it appeared in *Pravda*, May 5, 8, 9, 1923.
[2] *Izvestiya*, May 5, 1923.

strengthen the " link " between the proletariat and the peasantry, and a party congress had just referred, in this context, to the dangers of antagonizing the religious feelings of believers.[1] The persecution of Tikhon and the patriarchal church had been the subject of intensive propaganda abroad, where the Living Church was dismissed as a mere tool of the Soviet Government. In the mood of conciliation which followed the Curzon ultimatum, a policy less obnoxious to the outside world had its appeal. It seemed that the ends which the Soviet Government had in view could be achieved, not by the cooperation hitherto practised with the Living Church, but by a similar compromise with Tikhon and the patriarchal church. This, after a year of persecution and repression, was no longer unattainable. On June 26, 1923, Tikhon, whose impending trial had been several times announced, signed a confession of his " hostility to the Soviet authorities, and anti-Soviet acts ", admitting that these had been correctly stated in the charges brought against him and that the sentence on him had been in accordance with the criminal code. He expressed repentance for his actions, and petitioned to be set free. He declared that " henceforth I am no longer an enemy of the Soviet Government ", and that he had " completely and decisively severed all connexions with monarchists at home and abroad and with all counter-revolutionary white guard activities ".[2] On the strength of this confession Tikhon was released and allowed to resume his former patriarchal functions. Part of the understanding clearly was that Tikhon, in renouncing his hostility to the Soviet régime, was free to reassert his claims against the Living Church.[3] A fortnight after his release he made a public statement denouncing the leaders of the Living Church by name, and describing as " a lie and a deception " the pretence which they had made in May 1922 of acting with authority from him ; those who had acknowledged this illegal authority were

[1] See *The Interregnum, 1923–1924*, p. 17.
[2] *Izvestiya*, June 27, 1923.
[3] In an interview published in the *Manchester Guardian* on July 15, 1923, Tikhon said : " We, the members of the old church, are not now struggling against the Soviets, but against the Living Church ". Asked why he had been liberated, he replied : " I am persuaded that, having studied my case, the government became convinced that I was no counter-revolutionary. It was suggested that I should make a public declaration of the fact, and I wrote a letter to say so."

invited to "return into the saving bosom of the ecumenical church ".[1] Some of the leaders of the Living Church made their submission to Tikhon. The remaining members of the dissident groups now reorganized themselves into a single church, the " supreme church administration " being renamed the Holy Synod of the Russian Orthodox Church.[2]

The Soviet Government thus adopted a neutral position. The dissident church, generally referred to as " the synodal church " or " the renovators ", henceforth continued to exist side by side with the older body, " the patriarchal church ", but in much diminished strength and without direct Soviet support. On the exclusively secular character of the régime no compromise was to be thought of. Among the prescribed functions of village Soviets was " the supervision of the correct observance of the laws concerning the separation of the church from the state and of the school from the church ".[3] But, though anti-religious propaganda was not abandoned, the patriarchal church was no longer persecuted, and was recognized, in so far as an ecclesiastical institution could be recognized by a state whose official doctrine openly denounced religion. The period from 1923 to 1925, when conciliation of the peasant was in the forefront of party policy, was also the period of greatest toleration for the patriarchal church. The church, under Tikhon's leadership, took up the same attitude of qualified acceptance of the state. When Tikhon died at the age of eighty on April 7, 1925, his funeral was the occasion of a large religious demonstration which was rather ostentatiously tolerated by the Soviet authorities and reported in the Soviet press;[4] and a few days later a pronouncement was published which purported to have been signed by Tikhon a few hours before his death enjoining the faithful " to submit themselves loyally to the Soviet power, to pray to God to aid it in its efforts for the common good, and to organize the life of the parishes independently of the politicians ".[5] The circumstances of the publication threw some

---

[1] A translation of the statement is in W. C. Emhardt, *Religion in Soviet Russia* (1929), pp. 129-131.

[2] M. Spinka, *The Church and the Russian Revolution* (N.Y., 1927), pp. 271-272.

[3] *Sobranie Uzakonenii, 1924*, No. 82, art. 827 ; corresponding provisions from the decrees of the Ukrainian and White Russian republics are in P. Gidulyanov, *Otdelenie Tserkvi ot Gosudarstva v SSSR* (3rd ed. 1926), pp. 18, 19.

[4] *Pravda*, April 12, 13, 1925.          [5] *Ibid.* April 15, 1925.

doubts on the authenticity of the statement. But its content accorded with the policy pursued by Tikhon since 1923. The toleration by the state of a national church was conditional on ecclesiastical recognition of the secular power. A *modus vivendi* had been established between the revolutionary régime and an ancient national institution.

### (c) Literature

The party had made no pronouncement of its views on literature before the revolution. At the height of the revolution of 1905 the relaxation of the censorship prompted Lenin to write an article entitled *Party Organization and Party Literature* which afterwards gave rise to contested interpretations. Lenin insisted with some emphasis on the party character of literature :

> For the socialist proletariat the cause of literature . . . cannot in general be an individual concern independent of the common proletarian cause. Down with non-party *littérateurs*! Down with supermen *littérateurs*! The cause of literature must become part of the general proletarian cause, a " wheel and cog " in our single great social-democratic machine set in motion by the whole conscious vanguard of the whole working class.

Lenin anticipated the frenzied objection of " some intellectual, some fervent partisan of liberty " that it was impossible to bring about " the subordination to the collectivity of such a delicate, individual matter as literary creation ". Apart from the fact that the supposed liberty of the bourgeois writer was a myth, Lenin also pointed out that he was speaking only of " party literature and its subordination to party control ". Anyone would be free outside the party to write anything he pleased " without the slightest restrictions ". But the party was also free to exclude from its ranks anyone who expressed anti-party views.[1] It was afterwards claimed that Lenin in this article referred exclusively to political writing and not to belles-lettres at all. This was clearly not true. What was true was that neither Lenin nor any other party leader in 1905 contemplated a situation in which the party would have either the will or the power to establish a monopoly of literary

[1] Lenin, *Sochineniya*, viii, 387-389.

output.  He believed that the literary talents of party members, like their other talents, should be devoted to the service of the party and that whatever they wrote should conform to the party line.  But he assumed that non-party literature, which would be subject to no such obligations or restrictions, would continue to be written and published.[1]  Lenin was a reader of the Russian classics, but had no theory of literature.  When he wrote articles on Herzen and Tolstoy, he showed himself more concerned with their social than with their literary significance.  He took no interest in contemporary literary controversies.

When the revolution occurred in 1917 the centre of the literary stage in Russia was occupied by several schools or movements whose widely different theories converged on one point : all were in revolt against the view of nearly all nineteenth-century Russian literary criticism, which had treated literature as a manifestation of social thought and criticism as an instrument of ideological analysis and appraisal.  The new schools were at one in putting form before content.  Literature was based on the significant use of words ; and aesthetic criticism was concerned primarily with modes of expression.  This approach was shared by groups which had little else in common : Symbolists, Acmeists, Rhythmists, Futurists, and, finally, Formalists who became an organized movement only in 1916.  These groups purported to represent " advanced " thought in literature : some individual members of them supported the revolution.  Alexander Blok, who was a Symbolist and whose political affiliations were with the Social-Revolutionaries, wrote two famous poems which proclaimed his sympathy with the revolution.  The Formalists boasted the " revolutionary " credentials of their literary techniques.  Of all the groups the Futurists had the best claim to revolutionary status, partly because they had always made the bourgeoisie and bourgeois civilization a target for their shafts of ridicule and indignation, and partly because they had produced in Mayakovsky a considerable poet who found Bolshevism, at any rate in its destructive aspects, temperamentally congenial to him.  Mayakovsky not only wrote and recited in public a large amount

[1] This view was exactly parallel to Lenin's attitude to religion.  He believed that atheism, and even militant atheism, should be an obligation of party members, but that the state as such should tolerate religious activities, provided that these were not directed against public order.

of first-rate declamatory verse on revolutionary themes, but denounced all bourgeois art past and present in the coarsest and most unflattering terms.[1]  In the years between 1917 and 1920, when ordinary literary production and publication were almost at a standstill, and occasional poetry the main vehicle of literary expression, the revolution appeared to have found in Mayakovsky its poet laureate.

Yet it was difficult to see how the ideas of Futurists or Formalists, however advanced in their way, could be fitted into the doctrinal framework of Marxism or be made to serve the aspirations of the proletariat ;[2] and there were from the outset Bolsheviks who believed that the dictatorship of the proletariat must evolve its own literary movements and modes of literary expression.  Such views had been expressed before the revolution by Bogdanov, an independent Bolshevik who, in 1909, in association with Gorky and Lunacharsky, had founded a party school in Capri, and had crossed swords with Lenin in a famous philosophical dispute.   In 1910 he had incurred Lenin's disapproval by advocating a new proletarian culture, and by proposing " to develop proletarian science, . . . to work out a proletarian philosophy, and to turn art in the direction of proletarian strivings and experience ".[3]  But nobody had seriously thought of laying down a party line on these matters.  It was not therefore surprising that Bogdanov should have emerged as the moving spirit in a new Organization of Representatives of Proletarian Culture (henceforth known as Proletkult), which was set up on the eve of

[1] Brik, the Futurist critic, called bourgeois art " an exhalation from a swamp ", and Mayakovsky demanded that the firing-squad should give its attention to Raphael, Rastrelli, Pushkin and other " classical generals " ; the latter declaration provoked a protest from Lunacharsky against " the destructive tendencies in regard to the past and the attempt, while speaking in the name of a particular school, to speak at the same time in the name of authority ".  All these statements appeared in December 1918 and January 1919 in the semi-official journal *Iskusstvo Kommuny* and are quoted in V. Polonsky, *Ocherki Literaturnogo Dvizheniya Revolyutsionnoi Epokhi* (2nd ed. 1929), pp. 33, 249-251.

[2] In an article of February 1914 Trotsky had written : " The phenomenon of Futurism is the perfectly legitimate and in its way most finished crown of an epoch about which it can be rightly said : ' In the beginning was the word — and in the middle and the end as well ' " (Trotsky, *Sochineniya*, xx, 380).

[3] Lenin, *Sochineniya*, xiv, 297 ; Lenin's disapproval was evidently due in part to his suspicion that Bogdanov's " proletarian philosophy " would derive from Mach, his philosophical mentor.  For Bogdanov's career see *Literaturnaya Entsiklopediya*, i (1930), 526-530.

the revolution more or less independently of the party, and now
enjoyed the patronage of Bogdanov's old colleague, Lunacharsky,
first People's Commissar for Education.[1]  In the first months of
the revolution, and especially during the civil war, Proletkult·
recruited a large number of enthusiastic workers, founded local
branches, encouraged proletarian poets, founded journals for the
propagation of proletarian literature, and, in general, performed
important work in keeping culture alive and in disseminating it
among the workers.  It was not in itself a literary movement.  But
there emerged from it early in 1920 a group of proletarian writers
who called themselves the Forge or Smithy — a name calculated
to evoke the rôle of literature as a proletarian workshop — and
issued a manifesto which they described as " the red flag of the
platform-declaration of proletarian art ".  This group, after a
preliminary conference in May 1920, which mustered 150 supporters, was instrumental in convening in October 1920 an
All-Russian Congress of Proletarian Writers, which founded an
All-Russian Association of Proletarian Writers (VAPP).[2]

The views of Bogdanov, which dominated the activity of
Proletkult, formed a clear and consistent whole.  He conceived
the dictatorship of the proletariat as advancing on three parallel
but distinct lines, political, economic and cultural.  Its political
organ was the party, its economic organ the trade unions, its
cultural organ Proletkult.  Literature, like politics or economics,
was a class activity, but was sovereign in its own sphere : hence
it was inappropriate that Proletkult should be in any way subordinate to the party.  Bogdanov even maintained that Proletkult,
being exclusively proletarian, was more advanced than the party
which, as a political organ, was bound to take account of the
alliance with the petty bourgeois peasantry ; in a phrase which
was afterwards quoted against him, he described the proletarian
writers as " immediate socialists ".  Proletkult thus had the
positive rôle of acting as pace-maker of the revolution.  Bogdanov

[1] *Ibid.* ix (1935), 309-311, which names Polyansky, Pletnev and Kerzhentsev
as the other leading figures in Proletkult.  Polyansky was a historian and
literary critic, Kerzhentsev a party intellectual who was active in the Central
Institute of Labour (see *The Interregnum, 1923-1924*, p. 84) and was at different
times *polpred* in Sweden and Italy (*Literaturnaya Entsiklopediya*, v (1931), 187-
189) ; for Pletnev see p. 63 below.
[2] *Ibid.* v (1931), 703-707 ; the manifesto is quoted in V. Polonsky, *Ocherki
Literaturnogo Dvizheniya Revolyutsionnoi Epokhi* (2nd ed. 1929), pp. 52-53.

did not, like the Futurists, attack the culture of the past, but believed that the proletariat was capable of taking it over and assimilating it without the current aid of bourgeois writers. His position in this respect was analogous to that of supporters of workers' control in the factories who decried the employment of specialists.[1]

During the civil war, Proletkult and its supporters remained in the ascendant, partly because the prevailing political mood favoured a utopian faith in anything proletarian, and partly because the political leaders had little attention to give to anything not immediately related to the problem of survival.[2] But Lenin's disapproval of Bogdanov's doctrines was never in question. Lenin had no doubt that art and literature were part of the " superstructure " of society in the Marxist sense, and had social foundations which made it impossible to treat them as an autonomous activity divorced from economics and politics. Far from taking the lead, it seemed clear to Lenin that the cultural arm must necessarily lag behind : " the cultural task cannot be discharged as rapidly as the political and military tasks ".[3] Bogdanov's demand for independence, and his insistence on literature as an animating force in the dictatorship of the proletariat smacked of idealism. His claim that the proletariat was ripe to take over and develop by its own unaided efforts the heritage of bourgeois culture seemed as presumptuous as other utopian dreams of the period of war communism. To call the proletarian writers " immediate socialists " was a glaring example of the skipping of stages which was so contrary to the Marxist doctrine of revolution. Already in 1919 Lenin had proclaimed a " relentless hostility . . . to all inventions of intellectuals, to all ' proletarian cultures ' " ; [4]

[1] For a full account of Bogdanov's views, with references to his writings, see V. Polonsky, *Ocherki Literaturnogo Dvizheniya Revolyutsionnoi Epokhi* (2nd ed. 1929), pp. 56-71.

[2] Trotsky, seeking to explain the poverty of current literature, observed, in reply to the reproach that there were " no Belinskys ", that if Belinsky were alive he would probably be a member of the Politburo (L. Trotsky, *Literatura i Revolyutsiya* (1923), p. 155).

[3] Lenin, *Sochineniya*, xxvii, 51. Trotsky developed the same thesis in an article on *Proletarian Culture and Proletarian Art* ; the Russian proletariat had, through the circumstances of the revolution, come into power before it had time to assimilate bourgeois culture, and must now, first of all, make good this deficiency (L. Trotsky, *Literatura i Revolyutsiya* (1923), p. 144).

[4] Lenin, *Sochineniya*, xxiv, 305.

and, as soon as the military situation eased and victory was in sight, he quickly found an opportunity to reassert his disapproval of Bogdanov's pretensions. At a Komsomol congress in October 1920 he insisted that " we can build communism only from that sum total of knowledge of organizations and of institutions, with the store of human powers and resources, which have been bequeathed to us by the old society ". He reminded his audience that Marx had arrived at his conclusions by a thorough study of capitalist society and " by dint of making fully his own everything which earlier science could give " ; and he went on to define his attitude towards proletarian culture :

> Proletarian culture is not something that suddenly springs from nobody knows where, and is not invented by people who set up as specialists in proletarian culture. Proletarian culture is the regular development of those stores of knowledge which mankind has worked out for itself under the yoke of capitalist society, of feudal society, of bureaucratic society.[1]

Lenin instructed Lunacharsky to take steps, at a Proletkult congress held in the same month, to put Proletkult in its place as a subsidiary department of the People's Commissariat of Education (Narkompros) without independent status and powers. Lunacharsky failed to carry out these instructions, saying at the congress the opposite of what, according to Lenin, he had undertaken to say, and maintaining that " Proletkult must preserve its quality of independent activity ".[2] Lenin then brought the issue before the party central committee. A resolution was drafted, by the last paragraph of which the congress would " decisively reject as theoretically incorrect and practically harmful all attempts [of Proletkult] to invent its own special culture, to confine itself within its own particular organizations, . . . or to set up an ' autonomous domain ' of Proletkult within the institutions of Narkompros " ; and Bukharin and Pokrovsky were entrusted with the task of piloting it through the congress.[3] This was duly done ; and Bogdanov withdrew from the central committee of Proletkult. But, though Proletkult never recovered its former prestige, VAPP remained to uphold, in face of increasingly active opposition, the doubtful cause of proletarian literature.

[1] *Ibid.* xxv, 384-385, 387.        [2] *Izvestiya*, October 8, 1920.
[3] Lenin, *Sochineniya*, xxv, 409 ,636 637, note 197.

The year 1921 was a turning-point on the literary, as on the economic, front, and heralded the growth of a new outlook on the world of literary creation.   If NEP meant a retreat from the uncompromising assertion of proletarian principles and a compromise with the forces of capitalism, the way seemed open for a corresponding recognition of pre-revolutionary literary values and traditions.   In Soviet Russia, the month of February 1921 saw the foundation of a new literary movement of a different character from any of its predecessors.   Twelve young writers of bourgeois origin formed a group calling themselves the " Serapion brothers ". The name was borrowed from one of Hoffmann's tales, and indicated that they professed no common political allegiance, but only a common allegiance to art.   What united them and constituted their importance was that, far from rejecting the past, they were ready to model themselves on the classics of western and Russian literature, and regarded themselves as bearers of an existing literary tradition rather than as creators of a new one. In opposition both to the Futurists and to the Smithy, they stood for the principle of continuity.   Among the " Serapion brothers " who were destined to fame in Soviet literature were Vsevolod Ivanov, Fedin, Kaverin, Nikitin, Zoshchenko and the Formalist critic, Shklovsky.[1]   Their publications in the first year of their existence included an *Almanakh* and three numbers of a journal entitled *Literaturnye Zapiski*.   But the brotherhood would have exercised no great influence if its formation had not coincided with a fresh official initiative in the literary field.   In 1921 a State Publishing House (Gosizdat) was founded, though it was several years before it acquired a monopoly of publishing.   A decision of even greater immediate importance — it was a symptom of the general abandonment of unconditional hostility to the traditions of the past — was to establish in Petrograd two monthly literary journals on the lines of the " thick " journals of the pre-revolutionary period.   The first of these, edited by a party member Voronsky, began to appear in May 1921 under the title *Krasnaya Nov'*.   It was not originally an exclusively literary journal.   The first issue contained, in addition to Vsevolod Ivanov's story of the

[1] The main sources relating to the " Serapion brothers " are collected in *American Slavic and East European Review*, viii (1949), 47-64 ; the account in V. Pozner, *Panorama de la Littérature Russe Contemporaine* (1929), pp. 324-327, is by a former member of the group.

civil war *Partisans*, Lenin's article in defence of NEP, *On the Tax in Kind*, and articles by Radek and Krupskaya. But the literary items always came first ; and, as more new authors came on the scene and the public taste for literature of the familiar kind declared itself, the literary section came to predominate, and the major part of successive issues was devoted to prose fiction with an admixture of poetry and memoirs. Babel, Pilnyak, Vsevolod Ivanov, Kataev and Fedin were among those who in this way acquired fame as the new lights of Soviet literature. The second journal, entitled *Pechat' i Revolyutsiya*, and described as " a journal of literature, art, criticism and bibliography ", counted among its editors Lunacharsky, Pokrovsky the Marxist historian, and Polonsky, a Marxist literary critic. It followed the same general line, and appealed to the same public, as *Krasnaya Nov'*, though with less attention to current literary output.

The place in Soviet society of the contributors in these new literary journals was easy to define. The essence of NEP was not to reject or destroy capitalist forms, but to use them for the eventual advancement of socialism ; and this, too, had its literary application. Even under war communism, the employment of members of the bourgeoisie, first as military specialists, and later as specialists in administration and industrial management, had made great strides, which were rapidly consolidated under NEP. The argument for utilizing the services of bourgeois writers who were willing to work under the new régime, and in a spirit not unfriendly to it, became irresistible. It was Trotsky who, in the subsequent controversy aroused by their work, described them, in a phrase which stuck, as " not artists of the proletarian revolution, but its artistic fellow-travellers ".[1] The " fellow-travellers " were almost all young men between twenty and thirty. Having no pre-revolutionary past, they were moulded by the revolution and, while uncommitted to communist doctrine, accepted the revolution as an event in the history of the nation. Several of their best novels, beginning with Fedin's *Cities and Years* of 1924, wrestled with the problem of the adaptation of the young bourgeois intellectual to the revolution and its values. But the secret of their popularity lay in their treatment of revolutionary themes in traditional literary forms. In this sense they stood for the

[1] L. Trotsky, *Literatura i Revolyutsiya* (1923), p. 41.

continuity of Russian literature, and represented a reaction both
against the proletarian writers, who claimed to create a specifically
proletarian literature, and against stylistic innovators like the
Futurists and the Formalists, who regarded the literary methods
and techniques of the past as obsolete. In a society which had
begun to tire of the cult of innovation they enjoyed an immediate
success.

But the fellow-travellers represented historical continuity in
more than the formal or purely literary sense. In accepting the
revolution shorn of its communist doctrine and of its proletarian
basis, they insensibly transformed it into a national revolution in
the Russian tradition. The fellow-travellers, in Trotsky's analysis,
fell into the category of " Soviet *narodniks* " : they were all " more
or less inclined to look over the head of the worker and fix their
kaze with hope on the peasant ".[1] For Pilnyak, whose *Naked
Year*, published in 1921, was the first major work of a fellow-
traveller, the revolution was a disorderly tumult, an upsurge of
primitive peasant revolt in the style of Pugachev, sweeping away a
corrupt urban civilization. The national strain was, from the first,
strong in Pilnyak. His attitude was recorded in a " diary " of
1923 which he allowed to be published in a symposium in the
following year :

> I am not a communist and therefore do not acknowledge
> that I ought to be a communist and write as a communist. I
> acknowledge that the communist power in Russia is determined
> not by the will of the communists, but by the historic destinies
> of Russia ; and, in so far as I want to follow, according to my
> ability and as my conscience and mind dictate, these Russian
> historical destinies, I am with the communists — that is, in so
> far as the communists are with Russia, I am with them. . . .
> I acknowledge that the destinies of the Russian Communist
> Party are far less interesting to me than the destinies of Russia.[2]

[1] L. Trotsky, *Literatura i Revolyutsiya* (1923), pp. 41-42 ; elsewhere
(*ibid.* p. 164) Trotsky refers to " a peculiar neo-*narodnichestvo* " as " characteristic of all fellow-travellers ".

[2] *Pisateli ob Iskusstve i o Sebe*, No. 1 (1924), pp. 83-84 (no further numbers
of this publication are known to have appeared). The flavour of Pilnyak's view
emerges from a speech put into the mouth of an illiterate village elder in *The
Naked Year* : " Russia fell under the Tatars — there was the Tatar yoke ; Russia
fell under the Germans — there was the German yoke. Russia has a mind of her
own. The German has a mind, but his mind is foolishness — well-informed
about w.c.s. I say at the meeting : There is no such thing as the International,

Distinctively Slavophil motives appeared in nearly all Pilnyak's later novels. In *Three Capitals* he glorified pre-Petrine Russia and revived the familiar contrast between the decadent civilization of the west and wild, uncultivated, vital peasant Russia. In *Mother Earth* the native " Scythians " clearly have the best of the argument against " European " communists. Vsevolod Ivanov, one of the first Soviet writers to turn to Soviet Asia for his themes, was anti-rational as well as anti-western, exalting crude physical force above the sophistications of the intellect, and interpreting the revolution in terms of the healthy uncorrupted strength of the Russian peasant. The legacy of Bakunin seemed to have displaced the legacy of Marx. Leonov, a more sophisticated fellow-traveller, who drew his initial inspiration from Dostoevsky, depicted, in his novel *The Badgers* published in 1925, a group of peasant guerrillas who refuse to submit to communist rule and are ultimately put down by Soviet troops. But the leader of the badgers appears to have the last word :

> We are millions : we give bread and blood and strength. We are the land and we shall destroy the city.

At the height of NEP these were burning topical issues. The best of the fellow-travellers presented them with an ambivalence which was probably the product of their own divided minds [1] as well as of tactical discretion. But the fellow-travellers reflected the ideology of those who saw in NEP a salutary submission to the overwhelming resistance of the Russian peasant, and for whom the revolution seemed first and foremost a gesture of

but there is a Russian people's revolution, revolt and nothing else. Like in the days of Stepan Timofeevich [Razin]. ' And Karla Marxov ? ' they ask. A German, I say, and therefore a fool. ' And Lenin ? ' Lenin, I say, is of the peasants, a Bolshevik ; and you, I suppose, are communists ; therefore, I say, sound the alarm of freedom from the yoke. The land for the peasants ! Down with the merchants ! Down with the landowners, the fleecers ! Down with Constituent Assembly ! We want a Soviet of the land, where all may come who will, and decide under the open sky. Down with tea, down with coffee — they are small beer. Let there be truth and right. Moscow is our capital. Believe in what you like, in any blockhead you please. But the communists, too — down with them ! The Bolsheviks, I say, will make good by themselves."

[1] The critic Polonsky wittily compared the attitude of the fellow-travellers to the revolution with the attitude of Dostoevsky's Shatov to God. Shatov believed in Russia and in Orthodoxy, and believed that " Christ will come again in Russia ", but when challenged about his belief in God replied : "I . . . I shall believe in God " (V. Polonsky, *O Sovremennoi Literature* (1928), p. 73).

revolt against the intrusion of the west into the Russian national tradition. To accept the revolution while rejecting communism led inevitably to this conclusion.

About the same time as the fellow-travellers began to win recognition in Soviet Russia, a corresponding movement occurred among Russian *émigrés* abroad. These new bourgeois collaborators abroad differed from the fellow-travellers at home in having a past record of hostility to the revolution to be renounced and expunged ; and for this reason, unlike the fellow-travellers, they found it necessary to work out a theoretical justification for so paradoxical a step as a working compromise with the Soviet régime. In July 1921 a group of *émigrés* published in Prague a volume of essays entitled *Smena Vekh* (" A Changing of Landmarks "). The theme of the essays was the need for reconciliation between the Soviet régime and the Russian *émigrés* of former régimes ; and the argument was based on the essentially Russian character of the revolution and of the régime resulting from it. The leader of the group, Ustryalov, stated the argument in its most uncompromising form :

> No, neither we nor " the people " can properly evade our direct responsibility for the present crisis — for its dark, as for its bright, aspects. It is ours, it is genuinely Russian, it is rooted in our psychology, in our past, and nothing like it can or will happen in the west, even in the event of a social revolution copied in external forms from it. And if it is mathematically proved — as not altogether successful attempts are now being made to prove — that 90 per cent of the Russian revolutionaries are non-Russians, for the most part Jews, this does not in the least refute the purely Russian character of the movement. Even if " stranger " hands are harnessed to it, its soul, its " inner essence ", is all the same — for good or evil — truly Russian, a movement of the intelligentsia transmuted through the psychology of the people.
> It is not the non-Russian revolutionaries who govern the Russian revolution, but the Russian revolution which governs the non-Russian revolutionaries, who have assimilated themselves, externally or internally, to " the Russian soul " in its present condition.

In this interpretation, NEP became a vital turning-point in the history of the revolution. It was " the economic Brest of Bolshevism ", the adoption of " measures indispensable for the economic resurrection of the country irrespective of the fact that these measures have a bourgeois character ". Taking up the description of the Kronstadt rising in the émigré press as a " Russian thermidor ", Ustryalov argued that the thermidor had not implied a rejection of the French revolution, but its further progress by evolutionary means. In the same way, NEP meant that the Russian revolution had taken the path of evolution through " a transformation of the minds and hearts of its agents ". The revolution, he concluded, " is saving itself from its own excesses ". Klyuchnikov, another contributor, referring to the old charge that the Russian intelligentsia stood outside the nation and against the nation, openly invoked the tradition of Russian messianism as the basis of the reconciliation of the intelligentsia with the revolution :

> The Russian intelligentsia is seizing the principle of the *mystic* in the state, is being penetrated by " the *mystique* of the state ". Thus from being an extra-state or anti-state entity it will become a state entity, and through its mediation the state — the Russian state — will become that which it ought to be : the way of God on earth.[1]

In October 1921 a weekly journal bearing the same name, *Smena Vekh*, and preaching the same doctrine, appeared in Paris and ran regularly for several weeks. It referred with sympathy to events in Soviet Russia, and cautiously praised those intellectuals who had entered the party or the service of the Soviet Government. It drew a sharp distinction between Bolshevism and communism, and maintained that whatever the intentions of the Bolsheviks, the irresistible forces of NEP were carrying its authors along " the path of thermidor ".[2]

The initiative of the *smenovekhovtsy* provoked an ambivalent

---

[1] *Smena Vekh* (Prague, 2nd ed. 1922), pp. 50, 52-71 ; the theme of the reconciliation of the intelligentsia to the state, as well as the title of the volume, consciously recalled the famous volume *Vekhi*, published in 1908 by a group of Russian intellectuals who had embraced Orthodoxy, which attacked the Russian intelligentsia for its estrangement from the Russian nation. " In the Bolsheviks and through Bolshevism ", wrote Ustryalov shortly afterwards, " the Russian intelligentsia overcomes its historical apostasy from the people and its psychological apostasy from the state " (N. Ustryalov, *Pod Znakom Revolyutsii* (2nd ed. 1927), pp. 257-258).

[2] *Smena Vekh* (Paris), No. 3, November 12, 1921 ; No. 13, January 21, 1922 ; the last issue to appear was No. 20 of March 25, 1922.

response from the Soviet side. The original *Smena Vekh* was noticed on successive days, three months after its publication in Prague, by *Izvestiya* and *Pravda*; the latter observed with cautious satisfaction that its authors were " placing new landmarks on the path of the *rapprochement* of the intelligentsia with the revolution ", and that others would have to follow.[1] It was impossible to welcome whole-heartedly an acceptance of the revolution which so openly assumed that the revolution had abandoned its early ideals. An idealist view of the Bolshevik revolution as a unique expression of the Russian soul was utterly alien to everything believed and professed by the makers of the revolution ; and the interpretation of NEP as an evolution of Bolshevism in the direction of bourgeois moderation [2] was bound to be anathema to those who upheld NEP as a tactical manœuvre through which the aims of Bolshevism could be more surely achieved. Nevertheless, the breach in the anti-Soviet front of the Russian emigration was a bull point for the régime both at home and abroad, and would facilitate the reconciliation of former bourgeois intellectuals to their new rôle as loyal servants of the Soviet Government. The *smenovekhovtsy*, like the fellow-travellers, could not be ignored, and, while not admitted to the fold, could be used to further its ends. Bukharin dubbed them " friends in inverted commas ".[3]

The significant fact about the *smenovekh* movement was the immediate response which it evoked in intellectual circles in Soviet Russia. The volume of 1921 was reprinted in a Soviet edition, and two volumes of essays commenting on it appeared in the following year.[4] Lenin in 1922 admitted that the *smenovekhovtsy* " express the mood of thousands and tens of thousands of bourgeois of all sorts and Soviet officials who participate in our new economic policy ".[5] At the twelfth party congress a year later

---

[1] *Izvestiya*, October 13, 1921 ; *Pravda*, October 14, 1921.

[2] Ustryalov and his group, as Lenin indignantly said in March 1922 at the eleventh party congress, offered their support to the Soviet régime " on the ground that it has taken the path along which it is travelling towards ordinary bourgeois power " (Lenin, *Sochineniya*, xxvii, 243).

[3] N. Bukharin, *Proletarskaya Revolyutsiya i Kul'tura* (1923), pp. 5-6.

[4] V. Polonsky, *Ocherki Liternaturnogo Dvizheniya Revolyutsionnoi Epokhi* (2nd ed. 1929), pp. 291-292, lists these items under Nos. 23, 29 and 30 of his bibliography ; copies have not been traced.

Lenin, *Sochineniya*, xxvii, 243.

Stalin reiterated that the movement had " acquired a mass of
supporters among Soviet officials ".[1] Nor was the source of its
popularity in doubt. An early article in *Krasnaya Nov'* dubbed
the *smenovekhovtsy* " national Bolsheviks ".[2] Stalin directly
connected the movement with the growth of " Great-Russian
chauvinism ", which he treated as a sinister product of NEP.[3]
Bukharin denounced it as " Caesarism under the mask of revolution ", and quoted Ustryalov as admitting that his followers were
not socialists and were actuated first and foremost by the " patriotic
idea ".[4] In days when appeals to the continuity of Russian
history were still heretical, a writer could incur the sobriquet of
" crypto-*smenovekhovets* " by quoting precedents for current
policy from acts of Peter and Catherine the Great, or by saying
that Moscow was once more gathering the Russian lands round
her as in the sixteenth century.[5] The *smenovekhovtsy*, like the
fellow-travellers, were sometimes accused of being Slavophils ; [6]
and, though the charge was unjust (most of them were basically
western in outlook), it had its foundation in their eagerness to treat
the revolution as a specific episode in Russian history. As time
went on, they were less concerned to disown the socialist character
of the revolution than to assert its national character. Ustryalov
returned to the Soviet Union and settled in Harbin, where he was
employed in the education department of the Chinese Eastern
Railway. In the winter of 1925–1926 he visited Moscow and was
politely, though critically, received.[7] After the middle nineteen-
twenties the movement lost its importance and faded away. But
it had served its purpose, and helped to prepare the way for the
reconciliation of the revolutionary and the national tradition which
was a condition and concomitant of " socialism in one country ".

A third movement of the intelligentsia, taking shape primarily

[1] Stalin, *Sochineniya*, v, 244.
[2] *Krasnaya Nov'*, No. 3, September-October 1921, p. 271.
[3] Stalin, *Sochineniya*, v, 244-245.
[4] For Bukharin's article see pp. 309-310 below.
[5] *Planovoe Khozyaistvo*, No. 1, 1925, pp. 263-265 ; *Bol'shevik*, No. 5-6
(21-22), March 25, 1925, pp. 115-125.
[6] A contributor to *Russkaya Istoricheskaya Literatura v Klassovom Osvesh-
chenii*, ed. V. Polonsky, i (1927), 54, wrote that many of their articles could have
been signed by Ivan Aksakov.
[7] N. Ustryalov, *Pod Znakom Revolyutsii* (2nd ed. 1927), p. ix ; the visit to
Moscow was the occasion for a hostile critique of the *smenovekh* movement
in *Planovoe Khozyaistvo*, No. 6, 1926, pp. 215-233.

in *émigré* circles, but shared to a greater or less extent by groups in Soviet Russia, was what came to be known as the Eurasian movement. Alexander Blok in his poem *The Scythians*, written in Petrograd in January 1918, depicted the Russians as "Scythians" looking towards the "old world" of Europe with a mingled emotion of hatred and love, but ready to call in the hordes of Asia to redress the balance if their overtures were repulsed. The poem, inspired by a mood of defiance of the Germans at the time of the Brest-Litovsk negotiations,[1] had wider implications, and made an enormous impression. It was a mood which reflected familiar currents of Russian thought, ambivalence towards Europe and Slavophil faith in the primitive virtues of peasant Russia, in constructive anarchism, and in Russia's peculiar mission to revivify a decadent western world. After the publication of Blok's poem the name "Scythism" (*Skifstvo*) came to be applied, not to a literary movement, but to a tendency which inspired many writers in the first years of the revolution. Politically it was associated with the Left SRs as the modern representatives of the *narodniks*. It was reflected in two famous poems of 1918, Bely's *Christ is Risen* and Esenin's *Inonia*, and in the popularity among poets and writers of those years of Stenka Razin and Pugachev, the great leaders of Russian peasant revolt.[2] It was systematized by the SR literary critic Ivanov-Razumnik, and survived to influence fellow-travellers like Vsevolod Ivanov and Pilnyak.

But the most important theoretical development of "Scythism" occurred abroad. In 1921 an *émigré* group published in Sofia a collection of essays under the title *The Way Out to the East*, described in its sub-title as "A Declaration of the Eurasians". A short opening manifesto maintained that "Russia is not only ' west ' but ' east ', not only ' Europe ', but ' Asia ', and even not ' Europe ' but ' Eurasia ' ' ", and described revolutionary Russia as " a former European province " now in revolt against Europe. It concluded by asking whether the revolution portended the

---

[1] It was written between January 15/28 and 17/30, 1918, at the moment of Trotsky's second appearance at Brest-Litovsk (A. Blok, *Sochineniya*, v (1933), 21-24, 134-145) ; by the time it was published in *Znamya Truda*, the Left SR journal, on February 20, 1918, the negotiations had been broken off and the German advance resumed.

[2] The cult penetrated official circles : *Pravda*, January 27, 1925, published a long article by Pokrovsky on the 150th anniversary of the execution of Pugachev (January 10/21, 1775).

assimilation of Russia to western culture or the birth of a new
" Eurasian " culture. The revolution was condemned in so far
as it came from the west, but welcomed in so far as it cut off
Russia from the west. Another article attacked " Romano-
Germanic civilization " for its claim to represent universal culture
and for its " chauvinism " masquerading as " cosmopolitanism ".
Another, consciously or unconsciously borrowing from the geo-
political speculations of Mackinder, opposed the " continental "
idea of an economically self-sufficient Eurasia to the " oceanic "
idea of world-wide trade.[1]  The critic Polonsky had called
Scythism " the decadence of Slavophilism ".[2]  The Eurasians
inherited from the Slavophils their belief in the decadence of
western culture, and their dislike of the western elements in
Russian culture. They had the same affinities as the *smenovekh*
movement with Slavophilism, and preached the same indigenous
interpretation of the Russian revolution; and the *smenovekhovtsy*,
for their part, were quick to welcome the Eurasians as allies :

> In her revolutionary ideology [wrote Ustryalov], in this
> audacious, specifically eastern interpretation of western Marxism,
> Russia unexpectedly and miraculously realizes her immemorial
> historic " Eurasian " mission.[3]

The Eurasians differed from the Slavophils, whom they condemned as narrowly national, in appealing to the alliance of the
non-European world. But conceptions of a self-sufficient Russia
turning her back on Europe and relying on her firm foothold
among the peoples of Asia could be easily accommodated in the
strange amalgam of " socialism in one country ".

These movements for the qualified reconciliation of the
Russian intelligentsia, both inside and outside Soviet Russia, with
the Soviet régime were sufficiently important to be discussed in
August 1922 by the twelfth party conference which, on the
motion of Zinoviev, passed a resolution " On Anti-Soviet Parties
and Tendencies ". It attributed these " processes of collapse,

---

[1] *Iskhod k Vostoku* (Sofia, 1921) ; a further symposium entitled *Na Putyakh*
appeared in 1922 (these volumes both carried the sub-title *Utverzhdenie
Evraziitsev*). Several issues of a periodical entitled, first *Evraziiskii Vremennik*,
and later *Evraziiskaya Khronika*, as well as a number of miscellaneous publications, appeared in Prague between 1923 and 1930.
[2] V. Polonsky, *O Sovremennoi Literature* (1928), p. 52.
[3] N. Ustryalov, *Pod Znakom Revolyutsii* (2nd ed. 1927), p. 188.

disintegration and re-grouping in the anti-Soviet camp " to two
factors : " the hiving-off of certain groups of the bourgeois intelligentsia " and " the process of the partial restoration of capitalism
within the framework of the Soviet state, bringing about the
growth of elements of the so-called ' new bourgeoisie ' ". It
devoted a special paragraph to the *smenovekhovtsy* :

> The so-called *smenovekh* movement has so far played, and
> may continue to play, an objectively progressive rôle. It has
> welded together, and is welding together, those groups of the
> emigration and of the Russian intelligentsia which have " made
> their peace " with the Soviet power and are ready to work with
> it for the restoration of the country. *To this extent*, the *smenovekh*
> tendency has deserved, and deserves, a positive response. But
> at the same time it should not be forgotten for a moment that
> there are within the *smenovekh* movement strong bourgeois-
> restoration strains, that the *smenovekhovtsy* share with the
> Mensheviks and SRs the hope that economic concessions will
> be followed by political concessions in the direction of bourgeois
> democracy, etc.

While the resolution expressed apprehension of the dangers involved and continued to denounce foreign capitalists, SRs and
Mensheviks, its main practical recommendations were constructive. The party was to take advantage of " the splitting
process which had begun within the anti-Soviet groups " in order
to make a serious approach " to every group, formerly hostile to
the Soviet power, which now showed the slightest sincere desire
to give real assistance to the working class and the peasantry in the
restoration of the economy, the raising of the cultural level of
the population, etc." The resolution named " writers, poets,
etc." side by side with " representatives of technology, science and
the teaching profession " as worthy of " systematic support and
working cooperation " : every attempt was to be made " to
promote the crystallization of such tendencies and groups as display a real desire to help the workers' and peasants' state ".[1]

The qualified, but none the less decisive, encouragement by the
party conference of August 1922 of literary fellow-travellers did
not pass without opposition from those who still strove to uphold
the purity of proletarian art. Since the disgrace of Bogdanov,

[1] *VKP(B) v Rezolyutsiyakh* (1941), i, 463-467.

the leading figure in Proletkult had been one Pletnev, once a carpenter by trade and an old party member, who had become a writer of stories and plays — one of the few authentic proletarian writers.[1] In September 1922 *Pravda* published an article by Pletnev in which, while avoiding Bogdanov's error of demanding the independence of literature from the party, he once more pleaded that " the task of creating a proletarian culture can be carried out only by the forces of the proletariat itself ". This covert attack on the fellow-travellers incurred the displeasure of Lenin, who covered the page of *Pravda* containing Pletnev's article with disapproving annotations. A month later *Pravda* printed a reply to Pletnev by another party member Yakovlev, to whom Lenin's notes had apparently been communicated. Pletnev's thesis was roundly condemned on the basis of a comparison between fellow-travellers and specialists : " The mistake which comrades made in 1918–1919 about military specialists, and later about specialists in industry, is mechanically transferred by Pletnev to the sphere of culture ".[2] The snub to Pletnev was a further blow to Proletkult, which, though it continued to exist as a section of Narkompros, played no rôle in subsequent literary controversies. The employment, and integration into Soviet society, of former bourgeois intellectuals who were prepared to accept and serve the new régime was a natural and necessary corollary of NEP, and could no more be rejected in the name of proletarian culture than could the employment of specialists on the plea of workers' control. But, once this policy was adopted, a new doctrine, or at any rate a new emphasis in doctrine, gradually emerged. The conception of the " national " revolution, while it did not replace that of the proletarian revolution, proved a valuable supplement to it.

[1] *Literaturnaya Entsiklopediya*, viii (1934), 691-692.
[2] The dates of the articles in *Pravda* were September 27 and October 25, 1922 ; Lenin's annotations were published, with a facsimile of the page of *Pravda*, in *Voprosy Kul'tury pri Diktature Proletariata* (1925). Lenin continued in his last writings to drive home his case against Bogdanov : the persistence of illiteracy was " a menacing warning and reproach to those who were floating, and still float, in the empyrean of ' proletarian culture ' ", and very much had still to be done " in order to attain the level of the ordinary civilized state of western Europe " (Lenin, *Sochineniya*, xxvii, 387). Yakovlev renewed his attack in an article in *Pravda*, January 1, 1923, entitled *Menshevism in Proletkult Garments*.

Resistance to the intrusion of fellow-travellers and *smeno-vekhovtsy* into the preserves of Soviet literature did not end with the elimination of Proletkult, but was resumed in a new form. What finally disappeared in 1922 was the claim for an autonomous proletarian culture promoted by an organization outside the party. This was now replaced by a more insidious demand that the cause of proletarian literature should be espoused by the party itself, and vigorously asserted against fellow-travellers and other groups outside the party. The first move in this new process occurred in December 1922, when a group of young men broke away from the Smithy to found a new and more advanced group which they called October, and through which they hoped, having conquered the leadership of VAPP, to impose their literary policies on the party. From this point onwards, literary questions became a matter of controversy in the party itself and played a minor rôle in the party struggles of the ensuing period.[1] But these developments lay in the future. Down to 1924 or 1925 the fellow-travellers continued to dominate Soviet literature, and enjoyed the virtually unqualified confidence of the party leaders. It was through them that the ideals and policies of " socialism in one country " found popular literary expression.

Increased toleration for non-communist literary groups or individual writers sympathetic to the régime did not, of course, imply any relaxation of the ban on publications hostile to the régime. Lenin, having assured Clara Zetkin that " every artist, everyone who regards himself as such, has the right to create freely, in accordance with his ideal, independently of anything ", quickly added : " But, of course, we are communists ; we cannot sit with folded hands and let chaos develop as you please ".[2] Indeed it was at this time that the ban became absolute, and could be rigidly enforced. Censorship of the traditional kind was perhaps scarcely exercised or required ; for facilities were rarely available for the publication of works liable to incur official disapproval.

These events will be dealt with in Part III in the following volume.
[2] C. Zetkin, *Erinnerungen an Lenin* (Vienna, 1929), pp. 12-13. Trotsky defined his view of the relation of the state to literary groups at this time as follows : " While putting above everything the criterion for the revolution or against the revolution, to give them complete freedom on their own ground " (L. Trotsky, *La Révolution Trahie* (n.d. [1936]), p. 206). This appeared to exclude neutrality as a permissible attitude for the writer.

But Lunacharsky enunciated the principle, at the moment of the
introduction of NEP, in the first issue of *Pechat' i Revolyutsiya*,
with an outspokenness which left no room for doubt :

> We in no way shrink from the necessity of applying censor-
> ship even to belles-lettres, since under this banner and beneath
> this elegant exterior poison may be implanted in the still naïve
> and dark soul of the great mass of people, which is constantly
> ready to waver and, owing to the too great hardships of the
> journey, to throw off the hand which is leading it through the
> wilderness to the promised land.[1]

The year 1922 was apparently the last in which a few publications
of a non-popular character openly opposed to the Soviet régime
still saw the light — notably a theoretical economic journal
*Ekonomist*, which still professed the principles of *laissez-faire*
capitalism, and an almanac entitled *Shipovnik* (a revival of a pre-
revolutionary title), to which the philosopher-theologians Berdyaev,
Bulgakov and Stepun, the poet Khodashevich and the critic
Aikhenvald contributed, as well as some of the recognized fellow-
travellers.[2] Thereafter these hostile voices were silent in Soviet
Russia, and most of those who had raised them went into voluntary
exile.[3] Criticism henceforth would be couched only in the form
of divergent interpretations of the official line, not of open challenges to it. Another form of censorship which later became
frequent and important seems to have made its first appearance
about this time : the withdrawal from circulation of publications
which, though originally issued with full party or official approval,
had fallen out of date and represented views no longer accepted as
orthodox. A circular of 1923 from the propaganda section of the
party central committee to local party committees and sections of
the OGPU recommended the withdrawal from " small libraries
serving the mass reader " not only of " out-of-date, valueless or,
still more, harmful or counter-revolutionary books " but also of
" out-of-date agitational or informatory material of Soviet origin

[1] *Pechat' i Revolyutsiya*, No. 1 (May-June), 1921, pp. 7-8.
[2] V. Polonsky, *Ocherki Literaturnogo Dvizheniya Revolyutsionnoi Epokhi*
(2nd ed. 1929), pp. 132-136.
[3] According to M. Slonim, *Modern Russian Literature* (N.Y., 1953), p. 278,
Berdyaev, Bulgakov and others were placed under a ban as the result of an
article by Trotsky, entitled *Dictatorship, Where is thy Whip ?*, denouncing their
writings.

(1918, 1919, 1920) on questions which are at present regulated differently by the Soviet power (agrarian question, system of taxation, question of free trade, food policy, etc.) ".[1] Sharp reversals of policy were in the future to provide Soviet literary control with some of its most embarrassing problems.

### (d) Law

A change of attitude towards law is a natural sequel to any revolution. Revolution is a revolt against legal authority, and is directed to the overthrow of an existing legal order. But, once this order is destroyed, and the victorious revolutionaries have usurped the seats of power, they quickly experience the need to set up a legal authority of their own ; and they have to transform themselves from challengers and opponents of law into upholders and makers of it. The men of the French revolution sought to change the content of the law. But they accepted the principle of the authority and continuity of law ; and for them therefore the reversal of rôles was relatively easy. For the Bolsheviks the transition was complicated by the fact that they, as Marxists, were committed to a specific theory of law. Law was an emanation and instrument of the state, which was the instrument of a class. Hence, in the words of the *Communist Manifesto*, "your law is only the will of your class made into a law for all, a will whose essential character and direction are determined by the economic conditions of life of your class ". It followed from this that law, like the state, would die away in the future communist classless society. Marx allowed, however, in the passage of the *Critique of the Gotha Programme* in which he distinguished between the two stages of socialism, for a transitional period after the revolution during which " equal right in law is still in principle bourgeois right ". This was inevitable so long as full socialism (or communism) was not achieved ; for " law can never stand higher than the economic order and the cultural development of society conditioned by it ".[2] Thus, while the régime put in power by the victory of the revolution would continue to enjoy the support of law, this law would be in essence not a socialist creation, but a bourgeois survival, destined to die away as the new order established itself. For socialist law there was no more permanent place in Marx's scheme than for a socialist state. Engels, in an article written after Marx's death, had identified " the juridical view of the world " with " the classical bourgeois view of the world ", and described it as " the secularization of the theological view ".[1] Lenin fully endorsed these propositions, adding, in *State and Revolution*, the logical rider that not only the law, but the state, which temporarily survived the revolution, would be bourgeois, though " without the bourgeoisie ".[2] An early Soviet textbook referred coyly to " what we call Soviet law " and " so-called Soviet law ".[3]

The workers' and peasants' government established by the October revolution proceeded without question to exercise powers of legislation and enforcement of law. No body of men claiming to act as a government could do otherwise. But neither the first months of the revolution nor the civil war period which followed them left much leisure for the elaboration of theory ; and little that was said or done seemed incompatible with the silent assumption that law was a temporary expedient, borrowed for specific purposes from the defunct bourgeois order of society, and destined to die away as soon as socialism became a reality. The attitude of the new régime to pre-revolutionary law was not conclusively defined. An initial decree of November 1917, which abolished existing judicial institutions and set up local courts, elected or

---

[1] *Ibid*. xvi, i, 296 ; many years earlier, Herzen had coupled Roman law with the Catholic church and the rule of the bourgeoisie as a trinity of evil which Russia would never accept (*Polnoe Sobranie Sochinenii i Pisem A. I. Gertsena*, ed. Lemke, viii (1919), 151).

[2] Lenin, *Sochineniya*, xxi, 438. According to an earlier aphorism of Lenin (*ibid*. xiv, 212), " law is politics " (which may equally well be translated " law is policy ") ; in 1920 Lenin quoted a passage from an article written by him in 1906 : " The scientific concept of dictatorship means nothing else but completely unlimited power, restrained by no laws, by absolutely no rules, resting directly on force. The concept ' dictatorship ' *means nothing else but that* " (*ibid*. ix, 119 ; xxv, 441). In April 1917 he defined " revolutionary dictatorship " as " authority based on outright revolutionary seizure, on the direct initiative of the masses from below, and *not on law* given out by a centralized state authority " (*ibid*. xx, 94). As late as 1926 speakers in the TsIK of the RSFSR assumed that law was in principle " bourgeois law ", and that there was " nothing communist " about any law (*III Sessiya Vserossiiskogo Tsentral'-nogo Ispolnitel'nogo Komiteta XII Sozyva* (1926), pp. 134, 585).

[3] A. Goikhbarg, *Osnovy Chastnogo Imushchestvennogo Prava* (1924), pp. 8-9.

nominated by local Soviets, consisting of a judge and two lay assessors, laid it down that laws enacted by previous régimes should be treated as valid only in so far as they " have not been abrogated by the revolution, and are not in contradiction with the revolutionary conscience and revolutionary consciousness of right ".[1] A second and more elaborate decree on the courts of February 1918 prescribed that existing rules of procedure should be observed unless they had been specifically repealed or unless they contradicted " the consciousness of right of the toiling masses " (art. 8), and that existing codes of law should be applied unless they had been repealed or contradicted " socialist consciousness of right " (art. 36). The latter article added that civil courts should not be " limited by formal law ", but should be guided by " considerations of justice ", rejecting those of " a formal character ", and that the same principle should apply to criminal courts.[2] The most specific provision on this point was a direct prohibition, in a third decree of VTsIK on the constitution

[1] *Sobranie Uzakonenii, 1917-1918*, No. 4, art. 50. The decree also contained a provision for the creation of " revolutionary tribunals " to deal with cases of counter-revolution and profiteering : this was the beginning of the establishment of a separate system of jurisdiction to deal with political offences, which will be discussed in Part IV in the following volume.

[2] *Ibid.* No. 20, art. 420. The emphasis in the early period on " revolutionary consciousness of right " was apparently a reflection of the " intuitive " or " psychological " theory of law propounded by Petrazhitsky, a pre-revolutionary jurist of Kadet affiliations, which had a large following and was accepted by most Social-Revolutionaries and by some Bolsheviks. The epithet " revolutionary " was introduced in order to guard against any suspicion that an idealist conception of right was being smuggled into Soviet legal theory. Attempts were afterwards made to attribute the important place occupied by " consciousness of right " in the decrees of November 1917 and February 1918 to the fact that they were promulgated during the tenure of office of Steinberg, a Left SR, as People's Commissar for Justice. But this seems dubious. The most complete exposition of this theory of law by a Bolshevik jurist is in M. Reisner, *Pravo, Nashe Pravo, Inostrannoe Pravo* (1925), extracts from which are translated in *Soviet Legal Philosophy*, trs. H. W. Babb (Harvard, 1951) — see especially pp. 86-87. According to Reisner, Lunacharsky " with the support of Lenin " was responsible for the emphasis on " revolutionary legal consciousness " in early decrees on law : Lunacharsky, as a former follower of Bogdanov, was always suspect of leanings towards idealism. Stuchka in January 1918 had written : " We have taken our stand on the point of view [of the Petrazhitsky school] about intuitive right, but we differ profoundly from it about the basis of that point of view " (P. Stuchka, *13 Let Bor'by za Revolyutsionno-Marksistskuyu Teoriyu Prava* (1931), p. 10) ; later he added that it had been adopted in the decree of November 1917 " by necessity ", and that " we never declared this consciousness of right to be some mystical source of truth and justice " (*ibid.* p. 103).

of the courts in November 1918, on the citation of enactments
or judgments of former régimes : where no Soviet legislation
applied, recourse was to be had to " socialist consciousness of
right ".[1]  The result of these measures was afterwards described
by Stuchka, the influential Soviet jurist who played the largest
part in drafting them, as " the creation of the *proletarian court* —
*without bourgeois law*, but also without proletarian law ", though
he somewhat cryptically added that " we were sufficiently cautious
and did *not* come out *against law* in general ".[2]

New legislation in this period was mainly of an emergency
character, and often did not go beyond a solemn declaration of
principle or intention.  The land decree of October 26/November
8, 1917, provided legal cover for a spontaneous process of seizure
of land by the peasants : the decree of February 14, 1919, on
" the socialization of land " was a theoretical proclamation in
favour of collective agriculture.[3]  Two codes of law were promulgated in 1918 — a marriage code which secularized marriage and
made divorce automatic on the demand of either party,[4] and a
labour code which established the principle, applicable only to
former members of the bourgeoisie, of obligatory labour service.[5]
But these were thought of as pronouncements of policy rather than
as definitions of legally enforceable rights and obligations.  Lenin
was quoted as having taken, in the first months of the revolution, a
highly pragmatic view of law :

> Do not obey orders or decrees if they are harmful to the
> cause : do as your conscience dictates.  If as a result of the
> decree things turn out badly, but as a result of your actions

---

[1] *Sobranie Uzakonenii, 1917–1918*, No. 85, art. 889.
[2] *Vestnik Kommunisticheskoi Akademii*, xiii (1925), 236.
[3] For these decrees see *The Bolshevik Revolution, 1917–1923*, Vol. 2, pp.
35-36, 154-155.
[4] For the marriage code see p. 29 above.  A commentary on a version
of the code published in English in Moscow explained the current Soviet
philosophy of law :  " It is understood that in giving out its codes the
government of the proletariat engaged in implanting socialism in Russia does
not aim at making these codes such as might hold on for a long time.  It does
not wish to give birth to ' eternal ' codes, or codes which would last for
centuries. . . . It constructs them so that each day of their existence should
make less the necessity for their continuation as legislations of the state.  It
fixes for its laws one aim, namely that of making them superfluous " (*The
first Code of Laws of the Russian Socialistic Federal Soviet Republic* (Moscow,
1919), p. 4).        [5] See *The Bolshevik Revolution, 1917–1923*, Vol. 2, pp. 198-199.

well, nobody will blame you for that. But if you do not carry out the order or decree, and as a result of your actions things turn out badly, you will all have to be shot.[1]

And trained lawyers asked : " How can we work in the new peoples' courts when you have no law ? "[2]  Civil law, in the ordinary sense of the term, scarcely existed : " from November 1917 to 1922 ", wrote Stuchka, " law was formally lacking ".[3] Down to the end of 1922, when a civil code was introduced, " the number of civil cases before the courts was quite insignificant " ;[4] and in the universities it was proposed to abandon courses in branches of civil law, and to substitute courses on the corresponding branches of politics.[5]  Another authority of the period, referring to Engels's dictum on the identity between juridical and bourgeois, declared that to overcome the fetish of law was now even more important than to overcome the fetish of religion.[6]  Had not Marx written, in the preface to the *Critique of Political Economy*, of " relations of production . . . or, speaking juridically, property relations " ?  Once property in the means of production was abolished, it would be unnecessary to speak a juridical language at all.  Production would be regulated by administrative action.

If, however, the proletarian revolution, by abrogating property rights and private commercial enterprise, seemed to have made civil law immediately superfluous, the same cavalier attitude could not be adopted towards criminal law.  In the first days of the revolution Lenin impulsively exhorted the workers to " arrest and hand over to the revolutionary people's court anyone who dares to injure the people's cause ".[7]  The maintenance of order and the repression of crime were acute practical necessities which would monopolize the attention of the new courts for some time to come.  Moreover, reflections on the origin and nature of crime and plans for the reform of the criminal entered into the programmes of all Left parties.  The Bolshevik party programme

[1] *Sovetskoe Stroitel'stvo : Sbornik*, iv-v (1926), 88.
[2] *Ibid.* iv-v, 92.
[3] *Bol'shaya Sovetskaya Entsiklopediya*, xviii (1930), 74, art. Grazhdanskoe Pravo.    [4] A. Goikhbarg, *Kurs Grazhdanskogo Protsessa* (1928), p. 10.
[5] *Sovetskoe Gosudarstvo i Revolyutsiya Prava*, No. 11-12, 1930, pp. 48-49.
[6] A. Goikhbarg, *Osnovy Chastnogo Imushchestvennogo Prava* (1924), p. 9.
[7] Lenin, *Sochineniya*, xxii, 55.

adopted in the summer of 1919, which ignored questions of
civil law altogether, advocated the replacement " of privation
of freedom by compulsory labour with retention of freedom", "of
prisons by educational institutions ", and the establishment of
" comradely courts ", so that " measures of an educational character " might ultimately be substituted for punishment.[1]  Meanwhile, more orthodox penalties continued to be applied ; and in
December 1919, the People's Commissariat of Justice issued a
document entitled " Leading Principles of the Criminal Law of
the RSFSR ".  This was a hastily drafted document not free
from contradictions and obscurities.  Bourgeois codes of law, like
the bourgeois state, had, it declared, been destroyed, and should
be " placed in the historical archives ".  But the experience of
" the struggle with its class enemies " had " accustomed the
proletariat to uniform measures, had led to systematization, had
given birth to new law " ; and this — begging a question which
lay at the root of much subsequent controversy — was referred
to as " proletarian law ".  Law was, however, defined in unim-
peachable Marxist terms as " a system of social relations corresponding to the interests of the ruling class and secured by the
organized power of that class ".  Crime was defined as " any
infraction of the order of social relations protected by the criminal
law " ; and the function of criminal law was to protect that order
by the application of penalties for such acts.  The " leading principles " were permeated by the conception of all criminal law as
a measure of defence of the social and constitutional order.  Crime
was explicitly described as the product not of the personal guilt
of the criminal, but of the divided structure of a class society.
Criminal law was a provisional expedient adopted by a transitional
society until these divisions could be overcome, though "only
with the final destruction of the defeated hostile bourgeois and
intermediate classes, and with the realization of the communist
social order, will the proletariat abolish both the state as an organ
of coercion and law as a function of the state."[2]

Perhaps the most striking symptom of the original Bolshevik
attitude to law was the mistrust of professional judges.  This was
scarcely surprising in a situation which left so much discretion

---

[1] *VKP(B) v Rezolyutsiyakh*, i (1941), 288.
[2] *Sobranie Uzakonenii, 1919*, No. 66, art. 590.

to the " revolutionary consciousness " of the court, and where those who possessed expert legal knowledge were steeped in the traditions of the former régime, even if not open supporters of it. The original decree on the courts of November 1917 had prescribed that two assessors should sit with the judge, who acted as president of the court and was in practice primarily responsible for its proceedings. A regular system of " people's assessors " was worked out in the second decree of February 1918, which made it clear that the function of the assessors was to act as a check on the caprice, legal formalism or political unreliability of the judge : they received powers to remove the president of the court at any stage of the proceedings, to overrule the conviction of a defendant (though not apparently an acquittal), or to reduce a sentence. Nor was this all. The decree made provision for the establishment of a " supreme judicial control ", composed of delegates from lower courts, which had the right to quash any decision of a lower court, apparently on its own initiative, and was also invited to draw the attention of the legislative authorities to any contradiction between existing law (presumably the law of previous régimes) and the " people's consciousness of right ". This " supreme " control never seems to have functioned in the form in which it was devised. But it appeared in the third and major decree of November 1918 [1] in the form of provincial " councils of people's judges " elected by a provincial " congress of people's judges " to act as a court of appeal at the provincial level. Meanwhile the same decree extended the system of people's assessors. For major criminal charges (excluding, of course, those that came before the revolutionary tribunals) the court was constituted by a president with six assessors. The president could not be a member of a politically disqualified group ; [2] and he was expected to have had experience either in judicial work or in trade union organization. If these stipulations were literally applied by the Soviets or congresses of Soviets which elected the judges, courts trying important cases might easily contain no qualified lawyer. The principle was clearly stated in a report of the period by Kursky, People's Commissar for Justice :

[1] For the three decrees see pp. 67-69 above.
[2] For these disqualifications see *The Bolshevik Revolution, 1917–1923*, Vol. 1, p. 143.

The proletariat and the poorest peasantry, having conquered political power, were inevitably bound, in order to strengthen their power, to smash the whole juridical superstructure of the bourgeois state and, consequently, the courts. Henceforth the decisive voice in the courts must rest with the workers and poorest peasants in the person of assessors elected by the Soviets.[1]

" Our courts ", said Lenin in 1921, " are class courts, against the bourgeoisie ", just as " our army is a class army, against the bourgeoisie ".[2]  All these developments were consonant with the implicit assumption that law was a bourgeois expedient which was convenient and necessary in the period of transition but would be gradually eliminated with the growth of socialism.

Even before the end of the civil war and of the régime of war communism, a reaction had set in against this view of the character of law.  Every established régime needs to buttress its authority on law.  The essence of law is that its operation should be both comprehensible and predictable, and depend as little as possible on the personal idiosyncrasies of those who have to apply it.  Above all, law, in order to be effective, requires to be invested with a certain aroma of sanctity, which was conspicuously absent from the Marxist interpretation.  This gradually became apparent as the new régime established itself.  In March 1918, Lenin, in a draft article which, however, remained unpublished, explained that, while " new courts " had been essential to end the abuses of exploitation, it was also indispensable to organize the courts " on the principles of Soviet institutions, i.e. to promote the strictest development of the discipline and self-discipline of the toilers ".[3]  On the first anniversary of the revolution the sixth All-Russian Congress of Soviets in November 1918 issued a solemn declaration to the effect that " during a year of revolutionary struggle the Russian working class has evolved the fundamental laws of the RSFSR, strict observance of which forms a necessary condition for the development and strengthening of the power of the workers and peasants ".[4]  As a statement of fact,

[1] D. Kursky, *Izbrannye Stat'i i Rechi* (1948), p. 15.
[2] Lenin, *Sochineniya*, xxvi, 339.                    [3] *Ibid.* xxii, 424.
[4] *S"ezdy Sovetov RSFSR v Postanovleniyakh* (1939), p. 119 ; *Sobranie Uzakonenii, 1917–1918*, No. 90, art. 908.

this might have been difficult to justify ; Soviet legislation was still rudimentary.[1] But it betokened a new and hitherto unfamiliar recognition of the objective importance of law, undiminished by any suggestion of its bourgeois or transitional character or by any appeal to a subjective standard of revolutionary consciousness. The ground was thus prepared for a new conception of legality, which was to develop in the NEP period. A paragraph of the notes made by Lenin for a speech of October 1921 (though he did not develop the idea in the speech itself) gave a foretaste of the new turn of thinking about law :

> An increase of legality. . . . Learn to struggle in a cultured way for legality, while not forgetting the limitations of legality in revolution. The evil now is not in this, but in the confusion of illegalities.[2]

It was now for the first time clearly seen that an established régime, however revolutionary its origin, needed the support of a stable legal order ; and the sense of regularity and security inherent in law came to be exalted above the spontaneous deliveries of revolutionary intuition.

But the revival of law was also specifically connected with the economic practices revived and sanctioned by NEP, and was a direct outcome of them. " In order to put an end to doubts about the sincerity of the new course of economic policy ", a decree of August 25, 1921, laid down the rule that contracts could be invalidated only by a court decision, and that leasing agreements entered into by the Soviet authorities could be cancelled only by legislative action.[3] The conception of " due process of law " thus made its first appearance in Soviet jurisprudence in the wake of NEP. The party conference of December 1921 passed a resolution demanding " the establishment in all spheres of life of the strict principles of revolutionary legality ". A few days later Lenin echoed the phrase in his speech at the ninth All-Russian Congress of Soviets : " before us lies the task of developing private exchange — this is required by the new economic

---

[1] Kursky in 1919 enumerated a number of enactments which justified the phrase " a new criminal law " (D. Kursky, *Izbrannye Stat'i i Rechi* (1948), pp. 47-55).          [2] Lenin, *Sochineniya*, xxvii, 35.
[3] *Sobranie Uzakonenii, 1921*, No. 62, art. 455.

policy — and this requires more revolutionary legality ".[1]  In the civil war period, a commentator afterwards explained, Soviet organs had been bound to act on the principle of " revolutionary expediency ", and could not always conform to current legislative enactments.  Now the opposite principle of " revolutionary legality " was applicable.[2]  The change was far-reaching, even abrupt.  As late as February 1922, in the course of a campaign against legal formalism and red tape, Lenin wrote to Kursky, the People's Commissar for Justice :

> Broaden the application of state intervention in " private-juridical " relations, broaden the right of the state to annul " private " contracts, apply to " civil law relations " not the *corpus juris romani*, but our *revolutionary consciousness of right*.[3]

But such an attitude was flagrantly incompatible with the orderly conduct of trade and business which NEP sought to promote. " Revolutionary legality " meant the introduction of legal security into commercial relations, and proved an effective substitute for " revolutionary consciousness of right ".  The sphere in which this requirement weighed most heavily was that of foreign trade and of concessions to foreign firms.  Chicherin on the eve of the Genoa conference stressed the security which Soviet legislation offered to foreign trade ; [4] and this was the inspiration of a decree

---

[1] *VKP(B) v Rezolyutsiyakh* (1941) i, 410 ; Lenin, *Sochineniya*, xxvii, 140. This seems to have been the first recorded use of a famous catch-word ; *Entsiklopediya Gosudarstva i Prava*, i (1925), 1150, states that it was coined in 1920 because " some of our comrade-revolutionaries were shocked by the word ' legality ' ", but quotes no such early use.  Lenin anticipated the idea, but not the phrase, in August 1919, when, at the height of the civil war, he declared that, in order to destroy Kolchak and Denikin, it was " indispensable to maintain the strictest revolutionary order, indispensable to observe faithfully the laws and decrees of the Soviet power " (Lenin, *Sochineniya*, xxiv, 433).  P. Stuchka, *13 Let Bor'by za Revolyutsionno-Marksistskuyu Teoriyu Prava* (1931), p. 122, erroneously traced it to Lenin's memorandum of 1922 on the powers of the procurator (see pp. 81-82 below) ; the word " legality " appears again and again in the memorandum, but without the epithet.

[2] *Sovetskoe Stroitel'stvo : Sbornik*, iv-v (1926), 61-62.

[3] Lenin, *Sochineniya*, xxix, 419.  This seems to be the last recorded use — and not for publication — of the phrase " revolutionary consciousness of right " in a civil law context.  It was used again by Lenin three months later (*ibid.* xxvii, 296) in defence of terror ; and " socialist consciousness of right " appeared not very conspicuously in the criminal codes of the RSFSR of 1922 (art. 9) and 1926 (art. 45).  But after 1922 it was obsolescent.

[4] See *The Bolshevik Revolution, 1917-1923*, Vol. 3, pp. 360-361.

of May 22, 1922, " on the fundamental rights of private ownership as recognized by the RSFSR, protected by its laws and upheld by the courts of the RSFSR ", which proved to be a first step towards the adoption of a complete civil code in the following autumn.[1] The year 1922 was marked by the foundation of an Institute of Soviet Law with a monthly journal *Sovetskoe Pravo*, the purpose of which was described by Kursky, in an introductory article in the first issue, as " the construction of a contemporary *system of Soviet law* ".

It was thus no accident that the first two years of NEP were the great period of codification of Soviet law, seeing the birth of criminal, civil, agrarian and labour codes of the RSFSR. The character of the criminal code of May 1922 was clearly defined. It was enacted, in the words of the decree of VTsIK which introduced it,[2] " for the purpose of defending the workers' and peasants' government and the revolutionary legal order from those who would destroy it and from socially dangerous elements, and of establishing the foundations of revolutionary consciousness of right ". It followed the " leading principles " of 1919 in defining a crime as " any socially dangerous act or omission which threatens the foundations of the Soviet régime and the legal order established by the government of the workers and peasants during the period of transition to a communist order " (art. 6). It distinguished between crimes committed " in the interests of a restoration of bourgeois power " and those committed in the purely personal interests of the criminal, and between crimes against the state and against an individual person, and plainly regarded the former categories as more heinous than the latter. For the former, the code laid down minimum penalties which could not be reduced by the court, for the latter, maximum

---

[1] *Sobranie Uzakonenii, 1922*, No. 36, art. 423. One of the reasons given by Kursky, the People's Commissar for Justice, for the adoption of a civil code was the demand for " a recognized system of legal norms " put forward by Lloyd George at Genoa as a condition of regular relations with Soviet Russia (D. Kursky, *Izbrannye Stat'i i Rechi* (1948), p. 71) ; the Supreme Council at Cannes in January 1922 had required countries aspiring to foreign credits to undertake *inter alia* " that they will establish a legal and juridical system which sanctions and enforces commercial and other contracts with impartiality " (*Resolutions Adopted by the Supreme Council at Cannes, January 1922, as the Basis of the Genoa Conference*, Cmd. 1621 (1922), p. 3).
[2] *Sobranie Uzakonenii, 1922*, No. 15, art, 153.

penalties which could not be exceeded (arts. 25, 27).  The severest
normal penalties prescribed by the code were " expulsion from
the territories of the RSFSR ", " deprivation of liberty with or
without strict isolation " and " forced labour without taking into
custody ". . But the following article provided that, " until such
time as it may be abolished by the All-Russian Central Executive
Committee, in cases where the highest measure of punishment is
prescribed by articles of the present code, this is carried out by
shooting " (art. 32-33).  The " highest measure of punishment "
was reserved for crimes against the state.  But it was applied
by the code to a substantial number of such crimes, including
not only counter-revolutionary activities, but extreme forms of
" abuse of power " by officials, the perversion of justice for
interested reasons by judges, certain forms of bribery, and the
appropriation of public property by officials (arts. 110-111, 114,
128, 130).[1]  The main significance of the code was that it provided for the first time a specific list of acts which would be treated
by Soviet courts as crimes and of the penalties appropriate for
them, and thus substituted the precision of a code for the wide
competence of revolutionary consciousness.  It contained, moreover, an important innovation.  The leading principles had
assumed that, where the alleged crime had not been defined in
Soviet legislation, the gap would be filled by the revolutionary
consciousness of the court.  Article 10 of the code of 1922 instructed the court, in dealing with a form of crime not defined by
law, to apply by analogy the articles of the code " dealing with

[1] It should be noted that the rejection of any theory of the " rule of law "
i.e. of the legal limitation of the powers of the state as such, did not imply
any leniency towards officials exceeding the limits of authority conferred on
them by the law.  " Abuse of authority, or of an official position " was punish-
able with six months' imprisonment (art. 109), and " illegal arrest " with one
year's imprisonment (art. 115).  Crimes committed by officials always attracted
the special attention of the Cheka and later the OGPU ; for particulars of such
crimes dealt with by the Cheka in 1918 and 1919 see M. Latsis, *Dva Goda
Bor'by na Vnutrennem Fronte* (1920), pp. 68-69.  The first proclamation of
" revolutionary legality " by the party conference of December 1921 (see p. 74
above) specifically linked it with " strict responsibility both of organs and agents
of the government and of citizens for any infringement of laws enacted by the
Soviet power ".  According to Stuchka, one of the functions of revolutionary
legality was to overcome the reluctance of courts to deliver judgments against
official persons or institutions (*Vestnik Kommunisticheskoi Akademii*, xiii (1925),
246-247).

crimes most similar to it in importance and character ".[1] The significance of this change was the abandonment of " revolutionary consciousness " as a method of filling gaps in the legislative code and the substitution of what was, at any rate in form, a legal criterion.

It was a significant consequence of the new criminal code, or perhaps of the spirit in which it was administered, that increasingly severe penalties were imposed on crimes against property. In 1922 40 per cent of those convicted of such crimes received unconditional prison sentences ; in 1923 the proportion rose to 49 per cent, whereas the percentage of those receiving similar sentences for crimes against the person fell from 30 to 14.[2] In 1922 42 per cent of all prison sentences were for less than one year, and 10 per cent for over three years ; in 1923 the corresponding percentages were 30 and 28·5 respectively. These changes were attributed by some to " the influence of a petty bourgeois environment ".[3] It was noticed in particular that people's assessors who were peasants, when judging cases of theft, " try to discover some article under which the accused can be all but shot ".[4] This was only one example of a wide difference between sentences passed for different types of crime in town and country. Rural courts punished theft " seven times as harshly " as city courts, but were far more indulgent to illicit distilling of spirit or to offences against administrative orders.[5] The difference corresponded broadly to the distinction between crimes against the individual and crimes against the state. The new attitudes inculcated by NEP found their strongest support in the countryside.

---

[1] The introduction of the analogy principle met with a strong opposition : the first draft of the code had contained a clause based on the principle *nulla poena sine lege*, which would have limited the conception of crime to acts defined as such in the code (*III Sessiya Vserossiiskogo Tsentral'nogo Ispolnitel'nogo Komiteta IX Sozyva : Byulleten'*, No. 3 (May 17, 1922), p. 28) ; Krylenko defended the change as necessary, " particularly in our times when a large number of crimes are constantly changing their character " (*ibid*. No. 3 p. 34).

[2] *Ezhenedel'nik Sovetskoi Yustitsii*, No. 51-52, 1923, pp. 1191–1192.

[3] *V Vserossiiskii S"ezd Deyatelei Sovetskoi Yustitsii* (1924), pp. 242-243.

[4] *Ibid*. p. 244 ; the peasantry in general took the view that punishments for criminal offences were not severe enough (*Soveshchanie po Voprosam Sovetskogo Stroitel'stva 1925 g. : Yanvar'* (1925), pp. 64, 66).

[5] D. Kursky, *Izbrannye Stat'i i Rechi* (1948), p. 78.

The civil code [1] marked a more striking revival of legal conceptions. Its philosophy was defined in its first article : " Civil rights are protected by the law except in cases in which they are exercised in a sense contrary to the economic and social purposes for which they have been established ". Any notion of natural rights was ruled out as anathema to the Soviet conception of law. Soviet jurisprudence accepted no distinction between private and public law. " We do not recognize anything ' private ' ", wrote Lenin ; " for us everything relating to the economy is a matter not of private, but of public, law." [2]   Nevertheless, certain rights were conferred on individuals for " economic and social purposes " (these being further defined in art. 4 as " to develop the productive forces of the country ") ; and these rights would be protected by law. Within these limits the code was designed to establish " revolutionary legality " and to increase respect for the law. It declared the land and the means of production in nationalized industries removed for ever from the sphere of private ownership (arts. 21, 22). On the other hand, it guaranteed the right " to possess, enjoy and dispose of " property " within the limits fixed by the law " (art. 58). Enterprises might be leased to individuals for a maximum of six years ; for this period there was security of tenure without guarantee for what might come after. The treatment of inheritance in the code was a significant symptom of the change of outlook. In the first flush of revolutionary enthusiasm a decree had been passed in April 1918 to abolish the right of inheritance, though even here an exception had been made in favour of nearest relatives in respect of tools or implements used in personal labour or of other articles up to a value of 10,000 rubles. [3]   Under articles 416-418 of the civil code of 1922 the rights of inheritance and of testamentary disposition were recognized, but the potential beneficiaries were restricted to nearest relatives, and the total amount which might be bequeathed to 10,000 rubles. The fact that virtually no change was made in the practical position made the reversal of the theory all the more conspicuous and significant. In the civil code Soviet law appeared for the first time, not as the assailant, but as the protector, of

---

[1] For the civil code see *The Bolshevik Revolution, 1917–1923*, Vol. 2, pp. 342-343 ; its text is in *Sobranie Uzakonenii, 1922*, No. 71, art. 904.

[2] Lenin, *Sochineniya*, xxix, 419.

[3] *Sobranie Uzakonenii, 1917–1918*, No. 34, art. 456.

individual rights. In this sense, it was the embodiment of the spirit of NEP, and the charter of the nepman and of those who traded with him. An important, though less dramatic, aspect of the code was its endorsement of the principle of *khozraschet* for state enterprises and its assimilation of such enterprises to the status of juridical persons. While the fixed capital of state enterprises could not be made subject to private law, their working capital could, under article 18 of the code, be pledged as security for debts ; and they could sue and be sued in the courts in the ordinary way on contracts concluded by them.

The agrarian code and the labour code [1] were the counterparts of the civil code in their respective spheres. Neither the tenure of land nor the employment of labour was subject to unrestricted processes of exchange ; for this reason they could not properly find their place in the civil code. But under NEP they acquired much of the character of civil law relations. If the civil code was the charter of the nepman, the agrarian code was the charter of the land-holding peasant, and the labour code of the entrepreneur and industrial manager and of the free industrial worker. The agrarian code gave to the peasant a limited right of tenure over the land which he held, as well as the right, with certain reservations, to rent land and to employ hired labour. It was manifestly an expression of the new policy of concessions to the peasant, even at the expense of some return to the procedures of capitalism. The labour code similarly provided cover for the return to a free labour market. Though less clearly than the agrarian code and the civil code a concession to bourgeois forces, which did not occupy the same predominant place in industry as in agriculture and trade, it restored the contract between employer and worker as the basis of employment, placed the sanction of dismissal once more in the hands of the employer, and re-created the reserve army of labour in the shape of chronic unemployment. It was a code under which even the managers of state industry accepted the main capitalist presuppositions regarding the relations between employer and worker.

The same tendencies declared themselves in the new organization of the judicial system which was also undertaken in 1922.

[1] For these codes see *The Bolshevik Revolution, 1917–1923*, Vol. 2, pp. 330-333, 342-343.

It marked a strong reaction in favour of a professional judiciary and of strict observance of law. In May 1922, simultaneously with the adoption of the first criminal code of the RSFSR, the People's Commissariat of Justice introduced into VTsIK a draft decree instituting the office of public prosecutor or procurator, with powers not only to decide whether a prosecution should be instituted in a given case, but to recommend the annulment or amendment of any judgment or decision, whether of a court of law or of a department of the administration, which was in contravention of the law. Uniformity in judicial decisions would thus be assured. The procurator, who became the supreme custodian of " revolutionary legality ", was responsible only to the People's Commissar for Justice, and appointed procurators, who were subordinate to himself, in the autonomous republics, regions and provinces of the RSFSR. This proposal excited keen opposition in the party fraction in VTsIK. A majority alleged that the right of revision accorded to the procurator contravened the rights enjoyed by local authorities under the constitution of the RSFSR, and demanded that, at the least, the local procurators should be appointed under the system of " dual subordination " current in Soviet administration, i.e. that they should be responsible to the local Soviet authorities as well as to the procurator of the RSFSR.[1] At this point Lenin intervened on the side of the minority, and called for a reference of the issue to the Politburo. He argued that the system of "dual subordination" in administration was justified by differences in conditions between different regions (for example, the problems of agriculture in Kaluga were not the same as in Kazan), but that " legality cannot be of one kind in Kaluga and of another in Kazan, but must be one for the Russian republic, and indeed one for the whole federation of Soviet republics ". Rabkrin had the power to revise acts of the administration from the practical standpoint. It was the business of the procurator to see that " no single decision of a single local authority should part company with the law " ; and for this purpose a single central authority was required as a check on the ignorance or caprice of local decisions. " We live ", wrote Lenin, " in a sea of lawlessness ; and local influence is one of the greatest obstacles, if not the greatest, to the establishment of legality and

[1] See *ibid*. Vol. 1, pp. 218-219.

civilized behaviour." [1]   The debate in VTsIK was noteworthy
for an intervention by Skrypnik, the People's Commissar for
Justice of the Ukrainian SSR, who argued in vain that the rights
to be conferred on the procurator " would mean the abolition or
diminution of the power of the provincial executive committees,
and would paralyse the power of the whole Soviet system in the
localities ".   Krylenko, deputy People's Commissar for Justice of
the RSFSR, defended the project, and made a sweeping retort to
objectors who declared that the project made a "fetish" of the law :

> We suffer from an insufficiency of proper respect for written
> rules, for the law as such, from an insufficiency of this fetishistic
> attitude, not from an excess of it.[2]

The arguments of Lenin and of Krylenko prevailed, and the
decree was adopted in the form proposed by the People's Commissariat of Justice.[3]   A blow had been struck not only for the
unification of authority, but for the clothing of that unified
authority in strict legal form.

The organization of the judiciary was completed by a statute
of October 31, 1922, which received the approval of VTsIK at
the same session as the civil, agrarian and labour codes.   The
return to a professional judiciary was an important aspect of
the revived cult of legality.   As Krylenko said in submitting the
text of the new statute to VTsIK, "after five years of the existence
of Soviet power, especially in a period of development of civil
law relations, we must renounce the principle that anyone can be
a people's judge ".[4]   The aims of the statute were to establish
central control over appointments to higher courts and to ensure
a higher degree of professional competence in those appointed.
The statute moved cautiously along both these lines.   In the
lower courts the people's judge, who acted as president, was
elected for one year by the provincial executive committee and
could be re-elected.   But the elections took place "on the nomination of the provincial court or of the People's Commissariat of

[1] Lenin, *Sochineniya*, xxvii, 298-301 ; the circumstances are explained
*ibid.* xxvii, 544-545, note 142.
[2] *III Sessiya Vserossiiskogo Tsentral'nogo Ispolnitel'nogo Komiteta IX Sozyva :
Byulleten'*, No. 3 (May 17, 1922), pp. 5, 23.
[3] *Sobranie Uzakonenii, 1922*, No. 36, art, 424.
[4] *IV Sessiya Vserossiiskogo Tsentral'nogo Ispolnitel'nogo Komiteta IX Sozyva:
Byulleten'*, No. 1, October 25, 1922, p. 24.

Justice", and the candidate must have behind him the experience of not less than two years' "responsible political work" in public, trade union or party institutions or three years' " practical work in Soviet judicial institutions " (arts. 11-13). The people's assessors, who were the other members of these courts, were directly elected by local bodies, 50 per cent of them being workers, 35 per cent peasants, and 15 per cent Red Army men. But no assessor could sit for more than six days in the year, and a commission appointed by the county executive committee could object to anyone on the rota of assessors (arts. 15-28). In provincial courts the president and his two deputies (one for civil, one for criminal, affairs) were still formally elected by the provincial executive committee, but required the confirmation of the People's Commissariat of Justice which also had " the right equally to propose its own candidates " (arts. 59-63). The assessors in provincial courts were appointed by the provincial executive committee from lists drawn up by the judicial authorities ; they were required to have had the experience of not less than two years' work in public or trade union institutions (art. 64). At the summit a " supreme court of the RSFSR " constituted a final court of appeal : its president and his two deputies were appointed by the presidium of the TsIK of the RSFSR (arts. 95-96). Judges were subject to disciplinary action if they delivered " judgments in contradiction with the general spirit of the laws of the RSFSR and the interests of the working class ". (art. 112). The importance of the rôle of the procurator was emphasized : it was his function " to supervise the legality of all actions of the People's Commissariats, of all central authorities and institutions, and to recommend the abrogation or amendment of orders and decrees of such authorities if judged by him contrary to the law ".[1] Nor were the functions of the procurator restricted to criminal or administrative law. Under article 254 of the code of civil procedure of the RSFSR, he could reopen any civil case " if this is required by the defence of the interests of the workers' and peasants' state or of the toiling masses ".[2] The procurator became, as Lenin desired, the custodian not only of legality, but of the centralization of legal authority. The judicial system instituted for the RSFSR

[1] *Sobranie Uzakonenii, 1922*, No. 69, art. 902.
[2] *Sobranie Uzakonenii, 1923*, No. 46-47, art. 478.

by this decree was copied in the other Soviet republics, and served as the foundation of the judicial system of the USSR.

The return to legality and the restoration of a unified legal order were carried a step further when the USSR was founded in 1923. Both the framing of codes and the institution of courts of law remained formally the prerogative of the constituent republics, each of which had its own People's Commissariat of Justice ; but among the subjects reserved for the " supreme organs " of the USSR by article 1 of the constitution was " the establishment of the bases of courts of law and legal procedure as well as of the civil and criminal legislation of the union ". The principle of uniformity was thus safeguarded. Central control over the maintenance of " revolutionary legality " was assured by the establishment of a Supreme Court of the USSR and a procurator of the USSR ; these institutions were designed to supervise and co-ordinate the work of the corresponding organs of the union republics. The Supreme Court was set up for the purpose of " strengthening revolutionary legality on the territory of the Union of Soviet Socialist Republics ". Among its functions were " to give directive interpretations to the supreme courts of the union republics on questions of general union legislation ", and " to settle judicial disputes between union republics ". But care was taken to avoid the establishment of a right of "judicial review" which would have implied a non-Marxist conception of the supremacy of law. The Supreme Court was subject to the higher authority of TsIK and acted as its agent. It was " at the request of TsIK " that it was entitled to " give verdicts on the legality of decisions of the union republics from the point of view of the constitution"; and its most important function was "to examine and to protest, before the central executive committee of the Union of Soviet Socialist Republics, on the instance of the procurator of the Supreme Court of the Union of Soviet Socialist Republics, decisions and verdicts of the supreme courts of union republics on the ground of their incompatibility with general union legislation, or in so far as the interests of other republics may be affected by them". The procurator of the Supreme Court was appointed directly by the presidium of TsIK, and was attached to the court in the rôle of a mentor rather than a subordinate. It was his prerogative " to give rulings on all questions submitted for decision to the Supreme Court of the USSR, to undertake prosecutions before it, and, in the event of his non-agreement with a decision of the Supreme Court of the USSR in plenary session, to appeal against it to the presidium of the TsIK of the USSR ". The importance of the procurator resided in the fact that, since the People's Commissariats of Justice were re-publican commissariats, he was the highest, indeed the sole, judicial authority of the central government. While the con-stitution provided for no People's Commissariat of Justice of the union, and thus purported to leave the judicial power in the hands of the republics, the procurator of the USSR, in virtue of his power to overrule the procurators of the republics (who were often also the People's Commissars for Justice), exercised *de facto* the functions of a People's Commissar for Justice of the USSR. The authority of the law had been not only re-established, but centralized in this supreme office.[1]

The revival of law and the new cult of revolutionary legality were intimately connected with the need under NEP to provide the trader with the protection and guarantee of what he would regard as the normal processes of law ; and it was this aspect of the revival which made the strongest impression on those who set out to elaborate an up-to-date theory of law for the NEP period. On this basis it was possible to explain the revival of legality as being, like NEP itself, a retreat and a temporary compromise with capitalism. As Marx himself had said, the law of the initial stage of the transition to socialism would be in essence bourgeois ; to recognize its utility in the transitional period was not incompatible with the belief that it would die away with the coming of the socialist order. Thus Stuchka described civil law as " the result of the production of commodities for exchange ", and treated it as an expression of " that formal equality between persons which originates from the exchange of commodities on the basis of labour exchange values ".[2] According to this view, state-owned industry

---

[1] The controversy provoked by the constitutional aspects of this development will be discussed in Part IV in the following volume. The statute of the Supreme Court of the USSR adopted by TsIK in November 1923 is in *Sobᵛanie Uzakonenii, 1924*, No. 29-30, art. 278 ; the powers of the procurator of the USSR were defined in a decree of October 1924 (*Sobranie Zakonov, 1924*, No. 23, art. 203).

[2] *Bol'shaya Sovetskaya Entsiklopediya*, xviii (1930), 737.

and the planned sector of the economy would fall naturally outside the scope of law, being subject to other forms of regulation, though they might be artificially assimilated to the legal order through the principle of *khozraschet*. The civil code was " a genuine bourgeois civil code, being borrowed to the extent of nine-tenths from the best bourgeois civil codes of the west ".[1] Pashukanis, the other great jurist of the NEP period, carried Stuchka's thesis a step further, explaining all law as the expression of the bourgeois principle of commodity exchange between formally free individuals. " At the same moment when the product of labour acquires its quality as a commodity and becomes the bearer of value, the individual acquires his quality as a subject of law and becomes the bearer of rights." The legal relation was the expression of an economic relation. The same was true by analogy of public law : " the state machine really embodies itself as an impersonal ' general will ', as ' the rule of law ' etc., in so far as society represents a market " on which individuals exchanged values on formally equal terms. It was even true of criminal law which, though in the main it represented the crude repression of its adversaries by the ruling class, and did not deserve the name of law at all, nevertheless "enters as a component part into the juridical superstructure in so far as it embodies one of the varieties . . . of the form of equivalent exchange with all the consequences flowing from it " — the notion of equivalent retribution.[2] The essential point of the theories of Stuchka and Pashukanis was that, by explicitly associating the survival of law with the practices of NEP, they provided it with a temporary sanction, while leaving the way open for its eventual disappearance with the advent of socialism. The conception of law had the same ambivalent character as the general conception of NEP, which

[1] P. Stuchka, *13 Let Bor'by za Revolyutsionno-Marksistskuyu Teoriyu Prava* (1931), p. 106 ; elsewhere Stuchka described the Soviet civil code as " nothing but the formulae of bourgeois civil law, repeating in general the formulae of Roman law framed about 2000 years ago " (*ibid.* p. 121).

[2] E. Pashukanis, *Obshchaya Teoriya Prava i Marksizm* (3rd ed. 1929), pp. 70, 96, 125 ; this work, first published in 1924, is translated in full in *Soviet Legal Philosophy*, ed. H. W. Babb (Harvard, 1951), pp. 111-225. Pashukanis regarded imprisonment for a fixed term as a specifically bourgeois conception " profoundly connected with the conception of abstract man and abstract human labour measured by time " (*Entsiklopediya Gosudarstva i Prava*, ii (1925-1926), 917).

was simultaneously interpreted as a retreat from socialism and a necessary stage in the advance towards it.

It is unlikely that these rather far-fetched constructions of intellectuals made much impression on political leaders and administrators, who wanted to invest their authority with the sanctity of law, or on peasants and traders, who wanted legal security for the tenure of their possessions and for the transaction of their business, or even on ordinary citizens, who wanted to know where they stood and preferred hard-and-fast rules to the individual caprice of officials. Revolutionary enthusiasm had run out into the " sea of lawlessness " of which Lenin complained ; " revolutionary consciousness of right " had become too often an excuse for bureaucratic improvisation and petty tyranny. A certain parallel could even be established with the stabilization of the currency which, as Trotsky put it, was "indissolubly bound up with the restoration of 'norms of bourgeois law'".[1] In law, as in other matters, the conceptions of the early years of NEP were a reaction against the ideas of the period of war communism, when the dissolution of authority had been welcomed as a normal stage on the road towards the social utopia of the future. The return to legality was a spontaneous process reflecting both the need of an established government to rely on the prestige and sanctions of the law, and the need of the citizen to rely on the stability and regularity of a legal order. The epithet " revolutionary " prefixed to " legality " seemed at this time little more than a conventional mask for the reassertion of legal authority and legal continuity.

The return to the continuity of legal tradition was assisted by the personal factor. The initial " strike " of jurists which at first made it appear that " the transformation of an old bourgeois jurist into a Soviet jurist is an impossibility "[2] lasted no longer than the intransigence of other professional groups. Just as the revolutionary challenge to law had been accompanied by acute mistrust of the professional judge and the professional lawyer, whose affiliations were all with the old régime, so the revival of legal authority meant the reinstatement of the professional

---

[1] L. Trotsky, *La Révolution Trahie* (n.d. [1936]), pp. 85-86.
[2] P. Stuchka, *13 Let Bor'by za Revolyutsionno-Marksistskuyu Teoriyu Prava* (1931), p. 8.

exponent of the law. Codification in itself seemed to represent the very essence of fixity and permanence. Whatever the specific content of the codes, and whatever specific interests were served by them, they marked a victory for the principle of stability after the interlude of revolutionary turmoil, and established, however unwittingly, an element of continuity with the Russian past. Former Tsarist officials and lawyers were actively concerned in the preparation of the codes. The codes themselves often repeated the form, the ideas and the very phraseology of Russian pre-revolutionary codes,[1] and helped to create a familiar atmosphere of routine and regularity in which the representatives of the old order could accommodate themselves to the service of the new. Former judicial workers of the Tsarist régime found themselves charged with the interpretation and administration of Soviet law.

> If you open any text-book you like [said Stuchka in 1922], any work on Soviet law, complete disillusionment generally overtakes you. The cover is Soviet, but the inside gives off an ancient bourgeois smell.[2]

Gaps in Soviet law were more and more frankly made good by appealing to the provisions of Tsarist legislation. Thus, while official theorists continued to harp on the provisional status of law under NEP, the cult of legality was simultaneously preached in terms which reinforced the authority of law and treated it as an essential pillar of the national economy and the national state. The reversal of the initial hostility of the revolution towards law was one of the most striking symptoms of the change in the climate of opinion which paved the way for the doctrine of socialism in one country.

[1] N. Timasheff in *American Slavic and East European Review*, xii, No. 4 (December 1953), pp. 441-462, shows that the drafters of the 1922 criminal code "amply and willingly borrowed legal provisions from the pre-revolutionary law, especially from the code of 1903"; even the contested principle of analogy (see pp. 77-78 above) had figured in the Russian code of 1845. Kalinin, in commending to the third Union Congress of Soviets in May 1925 the principle of "administration on the basis of a code of laws", used the old Tsarist term *Svod Zakonov* which had been avoided in official Soviet terminology (*Tretii S"ezd Sovetov SSSR* (1925), p. 268).

[2] P. Stuchka, *13 Let Bor'by za Revolyutsionno-Marksistskuyu Teoriyu Prava* (1931), p. 81 ; the article originally appeared in *Sovetskoe Pravo*, No. 3, 1922, pp. 3-18.

CHAPTER 3

# CLASS AND PARTY

LENIN, in a speech delivered to an audience of workers a few days after the announcement of NEP, distinguished three classes in the Soviet social order : the proletariat, which, as the result of its superhuman exertions in the revolution and the civil war, was now "extremely weary and exhausted and extremely worn out"; the petty bourgeoisie, which he identified with the peasantry and described as " an independent class, that class which, after the annihilation of landowners and capitalists, remains the only class capable of resisting the proletariat " ; and the " landowners and capitalists ", who were " here . . . at present nowhere to be seen ", but still constituted a powerful enemy abroad.[1] After nearly two years' experience of NEP, in one of his last published articles, Lenin eliminated the " land-owners and capitalists " but introduced a new category : the social order in the Soviet republic was " based on the cooperation of two classes — workers and peasants — to which are now also admitted on certain conditions the nepmen, i.e. the bourgeoisie ".[2] From this time the " new bourgeoisie " or " new bourgeois strata " were constantly mentioned in party literature, being identified with "traders, private lessees of enterprises, various free professions in town and country, rural *kulaks* etc." or, more briefly, " *kulaks* in the country, nepmen in the town ".[3]  In the autumn of 1924 Zinoviev diagnosed the existence in the Soviet Union of " two classes and a ' fragment ' " ; the " fragment ", consisting of the new bourgeoisie and " remnants of the old bourgeoisie ", was said to " constitute, let us admit, a third class ".[4]  But by this time the structure of Soviet society had been

---

[1] Lenin, *Sochineniya*, xxvi, 287-291.        [2] *Ibid.* xxvii, 405.
[3] *VKP(B) v Rezolyutsiyakh* (1941), i, 463-464.
[4] G. Zinoviev, *Litsom k Derevne* (1925), p. 79.

further complicated by the emergence of another group, sometimes referred to as the "commanding staff" or "officer corps",[1]
and comprising specialists, technicians, administrators and professional men directly or indirectly in the service of the Soviet
Government or of state economic organs.  This group was distinguished from the "new bourgeoisie" by the fact that it was not
engaged in economic activities on its own account and controlled
no means of production.  The intelligentsia, in so far as it accepted
the Soviet régime and professed loyalty to it, fell within this
group.

Soviet society, after the elimination of the " landowners and
capitalists ", and after the introduction of NEP, thus consisted,
according to the diagnosis of its leaders and theorists, of three —
or rather four — groups or classes : the proletariat ; the peasantry
(including, no doubt, a considerable number of small independent
artisans whose natural affinities ranged them with the peasants) ;
the " new bourgeoisie ", i.e. the nepmen and the *kulaks* ;  and the
" officer corps ", i.e. the officials, managers, technicians and intellectuals of all kinds.  Of these groups the third, the nepmen and
the *kulaks*, lay outside the structure of Soviet society rather than
within it, being in the position of incongruous, and barely tolerated,
intruders.  As capitalists and employers of labour, they were
ineligible for official position or for party membership ;  under
the constitutions of the RSFSR of 1918 and of the USSR of 1923,
they did not even enjoy the franchise.  The other three groups
constituted between them what was still officially referred to
as the dictatorship of the proletariat.  The tripartite classification of " workers ", " peasants " and " employees and persons
engaged in intellectual work " appeared in official publications
at least as early as 1924.[2]  It was also used in party statistics
showing the social composition of the party.  Members were
classified as " workers ", meaning those whose primary occupation was " physical labour for wages in production or transport ",
" peasants " working either on their own account or in collective

---

[1] The phrase occurs with an enumeration of those belonging to the category
in a speech of Zinoviev of October 1924 (G. Zinoviev, *Litsom k Derevne* (1925),
p. 74) : the Russian term *komsostav* was military in origin, and would include
both officers and non-commissioned officers.

[2] See, for example, *Sovety, S"ezdy Sovetov i Ispolkomy* (1924), which classified delegates to the principal Soviet organs in these three categories.

forms of agriculture, and " employees ", including " intellectuals " (who were not recognized as a distinct category for
party purposes). The residual category of " others " included
workers in rural industries, independent artisans, housewives and
domestic workers and students ; few of these were party members.

The establishment of this tripartite classification was, however, the beginning and not the end of the difficulty of defining
the nature and function of class in Soviet society. The first
embarrassment was to decide on the distinguishing criterion of
class. Neither Marx nor Engels ever explained exactly what they
meant by class. But, in the familiar current usage which they
followed, it seemed clear that membership of a class was not
simply determined by the social and economic functions performed by the individual concerned, but that class was a durable
formation possessing a common ideology as well as common
interests, so that the term was applicable only when function had
been hardened by convention into something like status, and a
change of class for the individual did not automatically follow a
change of function, being rarely accomplished within a single
generation. Individuals behaved and thought and felt as capitalists
or workers not merely because they were at a given moment
occupied as such, but because they belonged by birth to the class
in question. It was natural that, when it became necessary to
classify Soviet citizens for party or governmental purposes, the
classification should have proceeded on the basis of social situation and not of present occupation.[1] In a stable society this might
have presented no great inconvenience. But, in a revolutionary
society subject to sharp changes of status and acutely conscious
of such changes, the practice became seriously misleading. At a
time when considerable numbers of workers and peasants had been
drafted into official and administrative posts in party, Soviet or
economic organs, published statistics of the three classes diverged
widely from the real situation, especially since the prestige
attaching to the status of " workers " and " peasants " encouraged
" employees and persons engaged in intellectual work " to lay
claim to this status on the slightest justification. This practice was

---

[1] Pre-revolutionary Russian society was legally divided into five " estates "
which also took no account of current occupation : virtually all Russian factory
workers before 1917 were classified as " peasants ".

resented in some party circles, and an attempt was made to distinguish authentic workers and peasants within the broader categories as " workers from the bench " and " peasants from the plough ". The " Lenin enrolment " of 1924 was confined to " workers from the bench ", and was designed to increase the proportion of actual workers in the party ranks.[1] But these restricted categories were never officially adopted for statistical purposes. A party circular of August 12, 1925, which gave explicit directions on the keeping of party records, made it clear that members should be registered in accordance with their social situation, not with their current occupation.[2]

In Soviet society the divergence between the two criteria of distinction between classes was considerable. In 1925 it was recorded that, while 74·8 per cent of the party members in the Leningrad province were returned as workers and 11·3 per cent as peasants, only 55·5 per cent were workers, and only 1·4 per cent peasants, by present occupation.[3] Of members of county, department or city district party committees 55·5 per cent were described as workers and 18·9 per cent as peasants. But only 16 per cent were " workers from the bench " and 8·8 per cent " peasants from the plough " ; and these low percentages were an advance on the previous year.[4] Molotov told the fourteenth party congress in December 1925 that, whereas 58 per cent of party members were returned in the statistics as workers, only 38 per cent were " workers from the bench ".[5] An official volume of electoral statistics for 1926 complained of " the usual methodological difficulties (i.e. whether to treat a worker now occupying an administrative post as a worker or as an employee) ", and drew attention to the fluctuating classification of " persons now employed in state institutions and public organizations " who were " formerly workers or peasants ".[6] Another element of confusion was revealed in some statistics of the composition of the Red

[1] See The Interregnum, 1923–1924, pp. 352-356.
[2] Izvestiya Tsentral'nogo Komiteta Rossiiskoi Kommunisticheskoi Partii (Bol'shevikov), No. 34, September 7, 1925, p. 8 ; Spravochnik Partiinogo Rabotnika, v, 1925 (1926), pp. 258-260.
[3] Leningradskaya Pravda, November 5, 1925.
[4] Partiinye, Professional'nye i Kooperativnye Organy i Gosapparat : k XIV S"ezdu RKP(B) (1926), p. 18.
[5] XIV S"ezd Vsesoyuznoi Kommunisticheskoi Partii (B) (1926), pp. 77-78.
[6] Perevybory v Sovety RSFSR v 1925–1926 godu (1926), i, 2 ; ii, 6.

Army at the beginning of 1927. At that date only 16 per cent of
Red Army men were returned as "workers"; but, if present
occupation had been the criterion, the proportion of workers
would have risen to 22 per cent.[1] Clearly, some workers of
peasant origin or peasant affiliations were at that time still being
registered as peasants. Nor were these uncertainties purely formal
or statistical. In a society still in process of crystallization, the
categories themselves were vague and undefined. It would often
have been difficult to determine, on any criterion of interest or of
conscious loyalty, the class affiliation of the peasant who went to
work in a factory or of a worker recruited for administrative work.

Uncertainties of classification as between different groups and
different individuals were, however, secondary to the major
difficulty of identifying in Soviet society the operation of those
class forces which, according to Marxist theory, provided the
dynamic of social action. Any analysis of the structure of Soviet
society in the NEP period is complicated by an incompatibility
between the objective conditions of the society and the terms in
which its leaders and its intellectuals, faithful to the Marxist
tradition, habitually thought and wrote about it. The revolution
which had occurred in Russia in October 1917 was recognized,
with some reservations as to the admixture in it of bourgeois
elements, as a fulfilment of the Marxist doctrine of the proletarian
revolution; and this implied acceptance of the Marxist class
analysis. On the other hand, it had occurred, contrary to Marxist
expectations, in a country where the proletariat was weak. The
Russian proletariat had achieved victory not by its own unaided
effort, but by invoking the assistance of the peasant and by accepting the time-honoured, but non-Marxist, goal of peasant revolutions — the seizure of land by the peasants. Lenin, like the other
Bolshevik leaders, did not believe that the régime could survive
unless the proletariats of other important countries made successful revolutions and came to its aid. But he also did not
believe that, without this aid, it was possible to create a socialist
economy in backward Russia. In the first place, the proletariat
was too weak to provide the industrial foundation of socialism;
secondly, any attempt to build socialism would bring a clash with
the proprietary ambitions of the overwhelmingly numerous

[1] K. Voroshilov, *Oborona SSSR* (1927), p. 184.

peasantry. By the end of 1920 the peasants, exhausted by the civil war and exasperated by the grain requisitions, were on the verge of revolt. The régime saved itself by the compromise of NEP, confirming the proprietorship of the peasants, restoring a free market in grain and thus opening the road to *kulaks* and nepmen — the new capitalists. It was admittedly a forced move, a " retreat ". Enemies of the régime called it a " thermidor " ; members of the " workers' opposition " in the party described it as a surrender of the proletariat to the petty bourgeois peasantry. This diagnosis could scarcely be avoided if the traditional class analysis were unconditionally applied. When it began to be claimed that NEP, while in one sense a retreat, was also a step on the road to socialism, and that, in spite of the failure of the pro-letarian revolution elsewhere, such an advance was in fact being made in Soviet Russia, a new diagnosis of class relations imposed itself. " The class struggle does not disappear under the dictator-ship of the proletariat," Lenin had written in 1919, " but merely assumes different forms." [1] The difference in form, however, seemed to carry with it a different conception of class. It involved serious theoretical embarrassments ; and these embarrassments, while they were inherent in the attempt to transpose the terms of the Marxist analysis from the nineteenth to the twentieth century and from western to eastern Europe, were also a reflexion of the fundamental problem, which had dogged the Bolsheviks ever since the victory of 1917, of bringing a proletarian revolution to fruition in a country where the proletariat was still a small and backward minority. What was happening in the Soviet Union could not be explained in terms of the traditional class analysis. The problem clearly emerges from an examination of the position, actual, theoretical and potential, of the three recognized groups or classes. [2]

The peasantry constituted the most serious and continuous preoccupation of Soviet statesmanship as well as the major

---

[1] Lenin, *Sochineniya*, xxiv, 513.
[2] The numerical strength of the respective groups cannot be precisely estimated, since the census figures and the statistics of population used by Gosplan distinguished between categories of employment, but not between social and professional groups within those categories. In the census of 1926

theoretical problem of the Marxist analyst of Soviet society. Marx, setting the peasantry apart from the two main classes of capitalist society, the bourgeoisie and the proletariat, treated it as a survival of pre-capitalist society, doomed to disintegrate under the impetus of progressive capitalism. But he recognized it, in less advanced countries, as a potentially revolutionary factor in alliance with the proletariat and as a battleground between the proletariat and the bourgeoisie.[1] Plekhanov, a good theoretical Marxist, firmly declared that the Russian peasantry was not a class but an " estate ", containing elements of two opposing classes.[2] Lenin, who rarely separated revolutionary theory from revolutionary tactics, fluctuated in his terminology. At the stage of the bourgeois revolution, the peasantry as a whole was the ally of the proletariat in overthrowing the power of the feudal landowner ; and in this context Lenin freely spoke of the peasantry

out of a total population of 147 millions, 82,700,000 were returned as being in civil occupation (children and other dependants, pensioners, unemployed, and members of the armed forces being excluded). Of this total 71,700,000 were engaged in agriculture, including forestry and fishing ; of 1,860,000 engaged in small and handicraft industry, a large proportion was engaged in rural industries, and was assimilated socially and politically to the peasantry. Those engaged in manufacturing (i.e. factory industry), mining, transport, trade and credit amounted to no more than 5,606,000 ; of these, manufacturing and mining accounted for 2,800,000. These figures do not distinguish between managers, technicians, employees and manual workers. The total of 2,030,000 engaged in public administration and social services includes all so employed from heads of department to door-keepers. (These figures are in F. Lorimer, *The Population of the Soviet Union* (League of Nations, Geneva, 1946), pp. 218-219.) It is interesting to compare these figures with those of party membership (for these see A. Bubnov, *VKP(B)* (1931), p. 615. The proportion of peasants in the population at this time on any estimate exceeded 80 per cent. The proportion of peasants in the party membership fluctuated in the years 1922–1926 between 28·8 and 25·7 per cent ; moreover, many of the so-called peasant members were rural party officials. The under-representation of peasants in the party was deliberate, and was justified by party doctrine. During the same period, the proportion of workers rose, almost entirely as the result of the " Lenin enrolment " of 1924, from 44 to 56·8 per cent, and the proportion of employees fell from 28·9 to 17·3 per cent. Since population figures do not distinguish between these two categories, comparison is impossible ; but it is certain that employees (i.e. the intellectuals) were, even at the end of the period, still heavily over-represented in the party.

[1] In *The Eighteenth Brumaire of Louis Bonaparte* Marx treated the French peasantry as a class which had lost its *raison d'être* with the overthrow of the landed aristocracy, and would be compelled to ally itself with the urban proletariat in order to protect its interests against the urban bourgeoisie (Marx i Engels, *Sochineniya*, viii, 408-409).

[2] See *The Bolshevik Revolution, 1917–1923*, Vol. 2, p. 12, note 2.

as a class.  At the stage of the socialist revolution, the proletariat sought the alliance of the " semi-proletarian " elements in the peasantry in opposition to its hostile bourgeois, and temporizing petty bourgeois, elements ; [1] and in this context the peasantry was not a class but a composite group drawn from different classes.  Once the attempt in 1918 to hasten the socialist revolution by exploiting the divisions in the peasantry through the committees of poor peasants had ended in failure, the retreat into the semi-capitalist régime of NEP became inevitable ;  and in this context it was once more relevant to speak of the peasantry as a class and of the importance of its link with the proletariat, though this usage was always tempered by consciousness that a renewal of the advance towards socialism depended on the possibility of exploiting the divisions in the peasantry which were momentarily in abeyance.[2]

By way of contrast with the depressed position of the industrial worker, the peasant appeared to have emerged under NEP as the main beneficiary of the revolution — a position due to his preponderant weight in the population and in the economy.  Retrospect confirmed the impression that he had supported the revolution on his own terms, and was powerful enough to enforce continued observance of those terms.  The land decree of October 26/ November 8, 1917, sealed the acceptance by the Bolsheviks of the peasant programme of land distribution, and secured the loyal support of the peasant throughout the civil war.  He fought to defend his newly won gains.  But he would not fight to carry the revolution to other countries ;  in this he was supremely uninterested.  The victory over the " whites " and the defeat of the Red Army before Warsaw in 1920 were alike symbolical of his attitude.[3]  Once the war was over, and the danger of a return of the landowners averted, the peasant could exact fresh terms for his continued support :  these were represented by NEP.  The French

[1] For this analysis, which dated from 1905, see *The Bolshevik Revolution, 1917–1923*, Vol. 1, pp. 54-55.

[2] Speaking at the third congress of Comintern three months after the introduction of NEP of the alliance between proletariat and peasantry as " an alliance of different classes ", Lenin explained that he meant an alliance between workers and poor peasants on the one hand and middle peasants on the other (*Sochineniya*, xxvi, 331) ;  but he was not always so careful to make these distinctions.

[3] See *The Bolshevik Revolution, 1917–1923*, Vol. 3, pp. 215-216.

elections of December 10, 1848 were described by Marx in *The Eighteenth Brumaire of Louis Bonaparte* as " *the reaction of the peasants* who had been compelled to carry the costs of the February revolution against the other classes in the nation, *a reaction of country against town* ". The same could have been said about NEP and its sequel in the Soviet Union.[1] Bukharin noted that the peasant had learned much in the army, and was "on a higher moral and intellectual level " than before the revolution :

> He says : We are the predominant force, and shall not allow others to treat us as silly children. We want to feed the workers, but we are the senior partners and demand our rights.[2]

When industry attempted to strike against the favours now shown to the peasant, and brought about the scissors crisis, Zinoviev and Kamenev at the twelfth party congress of April 1923 loudly proclaimed that agriculture was the foundation of the Soviet economy and a firm alliance with the peasant the key to Soviet policy ; and this conception continued to be proclaimed from party platforms, and to dominate party decisions, for the next two and a half years. " The peasant ", wrote Ustryalov from Harbin in 1923, " is becoming *the sole and real master of the Russian land.*"[3] These conditions made the weight of peasant influence strongly conservative. " The peasantry ", Lenin had noted in November 1922, " is satisfied with its present position."[4] The peasant accepted the revolution which had expropriated the landowners and distributed the land to the peasants under forms of tenure which approximated as closely as possible to peasant ownership : if this was socialism, he was a socialist. But this implied no ideological sympathy. The peasant, said Kalinin, was " more remote from the Soviet power than the intelligentsia ".[5] His horizon did not extend beyond the limits of his own economy and of the conditions necessary to make it prosper. " Socialism in one country " — provided he was allowed to choose his own interpretation of " socialism " — was a conception which fitted in perfectly with his interests and his aspirations.

---

[1] Marx i Engels, *Sochineniya*, viii, 339 ; the comparison was made in *Sotsialisticheskii Vestnik* (Berlin), No. 2-3 (120-121), February 11, 1926, p. 8.

[2] Extracts from speech of July 1921 in *The New Policies of Soviet Russia* (Chicago, 1921), pp. 52-54 ; the Russian original has not been traced.

[3] N. Ustryalov, *Pod Znakom Revolyutsii* (2nd ed. 1927), p. 148.

[4] Lenin, *Sochineniya*, xxvii, 347.          [5] *Pravda*, February 2, 1926.

This picture of the " peasantry " under NEP was, however, a deliberate and artificial simplification, which ignored the patent fact that the peasantry was not a homogeneous and undifferentiated mass, and that the beneficiaries of party and governmental favours were, on the whole, the enterprising and well-to-do peasants who knew how to look after themselves and to turn these favours to good account. Such a bias was, no doubt, inevitable. So long as conciliation of the peasantry remained the pivot of Soviet economic policy, the benefits of that policy would be reaped primarily by the well-to-do peasant, though the crumbs from his table might be gleaned by his poorer dependants. In Marxist terms, to encourage the development of agriculture, under whatever safeguards, on capitalist lines meant to tolerate capitalist inequality and capitalist exploitation in the countryside. The peasant whose influence, from 1921 onwards, made itself increasingly felt in the elaboration of party doctrine and of Soviet policy was mainly the well-to-do peasant. Eighteen months after Lenin had spoken of the satisfaction of the peasantry, Zinoviev admitted that it was the " prosperous *kulak* sector " of the peasantry which was most satisfied with the régime.[1]

In the first years of NEP this situation was accepted without undue alarm, and the reiteration of the " link " between the proletariat and the peasantry established by NEP remained the principal party slogan. The agitation which set in strongly after 1924 against a peasant policy whose effect was to favour the *kulak* marked a fresh turning-point. It meant, in theoretical terms, a reaction from the terminology which treated the peasantry as a class, and a renewed insistence on breaking up the peasantry into its class components. Not only was the peasantry no longer a class, but the different strata clearly belonged to different classes. The *kulak*, as a capitalist and an exploiter of labour, was the class-mate of the nepman. The class status of the poor peasant, or of the hired agricultural worker or *batrak* who sold his labour, ranged him with the proletariat, though in practice the *batrak*, who often retained a small plot of his own, was generally animated by the ambition, however hopeless, to acquire enough land to set up as an independent peasant, and felt no solidarity with the proletariat

---

[1] *Trinadtsatyi S"ezd Rossiiskoi Kommunisticheskoi Partii (Bol'shevikov)* (1924) pp. 100-102.

of the factories.[1] The middle peasant, an independent worker neither selling nor hiring labour, was in the same petty bourgeois category as the independent craftsman or artisan, the small shop-keeper or trader, in town or country. The characteristic of this group was that, while it owned or controlled the means of production with which it worked, and was to this extent capitalist, it did not generally exploit or employ the labour of others. The petty bourgeoisie had been described in the *Communist Manifesto* as a class balanced precariously between the bourgeoisie and the proletariat and a potential source of recruits for the latter. Hence the party was not unconditionally hostile to members of this group, and was continually concerned to prevent them from falling under the predominant influence of *kulaks* and nepmen, and to win their support for the proletariat : this was, in particular, the basis of the attitude towards the middle peasant adopted in 1919,[2] and strongly reasserted after 1924. The conflicting agrarian policies of the Soviet régime, which were for some time pursued side by side and simultaneously, were expressed in two different theoretical terminologies. When it was desired to emphasize the need to conciliate the well-to-do peasant, the peasantry was spoken of as a single class in alliance with the proletariat. When it was desired to curb the *kulak* and strengthen the hand of the middle and poor peasant against him, the peasantry was treated as a composite entity and dissolved into its class elements. It was no longer true that the class analysis determined policy. Policy determined what form of class analysis was appropriate to the given situation. The class analysis had been subordinated to the political issue.

The proletariat, like every group in Soviet society, had been profoundly affected by NEP, and had derived indirect material benefits from it. But its relative weight in the social order had declined. The conception of the leadership of the proletariat in the revolution, which had seemed effective in the heroic days of October 1917, proved unworkable once power had been seized. The idealized picture drawn by Lenin in 1918, in which " the whole mass of workers, not only its leading figures, but really the

---

[1] For the poor peasant and the *batrak* see p. 230, note 3, below.
[2] See *The Bolshevik Revolution, 1917–1923*, Vol. 2, pp. 161-165.

broadest strata, know that they themselves are building socialism with their own hands ",[1] remained a magnificent dream. Workers' control in the factories failed to turn the wheels of production ; the Red Guard was helpless in face of a disciplined army ; the administrative machine was not amenable to the simple prescriptions of *State and Revolution* ; above all, the proletariat was incapable of " leading " a peasantry which constituted 80 per cent of the population, and on which it was itself vitally dependent. The weakness of the proletariat, which had become apparent as soon as the need arose to consolidate the victory of October 1917, was further aggravated by the sequel. As early as March 1918 Bukharin diagnosed a " disintegration of the proletariat " ; with the chaos of the civil war, the collapse of industrial production and the flight from the hunger-stricken cities, the Russian proletariat seemed in danger of being completely reabsorbed into the peasant mass out of which it had so recently emerged, and with which it had retained so many links and affiliations.[2] It was no exaggeration when Lenin described the proletariat at the end of the civil war as " exhausted " and " worn out ". And, when at length NEP relaxed the tension and paved the way for recovery and reconstruction, it was the peasant rather than the industrial worker who took the lead and, for the next three years, occupied the chief place in the preoccupations of the party leaders. While agricultural production climbed slowly back towards its pre-war levels, industry lagged far behind ; moreover, it was light industry, where the workers had least skills and the worker's outlook and status diverged least from that of the peasant, which took the first steps on the road to recovery. In 1923 heavy industry, before the war the main occupation of the skilled and class-conscious worker, had still scarcely risen above the record low levels of 1920 and 1921. The proletariat had not only declined in numbers, but had lost its distinctive character.

This apparent decline was due in part to a misapprehension of the character of the proletariat. The proletariat did not, as a later Bolshevik commentator remarked, come into being " as a result of some ' immaculate conception ' ".[3] Many workers bore

---

[1] Lenin, *Sochineniya*, xxiii, 252.

[2] For a description of this process in the years of the civil war see *The Bolshevik Revolution, 1917–1923*, Vol. 2, pp. 193-195.

[3] *Bol'shevik*, No. 3-4, May 20, 1924, p. 18.

the hall-marks of a peasant or petty bourgeois origin. And this was truer of the Russian proletariat than of working classes of more advanced countries where lapse of time had detached the worker more completely from the soil, and long practice had bred habits of organization and concerted action.

> Very often [said Lenin in one of his last speeches] when people say " workers " they think that this means factory and workshop proletariat. It means nothing of the kind. Ever since the war people who are not proletarians have gone into our factories and workshops, have gone into them in order to escape from the war ; and are social and economic conditions such with us at present that genuine proletarians go into the factories and workshops ? Certainly not. That would be correct according to Marx, but Marx wrote not about Russia, but about capitalism in general, beginning with the fifteenth century. For 600 years that was correct, but in present day Russia it is incorrect. Again and again, those who go into the factories are not proletarians, but every sort of casual element.[1]

At the end of 1923 the Russian proletariat, dispersed and neglected, subjected to a long process of quantitative and qualitative deterioration, seemed to have touched the nadir of its prestige and influence.

This apparent reversal of fortune was a product of the conditions attending and following the victory of the revolution. " We are not in favour of seizure of power by a minority ", Lenin had declared in 1917 ;[2] and, even when at the critical moment he pronounced it naïve " to wait for a ' formal ' majority ", he was none the less confident that " a majority of the people are *for* us ".[3] The support of the peasants had, for the moment, supplied the condition of success. But the delay of the world proletariat in coming to the aid of the Russian revolution prolonged a situation which had at first seemed purely provisional. The Russian proletariat, unaided by the proletariats of the advanced countries and thrown back on its own resources, was unequal, in numbers, in organization and in experience, to the enormous burdens which the revolution had unexpectedly placed on it. The situation was one which Engels had foreseen in a different

[1] Lenin, *Sochineniya*, xxvii, 252.          [2] *Ibid.* xx, 96.
[3] *Ibid.* xxi, 193-194.

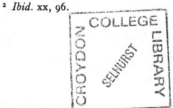

context, many years before, in his essay on *The Peasant War in Germany* :

> The worst thing that can happen to the leader of an extreme party is a conjunction of circumstances which compels him to take the administration into his hands in an epoch when the movement is not yet ripe for the rule of the class whose representative he is, or for the measures demanded by the rule of that class.  What he *can* do depends not on his will, but on the level of intensity reached by the clash of interests of different classes, and on the stage of development of the material conditions of existence, of the conditions of production and means of communication which always lie behind the development of class· contradictions.  What he *ought* to do, what his own party demands of him, depends not on himself and also not on the stage of development of the class struggle and of the conditions that lie behind it ;  he is bound by his former teachings and demands, which once again take their rise not out of the existing relation of social classes, and not out of the existing, more or less fortuitous, position of conditions of production and means of communication, but out of his more or less profound understanding of the general consequences of the social and political movement.  He is inevitably confronted by an insoluble dilemma :  what he *can* do contradicts his whole former behaviour, his principles and the immediate interests of his party ;  what he *ought* to do is impracticable.  In a word he is compelled to defend not his own party, not his own class, but that class for whose rule the movement is already ripe at the time in question.  He must in the interests of the movement itself defend the interests of an alien class, and put off his own class with phrases and promises, assuring it that the interests of this alien class are identical with its interests.  He who has fallen into this false position, is lost irretrievably.[1]

But no leader of a victorious revolution — and least of all Lenin — was likely to accept the view that he was irretrievably lost.

> The man who turns his back on the socialist revolution which is going on in Russia [he wrote] by pointing out the obvious lack of balance of forces is like a man who has got set in a mould, who does not see beyond the end of his nose, and forgets that there has never in history been a revolution of any magnitude without a whole series of examples of unbalanced forces.[2]

[1] Marx i Engels, *Sochineniya*, viii, 185-186.
[2] Lenin, *Sochineniya*, xxiii, 42.

And to those who denounced the Bolsheviks as dreamers Lenin replied : " How could one start a socialist revolution in such a country without dreamers ? "[1] The vision must take the place of the reality. What could not be realized in the present must be projected into the future. Lenin's chosen instrument, the party, must hold the fort and prepare the conditions in which the proletariat would mature and become ripe for the rôle assigned to it. The body of men gathered round Lenin in the party central committee and in the Soviet Government held sway in the name of the proletariat, of whose rights and interests they regarded themselves as trustees — trustees for the world proletariat in their capacity as prime movers in a world-wide proletarian revolution, trustees for the Russian proletariat in their capacity as provisional rulers of what had been the territory of the Russian Empire.

An unconfessed and unperceived modification thus began to creep into the doctrine of the dictatorship of the proletariat. The doctrine was constantly reiterated in party pronouncements.[2] But every attempt to translate it into practice revealed its illusory character. Lenin, who had made light in his pre-revolutionary writings of the tasks of administration, now spoke often and earnestly of the need to learn ; and Trotsky in 1919 was still more explicit on the administrative shortcomings of the proletariat :

> The conquest of power by itself does not transform the working class or bestow on it all the requisite merits and qualities : the conquest of power only opens before it the possibility of really learning, of developing and freeing itself from its historic shortcomings.[3]

This was the picture of a dictatorship of the proletariat *in posse* rather than *in esse*, a dictatorship not of the existing proletariat, but of the idealized proletariat of the future which had already

---

[1] *Ibid.* xxvi, 239.

[2] " This concept ", wrote Lenin of the dictatorship of the proletariat in 1920, " has meaning only when one class knows that it alone takes political power into its hands and does not deceive either itself or others by talk about popular, elected government, sanctioned by the whole people " (*Ibid.* xxvi, 286). The difficulty was, however, not the exclusion of other classes, but precisely the inability of the proletariat to " take political power into its hands ".

[3] Trotsky, *Sochineniya*, xxi, 97.

appeared in the pages of Engels and Plekhanov.[1] Such a conception raised the inevitable question by whom the dictatorship
should in fact be exercised in the interval while the proletariat
was being groomed for the eventual discharge of this function.
Nor was the answer in doubt. Thanks to the "low cultural level"
of the working masses, wrote Lenin in 1919, " the Soviets which,
according to their programme are organs of administration *by the
workers*, are in fact organs of administration *for the workers* by the
leading sector of the proletariat, not by the working masses ".[2]
In the same year he proclaimed the dictatorship of the party as
the working form of the dictatorship of the proletariat, explaining
that " the dictatorship of the working class is carried into effect
by the party of the Bolsheviks which since 1905 or earlier has
been united with the whole revolutionary proletariat ". In the
early years of the régime no embarrassment was felt about trans-
lating the dictatorship of the proletariat in current terminology as
the dictatorship of the party ; and a resolution of the twelfth party
congress of 1923 had declared that "the dictatorship of the working class cannot be assured otherwise than in the form of dictatorship of its leading vanguard, i.e. the Communist Party". Though
Stalin subsequently attacked the formula of the dictatorship of
the party,[3] the substance remained unchanged and uncontested.

[1] Engels, in the concluding sentence of his essay on Feuerbach, called the
German workers' movement " the heir of German classical philosophy "
because it was destined to provide " the key to an understanding of the whole
history of society " (Marx i Engels, *Sochineniya*, xiv, 678) ; Plekhanov looked
for " the appearance of the proletariat on the historical stage as the promised
Messiah " (*Sochineniya*, xv, 90), and noted in 1892 that " the proletariat in
Russia is growing and becoming strong, like the *bogatyr* in the fables, literally
not by days but by hours " (*ibid.* iii, 383-384).
[2] Lenin, *Sochineniya*, xxiv, 145.
[3] For these texts see *The Bolshevik Revolution, 1917–1923*, Vol. i, pp. 230-
232. Zinoviev reverted to this controversy in an article of 1924 : " The con-
sensus of opinion about the dictatorship of the proletariat can be expressed in the
following propositions. It is the dictatorship of a class if we look at the matter
from the social and class point of view. It is the dictatorship of the Soviet
state, a Soviet dictatorship, if we look at the matter from the point of view of
*juridical* form, i.e. from the specifically state point of view. It is the dictatorship
of a party if we look at the same question from the point of view of leadership,
from the point of view of the internal mechanism of the whole vast machine
of a transitional society " (*Pravda*, August 23, 1924). The first member of this
triple definition begs the issue. The dictatorship of the state and of the party
were, in fact, different facets of the same phenomenon ; but this could not be
identified with the dictatorship of the proletariat as a social class.

The party retained its directing rôle as " the basic lever of the dictatorship ".[1]

> The party [wrote Stalin in 1924] cannot be a real party if it limits itself to registering what the masses of the working class experience and think, if it drags along at the tail of the spontaneous movement, if it does not know how to combat the inertia and political indifference of the spontaneous movement.[2]

The party was no mere emanation or expression of the will of the proletariat. It was an organized class-conscious vanguard responsible for imparting will and purpose to the proletariat as a whole and giving it direction.

The change of emphasis in the first years of the revolution from the proletariat to the party, as the proletariat melted away and the party gathered strength, was a subtle and almost imperceptible process. In one sense it was a logical continuation of the rôle of the party before the revolution : Trotsky in 1904 had accused Lenin of seeking by his methods of organization to create a party which would " substitute itself for the working class ".[3] It would, however, be erroneous to regard it as a planned or purposeful development. In the days of desperate struggle before the revolution, the party had served as the trustee for the proletariat and the organizer of its forces. With the victory of the revolution this rôle should have become obsolete : the proletariat should have been ready to take over for itself the reins of authority. But this expectation was based on the firm belief in a world-wide proletarian revolution which would crown the Russian initiative, and enable its fruits to be garnered. When this hope was disappointed, when the aftermath of the revolution brought with it a struggle not less dire than that of earlier years, a struggle against enemies without and within, the party was quickly re-established in its old position and once more became the fighting vanguard of the proletariat, organizing and marshalling the masses behind it and speaking with authority in their name. Before the end of the civil war, Lenin had largely reverted to the moods and terms

[1] VKP(B) v Rezolyutsiyakh (1941), ii, 226.

[2] Stalin, Sochineniya, vi, 171.

[3] N. Trotsky, Nashi Politicheskie Zadachi (Geneva, 1904), p. 50 ; other passages from this pamphlet and Trotsky's own later comment on the controversy are quoted in The Bolshevik Revolution, 1917–1923, Vol. 1, p. 33.

of speech of the period before 1917, when the hopes of the revolution had been entirely centred in the devotion and organizing
capacity of the party. At the time of the introduction of NEP,
when industry sank to its lowest point, the substitution of the
party for the proletariat was to all intents and purposes an
accomplished fact.

The Bolshevik party, bearing the name of a workers' party,
had from the outset drawn a majority of its rank-and-file members
from the small body of Russian industrial workers. To win a
large following among the workers and to keep high the level of
proletarian membership had always been a professed party aim.
Yet, even in 1917, the proportion of workers in the party was no
more than 60 per cent ; and, after the revolution, with the collapse
of industrial production, and with the extensive recruitment of
administrators and intellectuals into the party, the proportion of
workers declined, falling as low as 41 per cent in 1921. At the
tenth party congress in that year Shlyapnikov, the leader of the
workers' opposition, complained that in Moscow only 4 per cent
of the metal workers, and in Petrograd only 2 per cent, belonged
to the party, the ranks of which were being swollen with peasants
and intellectuals. After 1921 conscious attempts were made to
redress the balance, first by the purges which fell more heavily on
non-worker members of the party than on the workers, and then
by the " Lenin enrolment " of 1924 and similar subsequent
campaigns to increase the influx of workers, though it was not till
1929 that the proportion of proletarian members once more
reached 60 per cent.[1] The recruitment into the party of carefully
selected workers, on whose loyalty the party leadership could
count, served the double purpose of countering an opposition
whose main strength lay among the party intellectuals,[2] and of
providing a nucleus through which the proletariat could be
influenced and won for party policies. At a later time the difference between party and non-party proletarians was largely effaced.
But in the middle nineteen-twenties those workers who were
members of the party not only enjoyed a privileged status which
was likely to ensure their loyalty to the party leadership, but

[1] See the tables in A. S. Bubnov, *VKP(B)* (1931), p. 615 ; Shlyapnikov's
remarks are in *Desyatyi S"ezd Rossiiskoi Kommunisticheskoi Partii* (1921),
pp. 29-30.        [2] See *The Interregnum, 1923–1924*, pp. 355-356.

served as agents by whom the authority of the party could be exercised over the proletariat as a whole. At this period about one in ten industrial workers were party members. The highest proportion of party members was to be found among the metal workers, the chemical workers and the printers, the lowest among textile workers, miners and timber workers.[1] Broadly speaking, the incidence of party membership was highest among the most skilled.

More significant than the mass recruitment of workers into the party was the promotion of worker members of the party to responsible posts in party or Soviet administration or in economic management. Lenin had insisted on the need for the workers to learn to administer if the dictatorship of the proletariat was to be realized. Even when the recruitment of bourgeois specialists to such posts had become normal practice, gaps still remained to be filled ; and the theory that potentially disloyal specialists should be checked and supervised by loyal and reliable workers was widely accepted and applied. In the early years the promotion of outstanding workers to managerial and other responsible posts was paradoxically hampered by opposition within the party itself, based partly on egalitarian prejudices against the creation of a so-called " workers' aristocracy ", and partly on fears that it betokened a revival of the syndicalist heresy of workers' control.[2]

[1] A. S. Bubnov, VKP(B) (1931), p. 617. According to a resolution of the thirteenth party congress at the time of the Lenin enrolment, " in some individual factories a majority of workers joined the party " (VKP(B) v Rezolyutsiyakh (1941), i, 571) ; but this was evidently exceptional.
[2] The idea of the promotion of workers to responsible posts was strongly pressed in 1919 and 1920 by one Goltsman, a leading member of the metal workers' union ; it was attacked by Bukharin, who, in an article in Pravda, September 14, 1919, entitled Workers' Aristocracy, or Unity of the Working Class ?, protested against the division of the proletariat into two groups (" regular factory workers materially bound up with urban industry " and " elements drawn into factory life at a comparatively recent date and bound up materially and ideologically with the petty bourgeoisie of country and town "), the former of which would dominate and discipline the latter, and by Zinoviev who prepared theses directed against Goltsman for the ninth party congress in March 1920 (G. Zinoviev, Sochineniya, vi, 344.) Goltsman defended himself in Pravda, March 26, 1920. The issue was never formally decided, though Lenin at the congress appeared to sympathize with Goltsman, mainly on the ground of his support of the principles of one-man management against collegiality (for this controversy see The Bolshevik Revolution, 1917–1923, Vol. 2, pp. 187-191) ; other echoes of the dispute were heard at the congress where Trotsky was accused of having formed a bloc with Goltsman, and Ryazanov

After the introduction of NEP any remaining objections in the party against the promotion of an élite of workers to responsible posts were finally overruled ; and the main limiting factor to the number of such appointments was a shortage of qualified candidates. This was, however, serious and persistent. Of workers occupying important posts in industry Nogin showed outstanding competence and rose to a position of unquestioned authority ; [1] but it is difficult to find any other name to match with his. Of a hundred communists in responsible positions, said Lenin at the eleventh party congress in 1922, 99 did not know the elements of business administration, and " will not understand that they do not know it and must begin to learn from the ABC ".[2] The congress resolution declared that the trade unions must become " a school of administration of socialist industry ", one of their " chief tasks " being " the promotion and training of administrators from the ranks of the workers and the toiling masses ".[3] From this time the promotion of party members to key positions became a familiar and an accepted practice, and an essential part of the system of party leadership. The word " promotion " (vydvizhenie) was introduced into the party vocabulary with the special connotation of the appointment of workers to important posts. A report submitted by the party central committee to the thirteenth party congress in May 1924 recorded the first results of a resolution passed in the previous November in favour of " promotion ", especially of workers. This was to take two forms : " promotion " by central party organs to " elective or other " posts in Soviet or economic organs, and " promotion " on the spot by local party organs. No particulars were available of the latter process. But of 788 appointments made in the intervening six months on decisions of the central committee, 173 were of " promotees ", i.e. persons specially promoted on grounds of merit. An analysis showed, however, that of 71 promotees appointed to local party posts, 20 were workers, 8 peasants, 33 employees or intellectuals

denounced the plan as " a kind of syndicalism which tries to promote trade unions of skilled workers to the rôle of managers of industry " (Lenin, Sochineniya, xxv, 120 ; Devyatyi S"ezd RKP(B) (1934), pp. 210-212, 247).
[1] See The Interregnum, 1923–1924, p. 278 ; a meeting with him is recorded in A. Ransome, 6 Weeks in Russia in 1919 (1919), pp. 94-99.
[2] Lenin, Sochineniya, xxvii, 251.
[3] VKP(B) v Rezolyutsiyakh (1941), i, 421.

and 10 " other ", and of 102 promotees to central party posts, 30
were workers, 3 peasants, 55 employees or intellectuals and 12
" other ".  It thus appeared that only 30 per cent of the promotees were workers, and more than half employees or intellectuals, and that the proportion of the latter was higher in
appointments to central posts than to local posts.[1]  Considering
that the category of " workers " doubtless included persons of
proletarian origin already engaged in administrative or intellectual
work, these figures are eloquent of the difficulties of finding any
large number of qualified " workers from the bench " to promote.

Progress was none the less made ;  and by 1924 the structure
of the boards of management of major industrial concerns had
fallen into a more or less regular pattern.  Statistics collected
from 88 of the large industrial trusts at the beginning of that year
showed that 91 per cent of the presidents of these trusts, and 48
per cent of all members of their boards, were party men ;  51 per
cent of the presidents, and 29 per cent of members of boards were
workers by origin, virtually all these being party men.  More
than half the proletarian presidents of trusts were former metal
workers.  The average age of presidents was lower than that of all
board members, 62 per cent of the presidents falling into the age
group 30-39.  Statistics of the same date from 639 large factories
showed that 48 per cent of the directors and 34 per cent of vice-
directors were party men :  details of origin were not given.  Of
the 18 largest factories (employing 5000 or more workers), all
the directors were party men, but only 31 per cent of the vice-
directors.  On the other hand, only 7 per cent of employees, i.e.
clerical staff, in the 88 trusts investigated were party members ;
the percentage tended to be higher in non-specialized administrative posts and lower in technical and financial posts.[2]  The
general picture which emerges from these figures is fairly clear.
It was a common, though not universal, practice, to place a party
member at the head of an important industrial trust or large
factory (or, perhaps, sometimes, to confer party membership on
a person holding such a post) ;  and a substantial proportion of
these industrial chiefs were former workers.  Among other

---

[1] *Izvestiya Tsentral'nogo Komiteta Rossiiskoi Kommunisticheskoi Partii (Bol'-
shevikov)*, No. 1 (59), January 1924, p. 66 ;  No. 4 (62), April 1924, pp. 49-51.
[2] *Komsostav Krupnoi Promyshlennosti* (1924), pp. 12-21, 31-32, 52-54 ;  this
was a party publication prepared for the thirteenth party congress.

directing staff, the proportion both of party members and of former workers was much smaller.  In large factories the combination of a party director and a technically qualified non-party vice-director was probably the most usual pattern.  Throughout all state economic organs and undertakings the conception prevailed of the party man who was also a worker at the top, responsible for major decisions and acting as watch-dog over the technically competent but politically suspect second in command.[1]

But, while some success may have been attained in imposing this system at the highest level, the practice of the promotion of workers to administrative appointments evidently continued to encounter difficulties.  Some of these were eloquently described in a report on " promotion " from the party organization in the Vyatka province in the spring of 1926.  Sometimes party cells refused to recommend good workers for promotion since this would mean their transfer elsewhere.  Sometimes the promotee himself " loses contact with the workers . . . and turns in the eyes of the workers into one of the 'bosses'".  Sometimes existing administrators and specialists cold-shouldered or bullied promotees appointed to work under them : this happened particularly on the railways, whose staffs adopted, " if not a hostile, at any rate an unwelcoming, attitude " to promotees.  Finally the worker himself was sometimes unwilling to be promoted to a Soviet, trade union or cooperative post where the rate of pay was lower than what he was receiving as an industrial worker.  Notwithstanding these handicaps, the report claimed that of 93 " workers from the bench " promoted to responsible posts during the past year, only 5 to 10 per cent had been failures.[2]  This experience was hardly borne out by other evidence.  Tomsky complained in the same year that of " thousands of workers " promoted to " all sorts of economic posts " many were failures owing to inability to familiarize themselves with the work, and

[1] An article and correspondence in *Leningradskaya Pravda*, May 5, 12, 13, 1925, suggest that by that time the system was taken for granted : a proposal was made to set up courses of instruction for Red directors, some of whom were ignorant of the elements of book-keeping.

[2] *Izvestiya Tsentral'nogo Komiteta Vsesoyuznoi Kommunisticheskoi Partii (B)*, No. 14 (135), April 12, 1926, pp. 5-7.  The same report contains information about the " promotion " of peasants to Soviet and cooperative work (see also *ibid.* No. 29-30, August 10, 1925, pp. 2-3) ; but this does not seem to have been taken very seriously.

were moved around from job to job in the hope of finding something which suited them.[1]  Yet, in spite of manifold shortcomings,
the availability of a ladder of promotion and the opening of the
party ranks to a relatively large number of workers sufficed to
create a nucleus of approval for the régime among the more
advanced and articulate sectors of the proletariat ; and the improvement, however gradual and halting, of material standards of
life kept the mass of workers docile, if not actively sympathetic to
the régime.    After 1923 few signs appeared of any widespread
proletarian discontent with the new order.

These arrangements were, however, far from constituting a
dictatorship of the proletariat.    Supreme decisions were exclusively in the hands of the highest party organs ;  and here the
representation of the workers was notoriously weak.    Few parties
bearing the title of " workers' " or " labour " parties in any
country have in fact been predominantly managed and controlled
by industrial workers :  members of other classes, acting on behalf
of the workers and in their name, have almost always played a
leading rôle.    In the workers' party of a country where the proletariat was so small, so recent and so backward in education and
organization as it was in pre-revolutionary Russia, this difficulty
was particularly acute.    At no period was the policy of the Russian
Communist Party shaped by industrial workers.    Lenin had
defended the predominance of the intellectuals in the initial stage
of its history when no workers at all were to be found in its
leading organs.[2]    In spite of vigorous and, after 1924, partially
successful attempts to redress the balance of the party membership
in favour of the workers, the supreme direction of the party
remained in the hands of educated professional revolutionaries.
In 1924 not more than four or five members of the party central
committee, and only one member of the Politburo, Tomsky, had
been factory workers ;  and these were not the real policy-makers
of the party.    The identification of the party leadership with the
government of the state affected its character in two different
ways.    It led to a further concentration of supreme power in a
few hands at the centre, and a further strengthening of party

---

[1] *XV Konferentsiya Vsesoyuznoi Kommunisticheskoi Partii (B)* (1927), p. 283 ;
the same passage confirms the statement that posts offered to workers on
promotion were sometimes less well paid than those which they already held.

[2] See *The Bolshevik Revolution, 1917–1923*, Vol. 1, pp. 16-17.

discipline, which was now reinforced by the needs of state security, so that to all appearances the party leadership became more autocratic than ever. On the other hand, the decisions to be taken were now of a kind which required for their execution the active cooperation of important groups of administrators and managers and the, at any rate passive, acquiescence of large masses of workers and peasants, so that the leadership was continually obliged, in the framing of policy, to take account of a vast complex of interests and opinions, not only in the ranks of the party, but in the population at large. In this sense it had become more responsible and less autocratic since the seizure of power. But it was not exercised by members of the proletariat.

The third group, which consisted of " employees and persons engaged in intellectual work ", was recruited mainly from the official or professional classes of the former régime. Its importance was no part of the original Bolshevik design, which had been to sweep away, together with the landowners and capitalists, all who had been directly or indirectly in the service of the Tsars. Every revolution by definition seeks to overthrow an existing political order and to eliminate the ruling group by which that order was upheld : every revolution is a social revolution in the sense that it seeks to alter the structure of society. In the Bolshevik revolution this process was particularly violent and bitter, partly because the divisions in Russian society were unusually deep, but partly also because the revolutionaries had a conscious theory of revolution which proclaimed the fundamental hostility of class to class and of the revolutionary class to all forms of the national state. This theory was expressed in the view of Marx, based on the experience of the Paris Commune, that the aim of the revolutionaries should be not to take over, but to smash, the machinery of the bourgeois state, and had been repeated and emphatically endorsed by Lenin in *State and Revolution*, written on the eve of the seizure of power. When the " workers' and peasants' government " established itself in Smolny on the morrow of the revolution, nothing was further from the thoughts of Lenin or the newly appointed People's Commissars than to take over and use the existing ministries of the Tsarist and Provisional Governments.

Trotsky set out to conduct the foreign relations of the new régime from Smolny, required nothing of the officials of the former Ministry of Foreign Affairs but the surrender of the " secret treaties ", and waited for the moment to " shut up shop ". The People's Commissar of Finance demanded the surrender of the funds of the State Bank, but was otherwise indifferent to the procedures of financial administration. Stalin's Commissariat of Nationalities had had no previous counterpart. Above all, the Red Guard, and the Red Army which grew out of it, were in no sense the successors of the old army, and were in many ways conceived as the direct antithesis of it. This attitude of the Bolsheviks was fully reciprocated in the boycott of the new revolutionary authority by the surviving officials of the old government which continued unbroken during the first weeks of the revolution. In June 1918 Lenin noted that " the intelligentsia is devoting its experience and its skills . . . to the service of the exploiters ".[1]

The problem of administration and management could not, however, be indefinitely evaded. It had been foreseen some years after Marx's death by Engels, who in a letter of 1891 to Bebel expressed the hope that, if only the revolution in Germany were delayed for another eight or ten years, the party would have time to train enough " young technicians, doctors, jurists and teachers " to provide staff for " the administration of factories and large institutions ". Engels went on to sound a note of apprehension :

> If as the result of a war we come to power before we are prepared for it, the technicians will be in principle our enemies and will deceive and betray us to the best of their ability : we shall have to get rid of them, and still they will deceive us. It was *always* like this with the French revolutionaries : they were obliged themselves to occupy the main posts in the administration, and gave the secondary, but still responsible, posts to old reactionaries, who obstructed and put the brake on everything.[2]

The difficulty was even graver than the parallel with the French revolution suggested. That revolution had occurred in a period when the machinery of political administration and economic

---

[1] Lenin, *Sochineniya*, xxiii, 90.
[2] Marx i Engels, *Sochineniya*, xxviii, 365.

management, the business of running the government and the
business of supplying the basic needs of the population, were
infinitely less complicated than they had become in the early
years of the twentieth century ; and the bourgeois class which
made the revolution was technically equipped to discharge these
functions.  In the Russia of 1917, the vision of a state and an
economy so simply contrived that they could be managed by the
untutored wisdom of workers and peasants, and of the handful of
intellectuals who constituted and supported the new régime, was
purely utopian.  Railway, postal and telegraphic communications
could be maintained only by those trained to operate them ;
factories taken over by the workers, from which managers and
engineers had been expelled or had retired in dudgeon, failed to
produce ; an efficient new army could not be built without drawing on the experience and military training of former Tsarist
officers.  The Bolshevik revolution took place in an age which
could no longer dispense with its technicians, its administrators
and its specialists, to whatever class they might belong and on
whichever side of the barricades they might have fought.  On
the eve of the October revolution Lenin at length noted that it
would be necessary to take " capitalists " and " *compel them to
work* in the new framework of state organisation . . . to put them
to the new state service ".[1]

Much time was spent in learning this unwelcome and unexpected lesson.  Military disasters drove it home in its crudest
and simplest form.  It was in process of creating the Red Army
that effective collaboration between the Bolshevik leaders and
" specialists " of the old order was first established ; this process
began immediately after Brest-Litovsk.  Less dramatically, but no
less significantly, a *modus vivendi* was found with the skilled
managers and engineers who were essential to the working of
industry ; workers' control was abandoned as an unpractical
aberration.  The same silent compromise was reached, especially
after the transfer of the capital to Moscow, with officials and
clerks of the old administration, of whom ever-increasing numbers
found themselves performing for the Soviet Government the same
routine services which they had rendered to its predecessors.
Away from the centre, the leaven of the new was in even scarcer

[1] Lenin, *Sochineniya*, xxi, 263.

supply and the mass of the old less rapidly and less deeply affected, though it is rare to obtain for this period such precise statistical evidence as is provided by the statement that out of 4766 Soviet officials in the town of Vyatka at the end of 1918, 4467 had occupied the same posts in the *zemstvo* administration under the Tsar.[1] In every politically important sphere, the highest posts and the formal direction of policy remained in Bolshevik hands. But, behind the party façade, more and more of the practical work was done by the same men who had previously done it for other masters.

The motives and dispositions of this large non-Bolshevik bureaucracy which turned the wheels of the Soviet régime in its formative years naturally varied. An early member of it, in reminiscences published a few years later in emigration,[2] distinguished three main categories. First came the army of bookkeepers, clerks of all grades, typists, etc., who form the staple of any office personnel, and who were generally free from any active political convictions : this group corresponded to those who, in economic administration, were described by Trotsky in an article of 1919 as " technicians without ideas ".[3] The second group was formed by more or less responsible former Tsarist officials, who " consisted almost exclusively of ' counter-revolutionaries ' of various kinds ", but justified their *volte-face* to themselves on the ground that, by entering the service of the Soviet Government, they were helping to " preserve values " and to " soften the régime ". Finally, there was a group which, though not previously associated with the party, was sympathetic towards the régime and worked for it willingly, though perhaps without much understanding : this was the group referred to by Trotsky, in the passage already quoted, as " representatives of the intelligentsia and semi-intelligentsia, who have sincerely adhered to the cause of the working class, but have not been inwardly transformed and have retained many qualities and habits of thought proper to a bourgeois milieu ". No doubt, habit and material need were more potent factors than sympathy in harnessing these bureaucrats and technicians to their appointed tasks. A contemporary

[1] Stalin, *Sochineniya*, iv, 216.
[2] *Arkhiv Russkoi Revolyutsii*, vi (Berlin, 1922), 304-311.
[3] Trotsky, *Sochineniya*, xxi, 99.

Bolshevik observer estimated that only one-tenth of the pre-war engineers in the service of the Soviet Government were favourable to the régime ; [1] and engineers were less likely to be hostile than some other categories. The sense of mutual mistrust between the bureaucracy and the technicians, on the one hand, and the party, on the other, persisted for many years and coloured all their relations.

The introduction of NEP, which appeared to sacrifice revolutionary policies and aspirations to immediate practical needs, appealed to this group hardly less strongly than to the peasantry, and was a powerful factor in winning and holding its loyalty to the Soviet power. Between 1921 and 1924 an extensive reconciliation took place between these " former people " and the Soviet régime. The comparatively small group of " specialists " of the days of war communism now swelled into an army of many thousands of former members of the administrative, industrial, mercantile or professional classes who made their peace, some grudgingly, some whole-heartedly, with the new order, and worked as Soviet officials, as " Red managers " in economic organs and institutions, as specialists in innumerable technical posts in industry and in administration, or as intellectuals in the professions, in education, in scientific research, literature or the arts. Under NEP the bureaucracy, the managers, the technicians and the intelligentsia — the " officer corps " of the new society — were predominantly, almost exclusively, made up of elements alien to the régime. Protests were heard from time to time against this predominance in particular fields — against the monopoly enjoyed by " fellow-travellers " in literature, against the numbers of former SRs in Narkomzem or of former Tsarist officials in Narkomfin.[2] But

[1] L. Kritsman, *Geroicheskii Period Velikoi Russkoi Revolyutsii* (n.d. [?1924]), pp. 148-149.
[2] The conservative attitude of these departments was sometimes attributed to the affiliations and sympathies of their personnel (for the special case of the Georgian Narkomzem see p. 287 below). The special reputation of Narkomfin in this respect went back to the employment there of Kutler in an influential position (see *The Bolshevik Revolution, 1917–1923*, Vol. 2, pp. 351-352 ; *The Interregnum, 1923–1924*, p. 134) ; Narkomfin also enjoyed the advice of a number of academic economists of the old school, currently known as " Narkomfin professors ". Special efforts were made in the NEP period to increase the very low proportion of party members in the personnel of Narkomfin (*Vestnik Finansov*, No. 1, January 1924, pp. 103-105 ; V. Dyachenko, *Sovetskie Finansy v Pervoi Faze Razvitiya Sovetskogo Gosudarstva*, i (1947), 241-242).

the phenomenon was universal and unavoidable, and its disappearance was a matter of time.  In 1929 former Tsarist officials still constituted 37 per cent of the personnel of Narkomfin, 27 per cent of that of Narkomtrud, and 26 per cent of that of Narkomtorg.  Of the staffs of local government offices in Leningrad 52 per cent were former Tsarist officials.[1]  The proportion may well have been higher in remoter provinces and provincial capitals.

These developments dictated a new attitude on the side of the party leaders.  When, in the days of war communism, Lenin had defended the use of specialists in industry against those who denounced it as a surrender to capitalism, and supported Trotsky's employment of former Tsarist officers on a massive scale to provide the cadres for the new army,[2] these were still the temporary expedients of a period of transition ; and, although the presence of these " hostile " and " counter-revolutionary " elements inevitably bred the dangers of bureaucracy and of deliberate or unconscious treason, the bourgeois specialists could still be regarded as isolated points in an ocean of proletarian and revolutionary endeavour which was sweeping everything before it.  But now, after the " retreat " of NEP, and the acceptance of capitalist ideas and practices over a large part of the economy, the position seemed reversed.  In a society which had restored money and the appeal to financial incentives, which accepted economic inequalities no longer as the exception, but a useful rule, and which had begun to display a qualified tolerance for the rights of property and inheritance, it was the workers who were isolated and in danger of being engulfed in a sea of resurgent capitalism.  At the eleventh party congress in April 1922, Lenin analysed the situation with a somewhat bewildered frankness :

> Wherein lies our strength ?  and what do we lack ?  We have quite enough political power.  I hardly think there is anyone who will assert that on such and such a practical question, in such and such a business institution, the communists, the Communist Party, lack sufficient power. . . . The economic power in the hands of the proletarian state of Russia is

[1] 15 Let Sovetskogo Stroitel'stva, ed. E. Pashukanis (1932), p. 255.
[2] See The Bolshevik Revolution, 1917–1923, Vol. 2, pp. 182-187 ;  Vol. 3, pp. 273-274.

quite adequate to ensure the transition to communism. What
then is lacking ? That is clear ; what is lacking is culture
among that stratum of the communists who perform the functions of administration. If we take Moscow with its 4,700
responsible communists, and if we take that huge bureaucratic
machine, that huge pile, we must ask : Who is directing whom ?
I doubt very much whether it can truthfully be said that the
communists are directing this pile. To tell the truth, they are
not directing, they are being directed. Something analogous
has happened here to what we were told in our history lessons
when we were children : sometimes one nation conquers
another, the nation that conquers is the conqueror and the
nation that is vanquished is the conquered nation. This is
simple and intelligible to all. But what happens to the culture
of these nations ? Here things are not so simple. If the con-
quering nation is more cultured than the vanquished nation
the former imposes its culture upon the latter ; but if the
opposite is the case, the vanquished nation imposes its culture
upon the conqueror. Has something like this happened in the
capital of the RSFSR ? Have the 4,700 communists (nearly a
whole army division, and all of the very best) become influenced
by an alien culture ? True, the vanquished give the impression
that they enjoy a high level of culture. But this is not the case
at all. Their culture is at a miserably low and insignificant
level. Nevertheless, it is higher than ours. Miserable and low
as it is, it is higher than that of our responsible communist
administrators, for the latter lack administrative ability. . . .

Very often the bourgeois officials know the business better
than our best communists, who are endowed with authority
and have every opportunity, but who cannot make the slightest
use of their rights and authority.

And Lenin went on, in an emphatic phrase twice repeated, to
defend the need to build a socialist economy " with the hands of
others " — " to build communism with non-communist hands ".[1]
Six months later, at the fourth congress of Comintern, he reverted
to the same theme :

We took over the old state apparatus, and that was our
misfortune. The state apparatus very often works against us.
It happened that in 1917, after we had seized power, the state

[1] Lenin, *Sochineniya*, xxvii, 244-246. On the eve of the twelfth party
congress Lenin's conception of " building communism with non-communist
hands " was strongly attacked by Osinsky in two articles in *Pravda*, March
24, April 15, 1923, each article being followed by a reply by Kamenev.

apparatus began to sabotage us. We had at that time a great
fright and asked them : " Please, come back to us ". And
they all came back and that was our misfortune. We have
now an immense mass of officials, but we have not sufficient
trained forces really to manage them. In practice it often
happens that here, above, where we hold the state power, the
apparatus manages to work, but there, below, where they take
decisions, they decide in such a way that they very often work
against our policies. Above we have, I do not know how many,
but I suppose at least some thousands, or at most some tens
of thousands, of our own people. But below are hundreds of
thousands of old officials inherited from the Tsar and from
bourgeois society, who work in part consciously, in part unconsciously, against us.[1]

Lenin's incessant preoccupation during the last twelve months of
his active life with the problem of bureaucracy was closely connected with his apprehensions of the growing influence of survivors of the pre-revolutionary social order in the working of the
Soviet administrative machine. One of his last articles on the
reform of Rabkrin contained a contemptuous reference to " those
who seek to maintain our apparatus in the same indecent pre-
revolutionary form in which it has remained up to the present ".[2]
Trotsky wrote at this time of " the class diversity of the Soviet
apparatus, and particularly the prevalence in it of aristocratic,
bourgeois and state-Soviet features ", as responsible for the
increase of bureaucracy.[3] In theory the communist administrator
was supposed to act as a watch-dog over his bourgeois and
potentially hostile colleagues and subordinates. In practice he
was often helpless in the hands of men more experienced than
himself, and was not always impervious to the seductions of pre-
revolutionary tradition.

In spite of these well-justified apprehensions, the Bolshevik
leaders recognized from a very early date the indispensability of
such non-proletarian props of the dictatorship of the proletariat.
These elements, though they defied class analysis, occupied a
crucial place in Soviet society, and the attitude of the party
towards them was the determining factor in their status. Though
they belonged by past affiliations to the aristocracy or to the

[1] Lenin, *Sochineniya*, xxvii, 353.      [2] *Ibid.* xxvii, 403.
[3] Trotsky, *Sochineniya*, xxi, 62.

bourgeoisie, they were neither exploiters of labour, engaged in extracting " surplus value " from the workers, nor owners of capital or of means of production in any form.¹ If they were bourgeois in background and tradition, they were no longer bourgeois in current situation and function, and were not therefore in principle irreconcilable class enemies like *kulaks* or nepmen. Their qualifications made them indispensable to the running both of the economic and of the political machine ; only with their help could the fatal handicap of the backwardness of the proletariat be overcome. Concessions to them were not, like concessions to *kulaks* or nepmen, mere temporary expedients. The economic status of the officials and the members of the intelligentsia now pressed into the service of the régime was such that the growth of loyalty to the régime on their part was not to be despaired of, though recalcitrance or even treachery was still to be expected in some cases. The employment of former Tsarist officers in the Red Army had justified itself in the civil war, and was not thereafter seriously contested. Prejudice against bourgeois specialists and managers engaged in industry was far more deep-seated and persistent ; but the party, from the ninth congress of 1920 onwards, struggled incessantly against it. Prejudice against bourgeois clerks and officials in Soviet departments and institutions was far less acute, since these formed a self-contained group and were not placed in authority over workers ; their rôle in Soviet society was quickly accepted and taken for granted.

Lenin's pronouncements at the eleventh party congress were the prelude to an active campaign to reconcile these groups with the party and the régime. The party conference of August 1922, which had greeted the emergence of the *smenovekhovtsy* among the *émigrés*,² also turned its attention to similar phenomena at home. It offered " systematic support and working collaboration " to those members of the intelligentsia " who have understood, at any rate in its fundamental features, the real meaning of the great revolution which has been accomplished ", and called on the party to work for " the crystallization of tendencies and groups which

¹ In the early NEP period a few members of the " free " professions — doctors, lawyers, writers, etc. — may have constituted a partial exception to this generalization. But they were never important, and soon became dependent on the state or the party, like other members of these groups, for their sources of income.    ² See p. 62 above.

show a real desire to aid the workers' and peasants' state ".[1]
A first All-Russian Congress of Scientific Workers was held in
Moscow at the end of November 1923.  It heard speeches from
Zinoviev and Bukharin, and a letter addressed to it by Trotsky
was read ;  and leading members of the old Academy of Sciences
professed allegiance to the Soviet power and willingness to col-
laborate with it in the interests of science.[2]  A year later, at an
All-Russian Congress of Engineers in December 1924, Rykov
spoke of the need to reconcile " men of learning and technicians "
with the workers' and peasants' state.  " The specialist, the
engineer, the man of science and technology must have full
independence and freedom to express his opinion on matters of
science and technology ", and must not be required to display
subservience either to " society " or to " the administration ".[3]
As the party began to turn its attention to the countryside, it
included within the scope of its interest the " rural intelligentsia "
of " teachers and agronomists " ;  Rykov in 1924 specially noted
the growing links between this group and the Soviet power, and
appealed to the party to " assimilate this intelligentsia ".[4]  Great
publicity was given to the first All-Union Congress of Teachers
which met in Moscow in January 1925.  Rykov spoke of " the
coming over to the side of the Soviet régime in the last two or three
years of the masses of the intelligentsia " ;  and Zinoviev, Bukharin
and Krupskaya harped on the same theme.[5]  In remote rural
regions, where Soviet officials were rarely, and party representatives
still more rarely, seen, teachers often formed an important link between the population and the régime.[6]  When in March 1925 the
TsIK of the USSR met for the first time outside Moscow, in
Tiflis, a delegation of Tiflis doctors appeared to present a loyal address, and Petrovsky, in a speech of thanks from the chair, referred

---

[1] *VKP(B) v Rezolyutsiyakh* (1941), i, 466.
[2] *Pravda*, November 30, 1923.
[3] *Izvestiya*, December 4, 1924 ;  the speech was printed under the title,
*Rôle and Tasks of the Intelligentsia in the Post-October Period.*
[4] A. I. Rykov, *Sochineniya*, iii (1929), 71 ;  in a further speech on the same
subject he included " doctors, hospital orderlies and veterinary workers " in
the rural intelligentsia (*ibid.* iii, 93).
[5] *Pravda*, January 16, 1925, and following days.
[6] *Sovetskoe Stroitel'stvo : Sbornik*, ii-iii (1925), 353 ;  in 1925 " special
attention was given to the reception of teachers into the party, and 3445 of them
were recruited in the first nine months of the year " (*K XIV S"ezdu RKP(B)
1925*), p. 5).

to the alliance between " labour and science ", and greeted the
Transcaucasian intelligentsia as helpers in the advance towards
communism.[1]  The third Union Congress of Soviets, meeting
two months later in Moscow, received a delegate from a conference
of rectors of universities which happened to be in session at the
same time, and referred in its main resolution to " the coming over
of broad strata of the intelligentsia to the side of the Soviet power ".[2]
The culmination of the process was the grandiose celebration in
September 1925 of the 200th anniversary of the Russian Academy
of Sciences, which was attended by men of learning from many
countries and obtained world-wide publicity, marking both the
reconciliation of the Soviet régime with the tradition of Russian
science and learning, and the establishment of new links between
Soviet science and that of the capitalist world.[3]

What therefore gradually came into being in the years of NEP
was an undeclared alliance between the party, speaking in the
name of the proletariat, and the group of "employees and persons
engaged in intellectual work ", still at this time composed mainly
of survivors of the pre-revolutionary régime or their immediate
descendants.  It was Gorky who had done much, before he left
Soviet Russia in 1921, to lay the foundations of such an alliance
in the field of literature, and, looking with a frank and disillusioned eye on the backwardness of the Russian peasant mass,
continued to insist on the necessity for giving it the widest possible
extension.

> The fundamental obstacle on the path of Russia's progress
> towards Europeanization and culture [he wrote in 1924] is the
> fact of the overwhelming predominance of the illiterate country-
> side over the town, the zoological individualism of the peasantry,
> and its almost total lack of social feelings.  The dictatorship of
> the politically literate workers in close alliance with the intelli-
> gentsia was in my view the only possible escape from a difficult
> situation, especially complicated by the war which brought still
> further anarchy into the countryside. . . . The Russian intel-
> ligentsia — the educated people and the workers — was, is,

[1] *SSSR : Tsentral'nyi Ispolnitel'nyi Komitet 2 Sozyva : 3 Sessiya* (1925),
pp. 143-146.
[2] *Tretii S"ezd Sovetov SSSR : Postanovleniya* (1925), p. 7.
[3] The speech of Zinoviev, who appeared as the principal official Soviet
representative, was fully reported in the press and published as a pamphlet
under the title *Nauka i Revolyutsiya* (1925).

and will long remain, the only cart-horse that can be harnessed
to the heavy load of Russian history.[1]

The obstacles to reconciliation were formidable, and not least on
the side of the party. But the resolution of the thirteenth party
congress of May 1924, at the height of the Lenin enrolment,
admitted a significant qualification when it demanded the purging
of " those elements of non-worker status who have not, during
their time in the ranks of the party, shown themselves true communists by improving the work of this or that state economic or
other organisation, or have not had direct contact with the worker
or peasant masses ".[2] Coming at this moment, the tacit admission
that the rendering of services to the régime made a man a " true
communist " and worthy of party membership was a striking concession to the specialist and industrial manager and to other
groups of administrators and intellectuals in the service of the
régime. Little by little, outstanding members of these groups
began to be admitted to the party as a reward for their loyalty
and as the token of the confidence which they had earned by
their services.[3] The ladder of promotion to positions of political
influence was now open to them as to outstanding members of the
proletariat. No doubt it was far more difficult for a manager, an
official or an intellectual than for a worker to rise in the party
hierarchy. But, except at the highest policy-making levels,
authority was within his grasp. Purged of his bourgeois origins
and associations, no longer branded as a capitalist or an exploiter of
labour, the individual specialist or intellectual was recognized, not,
like the nepman, as a temporarily tolerated class enemy, but as a
necessary and permanent pillar of the Soviet régime, and as such
was incorporated into the new structure of Soviet society. Indeed
the dictatorship of the proletariat, as organized by the party under
NEP, was exercised by a *de facto* alliance between the proletariat
and the officials, technicians and intellectuals who, abandoning

[1] *Russkii Sovremennik* (Berlin), i (1924), 235.
[2] *VKP(B) v Rezolyutsiyakh* (1941), i, 577.
[3] For the admission of industrial directors and managers to the party
between 1922 and 1924 see *The Interregnum, 1923–1924*, pp. 40-41 ; the
numbers involved were very small, but the trend none the less significant.
After 1924, with the " Lenin enrolment " of workers and subsequent recruitment of peasants, the proportion of " employees and others " in the party
membership declined (A. S. Bubnov, *VKP(B)* (1931), p. 615) ; but this did
not mean a decline either in their absolute numbers or in their influence.

their past bourgeois affiliations, loyally accepted service to the Soviet régime and aspired to party membership. It was this alliance, established under the direction and leadership of the party, which provided the ruling group of the society.

The picture which could be formed of Soviet society in the first years of NEP did not at first sight seem encouraging for the further progress of the revolution along the proletarian path. The proletariat, designated by Bolshevik theory as the artificers, leaders and main beneficiaries of the revolution, remained a small and ineffective minority. The peasantry, which still accounted for more than 80 per cent of the population of the Soviet Union, had ensured the success of the revolution, but had exacted as the price of its support a parcellation of the land in small peasant holdings which Marxist doctrine had always rejected as economically inefficient and politically retrograde. The peasantry had saved the revolution in the civil war. But it had revolted against the grain requisitions, and imposed on an unwilling government the retreat into NEP with the restoration of the grain market and with capitalist implications for other sectors of the economy. As a corollary to these forced concessions to class enemies and potential class enemies in country and town, the régime had been compelled to strike an alliance with a multitude of former bourgeois specialists, administrators and intellectuals, whose tradition and background seemed to preclude any sympathy with revolutionary aims, and many of whom had been drawn into cooperation with the régime in the belief that NEP marked the effective end of the revolution. It was not surprising that NEP, officially proclaimed by the party as a " retreat ", should have inspired fears in the Left wing of the party, and hopes among the survivors of the former régime, both in Soviet Russia and in emigration, that the advance would not be resumed, and that the country would settle down into a modified form of bourgeois capitalism on a Russian national pattern. Any conventional estimate of the social forces seemed to justify this expectation, which could be supported by arguments from the armoury of Marxist doctrine pointing to the impossibility of building socialism in a peasant country without the support of proletarian revolution elsewhere.

Had this constituted the whole picture or the main feature in it, then the deduction drawn from it of the imminence of a " Russian thermidor " would have been unescapable. Yet the deduction proved unsound. The proclamation of socialism in one country was followed, not by a stabilization on the basis of NEP with its favourable implications for the nepman and the *kulak*, or by a further retreat into capitalism, but by a feverish drive for the development of heavy industry — the traditional stronghold of the class-conscious worker, and an unprecedentedly rapid expansion of the proletarian sector of society, achieved by a progressive restriction of the market economy of NEP and a far-reaching intensification of planning. In other words, while socialism in one country made its concessions to nationalism, and thus seemed to diverge from the high road of Marxism, the proletarian or socialist element in it was also perfectly real. At a time when the industrial proletariat had been dispersed and weakened and its rôle in the economy catastrophically reduced, strength was none the less found to overcome the apparently irresistible forces interested in the conservation of the existing order and to carry forward the revolution along the lines laid down for it by Marxist doctrine. How was this paradox achieved — a paradox which falsified every current prediction and appeared to frustrate every attempt at rational analysis ? [1]

The answer is twofold. In the first place, the Russian revolution had gone much further than its English or its French predecessor in destroying the social and institutional framework of the old régime. This was in some measure due to the greater ruthlessness of the Russian Bolsheviks, to the insistence of both Marx and Lenin (inspired, in part, by the lessons of the Paris commune) on the need to " smash the bourgeois state machine ",

---

[1] One of the few Bolsheviks who asked this question at the time, the historian Pokrovsky, took refuge in a frankly mystical answer. Having confessed that " to put the accent on ' objective causes ' is no longer possible ", he went on : " ' Objective causes ' are now against us, and on this fact were founded the predictions both of our ' friends ', who are gradually losing hope that we shall ' reform ' and ' come to our senses ', and of our enemies, who are also gradually losing hope that we shall fail. The objective logic of the old ' economic materialism ' is against us — and we go forward. . . . This means that there is something in the very ' nature ' of the proletariat of our country which gives it the possibility to conquer even when ' objective causes ' are not for it, but against it " (*Istoriya Proletariata SSSR*, i (1930), pp. iii-iv).

but most of all to the weakness of the old Russian social and political order, which showed none of the tenacity and power for survival of its western counterparts. In France, and still more in England, old institutions and social groups, though maimed and disabled in the struggle, survived as recognizable features of the new landscape, and account had to be taken of them. In Russia, after the civil war, only individuals survived ; and the task of the Soviet régime was to take over, utilize or neutralize, not institutions or groups, but isolated and unorganized individuals whose capacity to resist or to influence was much more easily overcome. Specialists and administrators, nepmen and *kulaks* played their individual rôles. But they never succeeded in forming coherent opposition groups, and still less in cooperating with one another.[1] It was a special source of weakness that the Russian intellectuals, unlike the intellectuals of the west, had never been integrated into the ruling class, and remained alien, and in some degree hostile, to it. Their long tradition of opposition to the Tsars, while it did not issue in any consistent or coherent body of opinion, provided a background of support for the cause of revolution in the abstract ; and, while a majority of the intelligentsia was certainly inimical to the Soviet régime in its early stages, it was equally opposed to a restoration of the past and its sympathies and loyalties were not attracted to any alternative. The revolution continued to profit from the disunity and irresolution of its opponents. It had succeeded so easily in October 1917, not because the proletariat was strong, but because the bourgeoisie was weak ; and this correlation of forces continued to dominate its later history.

The second reason for the apparent paradox is more profound, and relates to the altered balance of social relations in the modern period. Marx had used the model of a perfectly free society of capitalist entrepreneurs ; and this model was sufficiently near to the current realities of western European, and particularly of British, society in the third quarter of the nineteenth century to provide a serviceable instrument of analysis. According to this conception, society was dominated exclusively by economic

---

[1] The success with which former Tsarist officers had been employed in the Red Army, apparently without serious risk of their combining against their new masters, was perhaps the most striking instance of this phenomenon.

relations, and the decisive economic relation was the absolute antagonism between two classes — the entrepreneurs, who could make the system work only through the progressively intense exploitation of the labour of the proletariat, and the proletarians, whose increasing numbers and increasing misery would inevitably impel them sooner or later to overthrow the system. The essential function of a class was to assert or defend its economic and social interests against those of another class : when such antagonisms ceased to exist, classes would lose their *raison d'être* and disappear. The life of classes was bound up with the class struggle. But " class relations " were the social expression of the economic concept of " relations of production " : " the sum total of relations of production ", wrote Marx in the preface to the *Critique of Political Economy*, " constitutes the economic structure of society ". And it was this economic structure which determined the political and ideological superstructure. It was not necessary to suppose that the individual member of the proletariat, any more than of any other class, was conscious of the nature or consequences of his rôle in the social order.

> The question is not [wrote Marx in *The Holy Family*] what this or that proletarian, or even the proletariat as a whole, considers as its aim at any given moment. The question is *what the proletariat is*, and what, in accordance with this *being*, it will be historically obliged to do. Its aim and action in history is patently, irrevocably preordained by its situation in life, as well as by the whole organization of contemporary bourgeois society.[1]

It was necessary, insisted Marx in *The Eighteenth Brumaire of Louis Bonaparte*, to distinguish " between the phrases and ideas of the parties and their real organism and real interests, between their conception of themselves and their real nature ".[2] This picture fitted in perfectly with the conception, shared equally by Adam Smith and Hegel, of a society in which innumerable individuals, engaged in the pursuit of their own interests, set in motion social processes of which they were themselves unconscious.

The Marxist conception of class was, then, a tool designed for the analysis of western capitalist society in the nineteenth century.

---

[1] *Karl Marx-Friedrich Engels : Historisch-Kritische Gesamtausgabe*, Ier Teil, iii, 207.          [2] Marx i Engels, *Sochineniya*, viii, 347.

It came into being in conditions where the division of society into
" castes " or " estates " had been abolished, and full legal equality
established, but where society remained divided into well-defined
strata distinguished by differences not of legal, but of economic
and social, status.   It was never easily or precisely applicable to a
society like that of Tsarist Russia, where legal " estates " still
existed, or perhaps to societies like those of the English-speaking
overseas countries, where no such divisions had ever existed, and
where classes therefore lacked the degree of traditional rigidity
assumed by the Marxist analysis.   But what is more important is
that by the end of the century it had ceased to be readily applicable even to the major countries for which it had been devised.
Even in western Europe " free " capitalist society was in process
of modification and transformation into " managed " capitalist
society.   The individual engaged in the private pursuit of his
interests was replaced as the essential unit of the economic and
social order by the collective group, taking the form now of the
trust or syndicate, now of the trade union, now of the nation state.
The enormous implications of the change were not all immediately
apparent.   But they included the widespread encroachment of
the sphere of collective action on that of individual action, with
serious consequences for the nineteenth-century conception of
freedom ;  the substitution of the conscious manipulation of the
economic process for the supposedly spontaneous operation of
economic laws, and of the flagrant irrationalism of collective
choices for the assumed (and, in part, real) rationality of individual
choice ;  and the intrusion of power in a more naked and violent
form into the ordering of social relations.   In a society moving
rapidly away from the basic assumptions of a *laissez-faire* economy,
the view of class as an objectively identifiable group whose common
action was the unconscious product of innumerable spontaneous
actions of individuals pursuing their own interests underwent
subtle and at first unperceived modifications.   By the turn of the
century the change of climate was beginning to make itself felt.

These results of the transformation of free into managed
capitalism may be briefly summed up by saying that the nineteenth-
century primacy of economics was succeeded by the twentieth-
century primacy of politics.   It would, of course, be misleading
to suggest that Marx regarded political action as unnecessary or

as incapable of affecting economic conditions. The writings of his earlier period were predominantly political ; " the proletariat ", declared the *Communist Manifesto*, " will use its political domin- ance in order to strip the bourgeoisie step by step of all capital, and concentrate all means of production in the hands of the state, i.e. of the proletariat organized as the dominant class ". Marx never at any time renounced the need for a political programme. But he did assume that class-consciousness developed spontane- ously within a class, that the active and class-conscious proletariat at the moment of its victory would constitute a majority, and that political action would be required merely to seal a victory already assured by its economic superiority. The state remained subordinate to the social and economic order.

> Only *political superstition* [wrote Marx in *The Holy Family* today supposes that social life must be held together by the state, whereas in reality the state is held together by social life.[1]

Fifty years later this hypothesis was open to serious question ; and Engels at the very end of his life, some years after Marx's death, had drawn attention to the way in which what Marx called the " superstructure " reacted on the basic economic structure, thus attenuating the unconditional primacy which Marx had appeared to assign to the economic factor. One specific development of managed capitalism was that political lines were beginning to cut across the economic lines of class solidarity, dividing classes against themselves and creating common interests within the nation between sectors of the proletariat and of the ruling bourgeoisie — a process which was to reach its culmination in 1914. Here too Engels, who was in closer touch than Marx with practical developments, qualified the unconditional applica- tion of Marx's model even before the publication of *Capital* by the half-serious remark that " the English proletariat is in reality becoming more and more bourgeois ".[2] By 1900 the Marxist assumption that the proletariat everywhere had a single deter- mining economic interest which made it the absolute " negation "

[1] *Karl Marx-Friedrich Engels : Historisch-Kritische Gesamtausgabe*, I$^{er}$ Teil, iii, 296.
[2] For the ambiguities of the theory of a " labour aristocracy " see *The Bolshevik Revolution, 1917-1923*, Vol. 3, pp. 182-183.

of the bourgeoisie was obsolescent. The classes were no longer either indivisible unities or unqualified opposites.

These were the conditions when Lenin in *What is to be Done ?* first attempted in 1902 to elaborate a programme of revolution. Firmly wedded to Marxist doctrine, he was none the less vividly conscious of the practical requirements of the movement and of the situation which he had to face. If political action was shaping the form of latter-day capitalism, political action must also shape the challenge to capitalism. Lenin took over the Marxist concept of a society polarized by the relations of production on which it was based, classes representing the element of antagonism inherent in these relations. Classes are " what allows one part of society to appropriate the labour of another part ".[1] But, in spite of this definition, Lenin from the outset insisted that the political consciousness of the working class could not be a spontaneous growth, and must be brought to the worker " from without ". It was Lenin who had first denounced " spontaneity " and " tail-endism ". It was Lenin who insisted on the political duty of the party to instruct the workers. The distinction between the party and the class, between the political and the economic unit, with the predominant and guiding rôle assigned to the party, was an essential feature of Bolshevik doctrine.[2] This already meant an emphasis on the political and voluntarist aspects of Marxism, as opposed to the Menshevik emphasis on its economic and determinist elements.

The unexpected victory of the Bolshevik revolution in Russia, and in Russia alone, not only brought this issue to a head, but put it in a new light. Either the revolution must be denounced, as it had been denounced by the Mensheviks, as non-Marxist, or it represented a vindication of the most extreme " political " interpretation of Marxism. The revolution had triumphed through political action in what were, from the Marxist standpoint, economically unripe conditions. As Lenin said, it was " the political half of socialism " which had been realized in Russia in

---

[1] Lenin, *Sochineniya*, xxv, 391. Lenin elsewhere gave a definition which also took account of the quasi-permanent character of class : " Classes are groups of people of such a kind that one group can appropriate the labour of another thanks to the difference of their position in the definite formation of the social economy " (*ibid.* xxiv, 337).

[2] See *The Bolshevik Revolution, 1917–1923*, Vol. 1 pp. 15-18.

the form of the dictatorship of the proletariat; [1] the economic half still lagged behind — a disconcerting reversal of the initial Marxist premiss. In the conditions created by the revolution, the strong political arm could alone make economic advance possible. " With the support of proletarian state power ", said Lenin in 1919, " the sprigs of communism will develop into full communism " ; [2] and in the controversy on the rôle of the trade unions in January 1921 he impatiently added : " Politics cannot but have precedence over economics ; to argue otherwise is to forget the ABC of Marxism ".[3] Lenin in one of his last writings argued that " the creation of socialism " could come only with the attainment of " a definite level of culture ", and went on :

> Why cannot we conquer by revolutionary means the pre-requisites for this definite level, and *then*, on the foundation of the workers' and peasants' power and of the Soviet order, move forward to catch up other nations ? [4]

In Krzhizhanovsky's words, written just after the fourteenth party congress of December 1925, " 300 years of monarchical rule created the society appropriate to it and the appropriate type of obedient subject, whose chief public virtue was silent submission to the ruling power " ; the revolution had not merely to sweep away the ancient patterns of life, but to " re-create the very type of citizen ", who would have the necessary energy and initiative for the " forthcoming battles ".[5] No longer could the individual be relied on to perform social functions of which he remained unconscious. The creation of consciousness was a vital task. The proletarian revolution in Russia had different prerequisites from the proletarian revolution whose inexorable advance Marx had charted in nineteenth-century western Europe, and raised

---

[1] Lenin, *Sochineniya*, xxii, 517.          [2] *Ibid.* xxiv, 345.
[3] *Ibid.* xxvi, 126.          [4] *Ibid.* xxvii, 400.
[5] G. Krzhizhanovsky, *Sochineniya*, iii (1936), 119. Bukharin had more precisely defined the type of working-class leader required : " We need a psychological type which possesses the good qualities of the old Russian intel-ligentsia in the sense of Marxist education, breadth of outlook and theoretical analysis of events, yet with an American practical stamp. . . . We need Marxism plus Americanism " (N. Bukharin, *Proletarskaya Revolyutsiya i Kul'tura* (1923), p. 49); Zinoviev told a Komsomol audience : " We must com-bine the best traits of Americanism with the best existing traits of the Russian people " (*Partiya i Vospitanie Smeny* (1924), p. 51).

issues that lay outside the perimeter of Marxist doctrine.  The task of building socialism in one country meant also the task of creating those who would build it.

It was this re-emergence of politics as the primary factor which made the party master of the situation, and exalted party above class.  No doubt, the party owed much, in the form which it ultimately assumed, to the foundations laid in the conspiratorial period of its existence before the revolution.  But it owed more to the peculiar position in which the victory of the revolution placed it.  The industrial proletariat which had played the heroic rôle in the achievement of October 1917 was a small and dwindling group.  So long as the victorious Russian revolution stood alone in a stubbornly capitalist world, its successful defence was a *tour de force* whose chances the Bolshevik leaders themselves had not at first rated high.  During the civil war the peasants had had a common interest with the workers in the defence of the régime. But once the " whites " had been finally defeated, the disappearance of this common interest put the régime in immediate jeopardy ;  and its hope of survival depended on the firm direction and iron discipline of the party.  Society had been split asunder and disintegrated by the revolution and the civil war ;  the party alone withstood the shock and emerged from the ordeal with the prestige and self-confidence born of victory.  The conception of NEP was based on belief in the effective control of the political dictatorship over the economic machine.  Its success depended on the efficacy of political coercions and sanctions in fields where capitalism had relied on the economic coercions and sanctions of the " hidden hand ".

After the introduction of NEP, however, even the party as a whole was no longer equal to this task ;  for expansion had resulted, as the tenth party congress recorded in March 1921, in " the entry into the party of elements which have not yet assimilated the communist view of life ".[1]  As the party had to mould and direct the society, so the party leadership — the central nucleus gathered round the Politburo, the Orgburo and the secretariat — had to mould and direct the party.  The tenth congress, which approved the introduction of NEP, also tightened up the disciplinary rules of the party.  The coincidence was calculated.  Never had discipline

[1] *VKP(B) v Rezolyutsiyakh* (1941), i, 366.

been more necessary in the party than at the moment when the bonds of society were being relaxed through the partial retreat into capitalism.  Three years later Stalin enumerated the three conditions which explained the rise of " groupings " in the party and made it imperative to prohibit them.  These were the mixed character of the Russian economy with its feudal survivals, the readmission of capitalist elements into the economy under NEP, and the mixed social composition of the party itself, which was made up of proletarians, peasants and intelligentsia.[1]  The dangers of social diversity could be countered only by the most rigid political discipline.  In 1921, it was party discipline which defeated the mutineers of Kronstadt and appeased the peasants. In the years that followed it was party discipline which warded off the more insidious advance of the nepmen and the *kulaks* and, in face of all contrary omens, carried on the revolution to its appointed conclusion.

The description and diagnosis of these developments was, however, complicated by the attempt to discuss them in strict Marxist terms.  The edge of the Marxist class analysis was seriously blunted when the attempt was made to apply it to a society where revolution had led to extreme fluidity and rapid changes in economic and social status, where the two basic classes were regarded no longer as antagonists, but as allies, and where a third class was postulated which could not be homogeneously defined in terms of the relations of production, and was not therefore in the Marxist sense a class at all.  Soviet society, as it began to take shape under NEP, presented a diversified picture of fluid social groupings which did not exhibit either the clearly defined social or economic function or the established stability of the Marxist class.  Cohesion was given to it by the organization of the party.  Party allegiance cut across social groups.  Social and economic divisions were intersected by a new line of political demarcation.  In order to determine the place of the individual in Soviet society, it was important to know not only to what social group he belonged, but whether he was or was not a member of the party, and what was the attitude of the party towards him and his group.  The political factor of the party enters into every valid analysis of the structure of Soviet society :  analysis in purely

[1] Stalin, *Sochineniya*, vi, 22.

social or economic categories becomes irrelevant. The party formed the new ruling class of this society out of a carefully balanced variety of social groups. It professed to act in the name of the proletariat ; and this profession, while it did not mean that the commanding positions in the party were occupied by workers, was not meaningless. The overriding aim of the party was to promote the advance to socialism through a process of industrialization which would increase the numbers, and ultimately the material well-being and political weight, of the proletariat ; and it was this aim, and the policies springing from it, which lent unity and substance to the party and to the ruling group. The purpose of the party was to carry on the revolution ; and, so long as this purpose remained effective, the party dominated the whole society and kept the social groups within it fluid and malleable.

The Bolshevik revolution, though it drew its impetus and inspiration from Marxist doctrine, revealed an interplay of political with economic factors more subtle and complex than had been allowed for in earlier versions of Marxism. Lenin and his comrades and successors were strongly impregnated with the ideals and purposes of Marxism, and sincerely believed that, by maintaining and extending the political power placed in their hands by the victory of the revolution, they could eventually create the kind of society which Marx had envisaged. But, except in this ideological sense, they could not be regarded as the agents of the proletariat, or of the proletariat and the peasantry ; neither the proletariat nor the peasantry, nor a combination of the two, could be said to constitute the ruling class in the sense in which the bourgeoisie could be said to constitute the ruling class in western countries. The dictatorship of the proletariat was a political, not an economic or social, phenomenon : it was the rule not of a class, but of a party or group. It did not rest on the strength of the proletariat, which was for the present extremely slight. Its declared political programme was to make the proletariat strong, and to create the conditions in which a dictatorship of the proletariat in the Marxist sense might become a reality. But these conditions included a remoulding, psychological as well as material, of the proletariat.

The rôle of the Bolshevik party and the hypertrophy of its

power were thus a direct outcome of the isolated victory of the proletarian revolution in a country whose retarded economic and political development failed to provide the conditions postulated by Marx and by early Marxists for such a victory.  On the morrow of revolution the Russian workers lacked even the elementary technical and political training which the advanced capitalism and democracy of western countries had provided.  In the absence of active support from the proletariats of the west, which remained passive, and of the Russian peasantry, which, having acquired the land, was potentially hostile to further innovation, the Russian proletariat was powerless to secure and consolidate the results of the victory, which in the years after 1917 were plainly threatened with collapse or defeat.  The party stepped into the gap.  Having planned and led the revolutionary *coup* of October 1917, it created a régime strong enough once more to emerge victorious from the civil war and to undertake the reorganization of the economic life of the country.  In so doing, the party succeeded to the automatic prestige, authority and sanctity attaching, in any established society, to its symbolic head — the kind of prestige, authority and sanctity formerly attaching to the person of the Tsar.  It was something both more and less than a ruling group.  The party became the society in miniature.  It reflected the tensions of the society and the distribution of power between social and economic groups, so that every major issue appeared in heightened perspective in the political microcosm of the party.  But the reflexion was not purely passive and representative.  The party line was no mere synthesis of these tensions and conflicts.  The party continued to carry the political programme of the proletarian revolution.  The history of the revolution consisted of the impact of this dynamic force on a society dominated by a backward peasant economy.  The coming of NEP had appeared to many to mean that the force of the revolution was spent, and that the party, as the bearer of this force, would be quietly reabsorbed into the traditional society.  Any social and economic analysis of the situation between 1921 and 1924 appeared to justify this conclusion.  In reality, the party leadership compromised far enough with the traditional society to ride the storm ;  this compromise was the essence not only of NEP, but of socialism in one country.  Yet in the sequel it had retained its revolutionary dynamic unimpaired, and imposed on the society the consummation of " revolution from above ". This was a political *dénouement* which constitutes a striking tribute to the infinite complexity of the factors that determine the course of history.

# CHAPTER 4

# PERSONALITIES

THE question of the rôle of great men in history is sometimes confused with that of the rôle of individuals in history. The two questions are logically separable. But certain analogies exist between them. The events of history are set in motion by the wills of individuals. But what individuals will is governed in part by the historical conditions in which they find themselves ; and these conditions impose still narrower limits on what they can effectively will. Hence the explanations of events given by the historian cannot be confined to simple statements of the will of the individuals concerned, and tend for this reason to create the illusion of " impersonal " forces at work in history, though the historian is well aware that the acts through which these forces find expression are the acts of individuals and are set in motion by the individual will. Similarly, conspicuous and memorable events in history are commonly said to be the work of great men. But the quality which makes the events conspicuous, and causes the actors in them to be hailed as great, appears to reside in factors that lie outside the events themselves. An event is significant on account of its causes or its consequences ; a great man is great because what he says or does represents not merely his own will, but the wills of large multitudes of his fellow-men, and perhaps not only of his own generation, but of generations yet unborn. The relation of great men to the world of history is reciprocal. The great man is great because he influences his contemporaries and posterity and helps to mould their destiny. But the great man is also the product of his environment, and is great because he embodies the wills and aspirations of his contemporaries, or anticipates those of their successors, on a more than ordinary scale. It is the business of the historian to explain these factors without either diminishing the achievement of the great man or depicting it as something that lies outside history and is not amenable to historical explanation. To ask whether the great man has shaped the course of events, or has himself been shaped by a course of events which is explicable in other terms, is to make an historically false distinction, to divide a single composite historical process.

This interconnexion between great events and great men explains why the appearance of great men commonly coincides with the turning-points of history. The name of Lenin is indissolubly linked, both for his contemporaries and for posterity, with the Bolshevik revolution. The accident of Lenin's illness and death at the moment when the revolution had completed its most turbulent phase, and was settling down to a post-revolutionary period compounded, in accordance with the usual pattern, of consolidation and of reaction, has served to throw his achievement into particularly dramatic relief. Lenin's career was cut off at the point where the revolutionary drama had reached its fifth act. The post-revolutionary epilogue was to be played out by men whose characters fitted, and reflected, the needs of the new period. The political qualities of Lenin's successors are an emblem of their time, and their political biographies are a part of its history. No historical period leaves a stronger impression than the middle nineteen-twenties in the Soviet Union of events that dominated its leading figures, dictating their opinions and determining their rise and fall. The qualities which made Trotsky an outstanding leader in the days of storm and stress unfitted him for leadership in the period of patient calculation and measured compromise which followed ; and he lacked all talent for self-adaptation. After his downfall, the other leaders struggled against one another for mastery. But at no time did the course of events — in the sense of the victory of this or that opinion or policy — seem to hang on the issue of the struggle. It cannot plausibly be maintained that the policies which ultimately emerged triumphant triumphed because Stalin advocated them. It is nearer the truth to say that Stalin rose to power through his skill in hitching his fortunes at precisely the right moment to policies that were about to win acclaim, and extricating himself in time from commitments to lost causes. Yet even this does not exhaust the complexity of the relation. The struggle between policies and the struggle for power between individual leaders were both real. But they proceeded independently and on different planes. The victory of Stalin over his rivals in the struggle for leadership can be simply explained by his immense superiority to them in almost every political art. But it cannot be said either that Stalin was victorious because he was the advocate of the policies which proved acceptable to the party, or that those policies were ultimately adopted as a result of Stalin's victory ; for most of these policies had been advocated by Stalin's opponents long before he, already well on the road to supreme power, made them his own. The historian who seeks to explain the major developments in the history of the Soviet Union in the nineteen-twenties will, for this reason, derive comparatively little help from the study of the characters of the principal leaders and of relations between them. These form, however, a part, though a minor part, of the story ; and some examination of the political opinions and public behaviour of these men will serve both to reflect and to illuminate the course of the events in which they were involved.

### (a) Trotsky

Lev Davidovich Trotsky (original name Bronstein) was born in 1879 at the village of Yanovka in the Ukraine, his father belonging to the not very numerous class of Jewish independent small farmers. At the age of nine he was sent to school in Odessa, living with relatives of his mother. His last year of schooling — 1896–1897 — was passed at Nikolaev, where he first began to read forbidden books and became politically conscious. Early in 1897 he joined a revolutionary group engaged in underground political work, and underwent conversion to Marxism. In the following year he was arrested, spent the next two years in a succession of prisons, and was sent to Siberia in 1900. In 1902 he made his escape and, travelling via Vienna, Zürich and Paris, joined Lenin and Martov in London. His literary talents, marked by a certain flamboyance of style, had already won him the party nickname of Pero or Pen : he quickly became a contributor to *Iskra*, earning the admiration of Lenin and the jealous disapproval of Plekhanov.

The second party congress of 1903 was an important turning-point in Trotsky's career. Differing from Lenin on the character of the party organization, he came down on the side of Martov and the Mensheviks. In the following year, in a pamphlet entitled *Our Political Tasks* and published under Menshevik auspices, he pronounced himself in favour of " opportunism in the organizational question " as against Lenin's " organizational rigorism ", and made a bitter and sweeping personal attack on Lenin, whom he denounced as a " Maximilian Robespierre " and a " slipshod attorney ", and accused of a desire to establish a "dictatorship over the proletariat ".[1] He soon broke with the Mensheviks, fell under the influence of Parvus, a German Social-Democrat of Russian origin, who inspired his theory of " permanent revolution ", and returned to Russia to participate actively in the 1905 revolution, becoming, at the age of twenty-six, the last president of the short-lived Petersburg Soviet, and demonstrating his capacity to sway and dominate an audience of workers. After a public trial which enhanced his reputation in revolutionary circles he spent another short period in Siberia, but escaped in time to attend the fifth party congress in London in 1907. From this time till 1917 he consistently attempted to occupy a position " outside the factions ", struggling to reconcile Bolsheviks and Mensheviks in the name of a " general party " line. Trotsky's view of the nature of the coming revolution was now far nearer to that of the Bolsheviks than of the Mensheviks. But, whereas Menshevism was always fluid and open to compromise, Bolshevism had hardened under Lenin's hand into a rigid core of doctrine which tolerated no dissent and treated as enemies those who rejected any item of it ; and this meant in practice that Trotsky found himself far more often at loggerheads with the Bolsheviks, and with Lenin in particular, than with the Mensheviks.[2] The fact that Lenin and Trotsky were at this time, in their different styles, already the two outstanding figures in the Russian Social-Democratic movement, and that there was no Menshevik of comparable stature to draw Trotsky's fire, merely deepened and sharpened the differences between them.

The years from 1907 to 1914 furnished that rich literature of

---

[1] See *The Bolshevik Revolution, 1917–1923*, Vol. I, p. 33.

[2] Trotsky himself afterwards noted this phenomenon : " The conciliatory line involved me in all the harsher opposition to Bolshevism since Lenin, in contrast to the Mensheviks, mercilessly rejected, and could not help rejecting, conciliationism ". (L. Trotsky, *Permanentnaya Revolyutsiya* (Berlin, 1930), p. 49).

controversy and mutual recrimination between the rival leaders which afterwards helped to build up the tradition of a fundamental incompatibility between the doctrine of Lenin and the doctrine of Trotsky. It was at this period that differences of opinion between them about " permanent revolution " and the rôle of the peasant, which, as the sequel showed, were never more than differences of emphasis, became inflated in the heat of controversy into differences of principle ; and this period also produced the abundant literature of mutual vituperation [1] which played so conspicuous a rôle in later controversies. The outbreak of war in 1914 brought about no immediate mitigation of the antipathy between them. Trotsky passed two years of the war in Paris, where in conjunction with Martov he edited a Russian anti-war newspaper *Nashe Slovo*, to which Lunacharsky, Ryazanov, Lozovsky, Chicherin, Radek and Rakovsky were regular or occasional contributors. Trotsky now stood on the extreme Left of the party. His views on the war differed in form rather than in substance from those of Lenin. But his eclecticism and willingness to cooperate with Mensheviks still estranged him from the Bolsheviks ; and his stern internationalism made him unsympathetic to the compromise which Lenin was prepared to make with the principle of national self-determination. At the end of 1916 Trotsky was expelled from France, and spent the first three months of 1917 in New York. He at once became a member of a Left-wing group in which Bukharin and Kollontai were leading figures and a contributor to the journal of the group, *Novyi Mir*. He evidently offended Kollontai, who wrote to Lenin that " Trotsky's arrival strengthened the Right wing at our meetings " and had delayed the endorsement of the Zimmerwald programme ; and this letter provoked the last of those personal outbursts of Lenin against Trotsky (" What a swine that Trotsky is ! ") which were afterwards so freely exploited by Trotsky's enemies.[2] On the outbreak of the

---

[1] See *The Bolshevik Revolution, 1917–1923*, Vol. 1, pp. 62-63.

[2] Lenin, *Sochineniya*, xxix, 290 ; the extract from Kollontai's letter is in *Proletarskaya Revolyutsiya*, No. 5, 1935, p. 39. Trotsky, in his own account of his stay in New York (L. Trotsky, *Moya Zhizn'* (Berlin, 1930), i, 310-312), writes condescendingly of Bukharin and Kollontai, but does not record any political disagreement. The fact that he and Bukharin returned to Russia by different routes may suggest that personal relations between them were not close.

February revolution he left for Russia, and, after a long detention by the British naval authorities at Halifax, Nova Scotia, finally reached Petrograd in May 1917. Lenin met him at first " with a certain restraint and hesitation ".[1] But, from the moment when Trotsky decided to join the Bolshevik party and to accept its organization and discipline, the difficulties melted away. In the critical months of 1917 Trotsky consistently saw eye to eye with Lenin : at this time there was, as Lenin recorded, " no better Bolshevik ".[2] Trotsky's experience and prestige as leader of the Petrograd Soviet of 1905 were invaluable, and he played the largest single part in organizing the *coup* of October 1917. Its brilliant success, and Trotsky's subsequent work in the recruitment and organization of the Red Army, made him in the eyes of the world the equal partner of Lenin : the names " Lenin and Trotsky " were coupled wherever the Russian revolution was spoken of. In the party the rôle of Lenin's principal lieutenant could not be denied him. It was true that Trotsky continued to have differences with Lenin in this period — on Brest-Litovsk, on the advance to Warsaw, on the relations of trade unions to the state, to name only the most famous. But it was also fair to recall the occasions on which he had stood with Lenin against other party leaders — in opposition to a coalition in November 1917, in support of the employment of " specialists " in the Red Army, in defence of the monopoly of foreign trade, in opposition to the coercion of Georgia in 1921–1922. Such alternations of agreement and disagreement were perfectly possible at this period between loyal party members. Lenin's criticism, in the testament, of Trotsky's " too far-reaching self-confidence " and " disposition to be too much attracted by the purely administrative side of affairs " was balanced by the recognition of him as " personally . . . the most able man in the present central committee ",[3] and implied no grain of doubt of his loyalty and devotion.

The position of Trotsky in the party was challenged only when Lenin was incapacitated and when Trotsky's rivals came together to block his potential claim to the succession. He himself remarked that " the beginning of the struggle with ' Trotskyism '

[1] L. Trotsky, *Moya Zhizn'* (Berlin, 1930), ii, 61.
[2] See *The Bolshevik Revolution, 1917–1923*, Vol. 1, p. 109, note 1.
[3] See *The Interregnum, 1923–1924*, p. 258.

coincides with the end of Lenin ",[1] though he failed to understand
the reason.  When Lenin disappeared from the scene, it quickly
became patent how much of the strength of Trotsky's position
had been due to Lenin's active support.  While he had a following
in the rank and file of the party, the other party leaders were his
implacable enemies.  What now brought about his downfall was
hostility not to his policies, but to his person.  It would be nearer
the truth to say that, between 1924 and 1927, Trotsky's policies
were discredited because he propounded them than that he was
discredited for propounding unacceptable policies.  He made mistakes ; but mistakes which would have been lived down and forgotten if committed by others proved fatal to him.  It was his
record, his outlook, his personality which were the real target of
attack and the real causes of his defeat.  An examination of these
will throw an indirect but significant light on the history of the
period.

Of all the Bolshevik leaders Trotsky was the most western
and the least specifically Russian.  Born in a Jewish family well
above the poverty line and with some intellectual ambitions, in a
part of Russia where anti-Semitism was rife in the period in which
he grew up, educated in a school which was a German foundation,
and where half the pupils in his day were still German, he can
hardly have escaped some perhaps unconscious prejudice against
things Russian.  By way of contrast, he conceived " an idealization
of the foreign world, of western Europe and America ",[2] whither
millions of his compatriots, including a high proportion of Jews,
were to migrate in the two decades before 1914.  He himself
came to western Europe at the impressionable age of twenty-three
— a refugee from the Russian police.  But, above all, the Russia
against which Trotsky reacted was the peasant Russia of his
youth.  The mature Trotsky was wholly urban.  The town was

    [1] *Byulleten' Oppozitsii* (Paris), No. 27 (March 1932), p. 2.
    [2] L. Trotsky, *Moya Zhizn'* (Berlin, 1930), i, 64, 114.  Trotsky disclaimed
any kind of national Jewish consciousness : " I am not a Jew but an inter-
nationalist ", he said on one occasion to a Jewish delegation (G. Ziv, *Trotsky :
Kharakteristika* (N.Y., 1921), p. 46).  But he was fully aware of the implications
of his Jewish origin in a Russian environment, and in 1917 rejected a suggestion
that he should be put in charge of the department of internal affairs on the
ground that the appointment of a Jew to such a post would " put an additional
weapon into the hands of our enemies " (L. Trotsky, *Moya Zhizn'* (Berlin,
1930), ii, 62-63).

the symbol of everything progressive : " the history of capitalism ",
he had written in 1906, " is the history of the subordination of
the country to the town ".[1]  Every Russian Marxist believed in
the economic superiority of western capitalist society and in the
backwardness of the primitive peasant Russian economy : every
Russian Marxist reacted against the Slavophil myth.  But Trotsky
showed particular zest in dwelling on the nullity of the Russian
contribution to civilization.  In statesmanship Russia had " not
got beyond third-rate imitations of the Duke of Alba, Colbert,
Turgot, Metternich or Bismarck ".  In philosophy and social
science, what Russia had given to the world was " nothing, a
round zero ".[2]  Even Trotsky's admiration of the classics of
Russian literature had a European flavour ; Karataev in *War and
Peace*, he remarked, was " the least comprehensible, or at any rate
the most remote from the European reader ", of Tolstoy's characters.[3]  The whole conception of the revolution was for Trotsky
inseparable from that of the impact of European civilization on
backward Russia :

> The revolution means the final break of the people with
> Asianism, with the 17th century, with holy Russia, with ikons
> and cockroaches, not a return to the pre-Petrine period, but on
> the contrary an assimilation of the whole people to civilization.[4]

To seek salvation in the west was Russia's revolutionary destiny.
In April 1916 Trotsky still deprecated " the national revolutionary
messianic mood which prompts one to see one's own nation-state
as destined to lead mankind to socialism ".[5]

The distinctively western cast of Trotsky's thought helps to
explain why before 1914 he found himself more at home with the

[1] L. Trotsky, *Itogi i Perspektivi* (2nd ed. 1919), p. 43.
[2] Trotsky, *Sochineniya*, xx, 330, 337.  The date of the article is 1912 ; it
contrasts markedly in tone with Lenin's article of 1914, *On the National Pride
of the Great Russians*: "We are full of a feeling of national pride, and precisely
for that reason *especially* hate *our* servile past " (Lenin, *Sochineniya*, xviii, 81).
[3] Trotsky, *Sochineniya*, xx, 252.
[4] L. Trotsky, *Literatura i Revolyutsiya* (1923), p. 68.
[5] *Nashe Slovo*, April 12, 1916, quoted in I. Deutscher, *The Prophet Armed*
(1954), p. 238 ; a phrase in Lenin's farewell letter to the Swiss workers of
March 1917 appeared to be a reply to this observation : " Not its special
qualities, but rather the special constellation of historical circumstances have
made the Russian proletariat, *for a certain, perhaps very brief, time*, the vanguard
of the revolutionary proletariat of the whole world " (*Sochineniya*, xx, 68).

more westernized Mensheviks than with the Bolsheviks. But he was also, of all Russian social-democrats, the only one who, during this period, achieved easy personal relations with the social-democrats of western Europe. The association with Parvus gave him his first entry into German party circles. Between 1907 and 1914 his position outside the two Russian factions made him the best interpreter of Russian party affairs to western European socialists, who shared his impatience of the doctrinal niceties of Russian party strife. In Berlin he was an assiduous visitor to the house of Kautsky, where he met the other German party leaders including the veteran Bebel ; and he was the only Russian whose contributions were welcomed by *Vorwärts* and *Neue Zeit*.[1] In Vienna he was on friendly terms with the Austrian socialist leaders.[2] He developed at this time a keen interest in the art, literature and intellectual movements of the west. Through Joffe, who was psychoanalysed by Adler, he had at any rate a superficial acquaintance with the work of Freud.[3] In Paris in 1915 and 1916 he came to know the French leaders of the extreme Left. After the foundation of Comintern he long remained, on the strength of these personal contacts, the main authority on relations with the French party. In the Russian party he made himself the champion of such supposedly western virtues as orderliness and punctuality.[4] If Trotsky struck Lunacharsky in 1905 as " unusually elegant, unlike the rest of us ",[5] if visitors to Moscow in the days of his greatness often noticed the " elegance " of his dress, and an American admirer described him as " highly bourgeois ",[6] this was another way of saying that Trotsky preferred

[1] Lenin in 1912 complained that " Trotsky is master at *Vorwärts* ", the foreign section of which was directed by Hilferding, " the friend of Trotsky " (Lenin, *Sochineniya* (4th ed.), xxxv, 11).

[2] Czernin relates that Victor Adler told him on the eve of his departure for Brest-Litovsk : " You will certainly get on all right with Trotsky " (O. Czernin, *In the World War* (1919), p. 234 : the passage is oddly omitted in the German edition published in the same year).

[3] Trotsky, *Sochineniya*, xxi, 423-432.

[4] In 1920 he apparently secured, with Lenin's support, the issue of a decree on punctual attendance at meetings and committees with fines for unpunctuality: the results were small (Trotsky, *ibid.* xxi, 700).

[5] A. Lunacharsky, *Revolyutsionnye Siluety* (1923), p. 20. This sketch appeared at a moment when it was already possible to criticize Trotsky and not yet obligatory to abuse him : it is the best in the volume.

[6] M. Eastman, *Portrait of a Youth* (1926), pp. 15, 31.

European conventions, affecting neither Lenin's proletarian cloth cap nor Stalin's Russian blouse. To describe Trotsky as the most European, and Stalin as the least European, of the early Bolshevik leaders is to state one of the underlying causes of the incompatibility between them. In the party where, after Lenin's death, men with little or no experience of the west were gradually coming to the top, the western quality of Trotsky's chosen ways of life and thought was an isolating factor. While it helped to account for the ready support which he at first obtained in most of the western communist parties, in the Russian party it was quickly turned against him. The resolution of the party central committee of January 1925 which passed judgment on him described Trotskyism as " a falsification of communism in the spirit of approximation to ' European ' patterns of pseudo-Marxism, i.e. in the last resort, in the spirit of 'European' social democracy ".[1]

Another quality put Trotsky at the opposite pole to Stalin. Of the original Bolsheviks only Stalin, and perhaps Zinoviev, were not pre-eminently intellectuals ; the rest (and the same was true of nearly all the Mensheviks) were men of ideas, men who resorted naturally to the written word and would have been uneasy about any course of action which could not be justified by theoretical argument. But in this respect Trotsky towered above them all.[2] For sheer force of intellect nobody in the party was a match for him. His anticipation of the dangers of personal dictatorship in the party in his pamphlet of 1904 ; his analysis of the future course of the revolution in *Results and Prospects* in 1906 ; his diagnosis of the different, but characteristic, chinks in the ideological armour of both Mensheviks and Bolsheviks in the

---

[1] *VKP(B) v Rezolyutsiyakh* (1941), i, 636. According to Stalin, *Sochineniya*, viii, 295, this resolution was drafted by Zinoviev ; Bela Kun, at this time a spokesman of Zinoviev, alleged in his article against Trotsky in December 1924 that Trotsky had always " tended towards ' western European Marxism ' in tactical and organizational questions " (*Pravda*, December 19, 1924).

[2] In exile Trotsky was to write of himself : " The desire for study has never left me, and many times in my life I felt that the revolution was pre-venting me from working systematically " (L. Trotsky, *Moya Zhizn'* (Berlin, 1930), i, 15). The words should not be taken too literally : they would scarcely have been written at the height of Trotsky's political activity. But they contrast strikingly with Lenin's *obiter dictum* on the contrast between " theory " and " life " (see *The Bolshevik Revolution, 1917–1923*, Vol. 1, pp. 24-25).

article of 1909 [1] — all these were extraordinary examples of penetrating acumen. The more testing conditions of political responsibility after 1917 revealed no falling off in intellectual power, though they brought out some of the defects of this quality in practical politics. The debate over Brest-Litovsk found Trotsky in the familiar posture of attempting to build a platform mid-way between two conflicting groups. The " no war, no peace " formula was a brilliant and ingenious improvisation. Its application was a gamble which nearly succeeded. But the verdict may be that to gamble in such a situation is not a mark of the highest statesmanship. In the succeeding years, Trotsky was, on a remarkable number of occasions, the first to elaborate and put forward policies which were eventually adopted, sometimes after he had been denounced for defending them. He was, so far as the record goes, the first advocate of NEP — at any rate in the party — a year before its acceptance. He was the protagonist of industrialization and planning at a time when these were denounced by the party leadership as destructive of NEP and of the " link " with the peasantry. The maintenance of labour armies and the " statization " of the trade unions, which were vehemently rejected when he proposed them, were realized, in substance though not in form, several years later. But this sequence of miscarriages — or of successes out of due time — suggests Trotsky's fundamental weakness as a responsible politician. He had an unfaltering, at times almost uncanny, perception of the social and economic trends of his time, and of the policies which would one day be demanded to take account of them. But he did not possess that supreme political sense of tact and of timing which is given to the great masters of statecraft. Once he had diagnosed the need for action, he lacked the patience to wait till the moment was ripe. The capacity to manipulate men, and to shape situations, in the interest of the course which he judged necessary eluded him. He had much of the common failing of the intellectual in politics : intolerance of the crude realities of the exercise of political power.

It was Trotsky's position in the party as the outstanding westerner and as the outstanding intellectual which, more than

---

[1] For the 1904 pamphlet see p. 140 above ; for the other two see *The Bolshevik Revolution, 1917–1923*, Vol. 1, pp. 58-59.

any specific issue of doctrine or policy, differentiated him from Lenin. Lunacharsky summed up the difference in the acute and provocative verdict that Trotsky was a more orthodox Marxist than Lenin.[1]  If Marxism is regarded primarily as a rigid analysis of the contradictions of the capitalist system and of bourgeois society, and only secondarily as a programme of action, if its economic and so-called " determinist " elements are exalted above its political and voluntarist aspects, then Trotsky was the better Marxist ; and this interpretation of Marxism, which can be supported by many passages of Marx himself, is on the whole the one which has prevailed in the west.  But this was not the interpretation which, under the leadership of Lenin, prevailed in the Bolshevik revolution.  Lenin brought to the interpretation of Marx's teaching a flexibility and an adaptability which were foreign to Trotsky's attitude, but which are probably essential to any application of theory to practice.  Both Lenin and Trotsky liked to invoke history.  But, while Lenin was fully alive to the necessity of moulding the course of history to his programme, Trotsky tended to treat history as an objective reality which was accessible to intellectual analysis, and was bound to justify that analysis in action if the analysis were correct.  The masses, by their spontaneous action, were the executors of the laws of history : the essence of the Bolshevik revolution was " the forcible entrance of the masses into the realm of rulership over their own destiny ".[2] From this court there was no appeal.  Trotsky relegated his defeated opponents to the dustbin of history.  But, in so doing, he deprived himself of any real answer when, in the hour of his own defeat, he found himself consigned to the same destination. His autobiography and many of his subsequent writings revolved round the tormenting question why he was defeated, why the masses failed to rise to his support — questions which for him could be answered only in terms of some error of analysis.  He patently failed to answer them, either to his own satisfaction or to that of the reader.  It is significant that in the concluding sentences of his autobiography he sought " consolation " in a quotation not from Marx, but from his old enemy Proudhon — not in an analysis of history, but in a gesture of defiance to it.

[1] A. Lunacharsky, *Revolyutsionnye Siluety* (1923), p. 27.
[2] L. Trotsky, *History of the Russian Revolution* (Engl. transl.), i (1932), 15.

The singularities of Trotsky's political destiny were closely interwoven with those of his personal character. The quality which Lenin called " self-confidence " and others bluntly branded as arrogance isolated him among his equals. An acquaintance of his early years, in a hostile but perceptive sketch, wrote of the desire " to rise above all, to be everywhere and always first " as the " fundamental quality " of his character ; and this gave his revolutionary convictions an austere and almost inhuman note which distinguished them from the equally intense but emotionally warmer convictions of Lenin :

> The revolution and his active " ego " coincided. Everything that was outside his " ego ", and therefore did not interest him, did not exist for him.
> The workers interested him as necessary instruments of his activity, of his revolutionary work ; his comrades interested him as a means with the cooperation of which he exercised his revolutionary activity. He loved the workers, he loved his comrades in the organization, because in them he loved himself.[1]

Between 1903 and 1917 he continued to play a lone hand ; and, when in 1917 the logic of the revolution and the magic of Lenin's personality made him a Bolshevik, they did not bring his isolation to an end. There was more than a grain of truth in Kamenev's later taunt that Trotsky " entered our party as an individualist, who thought, and still thinks, that in the fundamental question of the revolution it is not the party, but he, comrade Trotsky, who is right ".[2]  For Trotsky, even the Marxist sense of history seemed to take on a personal colour, and to centre round his own rôle on the historical stage. Unlike Lenin, wrote Lunacharsky, who " never looks at himself, never glances into the mirror of history, never even thinks of what posterity will say of him ", Trotsky " looks at himself often ", " treasures his historical rôle ", and coveted " the halo of a genuine revolutionary leader ".[3]

After Trotsky's downfall many who had once praised and flattered hastened to denigrate and condemn.  But there is contemporary evidence of the ambivalent attitude of the other leaders towards him and of their resentment of his authority and prestige :

[1] G. Ziv, Trotsky: Kharakteristika (N.Y., 1921), p. 12.
[2] Leninizm ili Trotskizm (1924), p. 47.
[3] A. Lunacharsky Revolyutsionnye Siluety (1923), p. 27.

indeed, nothing else could explain the rapidity and ease with which
the coalition was formed against him when Lenin withdrew from
the scene. " More feared than loved, perhaps — that is possible ",
wrote a French communist whose record of a visit to Moscow in
1921 appeared with a preface by Trotsky, " but his ascendancy is
prodigious ".[1]  " I love Trotsky, but am afraid of him ", wrote
the poetaster Demyan Bedny a little later.[2]  Angelica Balabanov,
an unsympathetic critic, passed a harsher judgment :

> His arrogance equals his gifts and capacities, and his manner
> of exercising it in personal relations creates very often a distance
> between himself and those about him which excludes both
> personal warmth and any feeling of sympathy and reciprocity.[3]

Lunacharsky referred to Trotsky's " nonchalant, high and mighty
way of speaking to all and sundry ", and noted that " a tremendous
imperiousness and a kind of inability or unwillingness to be at all
amiable and attentive to people " condemned him to " a certain
loneliness " in the party : he had " practically no immediate supporters ".[4]  A specialist without party affiliations who saw a good
deal of the leaders at this time acutely observed Trotsky's isolation :

> In any gathering of these old Bolsheviks Trotsky remained
> an alien. . . . Trotsky compelled them to respect him, to pay
> heed to every word he spoke.  Yet they resented it bitterly, or
> at least were dissatisfied and jealous whenever Lenin saw fit
> to defer publicly to Trotsky.[5]

It was easy for Lenin, the uncontested leader, to overlook Trotsky's
sudden and rapid promotion and to forget his past record in
admiration of his present deserts.  It was more difficult for those
jealous old Bolsheviks who felt that an intruder had supplanted
them both in authority and in Lenin's favour.  Trotsky never

    [1] A. Morizet, *Chez Lénine et Trotski* (1922), p. 108 ; three years earlier the
impressionable Frenchman Sadoul had referred to Trotsky's " Mephistophelean,
terrifying mask " (A. Sadoul, *Lettres sur la Révolution Bolchevique* (1919),
p. 396).                                        [2] *Pravda*, January 11, 1924.
    [3] A. Balabanov, *My Life as a Rebel* (Engl. transl. 1938), p. 176.
    [4] A. Lunacharsky, *Revolyutsionnye Siluety* (1923), pp. 20-21.  Gorky, in
his memoir of Lenin, compared Trotsky with Lassalle, and described him as
" with us, but not of us " (M. Gorky, *Days with Lenin* (Engl. transl. n.d.
[? 1932]), p. 57).  In view of Gorky's own position the passage reads somewhat
ironicaliy : it did not appear in the original version of the memoir in *Russkii
Sovremennik* (Berlin), i (1924), 229-244.
    [5] S. Liberman, *Building Lenin's Russia* (Chicago, 1945), p. 78.

seems to have realized the handicaps imposed on him by his late accession to the party. His behaviour accentuated them. His outstanding services to the party, and Lenin's ungrudging recognition of them, were a sufficient passport to pre-eminence ; he sought no other. He saw no reason to conciliate his enemies and rivals, and heedlessly added to their number.

It was doubtless this human shortcoming which Lenin had in mind when he wrote in the testament of Trotsky's addiction to " the purely administrative side of affairs ". His capacity as an administrator was second only to his intellectual power. The effortless success of the October *coup* of 1917 owed much to his organizing genius ; the creation of the Red Army was his supreme achievement ; and any department administered or supervised by Trotsky was a model of efficiency. Nor did this exhaust the astonishing range of his gifts. He was probably the greatest orator of the revolution. Before a limited and informed party assembly, his studied rhetorical effects were less effective than Lenin's direct simplicity ; and Stalin underlined the point when he missed the " simple and human " touch in Trotsky's exposition of Leninism.[1] But Trotsky's occasional flamboyance did not, like that of Zinoviev, mask an intellectual void or a weakness of inner conviction. It sprang from fierce, uncontrollable passion ; and in the ability to move a mass audience by the passionate sweep of his eloquence Trotsky stood out above any of his contemporaries. Yet the great intellectual, the great administrator, the great orator lacked one quality essential — at any rate in the conditions of the Russian revolution — to the great political leader. Trotsky could fire masses of men to acclaim and follow him. But he had no talent for leadership among equals. He could not establish his authority among colleagues by the modest arts of persuasion or by sympathetic attention to the views of men of lesser intellectual calibre than himself. He did not suffer fools, and he was accused of being unable to brook rivals. Where Lenin was supreme, Trotsky failed altogether.

Thus the political climate of the period, combined with his own weaknesses of character, sealed Trotsky's doom. Self-confident, haughty and aloof among his colleagues, secure in his own superiority and unconscious or contemptuous of the ruffled

[1] Stalin, *Sochineniya*, viii, 276.

emotions of those who felt themselves overshadowed by him,[1] he felt no need to defend himself against the powerful forces accumulated against him. Referring to the first attacks of the other leaders upon him in the winter of 1923-1924, he nonchalantly boasted that he had not read " any of these things ".[2] He made no attempt, till it was far too late, either to organize his friends or to divide his enemies. Trotsky had no political instinct in the narrower sense, no feeling for a situation, no sensitive touch for the levers of power. It was this defect which rendered him blind, in the years before the revolution, to the significance of Lenin's insistence on rigorous organization, and which, after the revolution, made him politically no match either for Lenin, whom he outshone in many spheres, or for Stalin, whom he eclipsed in almost all. But, even more than these personal shortcomings, the evolution of events contributed to his defeat. As an intellectual he lost his foothold in a time when theory was beginning to be at a discount, when political life revolved round the empirical solution of current practical problems, and the balance between conflicting factions and interests was maintained by clever political manœuvring. As a whole-hearted and impenitent westerner, he was out of place in a period when a return to Russian national tradition was being cunningly blended with the achievements of the revolution. As a revolutionary to the finger-tips, he was an incongruous figure in an age which seemed (though falsely seemed) to be set on a path of consolidation and stabilization. As an individualist, whose past recalcitrance to party discipline was unforgotten and unforgiven, he was suspect in a party which hymned the praises of collective leadership and was obsessed by the bogy of a Bonaparte. Trotsky was a hero of the revolution. He fell when the heroic age was over.

### (b) Zinoviev

Grigorii Evgenevich Zinoviev (original name Radomylsky) was born in 1883 in Elizavetgrad (renamed Zinovievsk in 1924) of a

---

[1] Trotsky contemptuously commented on the advantage which his indifference in this respect gave to Stalin : " Whenever I had occasion to tread on the corns of personal predilections, friendships or vanities, Stalin carefully gathered up all the people whose corns had been stepped on " (L. Trotsky, *Stalin* (N.Y., 1940), p. 289).

[2] M. Eastman, *Since Lenin Died* (1925), p. 94.

Jewish petty bourgeois family, his father being the proprietor of a small dairy farm. He was taught at home, and never attended school or university. He appears to have been the least highly educated of the Bolshevik leaders, except Stalin ; and he was the least successful of them all in his handling of men. From the age of fifteen he earned his living first as a teacher, then as a clerk in a business firm. At the turn of the century he became active in organizing strikes, and in 1902 he went abroad to Berlin, Paris and Berne. Early in 1903 he met Plekhanov and Lenin in Switzerland, and after the party congress of that year was sent back to Russia as a party worker. At this time his health began to give trouble ; a heart defect was diagnosed and he again went abroad. In 1906 he was back in Petersburg carrying on agitation among the metal workers ; he attended the fifth party congress in London in 1907 as their delegate, and was elected to the party central committee. In the following year he was arrested, but secured his release on grounds of ill health and returned to Switzerland.[1] From this time onward he became Lenin's intimate associate and disciple. He seems to have been the only Bolshevik to stand at Lenin's side in Paris in January 1910 in opposition to the policy of compromise with the Mensheviks and to the maintenance of the united party central committee,[2] and no doubt earned the leader's gratitude on that account. The Prague conference of 1912 made him a member of the new all-Bolshevik party central committee, and he moved with Lenin to Galicia in the following year. When Kamenev returned to Petersburg in 1914 to edit *Pravda*, Zinoviev remained behind as Lenin's principal collaborator, and followed his fortunes throughout the war. Zinoviev and Lenin appeared as joint authors on the title page of a pamphlet entitled *Socialism and the War*, published in Russian in Switzerland in 1915 and quickly translated into German and French; and of a collection of articles, *Against the Current*, published in Switzerland in German in 1916 and in a Russian version in Petrograd in 1918.

Zinoviev therefore already occupied a special position in the party when he returned to Petrograd with Lenin in the sealed

[1] The above particulars are taken from the authorized biography in *Entsiklopedicheskii Slovar' Russkogo Bibliograficheskogo Instituta Granat*, xli, i (n.d. [1927]), Prilozhenie, cols. 143-149.
[2] See *The Bolshevik Revolution, 1917–1923*, Vol. 1, p. 50.

train in April 1917. Like the other leading Bolsheviks, he was initially shocked by Lenin's April theses, but quickly rallied to them. He accompanied Lenin into hiding in July 1917, and returned with him to Petrograd in October. This proved a turning-point in Zinoviev's career. He associated himself with the cautious Kamenev in opposing Lenin's proposal for the immediate seizure of power. Less consistent than Kamenev, or more reluctant to burn his boats, he did not actually resign with him from the party central committee ; but, after the disclosure of the dispute in the non-party journal *Novaya Zhizn'*, he incurred Lenin's wrath equally with him. Ten days after the seizure of power, Zinoviev, with Kamenev and three other members of the central committee, resigned on the issue of Lenin's opposition to a coalition government with the SRs ; when confronted with an ultimatum from Lenin, Zinoviev, alone of the five, at once recanted and resumed his seat on the committee. These incidents were forgiven and, in large measure, forgotten by the party. But they left the impression of a basic timidity of character beneath a blustering exterior. Zinoviev shrank from the responsibility of decisive action, but shrank equally from the consequences of persisting in his opposition.

That Zinoviev held no important governmental office was a matter of accident and not a slur on his standing in the party. He was the chief party spokesman in the trade union central council and presided at the early trade union congresses. This was the beginning of his special association with Petrograd, the centre of the metal-working industries which furnished the Bolshevik core of the trade-union movement. In March 1918 he came out strongly against the project to move the capital and the party headquarters to Moscow ; [1] and, when this was carried over his head, he received the mandate to remain in Petrograd at the head of the party organization there. But soon new tasks awaited him. His record in the Zimmerwald organization during the war made him the obvious choice for the maintenance of contacts with Left-wing supporters abroad ; and his work in this field culminated in his appointment as president of the executive committee of the newly founded Communist International in March 1919 — a position of enormous prestige in Bolshevik circles at a

[1] L. Trotsky, *Moya Zhizn'* (Berlin, 1930), ii, 74.

time when world revolution seemed to hold the key to the future. Throughout this time Zinoviev was Lenin's loyal henchman and the unswerving supporter of his policies, thus effacing the memory of his momentary lapses in the autumn of 1917. When Lenin was incapacitated through illness, and Trotsky deliberately refused to be a candidate for the vacant chair, the triumvirate came almost spontaneously into being ; and, with Kamenev too modest, and Stalin too wary, to aspire to the highest position, Zinoviev emerged by common consent as its senior member. He made the principal report at the twelfth party congress of April 1923 and again at the thirteenth congress of May 1924 after Lenin's death.

Zinoviev was thus the leading figure in the party during the brief but important intermediate period which covered the last months of Lenin's illness and those immediately after his death ; and it is partly due to this circumstance that Zinoviev appears as the initiator of many sinister developments in party history. Zinoviev was more responsible than anyone for establishing the cult of Leninism and the convention that absolute fidelity to Lenin was the main and indispensable qualification for leadership of the party. This was natural, since his own close association with Lenin was his principal asset and the source of his prestige in the party. He invented, or was the first publicly to use, the word " Trotskyism " as a term of abuse. He unwittingly created a fateful precedent in party doctrine when, at the thirteenth party congress, he invited Trotsky not merely to submit to the decision of the majority but to confess himself in error ; and he initiated, no doubt in crude imitation of Lenin, the practice of denouncing as Menshevism any deviation from the path of current party orthodoxy.

The emergence of Zinoviev as *de facto* leader of the party and Lenin's potential successor threw his hitherto concealed weaknesses into relief.[1] He had no grasp of political or economic issues and preferred speech to action. The economic decisions taken in the autumn of 1923 were forced and generally belated ; and there is no evidence to associate Zinoviev with them. As the platform of the 46 showed, the party keenly felt the absence

---

[1] The contemporary verdict in A. Lunacharsky, *Revolyutsionnye Siluety* (1923), p. 32, that he " exceeded the anticipations of many " was a masterpiece of tact.

of economic leadership. The collapse of the German revolution in the same autumn was still more significant. Zinoviev may have been momentarily successful in unloading the blame on Radek and Brandler ; but the defeat could in the long run only lower the prestige of Comintern and of its president. More important, Zinoviev understood nothing of the nature of political power or of the management of men, and he lacked the native tact which sometimes goes with innocence. Clumsy in all his dealings, he revealed his cards before the time had come to play them. His ambition to assume the mantle of Lenin was so naïvely displayed as to make his vanity ridiculous. He brought on himself the principal odium of the campaign against Trotsky, and allowed Stalin to reap its advantages. Nor had he any gifts as an organizer. When he attempted to counter Stalin's rising power, the Leningrad party machine, of which he had hitherto been undisputed master, crumbled in his hand, leaving him helpless in face of an adversary infinitely astuter and better prepared for the fray.

No leading Bolshevik of this period incurred so much adverse personal criticism as Zinoviev, or appears to have been so widely disliked. None of them inspired so little personal respect. Zinoviev's intellect was nimble, but politically unschooled. Compared with his fellow-triumvirs he lacked the acumen of Kamenev or the application of Stalin. At the fourteenth party congress Molotov complained that, while Kamenev at least " tried to expound a complete system of opinions ", Zinoviev dealt in resounding phrases which offered " nothing new, nothing definite, no class content " ; [1] and the shaft went home. Stalin on the same occasion described his attitude as " wobbling " and " hysteria, not a policy ".[2] Trotsky wrote of his " incorrigible vacillations ".[3] Zinoviev never succeeded in attaining either depth of conviction or depth of understanding ; and this innate superficiality, among men who treated the subtleties of doctrine with passionate earnestness, won him an unenviable reputation for shiftiness and lack of scruple. It frequently appeared that there was no principle which he was not prepared to sacrifice on the

[1] XIV S''ezd Vsesoyuznoi Kommunisticheskoi Partii (B) (1926), p. 473.
[2] Stalin, Sochineniya, vii, 378.
[3] L. Trotsky, Moya Zhizn' (Berlin, 1930), ii, 273.

spur of the moment to the cause of political expediency or personal advancement. When attacked, he quickly abandoned his positions or defended them without courage or dignity. At the fourteenth party congress of December 1925, when he was fighting for his political life as leader of an opposition which challenged the fundamentals of party policy, he found it necessary to apologize for his temerity in stating his case :

> If our comrades in the central committee and the central control commission had said that, in the interests of peace, this should not be done, we should not have done it. They told us that there were no objections.[1]

And this lack of intrinsic seriousness was thrown into relief by a vein of vanity and self-importance which infected his literary style as well as his personal behaviour : Bukharin on the same occasion ironically taunted him with his " epoch-making books ".[2] Sukhanov, the journalist of the revolution, attributed to Zinoviev " the well-known qualities of the cat and the hare ".[3] Levi, the expelled KPD leader, described him as " an ass of European notoriety ".[4] A heckler at the fourteenth congress interrupted his protestations of innocence with the exclamation " Poor sheep ! " [5] Angelica Balabanov, writing many years later, recorded the verdict that Zinoviev was " after Mussolini . . . the most despicable individual I have ever met ".[6] No other Bolshevik leader was denounced, even by his worst enemies, in terms of such searing contempt.

These verdicts, not all of which were delivered after Zinoviev's fall, raise the question how Zinoviev was enabled, even for a short period, to play so conspicuous and important a rôle in party and state affairs. His record as Lenin's closest associate and disciple during the years of exile, his fame throughout Europe, among friends and foes alike, as president of the executive committee of Comintern, and his foolhardy willingness, after Lenin's breakdown, to claim responsibilities from which others shrank, provide

---

[1] *XIV S"ezd Vsesoyuznoi Kommunisticheskoi Partii (B)* (1926), p. 556.
[2] *Ibid.* p. 138.
[3] N. Sukhanov, *Zapiski o Revolyutsii*, iv (Berlin, 1922), 322.
[4] In a preface, dated December 28, 1924, to L. Trotsky, *1917 : Die Lehren der Revolution* (German transl. 1925).
[5] *XIV S"ezd Vsesoyuznoi Kommunisticheskoi Partii (B)* (1926), p. 556.
[6] A. Balabanov, *My Life as a Rebel* (Engl. transl. 1938), pp. 243-244.

a partial but inadequate explanation. An observer who saw him with Lenin during the war described his capacity to work day and night writing " newspaper articles, circulars to party friends, resolutions, brochures — everything that Lenin thought was required " ; and, while " what he wrote was neither deep nor original ", it was always serviceable. Zinoviev thus " moved in a world of verbal constructions ", and led a " bloodless paper existence " divorced from any real understanding of what was afoot.[1] It is no doubt unfortunate for Zinoviev's reputation that his outstanding excellence was one not easily transmitted to posterity. He was the possessor of " a powerful, extremely resonant, voice of tenor timbre ",[2] and was by common consent an impressive orator who could play on the emotions of a mass audience. He seems to have owed much of his authority in the party to this quality. Perhaps the crowning achievement of his career was his four-hour speech in German to the Halle congress of the German Independent Social-Democratic Party in October 1920 which won a majority of the party for fusion with the KPD.[3] But his style was flamboyant and repetitive, and even his best speeches lost their effect in print. His oratory seemed to require a background of applause and adulation. In the later years, when he was attempting to defend a minority opinion in the face of hostile audiences, his rhetorical genius deserted him. In adverse conditions he proved a far less formidable debater than Kamenev or Trotsky. After Lenin's death the hollowness of Zinoviev's fame was soon made apparent ; and the continued vacillations which attended his downfall deprived it of the dignity of tragedy.

### (c) Kamenev

Lev Borisovich Kamenev (original name Rozenfeld) was born in Moscow in 1883. His father was a skilled mechanic, educated in the Petersburg Technological Institute, who worked as an engine-driver on the Moscow-Kursk railway, moved shortly after his son's birth to the neighbourhood of Vilna, where he had a post

---

[1] O. Blum, *Russische Köpfe* (1923), p. 109.
[2] A. Lunacharsky, *Revolyutsionnye Siluety* (1923), p. 30 ; Emma Goldman an unfriendly witness, thought him " flabby and weak " and his voice " adolescent, high-pitched and lacking in appeal " (E. Goldman, *Living my Life* (1932) ii, 732).      [3] See *The Bolshevik Revolution, 1917–1923*, Vol. 3, p. 218.

in a local nail factory, and then in 1896 to Tiflis, where he was employed as a railway engineer. Kamenev began his education in Vilna, and passed through its later stages in the gymnasium at Tiflis. Though expelled from school in 1900 for contact with revolutionary groups and for reading illegal literature, he was able to enter Moscow University, where he studied law. He was arrested in 1902, was sent back to Tiflis " under police supervision " and went abroad. He joined a Russian Social-Democratic group in Paris and met Lenin. During this time he married Trotsky's sister, Olga. After the party congress of 1903 he was sent back to Russia as a party worker and was active in Petersburg, Moscow and Tiflis, where he was one of the organizers of the first " Caucasian committee " of the party. After more than one arrest and release, he went abroad again in 1908, and for the next five years remained, next to Zinoviev, Lenin's closest collaborator. Early in 1914 he was sent back to Petersburg to take charge of the party newspaper *Pravda*.[1]

What had hitherto distinguished Kamenev from other leading Bolsheviks was a mild and conciliatory temperament and a reluctance to go to extremes. The only occasion during this period on which he diverged from Lenin's views was when, at the last meeting of the united party central committee in Paris in January 1910, he had worked actively for the compromise which precariously preserved party unity.[2] Separated from Lenin by the war, he quickly displayed the same inclination to compromise in a more embarrassing context. In November 1914 the leading Bolsheviks in Petrograd were arrested *en masse* at a secret conference ; and at the trial in February 1915 Kamenev and some of the other defendants publicly dissociated themselves from Lenin's advocacy of national defeat. Their pliancy did not save them, and Kamenev spent the next two years in Siberia. Returning to Petrograd in company with Stalin in March 1917, he took up a position on the Right of the party, and came out as an advocate of

---

[1] Most of the above particulars are taken from the authorized biography in *Entsiklopedicheskii Slovar' Russkogo Bibliograficheskogo Instituta Granat*, xli, i (n.d. [1927]), Prilozhenie, cols. 162-168.
[2] For this compromise see *The Bolshevik Revolution, 1917-1923*, Vol. 1, p. 50 : Sukhanov's statement that Kamenev was once a Menshevik (N. Sukhanov, *Zapiski o Revolyutsii*, iv (Berlin, 1922), 322) seems untrue, and may be based on this episode.

national defence and a conditional supporter of the Provisional
Government. He persisted in this attitude after Lenin's return,
being the only Bolshevik who openly challenged the " April
theses " in *Pravda* ; and, with Zinoviev and Stalin now firmly on
Lenin's side, he continued his opposition at the April conference
of the party. His defeat there brought his recalcitrance to an end.
He faithfully followed the party line throughout the summer, and
was arrested with Trotsky in July when Lenin and Zinoviev went
into hiding. In every important issue which arose Kamenev
instinctively favoured moderation and compromise, and maintained his position long enough to absolve him from the charge of
mere weakness or opportunism. But he lacked any real independence of character or intellect, and always yielded in the end to the
weight of opinion of those about him.

This pattern was twice repeated in the autumn of 1917.
Kamenev, this time initially supported by Zinoviev, opposed the
seizure of power, resigned from the central committee and
endeavoured to put his views before the party. But, unable to
obtain further support, violently denounced by Lenin for the disclosure in *Novaya Zhizn'*, threatened with expulsion from the
party, and finally abandoned by Zinoviev, he came to heel, and
returned to his party allegiance. Little more than a week after
the seizure of power Kamenev and Zinoviev, followed by three
other members of the party central committee, opposed a decision
not to seek a coalition with other parties and, on November 4/17,
1917, confronted by an ultimatum from the majority, resigned from
the committee. Kamenev also resigned the post of president of
VTsIK, which now passed to Sverdlov. Two days later Zinoviev
recanted and was reinstated. But Kamenev and the other three
held firm for another three weeks, during which an agreement
was actually reached for the participation of Left SRs in the
government ; and, when they eventually applied for reinstatement, the party central committee on November 29/December
12, 1917, rejected their request. There seems, however, to have
been a general willingness to pass the sponge over this untoward
episode and to allow the offenders to resume party and governmental work. Kamenev was a member of the Brest-Litovsk
delegation during the first part of the negotiations ; according to
Trotsky, he " agreed with my formula at Brest, but joined Lenin

on his return to Moscow ".[1]  In January 1918 he was sent on a
mission to Great Britain and France and did not participate in the
later stage of the Brest-Litovsk negotiations or in the discussions
in the party on the ratification of the treaty.

From this time to the end of Lenin's active life, Kamenev
remained a faithful disciple and strayed no more from the party
line.  He was re-elected to the party central committee in 1919,
and during these years held several governmental posts.  But the
main symbol of his high rank in the party was his position as head
of the Moscow party organization and president of the Moscow
Soviet.  When Lenin was incapacitated, he was accepted without
question as a member of the ruling triumvirate.  While intellectually he stood above either of his colleagues, he proved in
action by far the least effective of the three, having neither the
ambition and self-confidence of Zinoviev nor the supreme political
skill of Stalin.  A strong personal antipathy to Stalin drew him
closer to Zinoviev, with whom he was linked by long-standing ties
of association.  He took a leading part in the campaign against
Trotsky, though perhaps with a certain sense of shame, since he
protested on one occasion that charges of "petty bourgeois deviation" should not be taken personally, or assumed to mean "that
we accuse this or that comrade whom we think mistaken of being
a representative of the petty bourgeoisie ".[2]  The strength of the
opposition in the Moscow party organization in the autumn of
1923 sapped Kamenev's prestige : clearly his leadership had not
been equal to the task of maintaining party discipline.  He probably supplied a large part of the intellectual ammunition for the
" new opposition " of 1925, and he alone had the courage to come
out openly against Stalin.  His speech for the opposition at the
fourteenth party congress was the finest of his career.  But
the conspicuous rôle devolved on Zinoviev; and Kamenev for the
rest of his career followed Zinoviev's lead to his own humiliation
and destruction.  The only occasion in these later years on which
he seems to have taken the initiative was also characteristic : he
was largely instrumental in hastening the reconciliation between
Zinoviev and Trotsky in the summer of 1926.

The authorities are agreed in depicting Kamenev as a highly

---

[1] L. Trotsky, *Moya Zhizn'* (Berlin, 1930), ii, 122.
[2] L. Kamenev, *Stat'i i Rechi*, x (1927), 257.

intelligent and cultivated man of amiable manners. He was an
excellent talker and an adequate, though not brilliant, public
speaker. He discharged with credit the task of supervising the
first collected edition of Lenin's writings ; Lenin is said to have
thought him " a clever politician ", while casting doubts on his
capacities as an administrator.[1] Kamenev was a man of sincerely
held beliefs, which were remarkably free from any admixture
whether of personal ambition or of political calculation. But these
qualities had their reverse side. Moderation was always Kamenev's
guiding star, even in the assertion of his beliefs. The point was
soon reached when it no longer seemed worth while to defend
them, partly through lack of conviction in the rightness of his own
judgment, partly through an amiable readiness to yield to the
importunities of his friends and associates. Molotov taunted him
with the habit of raising questions " by way of discussion " and
then abandoning them when he met with opposition, like a weakling
who does not stand up for his opinions.[2] Kamenev had neither the
desire nor the capacity to lead men : he lacked any clear vision
of a goal towards which he would have led them. He needed a
leader ; and this weakness ultimately linked his fate with that of a
man less intelligent, less upright and in every way less attractive
than himself.[3]

## (d) Bukharin

Nikolai Ivanovich Bukharin was younger than any of the other
recognized leaders of 1917, who never regarded him entirely as
their equal and treated him with a certain affectionate con-
descension. He was born in Moscow in 1888, both his father
and his mother being school teachers. The father was a mathe-
matician who also had a wide knowledge of literature, and is
described by his son as " a very unpractical person in daily life ".
Bukharin was by origin more distinctively an intellectual than
any other of the leading Bolsheviks. He was a brilliant pupil
at school, read illegal literature, was brought into contact with
Marxism, and joined the Russian Social-Democratic Workers'
Party in 1906. In the same year, in company with Ilya Erenburg,

[1] L. Trotsky, *Moya Zhizn'* (Berlin, 1930), ii, 216.
[2] *XIV S"ezd Vsesoyuznoi Kommunisticheskoi Partii (B)* (1926), pp. 484-485.
[3] N. Sukhanov, *Zapiski o Revolyutsii*, ii (Berlin, 1922), p. 243, has a good
character sketch of Kamenev.

he helped to organize a strike in a wallpaper factory. He studied in Moscow University, was arrested, released and re-arrested, finally escaping abroad in 1910. He engaged in party work, met Lenin in Cracow in 1912, and was in Vienna when war broke out in 1914. Expelled by the Austrian authorities, he spent some time in Switzerland, and went on in the autumn of 1915 to Sweden and Norway, and finally in October 1916 to the United States. Thence, after the February revolution of 1917, he returned via Japan and Siberia to Petrograd.[1]  With the other Bolshevik leaders proscribed and in hiding, Bukharin and Stalin played the principal rôles at the sixth party congress of August 1917. Bukharin now became a member of the party central committee — a position which he held continuously till 1929. In December 1917 he became editor of *Pravda*, and, after a brief interruption caused by his adherence to the Left opposition in the Brest-Litovsk period, resumed the post in the following year.

Bukharin won his reputation in the party as a theorist rather than as a practical politician ; and this preoccupation with doctrine tended throughout his career to make him the opponent of compromises dictated by expediency, and to drive him into extreme positions. The world war of 1914 inspired him to undertake an analysis of imperialism, for which Hilferding's *Finanzkapital*, published in 1909, served as the natural starting-point. Hilferding had portrayed the evolution of private enterprise capitalism into a system of national finance capitalism, in which expansive and self-assertive nations were the new units of power, and in which class contradictions within the nation had been eclipsed by conflict between nations. The war convinced Bukharin that this system represented a stage in capitalist development —'the new phenomenon of imperialism — in which capitalism had become incompatible with the further expansion of production, and had thus sealed its own doom. Competition for export markets, for raw material markets and for spheres of capital investment, were " simply three aspects of one and the same phenomenon : the conflict between the growth of productive forces and the ' national ' limitation on productive organization ".[2]  The moral was the

---

[1] The above particulars are taken from the autobiography in *Entsiklopedicheskii Slovar' Russkogo Bibliograficheskogo Instituta Granat*, xli, i (n.d. [1927]), Prilozhenie, cols. 51-56.

[2] N. Bukharin, *Mirovoe Khozyaistvo i Imperializm* (1918), p. 65.

inevitable breakdown of these national limitations, and the internationalization of capital as the final stage in the death-throes of capitalism. The article embodying these views appeared in a collective party volume published in September 1915 under the title *Kommunist*, to which Lenin also contributed, and apparently received Lenin's endorsement.[1]

Bukharin went on, however, to draw conclusions from his thesis which quickly brought him into conflict with Lenin. Bukharin's analysis of imperialism led him to adopt a position of unqualified hostility to the national state. If the nation was an obsolete and therefore reactionary political form, any kind of national policy was anathema to the true Marxist. Bukharin had on this ground accepted with reluctance Lenin's policy of "national defeatism"; and in November 1915 Pyatakov, Evgeniya Bosh and Bukharin, then in Stockholm, drew up a " platform ", accompanied by theses on the national question, which attacked Lenin's support of national self-determination as " utopian " and " harmful ".[2] In the following year he reverted in a further article to his analysis of national capitalism. Every " developed ' national system ' " under capitalism had now become " a state capitalist trust ". In spite of the fashion for describing this system as " state socialism ", it was really " state capitalism ", and it would bring into being " the finished type of the imperialist robber state " — a new Leviathan compared with which " the fantasy of Thomas Hobbes would seem a child's toy ". The workers had no option but to become " a simple appendage of the state apparatus " or to destroy it root and branch by the establishment of a proletarian dictatorship, the ultimate purpose of which was to abolish itself. When Bukharin submitted this article for publication in a volume of party essays, Lenin rejected it.[3] Relations

---

[1] Lenin wrote a preface for a revised edition of the article which was to be published as a pamphlet in Petrograd in the summer of 1917, but the preface was lost when the Provisional Government raided the party press. The pamphlet eventually appeared, without Lenin's preface but with a preface of Bukharin dated November 25, 1917, under the title *Mirovoe Khozyaistvo i Imperializm* (1918) ; the quotation above is from this edition.

[2] For a translation of the theses see O. H. Gankin and H. H. Fisher, *The Bolsheviks and the World War* (Stanford, 1940), pp. 219-223.

[3] It was eventually published in a truncated form (the conclusion having been lost) in *Revolyutsiya Prava: Sbornik*, i (1925), 5-32, with an explanatory note by Bukharin ; the quotations above are taken from this version.

became strained ; and, in October 1916, at the moment of his departure from Norway for the United States, Bukharin wrote a characteristic letter to Lenin in Switzerland :

> At any rate I ask one thing : if you must polemize etc., maintain such a tone as not to force a break. It would be very painful to me, more painful than I could bear, if joint work, even in the future, were to become impossible. I have the greatest respect for you and look on you as my revolutionary teacher and love you.[1]

When, however, after Bukharin's departure, an abbreviated form of the offending article appeared over a pen-name in the journal of the international youth movement,[2] Lenin made a sharp rejoinder. Bukharin, by denouncing the state in the abstract, and by ignoring the importance of the state as an instrument to be used for the overthrow of the bourgeoisie, had failed to distinguish between the Marxist and anarchist views of the state, and had fallen into a position bordering on anarchism.[3]

Already before the February revolution the differences between Lenin and Bukharin had narrowed. Lenin, brooding on the betrayal of the socialist cause by Kautsky and the German social-democrats, which he attributed to their worship of the national state, planned the essay which took shape some months later as *State and Revolution*, with its emphasis on the ultimate Marxist rejection of the state ; and this made him less unsympathetic to the anarchist leanings of Bukharin. In February 1917 he wrote to Kollontai, then with Bukharin in New York :

> I am preparing (have almost collected the material) an article on the question of the relation of Marxism to the state. I have come to conclusions that are even sharper against Kautsky than against Bukharin. . . . Bukharin is far better than Kautsky, but Bukharin's mistakes may spoil his " just cause " in the fight against Kautskyism.[4]

When Bukharin reached Petrograd in the early summer of 1917, Krupskaya's " first words " to him were : " V. I. asked me to tell you that in the question of the state he no longer has any

[1] *Bol'shevik*, No. 22, November 30, 1932, p. 88.
[2] *Jugend-Internationale*, No. 6, December 1, 1916, pp. 7-9 ; for this journal see *The Bolshevik Revolution, 1917-1923*, Vol. 3, p. 401.
[3] Lenin, *Sochineniya*, xix, 295-296.                    [4] *Ibid.* xxix, 291.

disagreements with you ".[1] Bukharin possessed, however, none of Lenin's suppleness of manœuvre in face of changing situations. At the sixth party congress in August 1917, at which (the principal leaders being in hiding) he was one of the main spokesmen, he foreshadowed " a holy war in the name of proletarian interests ", and declared that the only way out of the imperialist war was " an international proletarian revolution, however many victims it may cost us ".[2] In the long controversy in the party central committee during the Brest-Litovsk negotiations he was a fervent advocate of " revolutionary war ", being implacably opposed both to Lenin's advocacy of a surrender to the Germans and to Lenin's willingness to accept aid from the capitalist Powers of the west.[3] Bukharin's view of the state again came under fire from Lenin in the spring of 1918 when he rashly proposed to include in the party programme some description of " the developed socialist order in which there is no state " ; [4] and when, about the same time, he published an enthusiastic review of Lenin's State and Revolution, Lenin accused him of dwelling on all those passages which attacked the state and were no longer topical, and ignoring the passages which spoke of the need to create " the state of the commune " for the transition period.[5] Bukharin was the most influential figure in the group of " Left communists " who, in the spring of 1918, conducted a campaign against such concessions to bourgeois principle and practice as the formation of industrial trusts with the support of private capital, the employment of specialists and the establishment of one-man management in industry ; and he contested Lenin's conception of " state capitalism ", which he regarded as incompatible with the dictatorship of the proletariat.[6] He also rejected, once more in company

---

[1] *Revolyutsiya Prava : Sbornik*, i (1925), 5 ; the authority is Bukharin, but there is no reason to doubt the statement.

[2] *Shestoi S"ezd RSDRP(B)* (1934), p. 101.

[3] After Lenin's death Bukharin recalled that both these proposals " troubled our international conscience to the bottom of our heart " (N. Bukharin, *Ataka* (1924), p. 260).

[4] See *The Bolshevik Revolution, 1917–1923*, Vol. I, p. 246.

[5] Lenin, *Sochineniya*, xxii, 488 ; Bukharin's review appeared in *Kommunist*, No. 1, April 20, 1918, p. 19.

[6] *Ibid.* No. 3, May 16, 1918, pp. 8-11. For the controversy about " state capitalism " see *The Bolshevik Revolution, 1917–1923*, Vol. 2, pp. 88-95 ; its recrudescence in 1925 will be discussed in Part III in the following volume.

with Pyatakov, Lenin's compromise with the bourgeois principle of national self-determination, and proposed to substitute the slogan of " self-determination for the workers ".[1]

Bukharin was one of those who enthusiastically welcomed the policies of war communism not merely as emergency measures dictated by the needs of the civil war, but as milestones on the road from capitalism to socialism. This view was reflected in his major theoretical work of these years, *The Economics of the Transition Period*; and this, together with his popular text-books *The Programme of the Communists* and *The ABC of Communism* (the latter written jointly with Preobrazhensky and translated into many languages), gave him a lasting reputation as the leading party theorist. All these works were marked by a strong streak of utopian optimism. But before the end of the civil war period the optimism had begun to fade. The essence of war communism was the extraction of grain surpluses from the peasant by methods other than those of monetary inducement. By harvest time in 1920 it was clear that the only such method available was crude coercion, and that this method worked imperfectly. An article by Bukharin in *Pravda* of October 1, 1920, showed him for the first time vividly conscious of the magnitude and complexity of the peasant problem. On the other hand, belief in the impending achievement of a fully socialist society with the elimination of monetary incentives led him to accept the idea of a compulsory state labour service ; this fitted into the theory which he had propounded in *The Economics of the Transition Period* of " the self-organization of the working class ".[2] In the trade union controversy of the winter of 1920–1921 he found himself in alliance with Trotsky, whose influence seems to have been strong over him at this time.[3] He also came out at this time in favour of strict party discipline in opposition to the Democratic Centralist group (though many of its members were former Left communists),

---

[1] See *The Bolshevik Revolution, 1917-1923*, Vol. 1, p. 267.

[2] N. Bukharin, *Ekonomika Perekhodnogo Perioda* (1920), p. 151 ; for the application of this to labour service see *The Bolshevik Revolution, 1917–1923*, Vol. 2, p. 216.

[3] Trotsky spoke of " Bukharin's growing devotion to me " which began in New York in 1917 and continued to grow till 1923, when it " turned into its opposite " (L. Trotsky, *Moya Zhizn'* (Berlin, 1930), i, 311) ; in 1922 " Bukharin was devoted to me with a purely Bukharin-like, i.e. half-hysterical, half-childish, devotion " (*ibid.* ii, 207).

which advocated looser central control and a more " democratic " party organization.[1]

> The Moscow organization [he wrote] must be made healthy. . . . Organizationally it is necessary to remove from all groups the most factious elements, to send in new fresh forces of comrades not working in Moscow, and to set up a firm business-like Moscow committee, which would work and carry out the party line. It stands to reason that it is not at all necessary to exclude comrades even of the most extreme opposition, as certain hotheads wish. But it would be an excessive luxury for the party in the present difficult conditions to waste time and strength with arguments and disputes.[2]

Bukharin was not the only Bolshevik whose political views were in a state of disarray in the difficult period which followed the end of the civil war.

Bukharin, like the majority of the party, hailed the introduction of NEP as an escape from the impasse, both in policy and in political thinking, into which war communism appeared to have led. But he was now divided from most of his former associates of the Left, notably from Pyatakov and Preobrazhensky, who regarded NEP exclusively as a retreat and made no attempt to conceal their dislike of it. The transformation in Bukharin's attitude may be partly attributed to the influence of Lenin. " The whole succeeding period ", wrote Bukharin of the years after 1918, " is a period of the growing influence on me of Lenin, to whom, as to no one else, I am indebted for my Marxist educa-tion." [3] But Bukharin went characteristically further than his master. Having readjusted his ideas with his usual theoretical consistency, he found himself henceforth on the extreme Right of the party. An article written by him a few years later contained what was evidently intended as an apologia for this change of front in the post-NEP period :

> In the fire of this self-criticism the *illusions* of the period of childhood are consumed and vanish without a trace, real relations

---

[1] For this group see *The Bolshevik Revolution, 1917–1923*, Vol. 1, pp. 195-196.      [2] *Pravda*, November 6, 1920.

[3] *Entsiklopedicheskii Slovar' Russkogo Bibliograficheskogo Instituta Granat*, xli, i (n.d. [1927]), Prilozhenie, col. 56. Trotsky wrote long afterwards : " The naïve and ardent Bukharin venerated Lenin, loved him with the love of a child for its mother ; and, when he pertly opposed him in polemics, it was not otherwise than on his knees " (L. Trotsky, *Stalin* (N.Y., 1946), p. 380).

emerge in all their sober nakedness, and proletarian policy acquires in appearance sometimes a less emotional, but therefore more assured, character — one which clings closely to reality and therefore modifies this reality all the more faithfully.

From this point of view the transition to the new economic policy represented the collapse of our illusions.[1]

In the autumn of 1922 Bukharin joined Sokolnikov in advocating the abandonment of the monopoly of foreign trade, and incurred from Lenin the charge of " standing for the defence of the speculator, of the petty bourgeois, of the richest peasants, against the industrial proletariat ".[2] At the fourth congress of Comintern in November 1922 he abandoned another of the cherished convictions of his past, coming out as apologist of the national state and of the expediency of alliances between the Soviet Government and bourgeois Powers.[3] At the twelfth party congress of April 1923, out of loyalty to the sick Lenin, he ranged himself against the triumvirate on the Georgian question. But the same congress found him at the opposite end of the spectrum to Trotsky, defending the cause of the peasant and denouncing those who wished to press forward a policy of industrialization at his expense.[4] The scissors crisis reinforced his sympathies for the cause of the peasantry and confirmed him in a peasant orientation. At the beginning of the controversy on party democracy in the autumn of 1923 he had shown signs of being critical of the official line. But this marked the end of Bukharin's Leftist inclinations. In December 1923 he came out strongly and decisively against Trotsky.[5] From 1924 onwards he was the principal spokesman of the interests of the peasant, and especially of the well-to-do peasant who alone could be relied upon to produce the marketable grain stocks necessary to the development of the whole economy : the need to conciliate the peasant took precedence in his mind over the rapid development of industry. He adhered consistently to this view and remained the leader of the Right opposition till its defeat in 1929.

[1] *Bol'shevik*, No. 2, April 15, 1924, p. 1.
[2] See *The Bolshevik Revolution, 1917–1923*, Vol. 3, pp. 464-465.
[3] See *ibid.* Vol. 3, p. 447.
[4] *Dvenadtsatyi S"ezd Rossiiskoi Kommunisticheskoi Partii* (*Bol'shevikov*) (1923), pp. 173-174.
[5] For Bukharin's change of front at this time see *The Interregnum, 1923–1924*, p. 321.

A similar development occurred in Bukharin's opinions on world revolution. In July 1923, in company with Zinoviev, he was still eager to spur the German communists on the path of revolution.[1] But the shock of failure in Germany finally shattered the belief of Bukharin, as of so many others, in the imminence of European revolution and in the proletariat of the west. In his report on Comintern activities at the twelfth party congress in April 1923 he had dwelt extensively for the first time on the revolutionary potentialities of Asia.[2] The foundation of the Peasant International in the autumn of 1923 opened fresh vistas of faith and helped to kindle in Bukharin fresh hopes of the revolutionary potentialities of the peasant.[3] But this transfer of allegiance to new standard-bearers of revolution was accompanied by an important change in the time-table. From the time of the fifth congress of Comintern in May 1924 he became the principal theorist of the so-called " stabilization of capitalism ", admitting that " the picture is far more variegated than we used to see it ", and looking forward to a " transition period lasting perhaps for a considerable time ".[4] And when, in the next year, he appeared as the theoretical protagonist of " socialism in one country ", it was clear that for Bukharin, at any rate, this meant no longer revolution, but socialism by agreement with the peasantry — " a growing into socialism ". Even in this last phase of his intellectual development Bukharin remained in many ways, though on a different plane, faithful to the utopianism of his early revolutionary years. There was no element of ruthlessness in Bukharin's nature. He fervently believed in revolution by the spontaneous action of the masses — in " revolution from below ". When, after the introduction of NEP, he became disillusioned by the course of events, and perceived that further progress on the revolutionary path implied coercion and, above all, the coercion of the peasant — " revolution from above " — he instinctively shrank from the prospect, and was content to relegate revolution to a distant future rather than hasten it by such means. Bukharin was more

[1] See *The Interregnum, 1923–1924*, p. 186.
[2] See *The Bolshevik Revolution, 1917–1923*, Vol. 3, p. 231.
[3] For the Peasant International see *The Interregnum, 1923–1924*, pp. 197-199 ; for Bukharin's development of this theme see pp. 245-246 below.
[4] *Protokoll : Fünfter Kongress der Kommunistischen Internationale* (n.d.) ii, 520.

acutely conscious than any of the Bolshevik leaders of the cruel
dilemma. Incompatibility between ends and means.

Bukharin possessed most of the merits and defects of the
intellectual in politics. Lenin, while calling *The Economics of the
Transition Period* an " excellent book ", criticized it for lack of
factual foundation and concreteness, due to excessive philosophical
abstraction.[1] Lenin's testament described Bukharin as " the most
valuable and biggest theoretician of the party ", but qualified this
verdict by adding that " his theoretical views can only with the
very greatest doubt be regarded as fully Marxist ", that " there is
something scholastic in him ", and that " he has never learned,
and I think never has fully understood, the dialectic ". These
observations reveal the impatience of the working politician with
the unpractical rigidity of the intellectual ; and Bukharin's
opinions never appear to have carried serious weight in taking
decisions of policy. If a foreign visitor to Moscow in 1921 was
right in saying that Bukharin was " named in Russia as the
eventual successor of Lenin ",[2] this must have been the view of
outsiders who knew little of real relations in the party. On the
other hand, Bukharin's personal popularity was unrivalled : Lenin
in the testament justly described him as " the favourite of the
whole party ". At the fourteenth party congress, when his
extreme pro-*kulak* position was under general attack, Kamenev
and Orjonikidze both referred to him by the affectionate nickname
Bukharchik ; [3] and Stalin made one of his rare excursions into
rhetorical pathos when he declared that the opposition at the
fourteenth party congress in December 1925 " demand the blood
of comrade Bukharin " and that " we shall not give you that
blood ".[4]

The peculiar characteristic of Bukharin was a combination of
rigidity in ideas with malleability of temperament which made
him a ready tool in the hands of men less single-minded and
politically more astute. Once convinced by process of reasoning
of the rightness of a policy, he stuck to it with great tenacity and

[1] *Leninskii Sbornik*, xi (1929), 401-402.
[2] A. Morizet, *Chez Lénine et Trotski* (1922), p. 63.
[3] *XIV S"ezd Vsesoyuznoi Kommunisticheskoi Partii (B)* (1926), pp. 223, 269.
[4] *Ibid.* pp. 504-505 ; the passage is, for obvious reasons, omitted from the text of the speech in Stalin, *Sochineniya*, vii, 384.

without regard for its consequences to others or to himself. His honesty was transparent. Nobody could call Bukharin either opportunist or self-seeking. But these qualities, reinforced by a strong sense of personal loyalty, were at the service of anyone who could persuade him that the course of action proposed was consonant with his convictions. Lenin's verdict of December 1920 was indulgent, but decisive :

> Even big men, including Bukharin, have little weaknesses. If there is a catchword about with a twist in it, he cannot help falling for it.

And again three weeks later :

> We know how soft Bukharin is : it is one of the qualities we love him for and cannot help loving him for. We know that more than once he has been called in jest " soft wax ". It appears that any " unprincipled " person, any " demagogue ", can make an impression on this " soft wax ".[1]

Or, as Trotsky wrote in his autobiography, with more than a touch of bitterness :

> This man's nature is such that he must always lean on somebody, be dependent on somebody, attach himself to some-body. He becomes in these conditions nothing more than a medium through which somebody else speaks and acts.[2]

It was this pliability of temperament which enabled Bukharin to pass from one extreme of the party to the other after 1921, and eventually made him a ready tool in Stalin's hands. He appears at the outset to have had none of the repugnance for Stalin's methods felt, for example, by Kamenev, perhaps because the natures and interests of the two men were so utterly divergent that their paths did not cross or conflict. But Bukharin's weaknesses repeatedly led him into words and actions which are at first sight difficult to reconcile with the favourable verdict commonly passed on his character. That Stalin, who had no intellectual pretensions and was plainly indifferent to logic, should have felled his opponents with sophistical and dishonest arguments is less shocking than to find such arguments on the lips of Bukharin, who must have been

[1] Lenin, *Sochineniya*, xxvi, 68-69, 93.
[2] L. Trotsky, *Moya Zhizn'* (Berlin, 1930), i, 311.

well aware of their quality. He not only threw himself *con amore* into the campaign of the triumvirate against Trotsky, but in bitter controversy with his old associate Preobrazhensky replied to a measured and serious economic analysis in terms that were both evasive and crudely demagogic.[1] It was in the same spirit that he allowed himself to become in 1925 Stalin's mouthpiece and chief intellectual adjutant for the destruction of Zinoviev and Kamenev, and later of the united opposition.

But for the fact that these events were the prelude to Bukharin's own ruin, and made him the author of his own fearful punishment, it would be impossible to acquit him of an important share in any condemnation that may fall on Stalin's treatment of the opposition ; for he was a self-proclaimed accomplice in everything that Stalin did at this time. There are indeed some indications that he was not free, even at the moment of apparent triumph, of pangs of conscience and apprehensions about the future. When, at the fourteenth party congress in December 1925, Kamenev reproached Bukharin with turning against Zinoviev the weapons of distortion which he had previously hesitated to employ even against Trotsky, Trotsky broke the contemptuous silence with which he had followed the proceedings to mutter audibly : " He has acquired the taste ". After the congress Bukharin wrote a reproachful letter to Trotsky which contained the revealing phrase : " From this taste I tremble from head to foot ".[2] In July 1928, while still publicly supporting Stalin, he called Stalin in private conversation with Kamenev a " Genghis Khan ", and expressed the well-founded apprehension that he would destroy them all.[3] Bukharin was not one who sinned either unconsciously or without fear of retribution. He is one of the tragic figures of the revolution. His tragedy was not, however, a tragedy of greatness, but of a weak, amiable and keen-witted man caught up in the turmoil of events too vast for his moral stature.

[1] For this controversy see pp. 207-208 below. In 1921 Lenin had written : " There are people with such happy natures (Bukharin, for example) that even in the midst of the fiercest battles they cannot put venom into their attacks " (*Sochineniya*, xxvi, 121) ; this was no longer true of Bukharin in the controversies of the middle nineteen-twenties.

[2] Bukharin's letter is not extant, but the phrase is quoted in Trotsky's reply of January 9, 1926, of which a copy is preserved in the Trotsky archives, T 2926.

[3] A record of this conversation is in the Trotsky archives, T 1897.

## (e) Stalin

Iosif Vissarionovich Stalin (original name Djugashvili) was of humbler origin than any of the other Bolshevik leaders. He was born in the small Georgian town of Gori in 1879 of Georgian parents who had been born in serfdom ; he was the only one of their four children to survive infancy. His father worked as a cobbler in Gori, and was later employed in a shoe factory in Tiflis. Vissarion Djugashvili was addicted to alcohol, and died during his son's childhood. His widow was evidently a woman of some character. She is said to have maintained her boy by working as a washerwoman, and, though herself illiterate, secured his admission to the church school in Gori. Young Iosif (Soso was the Georgian form of the name) proved a brilliant enough pupil to be admitted, at the age of fifteen, to the theological seminary at Tiflis. The seminary had apparently had in the past a reputation for breeding subversive opinions. It was here that the future Stalin read his first forbidden books (including, perhaps, some Marxist literature), and graduated in the arts of dissimulation and intrigue.[1] What else he learned in the seminary is a matter for speculation. Some critics have attributed to its influence the taste for a flat formality of style and casuistry in argument which marked his later speeches and writings.

The story of Stalin's early career has been so overlaid with legend, adulatory and hostile, that no exact account of it will in all probability ever be recovered. Even the circumstances in which in his twentieth year he left, or was expelled from, the seminary are differently narrated. He became a Marxist, and a member of the embryonic and still undivided Russian Social-Democratic Party. He worked for a short time as a clerk in the Tiflis observatory. But he quickly joined the select, though now rapidly increasing, body of professional revolutionaries, dedicated entirely to the cause and dependent on the precarious and mysterious resources of the movement. His first article appeared in 1901

[1] In an interview with the German writer Emil Ludwig in 1934 Stalin referred to " the humiliating régime and jesuitical methods prevalent in the seminary ", and, when asked whether he found nothing good in the Jesuits, replied : " Yes, they are methodical and persevering in their work. But the basis of all their methods is spying, prying, peering into people's souls, to subject them to petty torment " (Stalin, *Sochineniya*, xiii, 114).

in a Georgian flysheet published illegally and intermittently in Baku. In the same year he moved from Tiflis to Batum ; and here, in April 1902, he suffered the first of several experiences of arrest, imprisonment and exile to Siberia. During the next ten years arrests and escapes alternated with spells of intense party activity. As a delegate of the Bolshevik organization in the Caucasus he attended the Bolshevik conference of December 1905 in Tammerfors, where he first met Lenin. He was present at the party congresses in Stockholm and London in 1906 and 1907. But his only sojourn of any length outside Russia was in the winter of 1912–1913, when he spent some weeks first in Cracow with Lenin, whose favourable notice he attracted by a painstaking essay on the national question, and then with the group of Bolsheviks in Vienna.

Stalin's solid but unspectacular talents and services to the party did not win immediate recognition. His rise in the party hierarchy began in 1912, when he was co-opted, presumably at Lenin's instigation, into the party central committee and sent to Petersburg to organize the publication of the new party newspaper *Pravda*. In the following year he was once more arrested and deported to Siberia ; and this time his exile continued till he was liberated, together with Kamenev and many other political exiles, by the outbreak of the February revolution of 1917. For a few weeks after his return to Petrograd he joined Kamenev, and a majority of Bolsheviks then in the capital, in a policy of qualified support for the Provisional Government. In the middle of April 1917 he rallied to Lenin's " April theses ", and through all the crises and controversies of the next few years remained a faithful and unswerving follower of Lenin. As People's Commissar for Affairs of Nationalities Stalin was still in the second rank of the leaders. In the civil war Lenin undoubtedly valued and used his devotion and his great organizing capacity ; and on more than one occasion he proved an active and effective check on Trotsky's policies. But his name was still scarcely known to the rank and file of the party, and not at all outside it. His appointment as a secretary-general of the party central committee in 1922 was a tribute to his reputation for practical efficiency among his colleagues in the party leadership, not to his popularity in the party in general. The post was not thought of as carrying political

significance or weight in public affairs.  That it served as the
perfect springboard for Stalin's rise to supreme power is evidence
of the peculiar and exceptional quality of his political genius.

The characteristic of Stalin which, in the light of later developments, most struck contemporary observers was his mediocrity,
his complete lack of distinction.  Sukhanov's verdict, which
referred to Stalin's activities in 1917 and was first published in
1922, is famous :

> The Bolshevik party, in spite of the low level of its " officer
> corps ", which in general was ignorant and collected by chance,
> disposed of a large number of powerful personalities and able
> leaders among its " general staff ".  Stalin, however, during his
> modest activity in the executive committee produced — and
> not on me alone — the impression of a grey blur, floating
> dimly across the scene and leaving no trace.  There is really
> nothing more to be said about him.[1]

In 1923 Lunacharsky's volume of popular sketches of Bolshevik
leaders omitted Stalin altogether.  Kamenev thought him " just
a small-town politician ".[2]  In 1929 Trotsky described him as
" the outstanding mediocrity of our party ".[3]  It is indeed plausible
to believe that Stalin's air of mediocrity was one of the factors
which contributed to his success.  The party feared a Bonaparte ;
and of all the leaders Stalin seemed the least likely — as Trotsky
seemed the most likely — to aspire to such a rôle.  In the years of
his slow rise to power, Stalin excited few jealousies.  He was
readily promoted because his promotion threatened nobody.  He
survived even Lenin's recommendation to oust him from his post
as secretary-general because nobody else felt so drastic a step to
be necessary.  Trotsky, when he began openly to denounce
Zinoviev and Kamenev in the autumn of 1924, left Stalin alone —
not because he had any desire to spare Stalin, but because it was
not worth while to expend his shafts on a secondary target.

But this immunity from attack, purchased by an apparent lack
of outstanding qualities, clearly does not by itself explain Stalin's
career.  More than almost any other great man in history, Stalin
illustrates the thesis that circumstances make the man, not the

[1] N. Sukhanov, *Zapiski o Revolyutsii*, ii (Berlin, 1922), 265-266.
[2] L. Trotsky, *Stalin* (N.Y., 1946), p. 393.
[3] L. Trotsky, *Chto i Kak Proizoshlo* (Paris, 1929), p. 25.

man the circumstances.   Stalin is the most impersonal of great
historical figures.   In the party struggles of the nineteen-twenties
he appears not to mould events, but to mould himself to them.
It is as difficult to define his opinions as to describe his personality.
Lack of definition, rather than the shiftiness of which he was
often accused, was the distinguishing feature of his position.   The
claim to be nothing more than a faithful follower and disciple of
Lenin was not altogether a pose.   He had no creed of his own.
He was content to be the favourite son of the revolution and the
man of the moment.   But this only makes his peculiar personal
qualities the more significant.   For the qualities which raised him
to greatness were precisely the qualities which mirrored the current
stage of the historical process.   They were the qualities, not only
of the man, but of the period.   " Every period has its great men ",
quoted Trotsky from Helvetius, " and if there are none it invents
them." [1]

Two characteristic features of Stalin's outlook, both of which
reflected his personal background and upbringing, were also conspicuous landmarks in the history of the revolution in the middle
nineteen-twenties.   The first was a reaction against the predominantly " European " framework in which the revolution had
hitherto been cast, and a conscious or unconscious reversion to
Russian national traditions.   The second was a turning away from
the highly developed intellectual and theoretical approach of the
first years of the revolution, and a renewed emphasis on the
practical and empirical tasks of administration.   This new attitude
had set in after the introduction of NEP, and was well established
at the time of Lenin's death.   It was altogether appropriate that
the major political figure of the ensuing period should have been
a man with few claims as a thinker, but an outstanding organizer
and administrator.

The absence of any significant western influence in the formation of Stalin's mind and character distinguished him sharply
from the other early Bolshevik leaders.   Alone among them he

---

[1] *Ibid.* p. 26.   Later Trotsky offered a more restrictive interpretation :
" Stalin took possession of power, not with the aid of personal qualities, but with
the aid of an impersonal machine.   And it was not he who created the machine,
but the machine that created him " (L. Trotsky, *Stalin* (N.Y., 1946), p. xv).
But it required something more than a machine to " create " Stalin and put him
in power.

had never lived in western Europe, and neither read nor spoke any western language. This peculiarity coloured his personal relations as well as his political opinions. He never seems to have felt entirely at ease with colleagues steeped in a European tradition and outlook : he particularly detested Chicherin and, according to Trotsky,[1] Rakovsky — both of them outstanding representatives of western culture. Those who stood closest to Stalin in later years — Molotov, Kirov, Kaganovich, Voroshilov, Kuibyshev — were as innocent as himself of any western background. Symptoms of a reaction against current assumptions of European pre-eminence might have been detected in Stalin even before the October revolution. When in August 1917 he observed at the sixth party congress in Petrograd that " it would be unworthy pedantry to ask that Russia should ' wait ' with her socialist transformation till Europe ' begins ' ", Stalin was merely re-formulating an idea first propounded by Trotsky and endorsed by Lenin. But, when he went on to speculate on the possibility that " Russia may be the country which points the way to socialism ", a new note of national fervour, unfamiliar at this time in Bolshevik doctrine, was added to the socialist creed.[2] Stalin remained a national rather than an international socialist. In the days when Comintern seemed a living organism, and engaged the constant and anxious attention of Lenin, Trotsky and Zinoviev, he remained apparently indifferent to it. He turned to it only in 1924 when it had ceased to be a potential instrument of world revolution, and had become a bureaucratic machine capable of impeding or furthering Soviet policy or his own political designs. Stalin's scepticism of the imminence of a German revolution, when this was assumed as a matter of course by almost every other leading Bolshevik, was an early example of his prescience.[3] By 1925, when he began to preach " socialism in one country ", his references to world revolution took on a casual and insouciant air which showed how little his heart was in it.

[1] Note on Rakovsky preserved in the Trotsky archives, where Rakovsky as " a genuine European " is contrasted with Stalin who " most fully represents the Petrine, most primitive, tendency in Bolshevism ".

[2] For these quotations see *The Bolshevik Revolution. 1917–1923*, Vol. 1, p. 92.

[3] See *The Interregnum, 1923–1924*, p. 187.

When international revolution will break out [he remarked early in that year], it is hard to say ; but, when it does break out, it will be a decisive factor.

Or again, a few days later :

The leading proletariat, the proletariat of the west, is the greatest strength and the most faithful, most important ally of our revolution and of our power.  But unfortunately the situation is such, and the condition of the revolutionary movement in the advanced capitalist countries such, that the proletariat of the west is not now in a position to render us direct and decisive help.[1]

Through all the apparent zigzags of Stalin's economic policy between 1923 and 1928, a single straight line was unwaveringly followed — the determination to make the Soviet Union powerful, and to make it self-sufficient and independent of the west. An unmistakable note of sincerity, often absent from his polemical utterances, was sounded in his denunciation of Sokolnikov for wanting the " Dawesification " of the Soviet Union, and in his own determination to make it " a country which can by its own efforts produce the equipment it requires ".[2]  Stalin could readily adapt his Marxism to a situation in which Marx's predictions of proletarian revolution in advanced capitalist countries had gone radically astray.  Unlike Lenin and Trotsky, or even Zinoviev and Bukharin, Stalin cared nothing for what happened in western Europe except in so far as it affected the destinies of his own country.  In pursuit of his aims he would imitate the west, borrow from the west, bargain with the west.  But everything was weighed in the scales of national policy.

It is, moreover, remarkable that Stalin's outlook, in spite of his Georgian origin, should have been not merely non-western, but distinctively Russian in the narrower sense.  It may be, as has often been suggested, that his character displayed some hidden

---

[1] Stalin, *Sochineniya*, vii, 21, 26.  *Byulleten' Oppozitsii* (Paris), No. 19, March 1931, p. 15, collected some remarks on this theme alleged to have been made by Stalin during the nineteen-twenties : Comintern, he said, " represents nothing and exists only thanks to our support " ; of the KPD : " They are all tarred with the same brush ; there are no revolutionaries among them any more " ; to someone who predicted world revolution within 40 or 50 years : " Revolution ?  Perhaps Comintern will make it ?  Look : it will make no revolution in 90 years ".          [2] Stalin, *Sochineniya*, vii, 355.

traits of a primitive Georgian tradition. It is more plausible to associate the frequent brutality and ruthlessness of his behaviour with the grinding poverty and harshness of his earliest environment. At a more conscious level, he seems to have reacted strongly against the predominantly Menshevik strain in Georgian social-democracy.[1] Politically, nothing that was Georgian seemed good to him. He was one of the engineers of the forced subjection of Georgia to Bolshevism in 1921, and throughout his career was notoriously opposed to all manifestations of Georgian nationalism. He was the most " Russian " of the early leaders not only in his rejection of the west, but in his low rating of the local nationalisms of the former Russian Empire. He became the protagonist not only of " socialism in one country ", but of a socialism built on a predominantly Russian foundation.

The reaction in Stalin's outlook against the intellectual and the theoretical was no less decisive than his reaction against the west, and was not unconnected with it. The tradition of the Russian intelligentsia was closely bound up with western Europe ; the familiar charge against it was that it drew its nourishment from foreign sources, and was divorced from the spirit of the Russian people or nation. All the original Bolshevik leaders, except Stalin, were in a sense the heirs or products of the Russian intelligentsia, and took for granted the premisses of nineteenth-century western rationalism. Stalin alone was reared in an educational tradition which was not only indifferent to western ways of life and thought, but consciously rejected them. The Marxism of the older Bolsheviks included an unconscious assimilation of the western cultural foundations on which Marxism had first arisen. The fundamental assumptions of the enlightenment were never questioned ; a basis of rational argument was always presupposed. Stalin's Marxism was imposed on a background totally alien to it, and acquired the character of a formalistic creed rather than of an

---

[1] The statement quoted in *Zarya Vostoka*, the Tiflis party journal, of December 23, 1925 (an extract from which is in the Trotsky archives), from a Tsarist police report, that Stalin had been active in the social-democratic party since 1902 " first as a Menshevik, then as Bolshevik ", has no great significance even if it is true. The split occurred only in 1903 and took some time to penetrate local groups ; Zhordania, the future Menshevik leader, was for some time the recognized leader of the whole party. It is certain that, from the moment when Stalin became conscious of the fact and implications of the split, he was whole-heartedly a Bolshevik.

intellectual conviction. The former seminarist was predisposed to regard faith as a more important virtue than reason.

Stalin's indifference or distrust for fine-drawn intellectual argument was displayed at an early stage of his party career. In 1911, in a letter to a Caucasian comrade, he called Lenin's famous dispute with Bogdanov on the philosophical premisses of Marxism " a storm in a tea-cup ".[1] Stalin never allowed doctrine to stand in the way of the demands of common sense. He was among the first of the Bolsheviks, at the fourth party congress in 1906, to support the distribution of land to the peasants. At the sixth congress in July 1917 he supported the thesis that " Russia may be the country which points the way to socialism " with a phrase which was so often repeated that it became a *cliché* :

There is a dogmatic Marxism and a creative Marxism : I take my stand on the latter.[2]

In the same spirit many years later, defending the policy of " socialism in one country " against an awkward quotation from Engels, he exclaimed that, if Engels were alive to see the present situation, he would only say : " Devil take the old formulae ! Long live the victorious revolution of the USSR ! "[3] In the long-standing debate between the determinist or " scientific " and voluntarist or " political " aspects of Marxism there was no doubt on which side Stalin would come down. In a curious unpublished draft essay of 1921 he distinguished the objective and subjective sides of " the proletarian movement ", identifying the former with the theory, and the latter with the programme, of Marxism, and added that " the sphere of action of strategy and tactics undoubtedly borders on the subjective side of the movement ".[4] " A stubborn empiricist, devoid of creative imagination ", was Trotsky's summing up.[5] From time to time, by way of vindicating his claim to leadership of the party, Stalin found it necessary to appear in the rôle of a theorist. But it was never in doubt that, in Stalin's conception of politics, doctrine was subsidiary to strategy and tactics.

[1] The letter was published in *Zarya Vostoka* (T flis), December 23, 1925 (see previous note).     [2] Stalin, *Sochineniya*, iii, 187.
[3] *Ibid.* vii, 303.     [4] *Ibid.* v, 62-63.
    L. Trotsky, *Chto i Kak Proizoshlo* (Paris, 1929), p. 25.

Distrust of intellectual processes seems to be reflected in Stalin's dislike of democratic procedures. " Power is exercised ", he remarked contemptuously in 1918, " not by those who elect and vote, but by those who govern." [1]    Railway transport in the civil war had been disorganized by " a multitude of collegiums and revolutionary committees ".[2]    At the thirteenth party conference in January 1924 he denounced those " intellectuals " who regarded the right to form fractions as a condition of democracy :

> The mass of the party understands democracy as the creation of conditions which guarantee the active participation of members of the party in the work of leading our country.  A few intellectuals of the opposition understand it as giving them the possibility of forming a fraction.[3]

And a few months later he contrasted " a formally democratic party " with " a proletarian party united by indissoluble bonds with the masses of the working class ".[4]    If, in the Politburo and in other bodies where policy was debated, Stalin had the reputation of being a man of few words, and was slow to commit himself to an opinion whether in speech or in writing,[5] his abstention was perhaps prompted not so much by a deliberate and calculated holding back as by a lack both of taste and of aptitude for such forms of expression.  What passed for cunning was, at any rate in early days, the product of diffidence.  The rise of Stalin was marked by an eclipse of democratic procedures in the party.  Decision by discussion, and if necessary by vote, in the central committee or in the Politburo was replaced by disciplined unanimity organized through the power of the secretariat.  Stalin never had any of that intellectual pleasure in argument which was so marked in Lenin, Trotsky and Bukharin.  Nothing that he said or wrote, at any rate after 1917, was divorced from some immediate political purpose.  Trotsky wrote of Stalin's " contemptuous attitude towards ideas ".[6]  Probably apocryphal utterances later attributed to him, such as " One Soviet tractor is

[1] Stalin, Sochineniya, iv, 37.              [2] Ibid. iv, 116-171.
[3] Ibid. vi, 40.                                        [4] Ibid. vi, 226.
[5] B. Bazhanov, Stalin (German transl. from French, 1931), pp. 17, 21.
[6] L. Trotsky, Stalin (N.Y., 1946), p. xv.

worth ten foreign communists " or " How many divisions has the Pope ? ", were framed to illustrate the low rating of ideological factors in Stalin's picture of the world.

It may well be that this anti-theoretical bias in Stalin affected his personal relations even more than his political opinions. In the first years after 1917 none of the Bolshevik leaders except Lenin appears to have treated Stalin as an important figure. Lenin recognized his outstanding gifts as an administrator and organizer ; the others saw only his commonplace and second-rate theoretical equipment. Yet it was a mistake to deduce from this intellectual shortcoming that Stalin had no gift for handling people. When he received a delegation of peasants in March 1925, he seems, from what looks like an authentic contemporary record, to have been remarkably successful in establishing easy relations with them. He " listened attentively like a *muzhik* and puffed at his pipe ", commented on practical points and exchanged artless jokes, so that " all were astonished at this simple, comradely attitude of comrade Stalin towards us, comparing it with the roughness and bureaucratic attitude of local party officials to the peasantry ".[1] In his dealings with colleagues, this ease of intercourse vanished altogether.[2] It was to them that Stalin exhibited the " rudeness " and lack of " loyalty " of which Lenin complained in the testament. Stalin smarted under their covert assumption of superiority, and met it with a constant sly depreciation of the party intellectuals. When attacking Trotsky, he recalled that Lenin at the second congress of the party had resisted Martov's demand to open the party to " non-proletarian elements " — an odd distortion of the famous dispute about the party statute — and quoted Lenin's rare criticism of the predominance of intellectuals in the party at the third party congress of 1905.[3] One of the frankest expressions of

[1] The interview which took place on March 14, 1925, was reported in *Bednota* (the peasant newspaper), April 5, 1925, by one of the participants ; though it shows Stalin in an unusually agreeable light, it was never utilized by any of Stalin's biographers, presumably because it contained an incautious remark about tenure of land which Stalin was afterwards obliged to disown (see pp. 247-248 below).

[2] One of Demyan Bedny's doggerel poems, intended as a friendly caricature, recounted an interview with Stalin at which he made all the correct remarks, while Stalin stroked his moustaches without uttering a single word, till he rose to end the interview with a hearty " Come again — it's pleasant to have a chat " (*Molodaya Gvardiya*, No. 9, September 1925, pp. 205-206).

[3] For this see *The Interregnum, 1923-1924*, p. 353.

Stalin's feelings appeared in a letter written to the German Communist Party leader Maslow in 1925 :

> We in Russia have also had a dying away of a number of old leaders from among the *littérateurs* and old " chiefs ". . . . This is a necessary process for a renewal of the leading cadres of a living and developing party.

And he named Lunacharsky, Bogdanov, Pokrovsky and Krasin among " former Bolshevik leaders who have passed over to a secondary rôle ".[1] Those whom he gathered around himself in later years were for the most part good party men whose theoretical pretensions were as few as his own. One of many interpretations of the great purges of the nineteen-thirties was that they were Stalin's final vengeance on the intellectuals who had despised him. He was particularly ruthless in forcing the intellectual life of the country into a narrow political strait-jacket. " We, Bolshevik practitioners," he was to say in the preface to the collected edition of his works in 1946.[2]

It has often been suggested that Stalin's background and education are reflected in his literary style. Lenin wrote and spoke plainly and easily with the air of one too completely preoccupied with what he is saying to pay much attention to the way in which it is said. Trotsky displayed the slightly mannered brilliance of an artist in words. Bukharin took evident pleasure, which communicated itself to the reader or hearer, in the lucidity and ingenuity of his argument. Neither the spoken nor the written word seemed to come easily to Stalin. His style had the workmanlike virtues of clarity and precision ; its vice was a total lack of imagination or of grace. When he wished to impress, he resorted to the schematic devices of enumeration, repetition and the rhetorical question, in which some critics detected liturgical echoes. But the form remained stiff, the content intellectually and emotionally trivial. Some of Stalin's earlier speeches made a favourable impression of moderation and caution. The applause that greeted his later denunciations of his enemies to packed audiences was no test. Stalin's victories were not won in the

---

[1] Stalin, *Sochineniya*, vii, 43 ; this version of the letter omits the name of the addressee and a few unimportant phrases preserved in the German version originally published in *Die Aktion*, xvi, No. 9, September 1925, pp. 214-217.
[2] Stalin, *Sochineniya*, i, p. xiii.

debating-chamber, and there is little evidence that he desired to shine there. The period of revolutionary oratory had passed with the day of the intellectuals.

If, however, Stalin, in his reaction against western influence and in his reaction against a theoretical approach to politics, was the product of his period, the dramatic element in Stalin's career and personality resides in the fact that it was he, above all, who carried forward the revolution to its appointed conclusion by bringing about the rapid industrialization of the country. By the irony of history it was Stalin, and not Trotsky, who became the effective champion of forced industrialization and comprehensive planning, and was prepared to sacrifice the peasant to this overriding purpose. It would be fanciful to ascribe this turn of events to any personal conviction or prejudice on Stalin's part ; nor is it necessary to convict him of hypocrisy when he attacked Trotsky for advocating measures less draconian than those which he himself would one day adopt. Nothing could better reveal the essentially impersonal character of Stalinist policy. If Stalin's methods often seemed to reflect characteristics derived from his personal background and upbringing, the aims which he pursued were dictated by the dynamic force inherent in the revolution itself. What Stalin brought to Soviet policy was not originality in conception, but vigour and ruthlessness in execution. When he rose to power in the middle nineteen-twenties, he became, and was determined to remain, the great executor of revolutionary policy. But the course of events makes it clear that he had at that time no vision of where that policy would lead.

Stalin's rôle in history thus remains paradoxical and in some sense contradictory. He carried out, in face of every obstacle and opposition, the industrialization of his country through intensive planning, and thus not only paid tribute to the validity of Marxist theory, but ranged the Soviet Union as an equal partner among the Great Powers of the western world. In virtue of this achievement he takes his undisputed place both as one of the great executors of the Marxist testament and one of the great westernizers in Russian history. Yet this *tour de force* had, when studied and analysed, a supremely paradoxical character. Stalin laid the foundations of the proletarian revolution on the grave of Russian capitalism, but through a deviation from Marxist premisses so sharp as to amount almost to a rejection of them.  He westernized Russia, but through a revolt, partly conscious, partly unconscious, against western influence and authority and a reversion to familiar national attitudes and traditions.  The goal to be attained and the methods adopted or proposed to attain it often seemed in flagrant contradiction — a contradiction which in turn reflected the uphill struggle to bring a socialist revolution to fruition in a backward environment.  Stalin's ambiguous record was an expression of this dilemma.  He was an emancipator and a tyrant ; a man devoted to a cause, yet a personal dictator ;  and he consistently displayed a ruthless vigour which issued, on the one hand, in extreme boldness and determination and, on the other, in extreme brutality and indifference to human suffering.  The key to these ambiguities cannot be found in the man himself.  The initial verdict of those who failed to find in Stalin any notable distinguishing marks had some justification.  Few great men have been so conspicuously as Stalin the product of the time and place in which they lived.

CHAPTER 5

# AGRICULTURE

## (a) The Harvest of 1924

THE thirteenth party congress in May 1924 had stressed the maintenance of Lenin's " link " between the proletariat and the peasantry, and commended the policy of generous concessions to the peasant. It had exhibited some uneasiness at the growing " differentiation " between different categories of peasant, and some divisions of opinion about the precise attitude to be adopted towards the *kulak*. But this issue had not seemed particularly urgent or important,[1] and cast no shadow on the prevailing mood of optimism. The first prognostications for the coming harvest were favourable, and it seems to have been taken for granted that the successful experience of 1922 and 1923 would be not only repeated, but surpassed. The sown area had been further increased, and reached more than 80 per cent of the pre-war figure. The smallest increase was in the area under rye, the largest in the area under cash crops, especially wheat, cotton, flax and sugar : this was evidence of growing prosperity and, in particular, of a growth in the number of well-to-do peasants who were not dependent on subsistence farming and could afford to grow for the market.[2] It was estimated that anything from 250 million to 400 million puds of grain, as against 200 millions in 1923, should be available for export.[3]

[1] See *The Interregnum, 1923–1924*, pp. 146-149.

[2] For detailed comparative figures see *Kontrol'nye Tsifry Narodnogo Khozyaistva na 1926–1927 god* (1926), p. 337 ; slightly different figures for 1924 are given in *Itogi Desyatiletiya Sovetskoi Vlasti v Tsifrakh, 1917–1927* (n.d.), pp. 168-171. According to a contemporary statement by Rykov, the sown area in 1924 exceeded that of 1913 in the consuming provinces of the RSFSR, but fell below it in the provinces which had suffered most from the famine of 1921, reaching an average of 88 per cent (*Shestoi S"ezd Professional'nykh Soyuzov SSSR* (1925), pp. 235-237).

[3] For the two estimates see *Sotsialisticheskoe Khozyaistvo*, No. 3, 1924, pp. 34-37 ; L. Kamenev, *Stat'i i Rechi*, x (1927), 274. The second estimate,

189

Early in June this optimistic view had to be abruptly revised. A serious drought threatened the harvest with ruin throughout the Volga basin and south-eastern Russia. Memories of the calamity of 1921 were still fresh ; and, within two or three weeks of his confident pronouncements at the party congress, Rykov, the president of Sovnarkom, was raising the alarm. The situation was now " so acute that it is essential for the party and the central committee to concern itself seriously with an examination of the peasant question ".[1] The harvest was likely to fail over five and a half million desyatins out of 77 million desyatins under cultivation ; and this would affect six million people, though Rykov denied, in an interview with anxious correspondents, that any comparison could be drawn with the disaster of 1921.[2] At the beginning of July 1924 Sovnarkom set up an emergency commission " for combating the consequences of the deficient harvest ".[3] On August 20, 1924, Rykov made an official report on the harvest to the party central committee. The total grain harvest was estimated at 2640 million puds, as against the 3000 millions which had been the expected yield of a good harvest : together with reserves in hand, this brought the total available stocks to 2800 million puds. Grain exports were now suspended till further orders.[4] In an economy where the margin between survival and catastrophe was so narrow and so precarious even this partial failure raised serious problems. A sum of 20

contained in a speech of June 9, 1924, was admittedly a desideratum rather than a prognostication, but marked the current mood. Kamenev went on to point the moral : " But who will give us these 400 millions. The poor peasant ? No ! We are bound to admit that the 400 million puds of grain, which we have to send abroad, will be produced by the middle peasant and, in part, by *kulak* elements."

[1] A. I. Rykov, *Sochineniya*, iii (1929), 120 ; the date of this speech was June 12, 1924.       [2] *Ibid.* iii, 169-175.
[3] *Sobranie Zakonov, 1924*, No. 1, art. 4 ; according to a pencil note of Krasin in the Trotsky archives, dated June 25, 1924, T 815, the decision of Sovnarkom was taken on the previous day. As late as July 3, *Leningradskaya Pravda* was still looking forward to extensive grain exports.
[4] A. I. Rykov, *Sochineniya*, iii (1929), 185-187 ; Rykov's speech was widely publicized in the press : an extract appeared in *Internationale Presse-Korrespondenz*, No. 115, September 2, 1924, pp. 1491-1493. *Izvestiya*, September 3, 1924, described a journey of Rykov by steamer down the Volga to inspect harvest conditions ; on August 30 he visited the Volga German autonomous SSR (*ibid.* September 9, 1924) ; Yagoda, the deputy chief of the OGPU, was one of those who accompanied him on this tour (W. Reswick, *I Dreamt Revolution* (Chicago, 1952), p. 84).

million rubles, later raised to 30 millions, was to be distributed in the form of agricultural credits to those who had suffered from failure of crops.[1] Special reliefs were granted from the incidence of the agricultural tax, which was now to yield only 340 million rubles for the year 1924–1925 in place of the estimated 400 millions.[2] The final results were less disastrous than had been feared.[3] Few, perhaps, actually starved after the partial crop failure of 1924. But its indirect consequences in the realm of price policy were felt throughout the economy, and had important political implications.

The scissors crisis of 1923 had been overcome when the scissors closed in the spring of 1924, and industrial and agricultural prices returned to approximately the same relation which had existed between them before 1914. The financial estimates for the 1924 harvest were based on the prices for grain current in May of that year. It was assumed that an average price of 75 kopeks for a pud of rye (with corresponding prices for other grain) would be paid to the grower, and that the crop would be marketed at 105 kopeks.[4] In July, when the partial failure of the harvest was known, prices began to soar. In August 1924 grain prices were 100 per cent above the low level of August 1923.[5] On August 23, 1924, *Ekonomicheskaya Zhizn'* pleaded for action " to reduce and stabilize grain prices ". The newly established People's Commissariat of Internal Trade (Narkomvnutorg), in accordance with the general policy of price control adopted at the end of

[1] A. I. Rykov, *Sochineniya*, iii (1929), 94, 120.

[2] *Sobranie Zakonov, 1924*, No. 3, art. 35 ; *SSSR : Tsentral'nyi Ispolnitel'nyi Komitet 2 Sozyva : 2 Sessiya* (1924), p. 141.

[3] According to later Gosplan figures compiled on a different basis from the current figures of Narkomzem, the total yield of grain in 1924–1925 was 3000 million puds as against 3360 million for the previous year ; the comparable figure for 1913 was 5450 millions (*Kontrol'nye Tsifry Narodnogo Khozyaistva na 1926–1927 god* (1926), p. 340). The failure was confined to grain crops ; cotton, flax, sugar beet and dairy and poultry products all increased, so that the total value of agricultural production at pre-war prices slightly exceeded that of 1923–1924.

[4] *Planovoe Khozyaistvo*, No. 5, 1925, p. 297 ; for tables showing the prices paid to the growers and the wholesale market prices for grain for each month of the financial years 1923–1924 and 1924–1925 see *ibid*. No. 11, 1925, pp. 114-115.

[5] *Ekonomicheskaya Zhizn'*, August 27, 1924, reporting a speech in which Kamenev referred to grain prices as " the central question of the moment in our internal economic situation " ; in the following month Kamenev spoke to the central committee of the Komsomol of " the fearful rise in grain prices " (L. Kamenev, *Stat'i i Rechi*, xi (1929), 104).

1923,[1] attempted to fix maximum prices for grain, and was so far
successful that the price to the grower of a pud of rye, which in
August 1924 stood at 99 kopeks, was forced down in September
and October respectively to 86 and 78 kopeks. The battle of the
grain was now joined. The peasant had learned in the years of
inflation that to hold grain was more prudent than to hold money.
The value of grain would not fall and might rise ; the value of
money would not rise and was only too likely to fall. Severe
" tax pressure " to enforce sales was applied in the form of
" strict time-limits for the payment of the single agricultural
tax ",[2] but without avail. The well-to-do peasants, " striving by
all means to keep the grain in their hands, and to pay the tax in
anything rather than in grain ", met their obligations out of cash
reserves or by selling animals or live-stock products or commercial
crops, and struck against the threatened price reduction by holding up their surplus grain ; and it was the well-to-do peasants
who had the surpluses.[3] More serious still, private traders
appeared on the scene to buy above the maximum price ; and this
meant, as one commentator remarked, a return to the situation
of 1918–1920 when there were " two markets and two purchasing-
powers for one ruble " — at free and at state regulated prices.[4]
" Private capital ", complained Pravda, " has thrown itself on
the grain market and disorganized it." [5]   Attempts at resistance
proved futile. At Rostov the authorities issued an order making
obligatory the delivery of 25 per cent of all flour milled in the
region to the state-purchasing authorities at a fixed price, and
prohibiting the transport of grain from the region. But the
result was a cessation of milling operations ; and the peasants
still preferred to hold their grain rather than to sell it at state
prices.[6] By December 1924 the state had collected only 118
million puds of grain out of the projected 380 millions ; [7] and the
grain stocks held by the state, which had amounted to 214 million

[1] See *The Interregnum, 1923–1924*, pp. 110-113.
[2] *Planovoe Khozyaistvo*, No. 10, 1925, p. 44 ; *Pravda*, October 21, 1924,
complained of " inadmissible slowness " in the collection of the tax.
[3] *SSSR : Tsentral'nyi Ispolnitel'nyi Komitet 2 Sozyva : 2 Sessiya* (1924),
pp. 50-51 ; L. Kamenev, *Stat'i i Rechi*, xi (1929) 280.
[4] *Sotsialisticheskoe Khozyaistvo*, No. 5, 1924, p. 101.
[5] *Pravda*, December 19, 1924.
[6] *Planovoe Khozyaistvo*, No. 2, 1925, p. 270.
[7] *Ibid.* No. 3, 1925, p. 275.

puds on January 1, 1924, stood at only 145 millions on January 1, 1925.[1] The situation was now critical. The estimate for the total collection was cut from 380 million to 290 million puds, the share of the Ukraine being reduced from 34 to 26 per cent of the total. All thought of grain exports went by the board, and an import of 30 million puds was authorized.[2] In November the official maximum price for rye had been raised to 85 kopeks a pud. The attempt to maintain the maximum prices was then abandoned. In December the price to the grower of a pud of rye rose to 102 kopeks, and thereafter rose by leaps and bounds till it reached 206 kopeks in May 1925.[3] The price-fixing policy had been defeated. The *kulak* had proved victorious. The cities were once more being held to ransom.

The rise in grain prices was alarming on two counts. It threatened to rekindle the discontents, so recently allayed, of the industrial proletariat, and to upset the delicately poised wages structure by irresistible demands for wage increases. Rykov put the case to a sympathetic audience at the sixth trade union congress in November 1924 :

An unlimited increase in grain prices would mean the collapse of our budget, since it would entail an increase in wages and an increase in the prices of manufactured goods, and the breakdown of our whole price policy and of the struggle with the " scissors ".[4]

But the rise in prices also threatened relations in the countryside. In the existing structure of rural society, the price question sharply divided the peasants themselves. Only the well-to-do peasants consistently had grain surpluses and were primarily interested in high prices. In the autumn of 1924 it was reported for the first time from the Ukraine that well-to-do peasants were buying grain from poorer peasants as " the most favourable commodity to insure their capital at the maximum rate of interest ".[5] To hold stocks of grain was not only a promising speculation, but

[1] *Vestnik Finansov*, No. 7, July 1925, p. 70.
[2] *Planovoe Khozyaistvo*, No. 5, 1925, pp. 298-299.
[3] See the tables in *Planovoe Khozyaistvo*, No. 11, 1925, pp. 114-115.
[4] *Shestoi S"ezd Professional'nykh Soyuzov SSSR* (1925), p. 246.
[5] *Planovoe Khozyaistvo*, No. 1, 1925, p. 47 ; Kamenev drew attention to the political implications of the phenomenon : " The buying of grain by private trading capital, and especially by *kulaks*, makes them a political force,

the best safeguard against inflation.  At the opposite end of the scale, the poor peasants who lived wholly or in part by hiring out their labour were normally on balance buyers, not sellers, of grain : these may have accounted at this time for something like one-third of the peasant population.[1]  Between the two extremes, the mass of middle peasants were buyers or sellers according to the failure or success of the harvest.[2]  High prices following a bad harvest tended therefore to benefit the well-to-do peasants, to press hardly on the poor peasants, and to drive more and more of the middle peasants into the category of poor peasants who could subsist only by hiring out their labour.[3]  Such was the situation which developed in the winter of 1924–1925 :

> When grain was at 60 kopeks, the poor peasant sold, and now that it is at a ruble, the *kulak* sells.  This is noticeable.  When a pud of grain stood at 60 kopeks, the poor peasant paid his tax while the middle peasant and the *kulak* held back ;  now the poor peasant is paying a ruble for a pud of grain and the *kulak* is selling it.[4]

the masters of the situation in the grain market, and, worse still, gives them the possibility of taking up the pose of benefactors in regard to the poor peasantry " (*Planovoe Khozyaistvo*, No. 1, 1925, p. 16).

[1] A formidable controversy on this point broke out at the time of the fourteenth party congress in December 1925.  Kamenev, on the basis of figures of the central statistical administration, had claimed that 37 per cent of the peasants were buyers and not sellers of grain ; Yakovlev, in an article in *Pravda* December 9, 1925, and at the congress itself (*XIV S"ezd Vsesoyuznoi Kommunisticheskoi Partii* (B) (1926), p. 305) attacked this estimate as absurdly exaggerated.  For the argument about the proportion of grain surpluses held by different categories of peasants see pp. 299, 306, 310-311 below.

[2] A calculation in *Sotsialisticheskoe Khozyaistvo*, No. 1, 1925, p. 140, purported to show that peasants holding up to 6 desyatins of land in the consuming provinces or up to 4 desyatins in the producing provinces bought more grain than they sold, and were therefore interested in low prices ; but the article was criticized (*ibid.* pp. 147-149) as being based on too small a sample.  In January 1925, in the agriculturally poor Leningrad province, after the bad harvest of 1924, 60 per cent of the peasants were said to be buying grain (*Soveshchanie po Voprosam Sovetskogo Stroitel'stva 1925 g. : Yanvar'* (1925), p. 131).

[3] A controversy on this issue had divided Russian Marxists as long ago as 1897, when a Marxist group in Samara had protested against high grain prices as beneficial to landowners and injurious to poor peasants : Lenin, who regarded the spread of capitalism in the Russian countryside as inevitable and desirable, attacked this view as a sentimental illusion (Lenin, *Sochineniya*, ii, 3-4 ; cf. Yu. Martov, *Zapiski Sotsial-Demokrata* (Berlin, 1922), pp. 328-330 ; N. Angarsky, *Legal'nyi Marksizm* (1925), pp. 100-107).

[4] *Soveshchanie po Voprosam Sovetskogo Stroitel'stva 1925 g. : Yanvar'* (1925). pp. 134-135.

High prices were thus readily tolerated in those official circles which supported the development of *kulak* agriculture, and looked without disfavour on the growing social and economic differentiation in the countryside.

At the outset of the harvest crisis, on July 30, 1924, both *Pravda* and *Leningradskaya Pravda* had featured an article by Zinoviev under the title *The Harvest Failure and Our Tasks*, the keynote of which was an italicized phrase :

> It is time, high time, to compel a number of our organizations *to turn their face more to the countryside*.

A few days later another article congratulated Zinoviev on having " launched the correct slogan ".

> It is necessary [declared the article] that our whole party should turn its face to the countryside, because this is demanded by the interests of the economy as a whole and therefore by the interests of the proletariat.[1]

From this time onwards, throughout the autumn and winter, the exhortation " Face to the countryside " was constantly reiterated in Zinoviev's speeches and articles and became the catchword of party policy.[2] On Zinoviev's lips the slogan served to emphasize his position as the heir of Lenin, faithfully occupied in continuing and extending the application of NEP ; it was a weapon in the campaign against Trotsky which centred round the charge of under-estimating the peasant ; and, above all, it expressed the wave of anxiety about the situation in the country which overtook the party leadership in the autumn of 1924. This anxiety, though primarily inspired by the shortcomings of the harvest and the rising grain prices, was also connected with two other symptoms of agrarian discontent which obtained wide publicity at this time.

For some time past the main Soviet newspapers, central and local, had enrolled a number of factory workers as regular contributors to their columns : the functions of the so-called " worker correspondent " (*rabkor*) were to report news of his factory,

---

[1] *Leningradskaya Pravda*, August 6, 1924.
[2] A volume of Zinoviev's articles and speeches, beginning with the article of July 30, 1924, was published under the title *Litsom k Derevne* (1925).

publicize achievements, draw attention to abuses and ventilate grievances. A conference of *rabkors* was held in the offices of *Pravda* in November 1923, and was addressed by Bukharin, as editor of *Pravda*, and by Ulyanova, Lenin's sister : it was decided to found a special journal, *Rabochii Korrespondent*, devoted to their work.[1] At the beginning of 1924, with the growing importance of the peasant in Soviet policy, this institution was extended to the countryside ; and the " village correspondent " (*sel'kor*) took his place beside the *rabkor*. Unfortunately, authentic peasants with even the minimum qualifications for such work were hard to find ; and it appears that many of the *sel'kors* were party officials or workers sent on duty to the country.[2] This initial handicap made their position particularly delicate. As good party men, they tended to espouse the cause of the poor peasants and *batraks* ; reviving the policy of the committees of poor peasants in 1918, some of them conceived it as their function to " kindle class-war in the villages ". The abuses to which they most eagerly drew attention were those committed by *kulaks*. Coinciding with the new official inclination to show greater indulgence to the *kulak*, this became a burning issue ; and a number of *sel'kors* were attacked or murdered, apparently at the instigation of *kulaks* and others whom they had denounced in the press. The first and most sensational of these affairs occurred on March 28, 1924, in a village named Dymovka, 50 versts from Nikolaevsk in the Ukraine. Two of the three members of the local party cell which controlled the village Soviet were said to have been in league with local *kulaks*, and to have connived at favours shown to them.[3]

[1] An account of these proceedings was given in *Internationale Presse-Korrespondenz*, No. 30, March 4, 1924, pp. 342-343 ; Bukharin's speeches at the conference were reprinted in N. Bukharin, *O Rabkore i Sel'kore* (2nd ed. 1926), pp. 33-47.

[2] It was afterwards stated that only 50 per cent of *rabkors* and 25 per cent of *sel'kors* were party members, and that " a majority of *sel'kors* are the more literate peasants drawn from the poor peasants and middle peasants and from Red Army men " (*Pechat' SSSR za 1924 i 1925 gg.*, ed. I. Vareikis (1926), p. 30) ; but this was hardly true in the early stages of the movement.

[3] The story of Dymovka as related at a party meeting by Zinoviev seems to have had little to do with *kulak* activities. According to this version, the two members of the party cell, Popandopulo and Postolati by name, were also OGPU agents : Popandopulo had gathered around him, by methods of intimidation and bribery, a personal following of 60 or 70 poor peasants, and with their support tyrannized over the community (G. Zinoviev, *Litsom k Derevne* (1925), p. 78). This is a picture of an unscrupulous party boss rather than of a *kulak*. But

The third member of the party cell, Malinovsky by name, in his capacity as a *sel'kor*, published the story in the press and was promptly murdered for his pains. In October 1924 six men were brought to trial in Nikolaevsk for their share in the crime. The trial was extensively reported in the press. Sosnovsky, a party publicist who had been active in unmasking the crime, made a long speech in support of the prosecution in which the issue of poor peasants against *kulaks* was heavily stressed. The accused were sentenced to death, though the sentences on three of them were afterwards commuted on the ground that they had been mere tools of the *kulaks*. After the trial inscriptions were said to have appeared on a wall in Dymovka threatening a similar fate to anyone else who sent complaints to the newspapers.[1]

It now transpired that the Dymovka murder was not an isolated incident. On October 3, 1924, *Pravda* reported a wave of such crimes throughout the country : the murder of *sel'kors* had become a " mass phenomenon " and marked " a recrudescence of counter-revolution in the country-side ". During the next two months a series of similar incidents was widely publicized in the press.[2] The party central committee, meeting at the end of October 1924, drew attention to the plight of the *sel'kors*, and pronounced it necessary to " take decisively under the protection of Soviet laws and Soviet organs any of those whose work of denunciation may provoke threats of violence from counter-revolutionary and *kulak* elements in the countryside ".[3] An article by Sosnovsky appeared in *Pravda* entitled *Dymovka not an Exceptional Phenomenon*;[4] and on November 11, 1924, the People's Commissariat of Justice issued an instruction to the courts " that the murder of a *rabkor* or a *sel'kor* should be treated as a counter-revolutionary act ".[5] One result of this publicity was to swell enormously the number of correspondents. At the

the incident was used in party propaganda to stimulate anti-*kulak* feeling ; and there is no doubt that the *sel'kors*, in general, conducted a campaign against the *kulaks*.

[1] *Pravda*, October 22, 23, 1924 ; *Izvestiya*, October 24, 28, 1924 ; G. Zinoviev, *Litsom k Derevne* (1925), pp. 77-78.
[2] See, for example, *Izvestiya*, October 29, November 13, 21, December 9, 13 ; the last two issues contained biographical notices of a number of murdered *sel'kors*.           [3] *VKP(B) v Rezolyutsiyakh* (1941), i, 633.
[4] *Pravda*, November 2, 1924.
[5] *Ezhenedel'nik Sovetskoi Yustitsii*, No. 45, 1924, p. 1092.

time of the thirteenth party congress in May 1924 there had been
only 15,000 in all. By January 1925 there were 60,000 *rabkors* and
80,000 *sel'kors*, by August 1925, 74,000 *rabkors* and 115,000
*sel'kors* ; and these figures were not exhaustive.[1]  In December 1924 a conference of *rabkors* and *sel'kors* was held in Moscow,
and listened to speeches by Zinoviev, Kamenev, Bukharin,
Krupskaya and others praising their work.  Cases were reported
to the conference of *rabkors* dismissed from their jobs on false
pretexts and of *sel'kors* openly intimidated or beaten up.  A list
of 20 murdered *sel'kors* was read, and this was said to be far from
complete.[2]  The publicity given to these incidents revealed a
growing sensitiveness in party circles to the favour now being
shown to the *kulaks* and to the discontent aroused by this policy
in the countryside.

Another untoward event engaged the attention of anxious
party leaders in the autumn of 1924 and sharpened the incipient
divisions in the party ranks.  A serious, though short-lived, revolt
occurred in Georgia at the end of August.  It was not the first
revolt against the Soviet power in that turbulent country.[3]  But
on this occasion an unexpected eagerness was shown to attribute
it to general rather than to local shortcomings.  At a conference
of rural party workers in October, Stalin called the events in
Georgia " indicative ", and unexpectedly concluded that " what
has happened in Georgia might be repeated all over Russia, if
we do not radically alter our whole approach to the peasantry ".[4]
The origin of the revolt was discussed a few days later at the
same meeting of the party central committee which had taken
cognizance of the murder of the *sel'kors*.  Stalin admitted that "in
certain counties . . . it indubitably had a mass character", that
its economic causes were the high prices of manufactured goods

[1] *Pechat' SSSR za 1924 i 1925 gg.*, ed. I. Vareikis (1926), pp. 29-30.
[2] The conference was reported in *Izvestiya*, December 6, 7, 1924, and
*Pravda*, December 13, 14, 1924 ; the speeches of Bukharin were reprinted in
N. Bukharin, *O Rabkore i Sel'kore* (2nd ed. 1926), pp. 51-63.  A dispute arose
at the conference on the form of organization which should be given to the
*rabkor* and *sel'kor* movement ; this will be discussed in Part III in the following
volume.
[3] The political aspects of the revolt will be discussed in a subsequent
volume.
[4] Stalin, *Sochineniya*, vi, 309 ; the conference was reported at length in
*Pravda*, October 23-26, 28, 1924.

and low price of maize, and that it was for this reason " indicative of the new conditions of the struggle throughout the Soviet land ".[1] Zinoviev, while he saw in the occurrence " a blend of Menshevism and nationalism ", refused to find more than 50 per cent of its causes in local conditions, and compared it with the Kronstadt and Tambov disturbances of 1921 as a warning signal of something wrong in the economy as a whole.[2] Three months later Stalin once more referred to the revolt as " a great warning ", and thought that " a new Tambov or a new Kronstadt are not in the least excluded ".[3] Another party commentator regarded the Georgian revolt as " a symptom of an acute stage in the relation of the peasants to the Soviet power ".[4] At the fourteenth party congress in December 1925 the Georgian disturbances were still referred to as a starting-point of the dissensions in the party on the agrarian question.[5]

These events did not immediately suffice to force issues of policy which the leaders themselves were still anxious to evade. At the meeting of the central committee at the end of October Kamenev made a report defending the policy of price intervention :

> Could we afford to take as the foundation of our economic policy submission to the spontaneous rise of grain prices? It was clear to us that that would be absolutely inadmissible. . . . We should cease to be masters of economic policy. The master would be he who could hoard grain at this price.[6]

But these brave words did not answer the practical question how to bring in the grain. Zinoviev, Bukharin and Sokolnikov, " under the fresh impression of the Georgian rising ", raised the question of further concessions to the peasantry. Zinoviev is even

[1] Stalin, *Sochineniya*, vi, 316-317.

[2] G. Zinoviev, *Litsom k Derevne* (1925), pp. 65-66. In a speech a few days later Zinoviev named the high price of bread, the low price of maize, and the veto on grain exports among the causes of the revolt (*ibid.* pp. 76-77) ; elsewhere he linked the Georgian affair (" a little subterranean shock ") with the Dymovka affair as warning symptoms (*Leningradskaya Pravda*, November 9, 1924).

[3] Stalin, *Sochineniya*, vii, 22 ; Stalin repeated the comparison and the warning a few weeks later (*ibid.* vii, 31).

[4] *Krasnaya Nov'*, No. 2, February 1925, p. 145.

[5] *XIV S"ezd Vsesoyuznoi Kommunisticheskoi Partii (B)* (1926), p. 190.

[6] L. Kamenev, *Stat'i i Rechi*, xi (1929), 175 ; the report was originally published in *Pravda*, October 31, 1924.

said to have proposed to " create official peasant non-party
fractions in VTsIK and in the Soviets, giving them the right to
issue their own newspapers, etc." But a majority of the committee thought that such proposals betokened an unnecessary
panic.[1] Stalin also quoted alarmist reports from different parts of
the country. In some places there had been " a mass refusal to
receive tax assessments " ; stormy mass meetings had been held
in others to demand from the government lower taxes and higher
grain prices. The inspirers of this campaign, as well as of the
Georgian revolt, had been " *kulaks*, speculators and other anti-
Soviet elements ". Stalin's conclusion that it was necessary to
" isolate the *kulaks* and speculators and to separate the toiling
peasantry from them ", though couched in conventional language,
might be interpreted as a cautious warning against a policy of
economic concessions which would inevitably bring most benefit
to the well-to-do peasants.[2] The committee spent a whole day
discussing the mood of the peasantry,[3] but, faced with these conflicting and hesitating views, could do no more than mark time.
The resolution adopted at the meeting confined itself to vague
recommendations on the strengthening of party work in the
country and the necessity of drawing more non-party elements
into the Soviets. It touched on the issue of policy in terms which
decided nothing at all, insisting on the need for " special instructions to local organizations (especially for the press) on questions
which, in the event of an incorrect approach to the moods of the
mass of the peasants, may lead to negative political results in the
country (such questions as the mutual relations of the working
classes and the peasantry, the *kulaks*, etc.) ".[4] Evidence of
embarrassment in face of the intractable agrarian problem was
provided by another act of the committee. Every year since 1917

[1] There is no formal record of these proposals, which rest on the later
evidence of hostile witnesses (*XIV S"ezd Vsesoyuznoi Kommunisticheskoi
Partii (B)* (1926), p. 190 ; M. Popov, *Naris Istorii Kommunistichnoi Partii
(Bil'shovikiv) Ukraini*, 2nd ed., Kharkov (1929), pp. 284-285).

[2] Stalin, *Sochineniya*, vi, 315-317.

[3] L. Kamenev, *Stat'i i Rechi*, xi (1929), 204.

[4] *VKP(B) v Rezolyutsiyakh* (1941), i, 632-633 ; according to Kamenev's
statement to the Moscow party organization immediately after the session, all
hopes of exporting grain had been abandoned, but it was still hoped to collect
300 million puds of grain at an average price of 80–85 kopeks (L. Kamenev,
*Stat'i i Rechi*, xi (1929), 198).

the party congress had been held in the spring. It was decided to postpone the congress of 1925 from the spring to the autumn, since vital decisions of policy could more easily be taken after, than before, the harvest. The congress in the spring could be replaced by a party conference ; this would precede the annual Union Congress of Soviets, which would be postponed from its normal date in January till April. Kamenev, who reported these arrangements to a meeting of the Moscow party organization, admitted that the postponement of the congress was a breach of the party statutes. Unless, however, strong objections were manifested in the party, the decision would stand.[1]

The crisis of the minor harvest failure of 1924 passed therefore without provoking important changes either in policy or in relations between the leaders. If, as is probable, Zinoviev was at this time more inclined than Stalin or Kamenev to a policy of further concessions to the peasant, all three members of the triumvirate were still deeply engaged in the latest and most acute phase of the struggle with Trotsky, which opened in October 1924 and ended only with his resignation from the People's Commissariat of War in January 1925.[2] In their speeches and articles of this period, they never failed to insist on the importance of the alliance with the peasantry in current policy and to censure Trotsky for his neglect of the peasant. But they sedulously avoided any public pronouncement on the proposal for further concessions, and refrained from taking sides in the increasingly embarrassing question of the attitude to be adopted towards the *kulak*. Whatever differences of opinion lurked beneath the surface, verbal compromises were possible, and would continue to be made, so long as the overriding need to hold the triumvirate together was still recognized by all its members.

The desire of the leaders to keep the issue of agrarian policy on safe and well-worn lines received adventitious assistance from an economic controversy which, side by side with the political controversy with Trotsky, attracted much attention in the autumn of 1924. Since the opposition of the " Left communists " in

[1] *Ibid.* xi, 189.
[2] The struggle will be described in Part III in the following volume.

March 1918 there had been no more persistent assailant of the
party line than the able and original economist Preobrazhensky.
Preobrazhensky had been the first to draw attention to the tendency of NEP to encourage the *kulak*,[1] and had incurred Lenin's
anger by attempting to raise this question at the eleventh party
congress in March 1922. Later he had criticized the deleterious
effects of NEP on planning.[2] He was one of the few to whom
Trotsky had confided his discussion with Lenin at the end of
1922 on the rise of bureaucracy in the party.[3] He was a leading
signatory of " the platform of the 46 " (much of it was probably
written by him) and a prominent member of the opposition
during the winter of 1923–1924 : he had crossed swords with
Stalin at the thirteenth party conference of January 1924, and
he spoke again for the opposition at the party congress four
months later.[4] Regarding NEP and all its implications and consequences with scarcely veiled suspicion, Preobrazhensky had
throughout this time steadily upheld the case for planning and for
increased support to industry, and had opposed all further concessions to the peasant.

In August 1924 Preobrazhensky read to the Communist
Academy a paper on *The Fundamental Law of Socialist Accumulation*, whose outstanding quality made it a landmark in the history
of Soviet economic theory.[5] Preobrazhensky's starting-point was
a comparison between the period of what Marx had called primitive

[1] A far-sighted article by Preobrazhensky in *Krasnaya Nov'*, No. 3,
September-October 1921, pp. 201-212, predicted that " the growth of the
*kulaks* in the new conditions must incontestably lead also to a new grouping of
forces in the countryside ", that the Soviet authority would be compelled to
intervene on the side of the poor peasant, that two or three years — or perhaps
more — of " peaceful cohabitation " between capitalist and socialist processes
of development in the countryside were probable, but that " the moment is
coming when a clash will be inevitable ". A further article in *Kommunisticheskii
Internatsional*, No. 25 (November 1922), cols. 6275-6290, anticipated some of
the main ideas propounded by him in the controversy of 1924–1926.

[2] See *The Bolshevik Revolution, 1917–1923*, Vol. 2, pp. 291-293, 379.

[3] See *The Interregnum, 1923–1924*, p. 297, note 3.

[4] See *ibid.* pp. 335-337, 364.

[5] It was originally published in *Vestnik Kommunisticheskoi Akademii*, viii
(1924), 47-116, and was republished under the title *The Law of Primitive
Socialist Accumulation* as the second chapter of E. Preobrazhensky, *Novaya
Ekonomika* (1926), pp. 52-126 ; in this second version some provocative phrases
were toned down, but without alteration of the sense. References below are to
the second version, except in passages where it diverged from the earlier version,
and where both references are given.

capitalist accumulation and the corresponding period in the advance to socialism.  Before the process of automatic accumulation proper to a mature capitalist economy could be set in motion, it had been necessary as a preliminary stage to go through a period of forced accumulation of capital in a small number of hands : this had been the stage of the open compulsion and exploitation of the worker who had to be drawn off the land into the factories, of " the separation of the producers from the means of production ". Similarly, " in order that the complex of the state economy may be able to develop all its economic advantages and establish for itself a new technical base ", socialism must pass through a preliminary stage of " primitive accumulation " : this accumulation meant " the accumulation in the hands of the state of material resources . . . coming from sources outside the complex of the state economy ", or, in other words, " the expropriation of the surplus product of the country for the broadening of socialist production ".[1]  Having enumerated the various forms of expropriation practised by capitalism in the process of primitive accumulation — the expropriation of the labour of small producers engaged in pre-capitalist forms of production, exploitation of colonies, expropriation by taxation, expropriation by state loans — Preobrazhensky rejected as unacceptable for a socialist government the method of " colonial robbery ".  On the other hand, the same objection did not apply to " the alienation for the benefit of socialism of a part of the surplus product of all pre-socialist economic forms " within the country itself : this method was, in fact, bound to play " an immense and directly decisive rôle in such peasant countries as the Soviet Union ".[2]  This brought Preobrazhensky to his diagnosis of the current situation :

> In the period of primitive socialist accumulation the state cannot *do without the exploitation of small-scale production, without the expropriation of a part of the surplus* product of the countryside and of artisan labour. . . .
> The idea that a socialist economy can develop by itself without touching the resources of the petty bourgeois, including the peasant, economy is beyond doubt a reactionary, petty bourgeois utopia.  The task of the socialist state consists not in taking from the petty bourgeois producers less than was taken by

---

[1] *Ibid.* pp. 52-58.          [2] *Ibid.* pp. 59-62.

capitalism, but to take more *out of the even greater* income which will be assured to the small producer by the rationalization of everything, including the small production of the country.[1]

In addition to taxation and loans, Preobrazhensky here pointed out that currency emission had also served as a form of taxation and " one of the methods of primitive accumulation ".[2]

Preobrazhensky then turned to " measures of primitive accumulation on an economic basis " (as opposed to taxation and loans which were administrative measures). After discussing various minor ways in which the private sector of the economy could be compelled to make its contribution to accumulation, he broached the vital issue of prices. At the thirteenth party conference in the previous January, he had already pointed to price policy as a method of extracting peasant surpluses " politically more advantageous " than taxation.[3] He now advocated " a price policy consciously directed to the exploitation of the private economy in all its forms ".[4] Preobrazhensky was writing an economic analysis, not a political pamphlet, and skirted round the most delicate point :

I do not speak here of the difficulties of a political nature which arise from the mutual relations of the working class

[1] *Vestnik Kommunisticheskoi Akademii*, viii (1924), 58-59 ; in the version in E. Preobrazhensky, *Novaya Ekonomika* (1926), pp. 62-64, the phrase " exploitation of small-scale production " was omitted, and " alienation " substituted for " expropriation ".

[2] *Ibid.* pp. 65-66. Preobrazhensky had drawn attention as early as 1920 to this virtue of the printing-press (see *The Bolshevik Revolution, 1917-1923*, Vol. 2, p. 261) ; in January 1924 he somewhat reluctantly approved the financial reform, " since we have by a spontaneous process arrived at the necessity of carrying it out " (*Trinadtsataya Konferentsiya Rossiiskoi Kommunisticheskoi Partii (Bol'shevikov)* (1924), p. 37).

[3] *Ibid.* p. 35. Sokolnikov in May 1923 had already looked forward to a time when the state budget, as well as investment for economic development, would be financed, not by direct taxation of the peasant, but from the profits of state industry; this meant that " the price of the product of nationalized industry will then include in a sense a certain rate of a kind of tax " (G. Sokolnikov, *Finansovaya Politika Revolyutsii*, ii (1926), 116).

[4] This was the original text in *Vestnik Kommunisticheskoi Akademii*, viii (1924), 79 ; in E. Preobrazhensky, *Novaya Ekonomika* (1926), p. 87, the phrase was toned down to read : " a price policy consciously directed to the expropriation of a *definite* part of the surplus product of the private economy in all its forms ".

and the peasantry, and make it often obligatory to speak of equivalent exchange, though that is even more utopian under the socialization of large-scale industry than under the rule of monopoly capitalism.[1]

In other words, the exploitation of the peasant (meaning the extraction from the peasant of his grain surpluses without making him a fully equivalent return) was a necessary condition of the initial stage of the advance to socialism.  It was the business of the politician to mask this disagreeable fact in decent language ; Preobrazhensky was no politician.

The position of the proletariat itself was, however, also ambiguous.  As a result of the revolution, " the working class is transformed from the object to the subject of exploitation ".  Nevertheless, it cannot be indifferent to its own health and to the conditions of labour, as was the capitalist employer ; and this puts " a certain brake on the tempo of socialist accumulation ".  Insistence on the eight-hour day was a case in point.[2]  But this merely made " the fundamental law of socialist accumulation " all the more irrevocably binding.  This Preobrazhensky now proceeded to formulate as follows :

> *The more economically backward, petty bourgeois and peasant in character is the country making the transition to a socialist organization of production, the smaller is the legacy which the proletariat of the country in question receives at the moment of the social revolution to build up its own socialist accumulation, and the more in proportion this socialist accumulation will be obliged to rely on the expropriation of the surplus product of pre-socialist forms of the economy.*

Only a more developed economy can " *rely on the surplus product of its own industry and its own agriculture* ".[3]  Preobrazhensky concluded his paper with some remarks on the struggle between two laws — the law of value and the law of socialist accumulation — in the Soviet economy.  Just as the law of value, which was

---

[1] *Ibid.* p. 88.          [2] *Ibid.* pp. 100-101.
[3] *Vestnik Kommunisticheskoi Akademii*, viii (1924), pp. 92-93 ; in the version in E. Preobrazhensky, *Novaya Ekonomika* (1926), pp. 101-102, the word " expropriation " has been replaced by the phrase " alienation of a part ".

essentially capitalist, still exercised its influence even over the socialized sector of the economy, so the law of socialist accumulation " extends its effect in a certain degree to the private sector of the economy only as to an alien territory ".[1]

The far-reaching implications of this analysis do not seem to have been immediately digested.[2] After its publication in the journal of the academy, Oganovsky, an official of Narkomzem and a former SR, wrote an article which, without mentioning Preobrazhensky by name, attacked those who sought to finance the development of industry, " not by way of expanding its output and its sales to the rural population, but by means of the direct extraction of surplus value from the peasant " : such a policy would be " to kill the goose that lays the golden eggs ".[3] But it was the development of the political situation which, at the end of the year, removed the argument altogether from the academic plane. Preobrazhensky's article proved a godsend to the leaders of the campaign against Trotsky. Preobrazhensky, who could reasonably be depicted as one of Trotsky's most loyal followers, had furnished ample material to justify the charge that Trotskyism was based on neglect of the point of view of the peasant and ran directly counter to Lenin's formula of the " link " between proletariat and peasantry, since it represented the interests of the two classes as fundamentally irreconcilable. Trotsky, though he never appears to have pronounced on Preobrazhensky's thesis, and was at any rate enough of a politician to put the case in less deliberately provocative terms, had already, at the twelfth party congress of April 1923, endorsed the appositeness of the phrase " primitive socialist accumulation " with all the implications which it carried.[4] Bukharin, the only economist among the party

[1] E. Preobrazhensky, *Novaya Ekonomika* (1926), pp. 116-117.

[2] Preobrazhensky later remarked that, when he first read his paper in August 1924, his opponents " were still more afraid of industrial over-accumulation and industrial over-production " (*Bol'shevik*, No. 15-16, August 31, 1926, p. 89) ; though the scissors had been closed, the danger of their reopening still seemed greater than any other.

[3] *Sotsialisticheskoe Khozyaistvo*, No. 5, 1924, p. 27 ; the five-year plan for agriculture issued by Narkomzem at the end of 1924 (see p. 491 below) was based on the hypothesis that " the development of industry presupposes the development of agriculture " (*Osnovy Perspektivnogo Plana Razvitiya Sel'skogo i Lesnogo Khozyaistva* (1924), p. 29).

[4] See *The Bolshevik Revolution, 1917–1923*, Vol. 2, p. 382 ; Preobrazhensky in his article also named V. M. Smirnov as the author of the phrase. In an

leaders of the first rank, replied at length to Preobrazhensky in an
article in *Pravda* entitled *A New Discovery in Soviet Economics,
or How to Ruin the Worker-Peasant Bloc*, and accompanied this
with another article in the party journal *Bolshevik* entitled *A
Critique of the Economic Platform of the Opposition*.[1]   The first
article opened with a reference to the controversy started in the
autumn of 1924 by the publication of Trotsky's *Lessons of October*,
and treated Preobrazhensky's article as representing " the economic foundation of Trotskyism " and " the economic side of the
anti-Leninist point of view ".  Bukharin stressed throughout the
basic incompatibility between Preobrazhensky's belief in progress
towards socialism at the expense of the peasant and Lenin's
conviction, embodied in NEP, that progress towards socialism
could be realized only in close alliance with the peasant and by
developing his resources and opportunities, through trade and
through the cooperatives, side by side with those of the proletariat.
Preobrazhensky wanted to introduce into socialism the oppressive
and restrictive procedures of monopoly capitalism.   The opposition view, concluded Bukharin, was " the workshop ideology
which has ' no time ' for other classes ", and which demanded
" increased pressure on the peasantry for the greater glory of the
proletariat ".  Like other similar " theories ", it would be rejected
by the overwhelming majority of the party.   The second article,
which bore the taunting sub-title *Lessons of October 1923*, covered
the same ground in greater detail and carried the war still further
into Trotsky's camp.   Trotsky, it was true, had never specifically
come out, like Preobrazhensky, for " a forced growth of industrial
accumulation ".   But his demand for a " dictatorship of industry "
amounted to the same thing.[2]   Once again the opposition had

unpublished note of 1927 preserved in the Trotsky archives Smirnov protested
that he scarcely remembered " the small article in which I was delivered of
this phrase ".   It was written in the days of war communism ; it had no relation
to NEP conditions.   He never used it again or built any theories on it ; and
" I profoundly regret my unhappy invention ".

[1] *Pravda*, December 12, 1924 ; *Bol'shevik*, No. 1, January 15, 1925, pp. 25-
57.   According to Bukharin, the second article was written " at the direct
order of the [party] central committee " (*Krasnaya Nov'*, No. 4, May 1925,
p. 267) ; this may have been a somewhat shamefaced apology for its demagogic
tone.   Both articles were later reprinted in a pamphlet, N. Bukharin, *Kritika
Ekonomicheskoi Platformy Oppozitsii* (1926).

[2] For Trotsky's demand that the " dictatorship " should belong to industry
and not to finance see *The Interregnum, 1923-1924*, p. 125.

demonstrated its incapacity to understand " *the problem of the worker-peasant bloc* " ; and this incapacity was " the fundamental vice of all Trotskyism ".

These crushing exposures of " the economic platform of the opposition ", having played their part in the defeat of Trotsky, proved too useful to be readily discarded. In vain did Preobrazhensky protest that Bukharin had treated " an attempt at a theoretical analysis of the Soviet economy " as if it were a description of " the economic policy of the proletarian state ", and offer to withdraw the invidious word " exploitation ".[1] For the next two years denunciation of the law of primitive socialist accumulation was the incessant theme of all who sought to defend the peasant cause and to resist or slow down the process of industrialization. But time did not make the fundamental issue raised by Preobrazhensky any less acute.[2] The slogans of the " peasant-worker bloc " and " Face to the countryside ", in the name of which the victory was achieved, were already obsolete, not only because any move to develop heavy industry made a practical answer to Preobrazhensky's problem imperative and urgent, but also because the peasantry was now less than ever an undifferentiated whole, and any decision to support the peasant which did not take account of divisions of interest within the peasantry itself was the beginning and not the end of the dilemma of agrarian policy. Differences about agrarian policy lay behind the discontents in the countryside in the autumn of 1924. These

---

[1] Preobrazhensky's reply appeared in *Vestnik Kommunisticheskoi Akademii*, xi (1925), 223-256 (reprinted in E. Preobrazhensky, *Novaya Ekonomika* (1926), pp. 210-255), with a note to the effect that Bukharin's counter-reply would be published in the next issue. In fact, it never appeared : a further reply was scarcely possible on the academic plane.

[2] It is worth listing Preobrazhensky's contributions to this controversy since they provide the most revealing available analysis, couched in strict Marxist terms, of the fundamental dilemma of the Soviet economy of this period : (1) *The Fundamental Law of Socialist Accumulation* (*Vestnik Kommunisticheskoi Akademii*, viii (1924), 47-116) ; (2) *Once Again About Socialist Accumulation* (*Reply to Comrade Bukharin*) (*ibid.* xi (1925), 223-256) ; (3) *Economic Notes I* (*Pravda*, December 15, 1925) ; (4) *The Law of Value in the Soviet Economy* (*Vestnik Kommunisticheskoi Akademii*, xiv (1926), 3-64) ; (5) Reply to the debate on (4), *ibid.* xv (1926), 157-160 ; (6) *Economic Notes II* (*Bol'shevik*, No. 6, March 31, 1926, pp. 60-69) ; (7) *Economic Notes III* (*ibid.* No. 15-16, August 31, 1926, pp. 68-83). Of these (1) was reprinted with slight modifications, and (2) and (4) without modifications, in E. Preobrazhensky, *Novaya Ekonomika* (1926).

differences provided the main theme for the new rift in the party leadership which gradually developed during 1925.

### (b)  The Issue in the Countryside

The development of Soviet agriculture in the middle nineteen-twenties forced on the attention of the party leaders the basic problems of the character and duration of NEP which they strove so studiously to evade.  Down to the spring of 1924 the serious issue in Soviet economics and politics had been the struggle between town and country, between the worker and the peasant. The thirteenth party congress in May 1924 marked the end of this period.  The closing of the scissors in favour of the peasant, and the adoption of a programme of support for heavy industry, had settled that issue for the time being.  A satisfactory balance had been achieved.  Recovery had gone far enough to enable the advance to be made simultaneously on both fronts.  Preobrazhensky's renewed assault on the peasant in the interests of industry was now either belated or premature.  It related to an issue which had been acute in 1923 and would once more become acute after 1926.  But for the moment a truce had been called.  Preobrazhensky's attack merely helped the party leaders, in the autumn of 1924, to concentrate attention on this issue, and thus maintain at full blast their campaign against Trotsky for his neglect of the peasant.  But, once Preobrazhensky had been refuted by Bukharin, and Trotsky decisively defeated and deposed from office in January 1925, this issue was relegated to the background, and was replaced by another.  The topical question was no longer whether, and in what proportions, to support and subsidize industry or agriculture, but, within the agreed policy of support for agriculture, what group of the peasantry was to become the main vehicle and recipient of that support, and what forms of land tenure were to be encouraged or tolerated.  The problem of the *kulak*, carefully thrust into the background throughout the greater part of 1924, now rose in all its stark and embarrassing stature. The climate had changed.  In terms of the party struggle, the change was expressed in the switch-over from the joint campaign of the triumvirate against Trotsky, which had occupied 1924, to the rift between Bukharin and Stalin, on the one hand, and

Zinoviev and Kamenev, on the other, which gradually widened throughout 1925.

The emancipation of the serfs left unsolved the major problem of the future organization of Russian agriculture : in so far as it hastened the dissolution of the large estates, it had complicated that problem. Among the peasantry it broke the equality of servile status, and set on foot the process of " differentiation " which had increasingly dominated peasant politics since that time. Agriculture could not flourish and develop on the basis of innumerable fragmentary holdings of individual peasants, the number of which multiplied with every increase of the population. The "black repartition " of the extreme wing of the *narodniks* and the equal division programme of the SRs — like the later Bolshevik incitement to unrestricted seizure of landowners' land by the peasants — were in plain contradiction with this fundamental economic truth. In these midget holdings, cultivated by peasants struggling to extract a bare subsistence from the soil, the primitive three-field rotation was still current. Hay and fodder crops were rare. Where the labour of man replaced the labour of animals, adequate ploughing and manuring became impossible. In particular, it was clear that the greater the number of small peasant holdings, the greater the proportion of the harvest which would be consumed by the grower for his own needs ; the larger the units of production, the greater the proportion of the harvest which would be marketed to feed a rising industrial population in the towns and to provide surpluses for export. The issue was stated in its simplest terms by the moderate Kamenev :

> We communists understand that we shall complete the building of socialism only when we draw into socialism these 20 million scattered peasant households, attract them into our common work of construction, and bind them into a single system.[1]

From this dilemma two ways of escape had divided opinions between them for the past fifty years. The first solution was to foster the survival of the *mir*, or peasant commune, and of the *dvor*, or peasant household, which was the unit of membership of the *mir*, and to make the collective elements of this ancient system

---

[1] L. Kamenev, *Stat'i i Rechi*, xii (1926), 514.

the starting-point for the erection of new forms of cooperative and collective agriculture. This, broadly speaking, was the old *narodnik* view ; and even Marx had given some encouragement to the belief that a direct transition could be made from the primitive peasant commune to the future socialist organization of agriculture.[1] The second solution was to abandon as rapidly as possible the collective principle, and to promote a competitive struggle among the peasants on individual or " capitalist " lines for the survival of the fittest, with the implication that the most enterprising would aggrandize themselves at the expense of the rest, and the least fit would move into the rapidly growing towns and factories or colonize the remoter regions of the broad Russian Empire. For many years the official policy remained ambivalent. The peasant commune struggled on side by side with the symptoms of rising capitalism, the most noteworthy of which was the division of peasantry between *kulaks*, poor peasants and middle peasants. The process of differentiation was described by Lenin in one of his earliest writings :

> On the one hand peasants were losing their agricultural inventory (animals as well as machines), while, on the other hand, some peasants were acquiring a modern inventory and mastering the use of machinery. On the one hand peasants were abandoning their land or selling or leasing their allotments, while, on the other hand, many peasants were renting additional plots and eagerly buying up privately owned land.[2]

It was not till 1906 that a decisive step was taken by the government in the shape of the Stolypin reform. Stolypin, who wholeheartedly and explicitly accepted the struggle for survival and the " wager on the strong ", worked to break up the peasant commune : the *kulak* became the hero and the hope of Russian agrarian policy.[3] It was a policy of production, not for subsistence, but for the market, domestic and foreign, of " the forced smashing of the commune in order to clear the way for capitalism

---

[1] See *The Bolshevik Revolution, 1917–1923*, Vol. 2, pp. 388-390.

[2] Lenin, *Sochineniya*, iii, 126 ; Lenin, however, criticized the word " differentiation ", which was a favourite term of the *narodniks*, on the ground that it covered only the phenomenon of inequality between different categories of peasant ; he preferred the word " disintegration ", which implied the gradual breakdown of the traditional rural economy.

[3] For the Stolypin reform see *The Bolshevik Revolution, 1917–1923*, Vol. 2, pp. 20-24.

in agriculture *at any cost* ".[1]   But it offered, at that cost, a capitalist solution of the future of Russian agriculture.   Considerable progress had been made, especially in the most fertile and densely populated areas of the Ukraine, the major source of Russia's grain exports, when war broke out in 1914.[2]

In the first years of the Soviet régime the same issues reappeared with all the persistence of basic economic factors.   The revolution passed a sponge over the recent past, confiscated the large estates of the landowners and the church, and redistributed the land all over the country on a professed basis of equality, though with many shortcomings in practice and some reservations about the interpretation of that ambiguous standard.   It largely increased the rôle of the peasant in the rural economy.   If before the revolution peasant holdings amounted to 240 million hectares, or 67·6 per cent of the agricultural land now included in the Soviet Union, the amount had risen ten years later to 314 million hectares, or 88·5 per cent of the whole.[3]   This heroic solution was, in the eyes of orthodox Bolsheviks, no more than a temporary expedient.   It had all the disadvantages of the state of affairs which Stolypin had set out to remedy : it restored cultivation for subsistence, and offered no adequate incentives for production for the market.   It provided the basis for Rosa Luxemburg's prompt criticism that " Lenin's agrarian reform has created for socialism a new and mighty popular stratum of enemies, whose resistance will be much more dangerous and stubborn than the resistance of the landowning gentry ".[4]   The decree of February 1918 on " the socialization of the land " pronounced " all forms of individual utilization of land " to be " transitory and obsolete ",

---

[1] Lenin, *Sochineniya*, iii, 123.

[2] An elaborate calculation, based on the scanty, scattered and unreliable material, appears to warrant the conclusion that of 13 or 14 million peasant households holding " allotment " land in 1905, some 5 millions were wholly unaffected by the Stolypin reforms, and 1,300,000 more practically unaffected. Some 5½ million households had had their independent titles recognized, and of these some 1,300,000 had been through the whole process of consolidation of their holdings before 1917.   The remainder in 1917 had reached different stages in the complicated transition from communal to individual holdings (G. T. Robinson, *Rural Russia under the Old Régime* (1932), pp. 226-227) ; but much of this work was undone by the revolution, or left uncompleted.

[3] *Itogi Desyatiletiya Sovetskoi Vlasti v Tsifrakh, 1917–1927* (n.d.), pp. 118-119.          [4] R. Luxemburg, *Die Russische Revolution* (1922), p. 87.

and listed first, Soviet farms and communes, secondly, *artels* and associations, and thirdly, individual holdings, in that order, as forms of land tenure.[1] But this preference for collective cultivation was linked with the compulsions of war communism ; and, when these failed, and the introduction of NEP restored the element of capitalist competition to the countryside, the issue between " socialist " and " capitalist " forms of land tenure, between collective and individual agriculture, was once more wide open, with the dice heavily loaded, as in the days of Stolypin, in favour of the " capitalist " solution. When, at the end of 1924, Narkomzem issued its five-year plan for agriculture, it plausibly argued that " the socialist form of agriculture " would be conceivable only when " the individual form of farming has become a form which puts the brake on development ". This stage had not yet been reached ; and in the present stage the correct formula was not " through collective agriculture to the development of productive forces ", but vice versa.[2] Nothing seemed to have been settled by the revolution. The problem had once again to be faced from the beginning.

Of forms of collective agriculture in the broadest sense of the term, the traditional peasant commune or *mir* was by far the most important. The essence of the *mir* was collective tenure, not collective working, of the land. It imposed on its members an obligatory rotation of crops, usually based on the primitive three-field system, and made provision for common use of pasture and water-rights. But the *dvor*, the peasant family or household, remained the unit of cultivation and of membership of the *mir* ; the main business of the *mir* was to allocate the land among the *dvors* belonging to it. This periodical redistribution of the land belonging to the *mir*, sometimes annually, more often at three-yearly or five-yearly intervals (according to the crop rotation in use),[3]

[1] For this decree see *The Bolshevik Revolution, 1917–1923*, Vol. 2, pp. 43-46.
[2] *Osnovy Perspektivnogo Plana Razvitiya Sel'skogo i Lesnogo Khozyaistva* (1924), p. 30 ; for this plan see p. 491 below.
[3] Annual redistribution seems to have become comparatively rare, though a detailed description of a region in the province of Yaroslav where it was still practised in the middle nineteen-twenties (*Na Agrarnom Fronte*, No. 1, 1926, pp. 166-174) does not suggest that it was exceptional. In some cases the *mir* had ceased to redistribute at all ; this was the first stage in the withering away of the *mir*.

made it the only form of land tenure under which the principle
of equality could be applied and maintained in perpetuity,
since redistribution took account of the changing numbers of
workers and "eaters" in each *dvor*.  In this respect, the insistence
in the first months of the revolution, under SR influence, on the
principle of equality helped to maintain the prestige of the *mir*
and to prolong its life.  In spite of the emphasis laid in party
doctrine and in the " socialization " decree of February 1918 on
collective cultivation, none of the Bolshevik leaders challenged
the rights of the *mir*.  The agrarian code of December 1922,
which embodied the principles of NEP, unconditionally accepted
the *mir* on the same footing as other current forms of land tenure.
The right of the individual household to secede from the *mir*
with an allocation of land and form an independent *khutor* or
*otrub* [1] was recognized, though hedged about with restrictions.
But it does not seem to have been widely exercised at this time.
The *mir* survived as the predominant form of tenure throughout
the European provinces of the Soviet Union, except the Ukraine
and White Russia, and in the cultivated areas of Siberia.  Of 233
million hectares of peasant land in the RSFSR on January 1,
1927, 222 millions were still held in communal tenure on the *mir*
system ; *khutors* and *otrubs* accounted for only 2 millions and 6
millions respectively, and various forms of kolkhozy for rather
more than 2 millions.[2]  The common assumption that the peasants
who seceded from the *mir* were the most enterprising, capable
and prosperous was probably in the main correct ;  and this sometimes led to attacks on the *khutor* as the stronghold of the *kulak*,
whereas the *mir* continued to uphold the interests of the poor
peasant.[3]  On the other hand, the contrary allegation was also

[1] See *The Bolshevik Revolution, 1917–1923*, Vol. 2, pp. 287, 296.  Both
*khutors* and *otrubs* were independent holdings separated from the *mir* ;  the
*khutor* was a farm on which the holder lived, the *otrub* a farm worked by a
holder who continued to live in the " village " with his former co-members of
the *mir*.

[2] *Itogi Desyatiletiya Sovetskoi Vlasti v Tsifrakh, 1917–1927* (n.d.), pp. 120-
121.  Figures for the Ukraine do not appear to be available ;  in parts of the
Ukraine the *mir* was obsolescent, and the Stolypin reform had made far more
progress here than elsewhere.

[3] An article to this effect appeared in *Leningradskaya Pravda*, November 4,
1925, as part of the current campaign against the *kulak* ;  according to this
article, there had been a strong move in the neighbouring provinces of Pskov
and Novgorod to form *khutors* after the beginning of NEP, but they had since

heard that some *kulaks* continued to cling to the *mir* as " the most convenient way for them to exploit the poor peasant and to conceal their excess of land ".[1] Throughout this period the *mir* had few defenders in the party. But the force of inertia, and the traditional resistance of the peasant to change, kept it in being over the greater part of the RSFSR.

The *dvor* as a unit of cultivation, whether in the *mir*, or in independent *khutors* or *otrubs*, typified the dilemma of Soviet agriculture even more dramatically than the *mir* ; for here too, and even more conspicuously, the collective principle was embodied in what was socially a primitive and backward form of organization. The framers of the agricultural code of 1922, fearful of an in- definite splitting of the land into small units of cultivation, had allowed provincial executive committees to fix limits of size, appropriate to each locality, below which separation from an existing *dvor* would not be sanctioned. The purpose and effect of the relevant articles of the code (arts. 85-89) was plainly to place obstacles in the way of the break-up of the *dvor* into still smaller units of individual cultivation. A decree of the White Russian SSR of April 1925 went so far as to prohibit altogether any further division of existing *dvors*.[2] Yet strict party doctrine could not help regarding the *dvor*, which in practice implied sub- jection of wife, family and other dependants to the paternal authority of the head of the household, as a " shameful relic of the time of serfdom ".[3] Peasant opinion itself was divided, and was governed by the circumstances of particular cases. It was commonly recognized that one of the main factors in the break-up of the *dvor* was the growing independence of women, and in particular the unwillingness of the young wife to submit to the

become unpopular, and many had been re-divided. A year later Sukhanov, a recent convert from Menshevism, was allowed to publish a " discussion article " extolling the *mir* as the last defence against the *kulak* (*Na Agrarnom Fronte*, No. 11-12, 1926, pp. 97-110).

[1] *Ibid.* No. 10, 1926, pp. 91-92 ; a similar allegation was made about the *kulaks* in the north Caucasus region (*Planovoe Khozyaistvo*, No. 11, 1925, pp. 70-71).

[2] *Zbor Zakonau i Zahadau BSSR*, *1925*, No. 24, art. 220 : it appears to have been a sequel to a decree of two months earlier which called for reorganiza- tion to do away with scattered holdings and strip holdings, and for the substitution of a multi-field system for the current three-field rotation (*ibid.* No. 8, art. 65).

[3] P. Stuchka, *13 Let Bor'by za Revolyutsionno-Marksistskuyu Teoriyu Prava* (1931), p. 208.

rule of her husband's parents in the family *dvor*.[1]  In 1926 the People's Commissar for Justice of the RSFSR declared that it was time to " put a cross over the reactionary utopia of seeking to preserve the patriarchal family and to check the process of division of peasant families into smaller units ".[2]  Whatever obstacles were placed in the way of further subdivision of existing *dvors*, it seems clear that the *dvor* was a dying institution, and that the process of disintegration, though gradual, was continuous.  Statistics showed throughout this period a net increase in the number of peasant households at the rate of at least 2 per cent per annum.[3]

The new experiments in collective cultivation sponsored or encouraged by the Soviet régime — the Sovkhozy and the kolkhozy in all their forms [4] — rejected periodical distribution, and were based on the principle of collective cultivation, departing in both respects from the time-honoured system of the *mir*.[5]  In theory they continued to enjoy the highest approval, and were applauded in every official pronouncement.  In practice the climate of NEP was inimical to forms of tenure associated with the first years of the revolution and the exactions of war communism ; and in 1925 it was estimated that not more than 2 per cent of the land of the USSR was under different forms of collective cultivation.[6] The Sovkhozy in particular continued to languish in the neglect and discredit into which they had fallen after the introduction of NEP.  The total area occupied by Sovkhozy fell from 3·4 million hectares in 1921 to 2·3 millions in 1926.[7]  Reports on the limited

[1] *Bednota*, May 1, 1925, printed a large number of letters from peasants for and against the perpetuation of the *dvor* :  the subject was frequently discussed in its columns.

[2] *III Sessiya Vserossiiskogo Tsentral'nogo Ispolnitel'nogo Komiteta XII Sozyva* (1926), p. 562 ; for some instructive comments on this problem, made in the course of a debate on the marriage code of the RSFSR, see *ibid.* pp. 560-561, 678-679.

[3] *Statisticheskii Spravochnik SSSR za 1927 g.* (1927), pp. 64-65 ; *Statisticheskii Spravochnik SSSR za 1928 g.* (1929), p. 82.

[4] For the early history of these experiments see *The Bolshevik Revolution, 1917-1923*, Vol. 2, pp. 151-156, 289-290.

[5] The agrarian code of December 1922 used the term " agrarian association " (*zemel'noe obshchestvo*) to cover both the old *mir* and the new kolkhoz ; the Sovkhozy, or state farms, were on a different footing (since those who worked on them were wage-earners), and were dealt with in another chapter.

[6] *Vestnik Kommunisticheskoi Akademii*, xiv (1926), 191.

[7] *Sovkhozy k xv Godovshchine Oktyabrya*, ed. Krylov (1932), p. 6 ; according to this account the year 1926 marked the nadir of the Sovkhozy, which thereafter began to make a gradual recovery.  *Itogi Desyatiletiya Sovetskoi Vlasti v*

number of Sovkhozy grouped under the authority of Gossel'-sindikat,[1] comprising 833 Sovkhozy in the RSFSR, 131 in the Ukraine, 164 in White Russia and 26 in the Transcaucasian SFSR, showed that by no means all the land occupied by these was being worked at all, and that of the part that was worked a considerable proportion (39 per cent in the RSFSR, 72 per cent in the Ukraine and 85 per cent in White Russia) was leased to peasants for cultivation.[2] Circumstantial evidence was quoted to show that in a considerable number of cases former landowners had retained the whole or part of their estates in the guise of Sovkhozy of which they were the managers. According to one report, some Sovkhozy did in fact fulfil their original purpose of serving as model farms for the surrounding peasants, supplied them with seeds, kept bulls and stallions and hired out tractors and agricultural machinery. But there is little evidence that these good intentions were often translated into practice ; and the conclusion of another contemporary report that " the Sovkhozy at the present time cannot serve as an example to the surrounding peasant population of correct and rational farming " was probably nearer the truth. It was admitted that the Sovkhozy were, in general, working at a loss and without adequate capital resources, and that, in the absence of state subsidies, which were not forthcoming, financial disaster had been avoided only by liquidating the less efficient Sovkhozy and transferring their property to the more efficient.[3] An account of the derelict condition of the few remaining Sovkhozy in the province of Smolensk in the autumn

*Tsifrakh, 1917-1927* (n.d.), pp. 120-121, gives a total area of nearly 2·3 million desyatins on January 1, 1927, in the RSFSR alone : figures for the RSFSR, the Ukraine and White Russia for 1924-1926 are given *ibid.* p. 164 (a breakdown of the figures of Sovkhozy in the RSFSR will be found in *Statisticheskii Spravochnik SSSR za 1927 g.* (1927), p. 60, which corrects an obvious misprint in the figure for Sovkhozy in the sugar trust).

[1] See *The Bolshevik Revolution, 1917-1923*, Vol. 2, p. 290.

[2] *Na Agrarnom Fronte*, No. 3, 1925, pp. 111-112. In the Ukraine, the Sovkhozy occupied only 145,000 desyatins as against 518,000 occupied by the Union Sugar Trust and 885,000 by other state enterprises ; over-population was acute, and there was " great pressure from the peasants to hand over this land to them " (*SSSR : Tsentral'nyi Ispolnitel'nyi Komitet 3 Sozyva : 2 Sessiya* (1926), p. 421).

[3] The reports are in *Na Agrarnom Fronte*, No. 1, 1925, pp. 57-68 (which is immediately followed (*ibid.* pp. 68-77) by a rather unconvincing apologia), and *Planovoe Khozyaistvo*, No. 4, 1925, pp. 27-41 ; a belated attempt to refute the second report was made in *Na Agrarnom Fronte*, No. 11-12, 1925, pp. 35-44.

of 1924 [1] appears to have been typical of prevailing conditions.
The rare description of a model Sovkhoz in the vicinity of
Chelyabinsk serves only to emphasize the exceptional character
of the phenomenon.[2]

Such evidence as is available goes to show that official policy
at this time took scant interest in the fate or in the future of the
Sovkhozy. In February 1925 the Orgburo of the party turned its
attention to their condition and issued a report. It admitted their
shortcomings, but thought that their already reduced number
should be maintained, " except for those which are obviously
non-viable economically ". For those that were retained, it
demanded an extension of the application of piece-work ; stricter
discipline ; a ten-hour working day ; and fuller employment of
labour.[3] A decree of the RSFSR of March 6, 1925, adopted a
less robust attitude, prescribing an improvement in the personnel
and organization of the Sovkhozy ; an increase in their capital ;
and more regular payment of wages and better conditions for
the workers.[4] But even these were utopian requirements in this
period. It was a more significant index of the prevailing state of
affairs when, in March 1925, the central administration of state
industry (a department of Vesenkha) advised industrial trusts to
liquidate all Sovkhozy under their control, since these yielded
losses and no longer served any useful purpose.[5] In October
1925 the TsIK of the White Russian SSR gave some attention to
the " grievous conditions " of the Sovkhozy in the republic, professed to detect some recent signs of improvement, and instructed
Sovnarkom to draw up a plan for their development.[6]

The neglect of the Sovkhozy was reflected in the miserable
plight of those who worked in them. A report on the status of
batraks [7] at the beginning of 1925 claimed in general terms that

---

[1] *Ekonomicheskaya Zhizn'*, October 12, 1924.
[2] P. Lezhnev-Fin'kovsky and K. D. Savchenko, *Kak Zhivet Derevnya* (1925),
pp. 34-44, where it is accompanied by a corresponding account of a bad Sovkhoz
in the same district ; the difference appears to have turned primarily on the
capacity and personality of the respective managers.
[3] *Izvestiya Tsentral'nogo Komiteta Rossiiskoi Kommunisticheskoi Partii
(Bol'shevikov)*, No. 11-12 (86-87), March 23, 1925, pp. 9-10.
[4] *Sobranie Uzakonenii, 1925*, No. 18, art. 121.
[5] *Pravda*, March 11, 1925.
[6] *Zbor Zakonau i Zahadau BSSR, 1925*, No. 48, art. 375.
[7] For the *batrak* see p. 230 below.

*batraks* employed in Sovkhozy were better off than those employed by individual peasants, but went on to attack a Sovkhoz
in the province of Tver which treated its workers with " intolerable roughness " and paid their wages in rations of the poorest
quality, and another near Semipalatinsk where " the discipline of
the stick " was in use.[1] Later in the same year *Pravda* complained
of the shameless exploitation of peasants working on a Sovkhoz
formed out of a previous landowner's estate in Bashkiria.[2] " If
we ask the peasants about the Sovkhozy ", said Bukharin to a
conference of Sovkhozy managers, " we shall in the majority of
cases get very unflattering answers." [3]

The collective farms formed on peasant land under voluntary
sponsorship (kolkhozy) fared at first a little better, having recovered
somewhat from the decline in numbers recorded after the introduction of NEP.  But the element of communal cultivation for which
the kolkhozy originally stood had substantially diminished.  The
original kolkhozy had been divided into two categories : agricultural communes whose members worked, lived and distributed
the proceeds of their work on a communal basis, the land being
registered in the name of the commune, and *artels* whose members
carried on cultivation and the marketing of products on a joint
basis, but lived separately and received their own share of the
proceeds, the land being registered in separate units in the
individual names of the members.  A third and still looser
category of kolkhoz had now grown up in the form of the so-
called TOZ (*Tovarishchestvo dlya obshchego zemlepol'zovanya*,
or Association for Common Cultivation), whose members merely
cooperated for the joint cultivation of certain tracts of land without
further mutual obligation.  But the credentials of this form of
association were suspect.  An investigation of a number of TOZy
in the province of Tambov in 1924 showed that " there is nothing
cooperative about them " ;  they served simply as cover for the
renting of land by rich peasants who found it expedient " to use
the cooperative trade mark ".[4]  Even those agricultural communes

[1] *Voprosy Truda*, No. 2, 1925, p. 34.
[2] *Pravda*, October 14, 1925.          [3] *Ibid.* March 6, 1925.
[4] Y. Yakovlev, *Nasha Derevnya* (1924), p. 57.  For the way in which industrial *artels* degenerated into " petty private concerns " see *The Interregnum*,
*1923–1924*, pp. 52-53 ;  the same thing was doubtless liable to happen to
agricultural *artels*.

which survived seem to have lost much of the original communal spirit.[1] Kolkhozy, like Sovkhozy, sometimes disguised the illicit retention by landowners of their former estates ; [2] and, though a decree was passed in March 1925 ordering the expulsion, before January 1 of the following year, of former landowners who, " under the guise of the organization of agricultural artels and communes, . . . exploit the peasants, using hired labour and adopting oppressive attitudes to the surrounding population ", the decree itself opened the door to exemptions in favour of former landowners who themselves worked, or had served in the Red Army, or " rendered special services to the state ". Even so, it seems to have remained a dead letter.[3]

It was, however, the general line of policy resulting from NEP, rather than the consciousness of these particular abuses, which accounted for official lukewarmness towards the kolkhozy. At the end of February 1925 a kolkhoz conference was held in Moscow at which kolkhoz peasants from all parts of the USSR recounted their achievements and their ambitions for the future, expecting, no doubt, a measure of official approbation and support. But cold water was quickly thrown on their ardour. The official *rapporteur* Kaminsky praised the agricultural cooperatives, which were in many respects the rivals of the kolkhozy and represented looser forms of cooperation ; " through the simplest form of agricultural cooperatives ", he told the conference, the path would

---

[1] P. Lezhnev-Fin'kovsky and K. D. Savchenko, *Kak Zhivet Derevnya* (1925), pp. 44-54, describe an agricultural commune in Siberia founded in 1920 by 56 peasants : only 9 of these remained in 1924, and made up for the deficiency in numbers by employing hired workers.

[2] *Na Agrarnom Fronte*, No. 3, 1925, p. 106.

[3] *Sobranie Zakonov, 1925*, No. 21, art. 136 (see also *ibid*. No. 44, art. 328) ; a year later a further decree (*Sobranie Zakonov, 1926*, No. 19, art. 118) retrospectively postponed the date for expulsion from January 1 (here misprinted August 1) to April 1, 1926. Cases certainly occurred of the retention, on one pretext or another, of relatively large holdings. A delegate at a conference on local government in January 1925 complained of old landowners with only five " eaters " in their family still holding 40-50 desyatins of land (*Soveshchanie po Voprosam Sovetskogo Stroitel'stva 1925 : Yanvar'* (1925), p. 138). In a Siberian district, where middle peasants, who formed the bulk of the population, occupied 286 farms of an average area of 12·8 desyatins, four rich peasants, holding 1400 desyatins between them and employing hired labour on a large scale, had apparently survived untouched since before the revolution (P. Lezhnev-Fin'kovsky and K. D. Savchenko, *Kak Zhivet Derevnya* (1925), pp. 27-28).

lead to " full collectivization " in the future.  Bukharin pointed the same moral and brusquely played down the rôle of the kolkhozy :

> We cannot begin collectivization from the angle of production, we must begin from the other angle.  The high road leads along the cooperative line. . . .
> Collective farms are not the main line, not the high road, not the chief path by which the peasant will come to socialism.

Krupskaya claimed that " collective farming is the best farming ", but stuck to generalities and apologized for being unable through illness to make a promised report on cultural work in the kolkhozy. In spite of protests by delegates from the Ukraine, where the kolkhoz movement seems to have been more highly developed than elsewhere,[1] the conference passed a resolution to the effect that " agricultural cooperatives, of which agricultural collectives (communes, *artels*, associations) are only one variety, represent in our conditions the only possible form of transition from small peasant production to a technically superior large-scale economy ", and condemned attempts to isolate the kolkhozy from the cooperatives and organize them in " independent unions of kolkhozy ".  The kolkhozy were to receive no further encouragement as a separate movement, but were to be merged in the general stream of agricultural cooperatives.[2]

---

[1] In the Ukraine some kolkhozy had the character, which they rarely possessed elsewhere, of genuine cooperative undertakings of small peasants for common defence against exploitation by *kulaks*.  " The poor peasants, the weak peasants who had received land, but had no possibility of working it with their own resources, entered kolkhozy " : some of these collapsed, but in 1925 there were 5578 kolkhozy in the Ukraine covering 286,600 persons (*Vserossiiskii Tsentral'nyi Ispolnitel'nyi Komitet XII Sozyva : Vtoraya Sessiya* (1925), p. 416).  In the Ukraine four-fifths of members of kolkhozy were described as poor peasants (*Na Agrarnom Fronte*, No. 2, 1926, p. 84).  In so far as the kolkhozy were really organizations of poor peasants, the move to subordinate the kolkhozy to the cooperatives was a move in favour of the well-to-do peasants who dominated the cooperatives (see pp. 277-279 below).

[2] The conference was reported in *Pravda*, March 5, 6, 1925.  More detailed accounts appeared in *Na Agrarnom Fronte*, No. 3, 1925, pp. 162-164 ; No. 4 1925, pp. 34-40 ; a further promised article on the conference did not appear, probably because the author was evidently unsympathetic to the policy decided on.  Bukharin's speech, which was delivered on March 4, 1925, was not recorded in these reports, possibly because it had appeared in full in *Pravda*, March 6, 1925 ; Bukharin made similar remarks in a later speech of April 1925 (see p. 264 below).  The proceedings of the conference were subsequently published as *Trudy Pervogo Vsesoyuznogo Soveshchaniya Sel'skokhozyaistvennykh Kollektivov* (1925), which has not been available.

In face of these pronouncements, the further development of collective cultivation was unlikely. The numbers of kolkhozy appear indeed to have increased in 1924 and 1925.[1] The kolkhoz of average size was composed at this time of about 50 " eaters ", including 20 or 30 " workers ".[2]  Especially favourable reports came from the Ukraine, where the kolkhozy secured credits from the cooperatives for the purchase of seeds and machines, and claimed to have produced substantially larger harvests than the surrounding peasants,[3] and the northern Caucasus, where kolkhozy were formed to cultivate land not previously worked.[4]  But this partial revival of the kolkhozy did not last.  From an alleged total in 1925 of 16,000 for the RSFSR and 5500 for the Ukraine they declined to a total for the whole USSR of 14,800 in 1927.[5]  Whether the figures were indicative of a real decline, or of the fact that it was no longer helpful to claim kolkhoz status, it is clear that from 1925 onwards collective agriculture was receiving little effective official support.

While all attempts to find a " socialist " solution of the problem of Soviet agriculture through the encouragement of collective units of cultivation remained in the doldrums, the rival forces of rural capitalism had registered an unmistakable advance. Since the coming of NEP, and with the greater freedoms accorded by the agrarian code of 1922, the reaction against equality had set in ;  the necessity of rewards for the enterprising and successful peasant was no longer contested.   When in 1924 the peasantry

[1] *Itogi Desyatiletiya Sovetskoi Vlasti v Tsifrakh, 1917–1927* (n.d.), p. 165.
[2] *Na Agrarnom Fronte*, No. 5-6, 1926, p. 121.
[3] *Planovoe Khozyaistvo*, No. 5, 1925, pp. 229-235 ; *Na Agrarnom Fronte*, No. 2, 1926, pp. 83-94 ; No. 5-6, pp. 129-138.   A large majority of Ukrainian kolkhozy appeared in the statistics as *artels*, though many TOZ are said to have registered as *artels*.
[4] *Planovoe Khozyaistvo*, No. 11, 1925, pp. 81-83.   Trotsky in a speech of November 1925 gave an unfavourable report on an agricultural commune which he had just visited in the Ukraine, where productivity was lower than that of the surrounding peasants (*Pravda*, November 28, 1925), but balanced this a few days later by a more favourable account of a commune in the northern Caucasus (*ibid.* December 5, 1925).
[5] *Na Agrarnom Fronte*, No. 5-6, 1926, pp. 121, 130 ; *Sdvigi v Sel'skom Khozyaistve SSSR* (Gosplan, 1931), p. 29.   The figures for 1925 and 1926 in *Itogi Desyatiletya Sovetskoi Vlasti v Tsifrakh, 1917–1927* (n.d.), p. 165, are substantially lower.

emerged victorious from the battle of the scissors, the only question was how far the reaction would be allowed to go. No fresh decision of public policy seemed to be required : the state had merely to stand aside, and allow the anti-egalitarian tendencies inherent in the situation to work themselves out. The case was frankly and cogently stated at the beginning of 1924 by a representative of Narkomzem :

> The rôle of the well-to-do peasantry in the growth of the production of grain and cattle is acquiring an exclusive significance in the national economy. On these strata of the peasantry, as well as on the agent who brings the commodities to the foreign or domestic market, rests the task of rebuilding the economy. All measures directed to economic recovery are, therefore, bound by the objective course of events to promote those conditions in which recovery is possible ; such measures will further the development of well-to-do farms and help to transform middle peasants into well-to-do peasants. There is no other way in a money-commodity relationship.
>
> The other side of the development of the domestic economy (industry) also pushes peasant agriculture along the path of differentiation in the near future. In so far as industry develops, the small, weak households will leave agriculture for industry and clear the ground for a deepening of class differences in the country.[1]

Another commentator described the process in the terms of Marxist class analysis :

> The process of the class differentiation of the peasantry and the process of the development of capitalism are synonymous, and are one and the same thing. The development of capitalism in the countryside means the growth of the capitalist strata of the peasantry on the one side and the growth of the proletarian strata of the peasantry on the other.[2]

Sufficiently wide diversities existed over the vast territory of the USSR to make generalization hazardous. But, broadly speaking, the process of differentiation was facilitated in three main ways : through the leasing of land, sanctioned, though with inconvenient restrictions which always enabled its legality to be called in question, by the agrarian code of 1922 ; through the loaning of draught animals and agricultural machines and implements, which had

---

[1] *Sotsialisticheskoe Khozyaistvo*, No. 2, 1924, p. 57
[2] L. Kritsman in *Vestnik Kommunisticheskoi Akademii*, xiv (1926), 191.

never been prohibited ; and through the hiring of labour, which had also been sanctioned, though with similar restrictions, by the code.[1] The leasing of land [2] took place in two conditions. The poor peasant not possessing enough land to feed his family could, if he had sufficient resources, rent a further plot to make up his holding to the requisite size.  A case was recorded from the north Caucasus in which *kulaks* rented 6000 desyatins of pasture from the state at 60 kopeks a desyatin, and leased it to poor peasants at a rate of three puds of wool per head of sheep grazed on it.[3]  But such transactions were rarely possible in the more densely populated regions, and can seldom have applied to arable land, since the poor peasant rarely possessed a horse or other draught animals and was unable, even if he could rent land, to cultivate it to yield a crop adequate for the needs of his family.  Hence the far commoner form of leasing was for the rich peasant, who possessed horses and implements in sufficient quantity, to rent land from his poorer neighbour, who, giving up the unequal struggle to cultivate his land on his own account, leased it to the *kulak* in return for a share of the harvest and, for the rest, lived by hiring out his labour.[4]   This practice was everywhere on the increase.

---

[1] For the provisions on leasing and hiring see *The Bolshevik Revolution, 1917–1923*, Vol. 2 pp. 288-289, 296-297.

[2] It is worth noting that what was leased was technically not the ownership of the land, but the right to utilize it.  By a strange anomaly or oversight, art. 27 of the agrarian code prohibited the buying, selling or mortgaging of land.  Since private ownership no longer existed, the provision, strictly interpreted, was meaningless ;  and the attempt was made to argue that this prohibition related not to the ownership, but to the utilization, of land.  But, according to Stuchka (*Vestnik Kommunisticheskoi Akademii*, xiii (1925), 244), the Supreme Court ruled that it must be held to relate strictly to ownership, and be regarded as an obsolete survival.

[3] *Na Agrarnom Fronte*, No. 9, 1926, p. 107.  The rate quoted seems, however, too exorbitant even for the greediest *kulak* ;  perhaps " three puds " should read " three pounds ".  The practice of sub-leasing was condemned in the resolution of the party central committee of April 30, 1925 (for which see pp. 266-269 below).

[4] The fullest review of conditions of leasing at this time is in *Ekonomicheskoe Obozrenie*, May 1925, pp. 1-17.  One point that emerges is that these were rarely cash transactions.  The rich peasant who rented land from the poor made payment in the form of a share of the harvest ;  the poor peasant who rented land from the rich (generally only a small plot for family cultivation) paid in the form of labour.  Similarly a *kulak* would advance grain for sowing·to the peasant to be repaid after the harvest, and exact from him a certain number of labour days by way of interest (*Vlast' Sovetov*, No. 11, March 15, 1925, p. 4). The habits of a monetary economy had still not penetrated the countryside.

Figures from two areas in the northern Caucasian region showed that, in one, two-and-a-half times as much land was leased in 1925 as in 1924, and in the other nearly twice as much.[1] The proportion of land leased rose in one case from 10 per cent to 20 per cent of the sown area, and in the other from 9 per cent to 15 per cent. On the other hand, since leasing agreements were seldom concluded for more than one year, the lessee had no interest in maintaining or improving the condition of the land held by him on lease,[2] and merely sought to obtain the maximum return from it in a single harvest. The number of poor peasants eager to dispose of land which they had not the capacity to cultivate was so large as to depress rents for such land to a very low figure. Where from 8 to 13 rubles a year had been paid for a desyatin of land in the northern Caucasus before the war, it was now worth only from 50 kopeks to 3 rubles a desyatin ; a case was quoted in which a widow holding 10 desyatins of land leased 9 of them to a *kulak* in return for an undertaking to plough and sow the remaining desyatin for her use with seed provided by her.[3] The proportion of farms which included rented land rose in the " consuming " provinces of the RSFSR from 5·5 per cent in 1925 to 10·8 per cent in 1926 and in the " producing " provinces from 11·2 per cent in 1925 to 18 per cent in 1926 :[4] this illustrated both the increase of leasing, and its greater prevalence in the more fertile regions. Everywhere the result of the development of leasing was a regrouping of the land and a tendency towards greater productivity through the creation of larger working units in the hands of the most enterprising and efficient peasants.

In view of the varied forms under which land was leased, and the aroma of discredit, if not illegality, surrounding the practice, it is unlikely that the full extent of these changes was reflected in official statistics ; nor are such statistics as exist strictly comparable

[1] *Planovoe Khozyaistvo*, No. 10, 1925, p. 38. Of the land in question less than 10 per cent was leased to small peasants to make up a deficiency ; about half was leased to peasants already holding 16 desyatins or more, the remainder to peasants holding between 6 and 16 desyatins (*ibid.* No. 10, 1925, p. 40).

[2] Under article 35 of the agrarian code, " the improvements which the lessee is obliged to carry out " were to be stipulated in the leasing agreement ; but this obligation was generally honoured in the breach.

[3] *Planovoe Khozyaistvo*, No. 10, 1925, pp. 36, 41.

[4] A. Gaister, *Rassloenie Sovetskoi Derevni* (1928), pp. 53-54.

with those of the earlier period. Down to 1921 the number of landless peasants and of peasants holding more than 4 desyatins of land were both declining; in 1920, 86 per cent of peasant households were returned as holding up to 4 desyatins of land.[1] Under NEP, and especially after 1923, these processes were partly reversed. More peasants lost their land; and more peasants had holdings in excess of 4 desyatins. In 1925, in the consuming provinces of the RSFSR, 22 per cent of the peasants were returned as holding either no land or less than 1 desyatin, 67 per cent from 1 to 4 desyatins, and 11 per cent more than 4 desyatins. In the producing provinces, the same three categories accounted for 13·5, 53 and 33·5 per cent of the peasants respectively.[2] In the Ukraine the proportion of landless peasants rose from 3·2 per cent in 1923 to 4·4 per cent in 1924, the proportion holding less than 2 desyatins falling from 37 to 33 per cent of the whole.[3] In the European provinces of the USSR the proportion of peasant holdings of from 6 to 10 desyatins rose from 5·6 per cent in 1922 to 10·2 per cent in 1925, and of holdings of over 10 desyatins from 1·2 per cent to 3·3 per cent.[4] In Siberia in 1925, the number of landless peasants and of peasants holding more than 4 desyatins was increasing, the number of holdings of less than 4 desyatins decreasing.[5] In 1926 a considerable proportion of small holders were lessors, and of large holders lessees, of land, the proportions varying closely in relation to the amount of land held; leasing was most prevalent in the central agricultural regions of the RSFSR, in the Ural region, in Siberia, in the Lower Volga region, in the North Caucasian region and in the Ukraine.[6]

The loaning of working animals and agricultural implements and machines was probably an even more important factor than the leasing of land in the growth of rural capitalism. The fall in

[1] See *The Bolshevik Revolution, 1917–1923*, Vol. 2, p. 168.

[2] *Statisticheskii Spravochnik SSSR za 1927 g.* (1927), pp. 78-79.

[3] *Na Agrarnom Fronte*, No. 2, 1925, p. 56; details of the proportion of landless peasants in different provinces of the Ukraine are given *ibid.* No. 4, 1925, p. 74.

[4] *XV Konferentsiya Vsesoyuznoi Kommunisticheskoi Partii (B)* (1927), p. 125.

[5] Statement of Lashevich reported in *Leningradskaya Pravda*, December 16, 1925.

[6] *Itogi Desyatiletiya Sovetskoi Vlasti v Tsifrakh, 1917–1927* (n.d.), pp. 144-151.

the number of working animals was a significant index of conditions in Soviet agriculture. Though the number of cattle and sheep in the USSR had by 1925 reached or almost reached 1916 levels, the number of draught horses in 1925 was only 20 millions as against 27 millions in 1916.[1] This total figure covered local diversities. In the " consuming " provinces of central and north-western Russia, the numbers of horses approximated to those of 1916. In the Ukraine, in the Volga provinces, in the northern Caucasus and in Central Asia figures well below the general average were registered ; and it was here that the pressure on the poor peasant was most acute.[2] While the result of the revolution had been to reduce by one-half the proportion of landless peasants, no corresponding reduction had taken place in the proportion of peasants not owning a horse. According to one set of figures, the percentage of peasants in the RSFSR not owning a horse fell from 29 to 27 between 1917 and 1920, and the percentage owning only one horse rose during the same years from 49 to 63.[3] According to later figures, the percentage of peasants in the RSFSR " without working animals " had fallen to 24 in 1922, and rose again to 27 in 1924.[4] In the Ukraine 52 per cent of the peasants had no horse in 1924, and 32 per cent only one horse.[5] In White Russia, in 1925, 19 per cent of peasants were said to have no horse [6] — a surprisingly low figure. Other figures for the RSFSR showed percentages of " horseless " peasants ranging from 44 in the

[1] Kontrol'nye Tsifry Narodnogo Khozyaistva na 1926–1927 god (1926), p. 338 ; the tables in Ekonomicheskoe Obozrenie, No. 3, 1929, p. 112, give a figure of 18,777,000 for 1925. Slightly higher figures are given in Itogi Desyatiletiya Sovetskoi Vlasti v Tsifrakh, 1917–1927 (n.d.), pp. 188-193, showing a steady increase in all regions from 1924 to 1926.

[2] Ekonomicheskoe Obozrenie, December 1925, pp. 94-97 ; approximately the same results are obtained from a table purporting to include all draught animals (ibid. January 1926, p. 13). In the Ukraine the numbers of cattle and sheep had risen in 1925 above the 1916 figures, but the numbers of horses and pigs had declined ; the total number of animals was said to be still falling (SSSR : Tsentral'nyi Ispolnitel'nyi Komitet 3 Sozyva : 2 Sessiya (1926), pp. 415, 422) ; in White Russia there had been " some increase " in the number of cattle in 1924–1925, but a continued decline in the number of horses (Na Agrarnom Fronte, No. 5-6, 1925, p. 177).            [3] Ibid. No. 2, 1925, p. 49.

[4] Planovoe Khozyaistvo, No. 8, 1925, p. 104.

[5] Na Agrarnom Fronte, No. 2, 1925, p. 54. In the Ukraine some peasants tried to use cattle for ploughing, but this was ineffective : the land required " strong horses " or tractors (Vserossiiskii Tsentral'nyi Ispolnitel'nyi Komitet XII Sozyva : Vtoraya Sessiya (1925), p. 419).

[6] Vlast' Sovetov, No. 20, May 15, 1925, pp. 14-15.

province of Olonets to 29 in the province of Kostroma, with
exceptionally low figures of 17 and 12 per cent respectively for the
province of Bryansk and the North-Western region.[1]  A popular
estimate for the whole of the USSR, frequently repeated at this
time, put the proportion of " horseless " peasants in 1924 at 40
per cent.[2]  In few regions of the USSR did the proportion of
peasants owning more than one animal rise above 10 per cent ;
only in Siberia and in the Ukraine did more than 20 per cent of
the peasants own more than one animal.[3]  The number of horses
owned was perhaps the most significant single index of peasant
prosperity or indigence, since the peasant not possessing a horse
or other draught animal could not, as a rule, make ends meet by
the cultivation of his own land without seeking supplementary
sources of income.  The possession of animals was also an important factor in maintaining the fertility of the soil.[4]

No general figures for agricultural implements are available.
But it may be assumed that those peasants who had most horses
were also best supplied in this respect, and that the " horseless "
peasants had none but the most primitive implements.[5]  The
possession of implements and machines varied with the size of
the holdings.  Of peasants farming less than 2 desyatins only 29
per cent in 1925 owned an iron plough ; only in farms of from 10

---

[1] *Na Agrarnom Fronte*, No. 3, 1926, p. 90.
[2] See, for example, Zinoviev's speech in *Pravda*, May 27, 1925 ; according
to statistics quoted in P. I. Lyashchenko, *Istoriya Narodnogo Khozyaistva SSSR*,
ii (1952), 279, the proportion of horseless peasants in 1912 was 31·4 per cent, as
against 29 per cent a decade earlier.  The division of land and the increase in
the number of peasant holdings after the revolution automatically increased the
number of " horseless " units.
[3] *Itogi Desyatiletiya Sovetskoi Vlasti v Tsifrakh, 1917–1927* (n.d.), pp. 136-
141 ; these statistics show a steady diminution of the number of horseless
peasants between 1924 and 1926 — a conclusion not borne out by other sources.
[4] " Give us cattle and all will be well ", the peasants of the German Volga
republic told Rykov when he visited the republic at the time of the harvest
failure in August 1924 (*Izvestiya*, September 9, 1924).
[5] A table in *Vestnik Kommunisticheskoi Akademii*, xiv (1926), 220, shows the
parallel increase in the percentage of peasants in the Ukraine without working
animals and without " inventory " :

|      | Without Working Animals | Without Inventory |
|------|------|------|
| 1921 | 19 | 24 |
| 1922 | 34 | 30 |
| 1923 | 45 | 34 |
| 1924 | 46 | 42 |

desyatins upwards could the possession of one or more ploughs be taken for granted. It was the exception rather than the rule to find a sowing or reaping machine on a farm of less than 10 desyatins.[1] Except in Siberia, where the greater size of farms and the scarcity of labour created special conditions,[2] agricultural machines were still scarcely known to the ordinary peasant ; and the replacement of the horse by the tractor was only on the horizon.[3]

These conditions once more divided the peasantry into three groups : those who, having more than enough animals and implements to cultivate their own land, were prepared to loan them on suitable terms to cultivate the land of others ; those who, having no draught animals and inadequate implements, were dependent on the first group, and, even if they owned land, were little better off than the landless ; and an intermediate middle group, which struggled in good years to rise into the upper group but more often fell in bad years into the lower. It was the classic " capitalist " pattern of a divorce between the ownership of the means of production and the ownership of labour power, and the exploitation of the latter by the former. The *kulak* was the man who, owning the means of production (in which category animals and implements were fully as important as the land itself), could dictate terms to the mass of his fellow-peasants.[4]   A description

---

[1] *Bol'shevik*, No. 15, August 15, 1928, p. 31.

[2] " The Siberian peasant . . . cannot do without harvesting machines. These machines are concentrated in the hands of ' strong ' peasants . . . and are the chief weapon for the exploitation of small holders.   Payment for the use of harvesting machines is made by the peasant chiefly in the form of labour and, of course, on oppressive terms. . . . In this way these ' strong ' peasants are on the one hand outstanding skilled farmers, ' pioneers of a renewal of the land ', . . . and on the other hand are indubitably exploiters of the village poor." (*Sovetskoe Stroitel'stvo : Sbornik*, ii-iii (1925), 352).

[3] See pp. 321-323 below.

[4] A detailed account of a settlement of Russian peasants in northern Kazakhstan on the borders of Siberia shows the 70 per cent who had no horse or only one horse becoming wholly dependent on the 30 per cent who had two or more (*Na Agrarnom Fronte*, No. 9, 1925, pp. 109-110) ; these conditions appear to have been typical, except that in more fertile regions, where cultivation was more intensive, the possession of one horse generally made a peasant self-sufficient at the current low subsistence level.   According to an article by A. P. Smirnov in *Pravda*, December 22, 1925, some well-to-do peasants owned no horses, and preferred to hire them when required ; but this must have been a rare phenomenon, confined to a few special regions (Smirnov was trying to depreciate the significance of the 40 per cent of " horseless " peasants).

of a rural district in the Tver province where " differentiation "
was said to have made little or no progress since the revolution
added the significant comment that the hiring of live or dead
inventory was virtually unknown there.[1]

The third factor in the process of differentiation — the hiring
of labour [2] — was a corollary and concomitant of the other two.
The increasing concentration in the hands of a well-to-do group of
peasants of the ownership or control of the means of production
meant, at the other end of the scale, an increasing number of poor
peasants whose only resource was to sell their labour. The *batrak*,
or hired agricultural worker,[3] was the counterpart of the *kulak*.
Both categories were, however, equally fluid and difficult to
define ; and estimates of the number of *batraks* are even less
precise and satisfactory than of the number of *kulaks*. The term
was used in common parlance to cover anyone working in the
countryside for some form of wages, whether in cash or in kind.
An official inquiry set on foot by Narkomtrud in the summer of
1924 investigated 1,600,000 peasants in this category. Of those
covered by the investigation, 117,000 were employed as agricultural workers, 190,000 in building, 145,000 as timber workers
and rafters, and 358,000 as miscellaneous unskilled workers.[4] This
inquiry seems, however, to have related mainly or exclusively to
male workers, while other sources show that about 40 per cent of
*batraks* employed on agricultural work were women.[5] Nor did it
include 100,000 workers on Sovkhozy, of whom 70,000 were
agricultural workers.[6] In 1926, when the absence of statistics
had become a scandal, and a delegate at the congress of

---

[1] A. Bolshakov, *Derevnya 1917–1927* (1927), p. 155.

[2] For Lenin's dictum that " the chief sign and indicator of capitalism in
agriculture is hired labour " see *The Bolshevik Revolution, 1917–1923*, Vol. 2,
p. 148.

[3] Poor peasants and *batraks* were differently defined, but were in practice
largely identical. The poor peasant was a peasant who, having insufficient land
to maintain him, was compelled to hire out his labour ; the *batrak* was a hired
agricultural worker who might or might not, but commonly did, supplement his
earnings from a plot of land of his own.

[4] *Sotsialisticheskoe Khozyaistvo*, No. 5, 1925, pp. 185-200.

[5] *Na Agrarnom Fronte*, No. 3, 1925, pp. 92-93 ; this is borne out by some
detailed figures, *ibid*. No. 5-6, 1925, pp. 93-95. The issues of this journal for
1925 and the first months of 1926 are the best source for the numbers and
conditions of work of the *batraks* at this period ; they contain a series of substantial articles on the subject with many quotations from otherwise inaccessible
material.                                     [6] *Ibid*. No. 1, 1925, p. 40.

Vserabotzemles, the trade union of agricultural workers,[1] had complained that " we count everything — cattle, sheep and pigs — but not *batraks* ", a total estimate was published of 3,500,000 hired workers in the Soviet countryside ; this comprised 250,000 workers on Sovkhozy, 1,600,000 agricultural workers (*batraks* in the narrower sense), 600,000 shepherds and herders, 600,000 timber workers and rafters, 100,000 foresters, 150,000 workers in cooperatives and rural handicrafts, and 50,000 in miscellaneous enterprises including fisheries and grain elevators.[2] These figures were admitted to be far from complete, since much rural employment was unrecorded or deliberately concealed in false guises.[3] A more serious defect was failure to distinguish between temporary and permanent workers. The Narkomtrud inquiry of 1924 reported that the average period of employment of a *batrak* was six months.[4] Much of this labour was in its nature casual : there were few parts of the Soviet Union where climatic conditions made it useful to employ hired labour all the year round. Few *batraks* — according to one estimate, not more than 20 per cent — regarded hired work as a regular occupation. Some of them had been independent peasants in the past ; many still had some land of their own ; most of them dreamed of regaining or acquiring independent status.[5]

Between 1922 and 1924 the lot of the *batraks* steadily worsened.

[1] See *The Interregnum, 1923-1924*, p. 7, note 1.
[2] *Na Agrarnom Fronte*, No. 5-6, 1926, pp. 47-48 ; *Itogi Desyatiletiya Sovetskoi Vlasti v Tsifrakh, 1917-1927* (n.d.), p. 162, gives a total of 2,083,000 hired agricultural workers in August 1926, excluding workers on Sovkhozy.
[3] Among these disguises were fictitious adoptions, adoptions of orphans and homeless children, bogus kolkhozy, fictitious marriages " for the season ", and, in the Asiatic regions, open polygamy (*Na Agrarnom Fronte*, No. 5-6, 1925, p. 93 ; see also L. Kritsman, *Klassovoe Rassloenie v Sovetskoi Derevne* (1926), pp. 163-164). *Pravda*, April 11, 1925, described the " wife for the working season " as " an everyday phenomenon " ; *Voprosy Truda*, No. 3, 1925, pp. 37-38, quotes a contract by which an employer undertook to treat a woman working for him as his wife for three years.
[4] Detailed figures in *Na Agrarnom Fronte*, No. 5-6, 1925, pp. 93-97, show that the largest number of engagements were for six months or for " the season " ; the average length of employment was six months. In White Russia, by way of exception, annual engagements seem to have been the rule.
[5] *Ibid.* No. 3, 1925, p. 94 ; according to A. Gaister, *Rassloenie Sovetskoi Derevni* (1928), p. 93, 46 per cent of *batraks* in 1926 had no land, 52 per cent had plots of land, but no animals (this volume is a report of a group set up in 1926 by the Communist Academy under the presidency of Kritsman to investigate the problem of " differentiation ").

As the number of potential workers grew under pressure of the natural population increase, with declining reserves of land and animals, and the closing, through unemployment in industry, of the most obvious avenue of escape, the conditions of employment deteriorated. " An immense excess of free working hands in the countryside and an acute shortage of draught animals " [1] was a familiar symptom ; and the exploitation of the poor peasant by the *kulak*, far from being an invention of the demagogues, was a stern reality of the rural scene. The commonest form of hiring was for the *batrak* to live with his employer and receive wages in kind, mainly in the form of food, lodging, and sometimes clothing, with the occasional addition of a few puds of rye. Sometimes, where the *batrak* had retained land of his own, the employer, under arrangements of the kind already described, discharged his debt by sending horses to plough the *batrak*'s land. In western Siberia, where population was sparser and conditions of employment more favourable, the *batrak* was reported to be receiving from 2·50 to 12 rubles a month in cash in addition to food, lodging and clothing.[2] But such rates, if genuine, were exceptional. In general, the total monthly wage of the agricultural worker rarely exceeded 12 rubles a month, or half that amount for a woman,[3] and was hardly ever paid in cash. As usual in agricultural work, no limit was placed on the number of hours worked. Rural employers were said to prefer workers of the lowest grade, since these were the most docile and least demanding. In the Ukraine 80 per cent of the *batraks* in 1924 were reported as illiterate ; in the Soviet Union as a whole, 38 per cent of *batraks* belonging to

---

[1] *Planovoe Khozyaistvo*, No. 11, 1925, p. 67 ; in the words of a report of Narkomzem, " the saturation of the rural economy with population and labour had increased, its saturation with capital, implements and means of production had decreased " (*Osnovy Perspektivnogo Plana Razvitiya Sel'skogo i Lesnogo Khozyaistva* (1924), p. 25).

[2] P. Lezhnev-Fin'kovsky and K. D. Savchenko, *Kak Zhivet Derevnya* (1925), pp. 17-18.

[3] *Voprosy Truda*, No. 2, 1925, p. 34 ; *Planovoe Khozyaistvo*, No. 11, 1925, p. 66 : this was said to be less than half the pre-war rate for such work. A table in *Na Agrarnom Fronte*, No. 1, 1926, p. 49, purports to show that the real wages of a *batrak* in 1924 were from 50 to 60 per cent of the 1913 rate. The lowest wages were paid in the black earth region, where population pressure was most acute. According to A. Gaister, *Rassloenie Sovetskoi Derevni* (1928), p. 95, the average monthly wage of the adult *batrak* under collective agreements concluded in 1925-1926 was 21·6 rubles ; but such rates were certainly not effective.

Vserabotzemles (and these were presumably the least backward members of their class), were illiterate, and another 16 per cent semi-literate.[1] For the same reason preference was given in some places to youths between fourteen and sixteen.[2] A contract was recorded for the employment of a child of seven in return for food and certain specified articles of clothing.[3] In the Archangel province poor peasants hired out their children to rich peasants as shepherds : the children received from 3 to 5 rubles (in addition, presumably, to their keep) for looking after the flocks for four months.[4] Stories were told of the use of " threats of dismissal, various kinds of intimidation and even blows " against *batraks* who attempted to assert their rights or to join the trade union.[5] The plight of the *batraks* was not unfairly summed up in a sympathetic article :

By and large *the overwhelming mass of our* batraks *today are outside the protection of the law and organization.* They do not know the laws, there is no way of creating an organization, and the assistance of the state *which should be given by the agricultural labour inspectorate* is lacking, since this inspectorate in fact does not exist. The *batrak* is in the majority of cases wholly at the mercy of " spontaneous market regulation " of his relations with his employer.[6]

The hiring of labour was least prevalent in the European provinces of the RSFSR, most prevalent in the Ukrainian and Crimean republics, in the North Caucasian region, and in all Asiatic regions of the RSFSR.[7]

The organization of these workers, when it was at length attempted, was a formidable task. The peasant committees of mutual aid, whose status had been so hotly debated at the thirteenth party congress,[8] proved as ineffective as the sponsors of

[1] *Na Agrarnom Fronte* No. 3, 1925, p. 93 ; *ibid.* No. 7-8, 1925, p. 87.
[2] *Ibid.* No. 4, 1925, p. 70.
[3] *Voprosy Truda*, No. 3, 1925, p. 37.
[4] *Pravda*, November 20, 1924.
[5] *Na Agrarnom Fronte*, No. 3, 1925, p. 97.
[6] *Ibid.* No. 1, 1925, p. 37.
[7] A. Gaister, *Rassloenie Sovetskoi Derevni* (1928), p. 98 ; a table (*ibid.* p. 100) shows the close correlation between the hiring of labour and the leasing of land.
[8] See *The Interregnum, 1923–1924*, pp. 148-149.

the *kulak* interest at the congress had intended them to be ; in the words of a party report of 1925 they were " not hitherto in a position to give the poor peasant population the real help which would have enabled it to avoid becoming an object of exploitation by the possessing groups in the countryside ".[1]  Vserabotzemles had existed since 1920, but did little to help even workers on the Sovkhozy and in the timber industry, for whom it was supposed primarily to cater.   In April 1923 the twelfth party congress suggested that, " with the growing use of hired labour in the countryside ", it was time for Vserabotzemles to " work out sufficiently flexible forms and methods for counting, organizing and protecting hired workers in the peasant economy ".[2]   This, however, produced little result ;  and a year later the thirteenth party congress, which first became anxious about the process of differentiation and the advance of capitalism in the country- side,  decided that, for the delicate task of organizing the scattered and backward *batraks*, " a substantial body of paid workers " was required, and that funds should be provided for the purpose.[3]   After this Vserabotzemles received " a few hundred thousand rubles " for its work, and was authorized to levy contributions from employers of labour.[4]   But the enormous practical difficulties of work among the *batraks* were increased by persistent obstruction from the well-to-do peasants who dominated the rural Soviets and their organs ; [5]  and the trade-

---

[1] *Na Agrarnom Fronte*, No. 9, 1925, p. 14.

[2] *VKP(B) v Rezolyutsiyakh* (1941), i, 518.                      [3] *Ibid.* i, 592.

[4] *Na Agrarnom Fronte*, No. 3, 1925, p. 96.   A report of a trade union organizer from the province of Tula quoted in *Sotsialisticheskii Vestnik* (Berlin), No. 9 (103), May 15, 1925, p. 8, gave a cynical account of these proceedings : " Every peasant who hires labour has to register it at the rural district executive committee.  The representative of Vserabotzemles then concludes a contract with the employer who undertakes to deduct 4 per cent from the wages of the *batrak* and pay it to Vserabotzemles.   In this way 1000 *batraks* have been organized."

[5] The political position will be discussed in Part IV in the following volume. According to the trade-union journal, " the rural district executive committees and village Soviets in most cases stand aside and do not interfere in ' a private question ', considering, apparently, that their functions do not include the protection of the professional interests of rural labourers " (*Trud*, February 8, 1924).   In October 1924 Kalinin resisted a proposal to make the rural district executive committees responsible for the inspection of rural labour (*Vseross- iiskii Tsentral'nyi Ispolnitel'nyi Komitet XII Sozyva : Vtoraya Sessiya* (1925). p. 75 ; for Larin's comment see *Na Agrarnom Fronte*, No. 3, 1925, p. 96).

union movement as a whole regarded Vserabotzemles and its
efforts with unconcealed contempt.[1]

The picture of Soviet agriculture which emerged from these
scattered and disjointed fragments of experience was one of
enormous complexity. Not only was generalization hazardous,
but nothing was quite what it seemed. The classic relation of
landlord and tenant was frequently reversed. The wealthy peasant
holding land and leasing it out in parcels to poor peasants at an
exorbitant rent was a less common phenomenon than the poor
peasant leasing to the wealthy peasant the inadequate holdings of
land which did not suffice to feed him and his family, or which,
for lack of live-stock and implements, he was unable to cultivate.
But the relation of employer and employed was equally ambiguous
in the new context. The wealthy peasant could take over land
from the poor peasant without even resorting to processes of rent.
A common alternative arrangement was an undertaking by the
rich peasant to cultivate the poor peasant's land in return for a
major share of the harvest, so that the rich peasant, in virtue of
his ownership of animals and implements, became the principal
beneficiary from land which he neither owned nor rented. Theoretically, the rich peasant had been hired to work the land for the
poor peasant ; and through this fictitious legal status he hoped to
avoid the opprobrium of being branded as a *kulak*. Attention
was first drawn to the anomaly of these developments in the
resolution of the thirteenth party congress in May 1924 :

> The peculiarity of the process of differentiation which is
> going on in the countryside consists in the fact that its funda-
> mental element at the present time is not so much land as trade,
> animals and inventory, which are converted into an instrument of
> accumulation and a means of exploiting the weaker elements.[2]

---

[1] The sixth trade union congress in November 1924 received greetings
from several delegations of *batraks*, but showed so little interest in Vserabotzemles that it refused to pass a special resolution supporting its work (*Shestoi
S"ezd Professional'nykh Soyuzov SSSR* (1925), pp. 210-215, 619) ; according
to Tomsky a year later, Vserabotzemles was run by trade union-officials from the
cities, who went into the country " carrying a labour code in one hand and
collective contracts, statutes and instructions in the other ", and failed to
arouse any enthusiasm among the *batraks* or the middle peasants (*XIV S"ezd
Vsesoyuznoi Kommunisticheskoi Partii (B)* (1926), pp. 739-740).

[2] *VKP(B) v Rezolyutsiyakh* (1941), i, 589.

And the complexities were acutely summarized by a contemporary commentator :

> The basic form of capitalist economy in the Soviet country-
> side, and one which is growing in importance, is a capitalist
> (predominantly *small capitalist*) economy founded on the loan-
> ing of working animals and agricultural implements, in which
> the concealed *capitalist* appears *in the guise of a worker* working
> on another man's farm with his own animals and implements,
> and the concealed *proletarian* appears *in the guise of a proprietor*
> who is without animals or implements . . . , and hires these
> indispensable means of production.[1]

These conditions frustrated any attempt to estimate the relative strength of the different groups in the countryside. Legal forms did not correspond to economic realities. Statistics were not so much inaccurate as basically misleading. The workers of Gosplan reported in 1926 that " the character and tempo of the process of differentiation in the countryside " could not be accurately computed " for lack of data ".[2]

In these circumstances it is not surprising that the classification of the peasantry into the traditional three categories — *kulaks*, middle peasants and poor peasants [3] — gave rise to endless terminological and statistical controversy. Kamenev on one occasion coyly declined an invitation to define a *kulak*, and on another remarked that, though a great addict of diagrams, he had never been able to get a sufficiently precise picture of " differentiation " to be able to represent it diagramatically.[4] The generally accepted distinguishing marks of a *kulak* were the systematic employment of hired labour ; the acquisition of land, whether by rent or by other processes ; the ownership of means of production (animals and inventory) on an extensive scale ; and the enjoyment of income from commercial or financial operations not directly connected with agricultural production.[5] The *kulak* was

[1] L. Kritsman, *Klassovoe Rassloenie Sovetskoi Derevni* (1926), p. 161.
[2] *Kontrol'nye Tsifry Narodnogo Khozyaistva na 1926–1927 god* (1926), p. 9.
[3] For the three groups see *The Bolshevik Revolution, 1917–1923*, Vol. 2, p. 160.
[4] L. Kamenev, *Stat'i i Rechi*, xii (1926), 198, 355.
[5] For a description, rather than a definition, of the *kulak* on these lines see *Planovoe Khozyaistvo*, No. 12, 1925, p. 37. The first official attempt to define the three categories of peasant in precise terms was made by the central statistical

essentially a small rural capitalist. *Kulak* was, however, a word of
strong emotional content ; and those who supported the *kulaks*
generally preferred to speak of them as " well-to-do " (*zazhit-
ochnye*) or " rich " (*bogatye*) peasants, sometimes distinguishing
between the two words and claiming that, whereas " rich "
peasants were *kulaks*, " well-to-do " peasants were not. Everyone was agreed on the hatred felt by the poor peasant (and the
middle peasant) for the *kulak*. But there were differences of
opinion on the question who was regarded by the middle or poor
peasant as a *kulak* ; and here, too, defenders of the *kulak* policy
tended to restrict the category. According to one observer,
peasants regarded as a *kulak* "the peasant-exploiter, the man who
keeps hired workers on his farm and lives on them, all those who
do not live by their own work " ; the peasant who himself worked
was not a *kulak* in the eyes of other peasants merely because he was
more successful and lived better.[1] The same ambiguity prevailed
in the classification of middle and poor peasants, and opinions on
this point equally reflected differences of policy. A characteristic
argument between Larin and Kalinin occurred in the TsIK of
the RSFSR in October 1924, the former maintaining that 64 per
cent of all peasants were poor peasants, the latter that 80 per cent
of them were middle peasants.[2] An official estimate of 1925, made
before the controversy had reached its acute stage, put the number
of poor peasant households at 10 millions or 45 per cent of the
total ; of middle peasant households at between 11 and 12
millions or 51 per cent ; and of *kulak* households at about a
million or 4 per cent, though it was admitted that the proportion

administration in its handbook *Statisticheskii Spravochnik SSSR za 1928 g.*
(1929). " Agricultural *entrepreneurs* " (the synonym for well-to-do peasants or
*kulaks*) were defined by an elaborate series of calculations taking into account
degree of dependence on agricultural or other income, value of means of production owned and amount of hired labour employed. Poor peasants were
defined as those " without working animals and with sown land up to 4 desyatins " or " with one working animal and sown land up to one desyatin ".
The remainder were classified as " middle peasants ". Larin distinguished
four types of *kulak* (the same individual could, of course, belong to more than
one type) : the employer-*kulak*, the speculator-*kulak*, the merchant-*kulak*,
and the usurer-*kulak* (Yu. Larin *Rost Krest'yanskoi Obshchestvennosti* (1925),
pp. 53-57).
[1] P. Lezhnev-Fin'kovsky and K. D. Savchenko, *Kak Zhivet Derevnya*
(1925), pp. 30-31.
[2] *Vserossiiskii Tsentral'nyi Ispolnitel'nyi Komitet XII Sozyva : Vtoraya
Sessiya* (1925), pp. 58-59 73.

of *kulaks* was increasing and had in some areas reached 10 or 12 per cent.[1]

The process of differentiation was not only complex, but presented wide diversities in different regions of the Soviet Union. A detailed account of a single rural district in the province of Smolensk described the process at work on what would appear to have been normal lines among the middle peasants who were said to form three-quarters of the population. The transformation of the middle peasant into a poor peasant was naturally a far commoner process than his rise to the status of a *kulak* : his reserves being non-existent, any natural calamity or crop failure was liable to throw him on to the labour market. Industrial unemployment prevented any large flow of workers to the towns and increased the pressure of over-population in the country, where the poor peasants were shamelessly exploited by the *kulaks*, being paid mainly in kind and often receiving no fixed wages. This authority added that the influence of the *kulak* was increasing and that "the attitude of the *kulaks* to the Soviet power is satisfactory ".[2] The greatest friction occurred in the Ukraine. The fertile soil of the Ukraine had long made it especially propitious to large-scale holdings and cultivation for the market. It had been the main source of Russian grain exports before 1914 and was the centre of the Russian sugar industry : since the Stolypin reform it had been the home of the largest and most prosperous individual peasant holdings — in other words, of the most successful *kulaks*. It was also subject to the most extreme rural over-population, and had the largest number of landless agricultural workers living near the hunger line. Differentiation within the peasantry showed itself in the Ukraine in its most acute form. It was no accident that the committees of poor peasants (komnezamozhi) lingered in the Ukraine, and kept alive the tradition of the class struggle, long after they had become a memory elsewhere in the Soviet Union.[3] But here, too, in 1925 the poor peasant was on the defensive. In the northern Caucasus well-to-do peasants told a

---

[1] *Na Agrarnom Fronte*, No. 5-6, 1925, p. 8 ; Lenin in 1918, not using precise figures, had written of 10 million poor peasant households, 3 million middle peasant households and 2 million *kulak* households (*Sochineniya*, xxiii, 207) ; for another estimate see *The Interregnum, 1923–1924*, p. 6, note 3.

[2] A. Gagarin, *Khozyaistvo, Zhizn' i Nastroeniya Derevni* (1925), pp. 32-43.

[3] See *The Bolshevik Revolution, 1917–1923*, Vol. 2, p. 159, note 4.

government commission that they should be left to " settle " the poor peasants, and that every " strong " peasant possessing horses should have a poor peasant attached to him as a worker : then grain would be grown in plenty — " millions of *puds* as we used to grow " — and the poor peasant would be fed. Other rich peasants denounced the poor peasants as " loafers who are always coming to you for help, but cannot work on the land ". It was waste of money to help them : " better leave it to us to carry on ' civilized ' farming ".[1]

The issue thus began to present itself in the spring and summer of 1925 in terms which could no longer be evaded. The problem which the Russian Marxists purported to solve by their advocacy of large-scale collective agriculture was not a matter of doctrine : it was the same problem which had inspired Stolypin's " wager on the strong " — the problem of the efficient unit of production in Russian agriculture. But its solution now seemed both more urgent and more remote than ever. In 1913, 20 per cent of the grain harvest had been marketed, and 22 per cent of all agricultural production. In 1924–1925 these already low percentages had fallen to 14 and 16 respectively.[2] The revolutionary redistribution of the land after 1917 had merely aggravated the fundamental difficulty confronting Russian agriculture ever since the emancipation. The individual peasant, left to himself with the small holding which any approximately equal distribution of the land must inevitably produce, had relapsed into subsistence farming to feed himself and his family, and met his other needs by barter with his neighbours and by seeking employment for himself or members of his family in casual labour or in rural handicrafts. It was beyond his capacity, and did not enter into his view of life, to produce food for sale to the cities, much less for export. The coercion of the tax-gatherer, so soon as it reached a point where it threatened the peasant with hunger, provoked violent mass resistance, as the experience of 1920–1921 had shown. The production of the more specialized crops which had a higher market value and were required as raw material for Soviet industries (beet,

[1] *Planovoe Khozyaistvo*, No. 10, 1925, pp. 36-37.
[2] *Ekonomicheskii Byulleten' Kon'yunkturnogo Instituta*, No. 11-12, 1927, p. 52.

cotton, flax, oil-seeds) was still further beyond the reach of the small peasant ; these crops required to be grown in larger areas and with equipment which the peasant did not possess.  The small peasant had neither the animals nor the manure nor the machinery required for the intensive forms of cultivation which alone would yield large crops for the market, and increase the output of meat and dairy products.  If Soviet agriculture was to feed Soviet cities and factories, to furnish raw material to Soviet industries, to provide a surplus for export and to accumulate reserves of capital for industrial development, some other form of organization was imperatively demanded.  To this problem there were in theory two alternative solutions : the socialist solution of large-scale collective agriculture, which none of the Soviet leaders at this time seriously regarded as practicable,[1] and the capitalist solution of removing restrictions on the free play of economic forces and opening peasant agriculture to the processes of competition.  The logical expression of this policy would be to applaud the efficient and enterprising peasant who increased his holdings and his equipment, hired labour and produced valuable cash crops, and to allow the weak and inefficient peasant to lease or abandon his land and become a hired worker, either in agriculture or in the factories.  Shorn of disguises and embellishments, this was a policy of supporting the *kulak*, and encouraging him to rise on the shoulders of the weak and inefficient.  In an expanding economy it might be possible to achieve the enrichment of the *kulak* without impoverishing the *batrak*.  In 1925 it was not easy to see any other course for Soviet agriculture which promised to deliver the required results.

### (c)  The Wager on the Kulak

At the thirteenth party congress in May 1924, Kalinin had been the only public advocate (and even then in veiled terms) of concessions to the *kulak*.[2]  But, though he had obviously enjoyed a good deal of support behind the scenes, the weight of opinion

[1] Even Larin wrote at this time : " Of course, we do not propose to drive 50 millions of the population of the USSR into socialism by force " (Yu. Larin, *Rost Krest'yanskoi Obshchestvennosti* (1925), p. 37).
[2] See *The Interregnum, 1923–1924*, pp. 147-149.

was against him ; and a writer in the party journal caustically
described the " *kulak* deviation " as a " Stolypin-Soviet policy ".[1]
The *kulak* still had a bad press. *Leningradskaya Pravda* of July
2, 1924, issued the first number of a supplement for peasants
under the title *Nasha Derevnya*, which contained several stories
of *kulak* abuses and a cartoon depicting a fat *kulak* — the typical
capitalist of the Soviet political cartoon — surrounded by a group
of thin, empty-handed poor peasants.[2]   But, with the troubles of
the harvest, opinion became more fluid ; and in August 1924,
at the very moment when Preobrazhensky was delivering his
broadside from the Left, an ingenious attempt was made from
the opposite angle to appease those party consciences which were
still obstinately wedded to the conception of large-scale collective
agriculture as an essential condition of socialism, and stubbornly
resisted the *kulak* solution.   Varga, the economist of Comintern,
published a book entitled *Outlines of the Agrarian Question*
(sections of it also appeared in *Pravda*), in which he recalled the
famous controversy on the agrarian question in the German
Social-Democratic Party of the early nineteen-hundreds.   A
section of the party had then maintained that agriculture need not
necessarily follow the same course of development as industry,
and that individual peasant agriculture could be made both
efficient and compatible with the realization of socialism.   The
" orthodox " Marxist predilection for large-scale collective agriculture had been defended at that time by Kautsky, who denounced
the advocates of the contrary view as " revisionists ".   Kautsky's
subsequent defection now made it easier to discredit a cause with
which he had once been prominently associated ; and Varga did
not hesitate to employ this weapon, marshalling the arguments of
Kautsky's old opponents in support of the new thesis that, in
capitalist conditions such as prevailed under NEP, individual
peasant enterprise was more efficient than large-scale collective
cultivation, which should be relegated to a later stage in the
ultimate realization of socialism.   This argument produced a
sharp reaction.   Orthodox Marxists did not fail to expose the

[1] *Bol'shevik*, No. 3-4, May 20, 1924, pp. 23, 25 ; the writer was Slepkov,
afterwards closely associated with Bukharin.
[2] The supplement is dated June 2, 1924 — no doubt in error. No further
issue has been traced, but *Leningradskaya Pravda*, November 2, 1924, announced that the supplement would appear twice a month.

" revisionist " character of Varga's thesis.[1] A new monthly journal, *Na Agrarnom Fronte*, issued under the auspices of the Communist Academy, started publication in January 1925, and became primarily the organ of party intellectuals who viewed current " bourgeois " tendencies in agrarian policy with a critical eye. The controversy continued ; and, while Varga secured no official support, he had done something to shake the confident assumption of the superiority of large-scale agriculture.

While the party leaders showed the utmost reluctance to plunge into these troubled waters, concrete issues arose and decisions had to be taken. A first minor clash occurred over the aftermath of the Dymovka scandal. This occurrence had been actively used to discredit the group within the party which favoured support for the well-to-do peasant ; and this group, increasingly numerous and influential, now attempted to strike back by attacking Sosnovsky, who had led the campaign against Dymovka and was alleged to have described the whole Soviet apparatus in the countryside as rotten. The personal issue was raised in the Orgburo at the end of January 1925, when Stalin came to the defence of Sosnovsky :

> The present course of our policy is a new course marking out a new line in our policy in relation to the countryside, in the building of socialism. This comrades fail to understand. If they do not understand this fundamental fact, none of our work will succeed, and we shall have no socialist construction. . . . We must tell the comrades not to be afraid of dragging fragments of life, however unpleasant they may be, into the light of day.

---

[1] The first hostile critique appeared in *Bol'shevik*, No. 10, September 5, 1924, pp. 66-80. A long attack by Milyutin, under the title *Agrarian Revisionism*, appeared in *Pravda*, October 25, 1924 ; in a reply published *ibid*. December 11, 1924, Varga quoted the precedent of the Hungarian revolution of 1919, when Bela Kun had antagonized the peasants by attempting to collectivize agriculture (Varga had criticized the policies of war communism on the same grounds — see *The Bolshevik Revolution, 1917–1923*, Vol. 2, p. 169, note 3). The interest excited by the controversy is shown by the reprinting of the two articles in *Internationale Presse-Korrespondenz*, No. 159, December 9, 1924, pp. 2179-2181 ; No. 169, December 30, 1924, pp. 2327-2330. Milyutin again replied in an article in *Na Agrarnom Fronte*, No. 2, 1925, pp. 1-10 ; No. 3, 1925, pp. 7-19. On the other hand, an article in *Ekonomicheskoe Obozrenie*, March 1925, pp. 72-89, which appeared with a note dissociating the editors from the views expressed, repeated the argument for regarding the issue of large-scale agriculture as at any rate open ; Kautsky's association with the impugned doctrine was once more quoted as a bull point.

The *sel'kors*, he concluded, should be regarded as " one of the main levers in the process of bringing our shortcomings to light and of correcting our party and Soviet constructive work in the localities ".[1] The Dymovka affair assumed sufficient importance to become the subject of a joint resolution of the party central committee and central control commission, where it was described as revealing the dangers of " a deviation from the correct line in Soviet and in party organization in conditions of increasing capitalism ". Popandopulo and Postolati, the resolution concluded, had acted " under the direct influence of *kulaks* ". Party organizations were censured for failing to recognize " the political significance of the murder of the *sel'kor* Malinovsky " ; and Sosnovsky was praised for his assiduity in bringing these abuses to light. The issue of *Pravda* which printed this resolution drove home the moral in a leading article on *The Lessons of Dymovka* : the Dymovka affair was a " typical phenomenon ", illustrating the danger of capitalism in the countryside, " where the weakest flank of the Soviet front faces *kulak* capital ".[2] A Moscow provincial party conference produced some more plain speaking in high quarters on the same theme. Uglanov, the secretary of the Moscow party organization and now a supporter of Stalin, spoke of the murder of the president of a village Soviet because he had tried to insist on punctual collection of the agricultural tax, of a threat to kill the secretary of a local Komsomol organization, and of stormy meetings of rural district congresses of Soviets which *kulaks* tried to control by organizing other peasants.[3] Stalin seized the occasion for a hedging speech which, none the less, showed a strong scepticism about the advantages of a pro-*kulak* policy. He deprecated the inclination to treat the slogan " Face to the countryside " as if it meant to turn one's back on the towns ; and, after dwelling on the fact that the peasantry was at the present time the only available ally for the proletariat and therefore indispensable, he went on :

> This ally is, as you know, not very firm ; the peasantry is not such a dependable ally as the proletariat of the capitalistically developed countries.[4]

[1] Stalin, *Sochineniya*, vii, 22-23.
[2] *Pravda*, February 6, 1925.    [3] *Ibid.* January 28, 1925.
[4] Stalin, *Sochineniya*, vii, 25-28.

This was an unusual line for Stalin to take, and showed a marked
desire to remain uncommitted for as long as possible in the current
controversy. If anyone in January 1925 had been acute enough
to predict an imminent break between Stalin and Zinoviev on
this issue, he would almost certainly have seen in Zinoviev the
prospective champion of a peasant policy and Stalin as its
opponent.

The trend soon became too strong to be resisted. A peasant
delegate to the conference on local government held in Moscow
in January 1925 used an argument which was to be frequently
repeated in varying forms in the next few months :

> You call on us to improve our cultivation, to mechanize it.
> We shall grow, our revenues will grow, we shall have more
> horses, cattle and machines, and what will you then do ? The
> representative of the executive committee of the department
> answers : " We shall dekulakize you ".

And Rykov answered reassuringly that one who, by more
efficient cultivation or by introducing new crops, produced 40
puds where only two puds grew before deserved to be rewarded,
not condemned :

> The man who looks after his land well is not a *kulak*, but
> a Soviet worker.[1]

A variety of reasons strengthened the case of those who, in
the first months of 1925, demanded further concessions to the
peasantry — meaning, in practice, to the well-to-do peasants or
*kulaks*. Some of these reasons were purely psychological. In the
spring of 1924, optimism had prevailed about the coming harvest
on no better ground than that of the excellent harvests of 1922
and 1923 ; these hopes had been disappointed, and an equally
ungrounded pessimism now set in about the prospects of the 1925
harvest. It was reinforced by the exceptionally mild weather and
light snowfall of the winter of 1924–1925, which was thought to
augur ill for the sowing season. The well-to-do peasants had
held up surpluses after the 1924 harvest, and were still holding
them in an attempt to defeat the efforts of the government to force

[1] *Soveshchanie po Voprosam Sovetskogo Stroitel'stva 1925 g. : Yanvar'*
(1925), pp. 157, 192-193.

down grain prices. In these conditions they lacked any incentive to extend the sown area in 1925 and might even contract it, with catastrophic effects in the event of a bad harvest. Visions of a strike of peasants, and of a return to the process of holding up the cities and factories to ransom, began to haunt the party leaders. The fate of the country seemed to depend on the good will of larger peasants who alone could produce and deliver the indispensable surpluses of grain. It had become necessary to appease them at all costs.

The movement now only required a leader, and Bukharin, almost by accident, found himself cast for the rôle. In his polemic against Preobrazhensky,[1] he had in the main been concerned to make out a political case, and to discredit his adversary for hostility to the claims of the peasant. But he had also, in passing, attempted to answer Preobrazhensky's searching economic argument and to explain by what means, other than those propounded by Preobrazhensky, socialist accumulation could take place. The proletarian state, according to Bukharin, would obtain what it required " on the basis of *the growing rationalization and the growing profitability* of the peasant economy ", the main instrument of rationalization being agricultural cooperatives.[2] No other answer was possible. Once a policy of forced accumulation was rejected, no alternative remained but to count on the growth of a prosperous peasantry which would provide both a lucrative market for industrial goods and capital out of its savings for the further development of industry. The success of this policy, problematical in any event, was, however, conceivable only on the hypothesis of the emergence of a well-to-do group of peasants, since they alone could perform the required function of providing for future capital accumulation. At the beginning of 1925 the exigencies of the struggle against the Trotskyite opposition, as well as material conditions in the country, pointed the way not merely to concessions to the peasantry in general, but specifically to the encouragement of differentiation and the toleration of the *kulak*. An article by Bukharin in the party journal in February 1925 was a sustained eulogy of the peasant as a motive force of the revolution. It attacked the " bourgeois " theory of the innate

---

[1] See pp. 207-208 above.
[2] N. Bukharin, *Kritika Ekonomicheskoi Platformy Oppozitsii* (1926), pp. 25-26.

conservatism of the peasant. The peasants formed one-half of the population of continental Europe ; and the coming agricultural crisis would compel the capitalist states also to " face the countryside ". The peasantry " can become, and will become, under the leadership of the proletariat, the great liberating force of our time ".[1] An article by Kritsman, the historian of war communism, on *The Alliance of the Proletariat and a Majority of the Peasantry in the USSR after the Victory of the Revolution*, was printed in *Bol'shevik* as a " discussion article ", and followed in a later number by two articles refuting it ; [2] the splitting of the peasantry to spearhead an attack on the *kulak* was an obsolete and unwelcome slogan.

Throughout the spring of 1925 the movement gathered fresh impetus. In March 1925 TsIK insisted that " only the development and expansion of the peasant market can serve as a foundation for the further development of industry and transport ", and that the interests of industry and transport were dependent on " the development of agriculture and of the marketability of its products " ; [3] and the " peasant market " meant primarily the well-to-do peasant. Rykov, when told by a party member that any peasant having two horses and two cows was a *kulak*, retorted that he hoped that in two years' time every peasant would be just such a *kulak*.[4] An article in the party journal on class differentiation explained that " *kulaks*, as a specific category of exploiter-landowners, . . . cannot in our conditions develop in their most garish colours ".[5] It was the name rather than the function of the *kulak* which now appeared to constitute the stumbling-block. A. P. Smirnov, in an article in *Pravda* on the need for a " strong working peasantry ", deprecated the habit of applying the " *opprobrious nickname* " of *kulak* to " every strong and hard-working

[1] *Bol'shevik*, No. 3-4, February 25, 1925, pp. 3-17 ; Bukharin in his speech at IKKI early in April 1925 spoke of the struggle between bourgeoisie and proletariat on a world scale for the control of the peasantry, denounced a " workshop-proletarian ideology (Trotskyism) ", and, in general, emphasized the importance of conciliating the peasant (*Rasshirennyi Plenum Ispolnitel'nogo Komiteta Kommunisticheskogo Internatsionala* (1925), pp. 304-328).

[2] *Bol'shevik*, No. 2 (18), January 31, 1925, pp. 40-55 ; No. 5-6 (21-22), March 25, 1925, pp. 73-100.

[3] *SSSR : Tsentral'nui Ispolnitel'nyi Komitet 2 Sozyva : 3 Sessiya : Postanovleniya* (1925), pp. 7-8. [4] *Izvestiya*, February 10, 1925.

[5] *Bol'shevik*, No. 5-6 (21-22), March 25, 1925, p. 25.

household " ; and Kamenev declared that, if every hard-working peasant were branded as a *kulak*, " we should be chopping off the branch on which we are sitting ".[1] Larin, taking up the cudgels for the opposition, discerned " a sort of '*de jure* recognition ' of the well-to-do proprietor as a legitimate species in Soviet rural life ", and complained that some people refused to call a man a *kulak* even if he employed ten hired workers for the season ; a reply, which appeared in the party newspaper for peasants, *Bednota*, accused him of ridiculing the middle peasant and hypocritically flattering the poor peasant.[2]

A curious and significant record of the mood prevailing in the spring of 1925 is contained in the interview given by Stalin on March 14 to a delegation of an All-Union Congress of *Sel'kors*, consisting of peasants drawn from all parts of the USSR. The conversation ranged widely and extended over two long sessions. Criticism was directed against the *kulak* and against the incidence of the agricultural tax : " the well-to-do enjoy the rebates, but from the poor peasant the last penny is taken ". But the opposite case was also heard. A woman delegate from the Bashkir republic alleged that any peasant with three horses and three cows was called a *kulak*, even if he did not exploit labour ; and Stalin agreed that " it happens that, if a peasant puts on a new roof, they say he is a *kulak* ". The most radical demand related to the tenure of land. A delegate from the province of Tula complained that without security of tenure the peasant had no incentive to improve his land. He suggested that the land should be divided into small holdings and allocated to the individual peasant irrespective of the size of his household, " so that there may be unchangeable boundaries, so that there may be stable utilization of land ". Stalin agreed that without security the peasant would not manure his land and asked for how many years the land should be allocated; and, when the peasant replied, " For 20 years ", Stalin is said to have inquired : " And suppose for longer, for 40 years, or even for ever ? " This boldness evidently surprised the peasant, who replied: " Perhaps for longer, perhaps for ever, but this would need thinking over by more than one head ". Stalin then wound up the discussion by saying that this would not be ownership,

---

[1] *Pravda*, April 5, 6, 1925 ; *Ekonomicheskaya Zhizn'*, April 14, 1925.
[2] *Pravda*, April 8, 1925 ; *Bednota*, April 29, 1925.

since the land could not be sold, but that " it will be possible to utilize it with confidence ".[1]

The conversation committed nobody. But it reveals the freedom with which such far-reaching projects were being canvassed in the spring of 1925. In effect what lay behind the current demand for freedom to acquire land by leasing and to cultivate it with hired labour was the ambition of the enterprising and successful peasant for security of tenure. The " fundamental law " of May 1922 and the agrarian code of the same year [2] had whetted this ambition but not satisfied it. They had recognized the right of the peasant to leave the *mir* and establish an independent holding in the form of an *otrub* or a *khutor*. But the land so allocated remained in principle subject to the *mir* to which it had belonged, and the *mir* was at liberty by a vote of its members to revise the boundaries of such holdings at any time. Since the sale of land was excluded altogether, and the leasing of land and hiring of labour were allowed only in special circumstances and for limited periods, some authority to revise or withdraw holdings was necessary in order to take account of changes in the size and working capacity of households. What the well-to-do peasant now wanted was not merely the temporary right to obtain land on lease and hire labour, but a security of tenure which would have amounted (except perhaps in respect of the right to sell) to a restoration of private property in land.[3] Such a step would, however, have been in direct contradiction both to the letter of the law and to cherished party prejudices. Article 2 of the agrarian code recognized all land as " the property of the workers' and peasants' state " ; and Lenin in one of his last articles had

---

[1] For the report of this interview see p. 183 above. Stalin stated at the fourteenth party congress in December 1925 that he saw the report for the first time in October and at once denied its authenticity (Stalin, *Sochineniya*, vii, 362-363). By this time it was compromising to have toyed with proposals which looked like a restoration of private property in land ; but the denial does not carry conviction. On the other hand, Stalin was clearly thinking aloud rather than making a statement of policy.

[2] See *The Bolshevik Revolution, 1917-1923*, Vol. 2, pp. 289, 296-297.

[3] In the Ukraine, where independent peasant holdings were the predominant form of cultivation, the fifth Ukrainian congress of Soviets had adopted in December 1920 a decree guaranteeing security of tenure for a maximum of 9 years (M. Popov, *Naris Istorii Kommunistichnoi Partii (Bil'shovikiv) Ukraini* (2nd ed. 1929), p. 252 ; *Zbir Zakoniv i Rosporyadzhen', 1921*, No. 3, art. 94) ; but no corresponding provision seems to have been enacted elsewhere.

specifically referred to the state ownership of land, as of other means of production, as a safeguard against the return of capitalism.[1] This testimony could not be lightly disregarded. For the present the aspirations of the would-be peasant proprietor would have to be confined within the safer limits of the right to lease and the right to employ hired labour.

A decision could now no longer be delayed on three current issues of agrarian policy; all of them turned on increasingly insistent demands of the well-to-do peasant, which would have to be satisfied in some form or other if his good-will was to be retained. The first was for a reduction in the burden of the agricultural tax. The second was for an unequivocal recognition of the right to employ hired labour, and a removal of the conditions and restrictions with which it had been hedged about in the agrarian code. The third was for the unrestricted right to acquire land by leasing. These three demands had one characteristic in common. All of them would increase the differentiation in the countryside and help the well-to-do peasant to better himself at the expense of the poorer peasant, who would more and more be driven off the land and find himself working as a *batrak* for his richer neighbour. The party leaders, having triumphed over Trotsky, were committed by the logic of the situation to the only course which seemed to hold out hopes of increased agricultural production — the appeasement of the *kulak*.

The first contested question to require settlement in the spring of 1925 was that of the agricultural tax. Since this tax was one of the main fiscal resources of the Soviet Union, and the only direct tax borne by the peasant, a decision on the total amount to be levied and on the method of its assessment for the coming financial year was not only an important act of agrarian policy, but a necessary preliminary to the framing of the budget. The tax in kind, which in 1921 replaced the requisitions of war communism, was converted in 1923 into the " single agricultural tax ", computed in terms of money and assessed on the following basis.[2]

[1] Lenin, *Sochineniya*, xxvii, 395.
[2] *Sobranie Uzakonenii*, *1923*, No. 41, art. 451 ; for the decision of the twelfth party congress on which the decree was based see *The Interregnum*, *1923–1924*, pp. 18-19.

On the completion of the harvest, Narkomfin fixed the total amount of the tax for the year, and laid on each province an assessment based on the estimated yield of the harvest for that province. The provincial authorities divided the assessment on the same principle between the counties, and the county authorities between the rural districts. The rural district authorities then divided up the sum required of them between the peasants of the district in proportion, not to the value of the crops harvested by them (this it would have been impracticable to verify), but to the amount of land cultivated by them : for the purposes of the assessment animals owned by the peasants were converted into units of arable land at rates laid down in the decree. The tax was progressive, holdings being divided into grades according to the amount of land held. But the grading was not steep ; and much evidently depended on the discretion of the rural district authorities. In the financial year 1923–1924 the tax was collected partly in money and partly in kind at the choice of the peasant. In 1924–1925 it became exclusively a money tax. But the basis of assessment and collection, with some minor changes, mainly in the form of exemptions for special crops, remained the same. Recognition was also given to special conditions of different regions of the USSR. In regions where cultivation was less intensive (the northern Caucasus, Siberia), a desyatin of sown land (and not, as elsewhere, of ploughed land) was taken as the basis of assessment. In regions where animal husbandry predominated, one head of large horned cattle (instead of one desyatin of land) was taken as the unit : one horse or camel, two donkeys, one mule, three sheep or three goats (not exempt in these special regions) were equivalent to one head of cattle.[1]

Common criticisms of the system were that the incidence of the tax on individuals varied from district to district, since the calculations on which the burden was distributed between

<hr>

[1] *Sobranie Uzakonenii, 1924*, No. 58, art. 570. In the previous year in stock-breeding regions a pound of meat had been taken as the unit (*Sobranie Uzakonenii, 1923*, No. 42, art. 452) : this had evidently not been a success. As an instance of variations in the scale of assessment in different regions, the rates of conversion in the Turkestan Autonomous SSR for 1924–1925 were one sheep or goat for $\frac{1}{10}$ desyatin of irrigated land, one camel, ox, mule or donkey for $\frac{2}{10}$ desyatin, one horse or head of large horned cattle other than oxen for $\frac{3}{10}$ desyatin (*Sbornik Dekretov, Postanovlenii, Rasporyazhenii i Prikazov po Narodnomu Khozyaistvu*, No. 10, July 1924, pp. 72-73).

provinces, counties and rural districts were only rough-and-ready estimates ; that the assessment of the individual on the amount of land cultivated and number of animals owned by him discouraged him from increasing his holdings ; and that the peasant received his individual assessment only at the moment when payment was demanded, and had no means of knowing in advance what would be required of him. The gravest criticism of all, however, was that the system, in spite of the supposedly progressive basis of assessment, favoured the well-to-do peasant at the expense of the poor peasant.[1] Only cultivated land, cattle and horses were now assessed for tax ; poultry had never been included ; sheep and goats were included only in regions where they were a main element in the economy ; and pigs and calves had been excluded since 1923–1924. " Technical " crops, which were generally grown for the market by the well-to-do peasant, were also excluded. Income from trade and rural industries was exempt from tax. It was calculated that such income constituted, in different regions, from 20 to 45 per cent of total peasant incomes; and it was common knowledge that it was mainly the well-to-do peasant in the economically advanced regions who enjoyed these untaxed sources of income.[2] It was generally admitted, though precise comparisons were impossible, that the peasant was less heavily taxed than he had been before the revolution, and that the burden of taxation fell more lightly on him than on the industrial worker.[3]

[1] A decree of August 24, 1923, purported to grant exemption to peasants owning no animals and holding not more land than ⅜ desyatin for each member of his family (*Sobranie Uzakonenii, 1923*, No. 97, art. 969). But this decree does not seem to have been brought into effect till the year 1924–1925, when 20 per cent of peasant households were exempt from the tax as against only 2 per cent in 1923–1924 (*Finansy i Kredit SSSR*, ed. V. Dyachenko and G. Kozlov (1938), p. 120) ; Sokolnikov in his speech at the twelfth party congress in April 1923 admitted that the tax imposed often did not correspond to ability to pay (G. Sokolnikov, *Finansovaya Politika Revolyutsii*, ii (1926), 109).

[2] For detailed calculations see *Ekonomicheskoe Obozrenie*, March 1929, pp. 34-36. According to an estimate in *Bol'shevik*, No. 2, January 30, 1926, p. 90, 21 per cent of peasant income was untaxed ; in the provinces of Kursk and Smolensk respectively, 72 and 68 per cent of peasant households had subsidiary agricultural earnings, and 10 and 15 per cent non-agricultural earnings, which were exempt from tax in 1925–1926 (*ibid.* pp. 91-192).

[3] On the first point, a source favourable to the peasant calculated that, whereas the average income of the peasant in 1925 was only 82 per cent of the pre-war figure, his effective income reached 95 per cent owing to lower taxation

Ever since the introduction of a graded income-tax in the cities in the autumn of 1922,[1] the idea had been canvassed from time to time of converting the agricultural tax to an income basis. The twelfth party congress of April 1923 pronounced in general terms that " our legislation (first and foremost, our tax legislation) should take account of class divisions in the countryside, appropriately placing the chief economic burdens on the most well-to-do farms ".[2]  In the following year, even the conservative Rykov discovered an argument for making the tax progressive :

> If the peasant who has ten or twelve animals sells two, his farm will be a little weaker, but it will be kept going.  If a peasant with one horse sells his horse, the farm will collapse.[3]

The second Union Congress of Soviets decided that " the tax should be progressive, with the closest possible approximation to an income tax and with exemption for the poor peasant ", and that it should be levied on all income of peasants, not only on agricultural income.[4]  This was, however, easier said than done. In the autumn of 1924 a definite proposal emanated from Narkomfin for the placing of the agricultural tax on an income basis, and was ventilated by Sokolnikov at VTsIK ; [5] and it seems to have had the backing of Rabkrin and the central control commission of the party.[6]  But even this powerful advocacy did not avail.  Apart from the well-founded objection that the peasant's income could not be accurately computed in monetary terms, the proposal

---

(*Ekonomicheskoe Obozrenie*, January 1926, p. 13) ;  on the second point, a table in the official *Vestnik Finansov*, No. 10, October 1925, p. 65, showed that the peasant in 1924–1925 paid in all taxes, direct and indirect, an average of 4·41 rubles a head (for a variant figure of 5·43 rubles see *ibid.* p. 47), the worker 14·2 rubles and other town-dwellers 40·26 rubles.  It is fair to add that the advantage enjoyed by the peasant in the matter of taxation was compensated by a price policy which discriminated against him : for taxation and price policy as alternative methods of extracting surplus value from the peasantry see pp. 203-204 above.

[1] See *The Bolshevik Revolution, 1917–1923*, Vol. 2, pp. 354-355.

[2] *VKP(B) v Rezolyutsiyakh* (1941), i, 518.

[3] *Izvestiya*, September 9, 1924.

[4] *II S"ezd Sovetov SSSR : Postanovleniya* (1924), p. 13.

[5] *Sotsialisticheskoe Khozyaistvo*, No. 5, 1924, p. 56 ; *SSSR : Tsentral'nyi Ispolnitel'nyi Komitet 2 Sozyva : 2 Sessiya* (1924), p. 143.

[6] *Chetyrnadtsataya Konferentsiya Rossiiskoi Kommunisticheskoi Partii (Bol'-shevikov)* (1925), p. 73 : a published report of Rabkrin advocating this solution was sceptically reviewed in *Vestnik Finansov*, No. 7, July 1925, p. 250.

continued to encounter the opposition of the well-to-do peasants.[1] Another current proposal — a curious reversion to capitalist ways of thought — was to levy the tax on the estimated annual rental value of the land.[2] But this project presupposed a vast machinery for surveying and estimating land values which did not exist.

In the new climate, so mild and propitious for the *kulak*, of the spring of 1925, the question of the agricultural tax for 1925–1926 was discussed at the session of TsIK in March. Complaints were heard that the well-to-do peasants were evading their full share of the burden, that the earnings of the *kulak* from trade and industry escaped taxation, and that the exemptions were more readily accorded to the *kulak*, who knew how to get round the authorities, than to the poor peasant.[3] But these isolated voices appear to have carried little weight against the preponderant opinion. The question was referred to a commission of peasant members of TsIK which, as was to be expected, reported against any change in the basis of the tax.[4] The resolution adopted at the end of the session recommended reduction of the total assessment from 470 million rubles in 1924–1925 (of which 400 millions were collected) to 300 millions for the coming year ; moreover, the 100 million rubles required for the budgets of the local authorities, which had in previous years been collected as a supplement to the main tax, were in 1925–1926 to be taken out of the total. Nor did these drastic reductions exhaust the proposed concessions. Strong objections had been raised to the rate of assessment of animals for tax. There had been large slaughterings of cattle in the autumn of 1924 ; and these were attributed to the high rate at which they were assessed, or to fears of the introduction of an income-tax which would bear on them still more heavily.[5] It was now

---

[1] *XIV S"ezd Vsesoyuznoi Kommunisticheskoi Partii (B)* (1926), p. 328.

[2] This proposal was tentatively adopted by Tsyurupa at the fourteenth party conference (*Chetyrnadtsataya Konferentsiya Rossiiskoi Kommunisticheskoi Partii (Bol'shevikov)* (1925), pp. 67-68), and in an article in *Planovoe Khozyaistvo*, No. 11, 1925, pp. 60-63.

[3] *SSSR : Tsentral'nyi Ispolnitel'nyi Komitet 2 Sozyva : 3 Sessiya* (1925), pp. 124, 238-239.

[4] *Ekonomicheskoe Obozrenie*, April 1925, pp. 69-70 ; the appointment and report of this commission do not figure in the official record of the session.

[5] *Ibid.* pp. 230, 233-234 : one reason, which was not mentioned, for the slaughterings was that the peasants sold their cattle in order to pay the tax, preferring to hold back their grain for higher prices.

decided to reduce the rating of animals for assessment by one-third, so that a horse would be taken as the equivalent of half a desyatin of cultivated land, not three-quarters as hitherto : this was definitely a concession to the well-to-do peasant, who alone possessed animals in any quantity.[1]  No recommendation was adopted to make the tax more progressive, and an amendment pronouncing in favour of a transition to income-tax in 1926–1927 was rejected.[2]  The decision seems to have been ill received in some party circles.  It was Stalin, once more steering a cautious middle course, who urged, at a meeting of the Orgburo on April 6, 1925, that " the poor peasant should not be oppressed, and the *kulak* not freed from tax burdens ".[3]

As the time drew near for the fourteenth party conference to assemble at the end of April 1925, the question of the agricultural tax was the only one of the three topical issues of agrarian policy on which progress had been made ; and it was the only one of the three which, by decision of the January session of the party central committee, was formally on the agenda of the conference.[4]  The other two issues — the hiring of labour and the leasing of land — were under examination by a committee of the Politburo presided over by Molotov.[5]  No serious difference of opinion about the course to be pursued seems to have existed between the party leaders at this time.  Kamenev spoke frankly and emphatically to the Moscow provincial congress of Soviets which met in the first half of April 1925 :

> *We shall also have to review our legislation about the utiliza-*
> *tion of land, about hiring of labour and about leasing, since we*
> *have many juridical restrictions which are in fact of a kind to hold*
> *back the development of productive forces in the countryside,*

[1] The resolution is in *SSSR : Tsentral'nyi Ispolnitel'nyi Komitet 2 Sozvya : 3 Sessiya : Postanovleniya* (1925), pp. 8-10, and in *Sobranie Zakonov, 1925,* No. 17, art. 124.

[2] *SSSR : Tsentral'nyi Ispolnitel'nyi Komitet 2 Sozvya : 3 Sessiya* (1925), pp. 254-255.  An illustration of the complexities of tax-gathering in the rural areas of the USSR was furnished by a proposal to include in the resolution special provisions for the nomad and semi-nomad peoples ; the spokesman of Narkomfin successfully pleaded that nothing should be laid down in the resolution, and that it should be left to the presidium of TsIK to sanction any necessary adjustments (*ibid.* pp. 235-236, 254).   [3] Stalin, *Sochineniya,* vii, 80.

[4] *VKP(B) v Rezolyutsiyakh* (1941), i, 635.

[5] *Leningradskaya Organizatsiva i Chetyrnadtsatyi S"ezd* (1926), p. 108.

exacerbating class relations instead of leading them into the right channel. . . .

We are for the development of productive forces, we are against those survivals which impede the development of productive forces. . . . We are for peasant accumulation — the Soviet power must take its stand on this point of view — but we are for the regulation of this accumulation.[1]

Bukharin, in his speech at IKKI early in April 1925, invoked the somewhat sophistical argument that " the poor peasants, not finding employment for their labour power, are sometimes against us on the ground that we prohibit hired labour in agriculture ".[2] And Rykov, in the Leningrad provincial congress of Soviets held a few days later, declared that production in the countryside must be expanded " even if strong households resort to the hiring of labour and, in certain cases, to the renting of land ".[3]  But, while the leaders were unanimously in favour of the concessions which the well-to-do peasants demanded, they were conscious of a widespread hostility to the proposals in the rank and file of the party and of the danger that this would be voiced by a significant number of delegates at the forthcoming party conference.  The Politburo met on April 16, 1925, ten days in advance of the conference, to consider the tactics to be adopted.[4]

The least keenly contested of the three issues was that of the agricultural tax.  Here the main work had already been done by TsIK at its session in the previous month.  The question was already on the agenda of the conference ; and it was only necessary to stave off a renewed offensive in favour of an income-tax.  The Politburo recommended that the party central committee should be invited to re-examine the current proposals to place the agricultural tax on an income or rental basis and submit a report to the next party congress.  This decision guarded against the danger

[1] L. Kamenev, *Stat'i i Rechi*, xii (1926), 132-133 ; it is perhaps significant that a brief report of the speech in *Vlast' Sovetov*, No. 17-18, May 3, 1925, p. 29, omitted the reference to legislation about hired labour, and quoted Kamenev as referring to *kulaks* as " the bourgeois top stratum in the countryside which utilizes hired labour ".

[2] *Rasshirennyi Plenum Ispolkoma Kommunisticheskogo Internatsionala* (1925), p. 370.

[3] *Vlast' Sovetov*, No. 19, May 10, 1925. p. 18.

[4] The resolution adopted at this meeting, like most resolutions of the Politburo, was not published, but was quoted piecemeal in subsequent speeches and resolutions.

of pressure from the anti-*kulak* group for a quick decision in favour of an income-tax, and allowed the question to be safely and inconclusively ventilated at the conference.

The second issue, that of hired labour, required more cautious handling. The Politburo decided on urgent governmental action in advance of the conference. On April 18, 1925, two days after the meeting of the Politburo, Sovnarkom adopted a decree under the title of " temporary rules " (a hint to objectors that the system was not intended to be permanent) which was designed as a charter for *batraks*. According to the rules now laid down, hired workers must be covered by a written agreement signed by themselves or by a trade union on their behalf. More than eight hours a day could be worked only " with the consent of the parties, depending on the character of the work at different agricultural seasons " ; there was to be one rest day in a week. Children could not be employed on heavy work under the age of fourteen, or on light work under twelve. Pay must not be below the statutory minimum wage, and the worker could not be required to accept payment in kind. Social insurance was obligatory if three or more workers were employed permanently.[1] Given the conditions of agricultural work, the level of intelligence and initiative of most of those working as *batraks*, and the weakness of Vserabotzemles, the agricultural trade union, these " rules " can have had little practical effect. But at the present juncture they served two purposes. By making formal provision for the protection of those hired as agricultural workers, they gave clear legal sanction to a practice hitherto regarded as having an exceptional and somewhat equivocal character ; and they provided an answer to critics who accused the party leaders of being indifferent to the fate of the *batraks*. A legal commentator described

---

[1] *Sobranie Zakonov, 1925*, No. 26, art. 183. Doubt may have been felt by constitutional purists whether the enactment of some of these provisions fell within the competence of organs of the USSR ; supplementary decrees giving effect to them were issued by several of the republics two or three months later. For example, the Transcaucasian SFSR issued a decree of its own in similar terms on July 30, 1925 (*Sobranie Uzakonenii Zakavkazkoi SFSR, 1925*, No. 8, art. 475) ; the Tatar autonomous SSR reprinted the " temporary rules " among its own decrees (*Sobranie Uzakonenii Tatarskoi ASSR, 1925*, No. 27, art. 189). Many of the republics later issued supplementary decrees tightening up the regulations (e.g. the Ukrainian SSR in *Zbirnik Uzakonen' ta Rosporyadzhen', 1925*, No. 62-63, art. 350 ; *id. 1926*, No. 4, art. 35 ; the White Russian SSR in *Zbor Zakonau i Zahadau BSSR, 1925*, No. 42, art. 351).

them as "a first partial step" towards the extension of the labour code to the peasantry.[1] The "temporary rules" were given maximum publicity, being printed in full in the leading newspapers. The official economic daily, *Ekonomicheskaya Zhizn'*, welcomed them on the ground that they introduced "clarity into class relations in the countryside", and made it plain that a process of accumulation of wealth was no longer "economically and socially discredited".[2] The party peasant newspaper, *Bednota*, struck a slightly apologetic note, explaining that "an increase in the number of strong and well-to-do peasant households employing hired labour is an unavoidable phenomenon, if we are to speak seriously of a new economic policy in the countryside as well as in the town".[3]

The third issue, that of the right to lease land, was the most delicate of all, since it involved the hotly contested question of tenure; and what happened was clear proof of the embarrassment of the Politburo. On April 21, 1925, the presidium of the TsIK of the RSFSR, acting on behalf of the TsIK which was not in session, adopted an addition to art. 28 of the agrarian code, which accorded to households "temporarily weakened in consequence of some natural disaster (bad harvest, fire, cattle disease, etc.)" or not possessing, for any reason, the necessary inventory or labour power to work their land, the right to lease the whole or part of their land for payment in cash or in kind.[4] Just as the employment of hired labour had been sanctioned by the "temporary rules" as a provisional expedient, so the leasing of land was allowed under the guise of a concession to temporary emergencies. But, whereas the temporary rules were given full publicity in advance of the conference, the decree of the RSFSR on leasing, though adopted on April 21, was published for the first time in *Izvestiya* on May 1, 1925, two days after the conference had dispersed. Whether the difference was due to a greater sense of

---

[1] *Vestnik Kommunisticheskoi Akademii*, xiii (1925), 245.

[2] *Ekonomicheskaya Zhizn'*, April 22, 1925.

[3] *Bednota*, April 23, 1925. Later *Bednota* became bolder, and in a leading article of June 2, 1925, attacked the principle laid down in art. 40 of the agrarian code (which was not amended) that the employment of hired labour was permissible only if the employer worked himself (since labour was the only basis of the right to hold land): "This demand is hardly viable in the conditions of NEP; in any case, this demand is extremely difficult to enforce in real life".

[4] *Sobranie Uzakonenii, 1925*, No. 27, art. 191.

embarrassment, or a stronger fear of opposition in the party on the second issue, can only be guessed. But it can hardly be doubted that the purpose of both decrees, issued at this moment, was to blanket discussion of these questions at the conference.

Thus, when the fourteenth party conference met on April 27, 1925, the plan to prevent the crucial agrarian question from becoming a bone of contention had been carefully worked out by the Politburo. These cautious arrangements might have succeeded but for an indiscretion on the part of Bukharin, who chose this embarrassing moment to come out as a thinly disguised champion of the " Stolypin-Soviet " line. The occasion was a speech at a mass party meeting in the Bol'shoi Theatre in Moscow, held on the day after the Politburo had taken its major decisions on the hiring of labour and the leasing of land. When Bukharin spoke on April 17, 1925, the decisions had not been formally announced. But it was clearly his function to prepare the ground for them ; and in so doing he developed, in far plainer terms than had yet been heard from a party platform, the philosophy of the wager on the *kulak*. He started [1] with the picture of Soviet Russia surrounded by a capitalist world which had succeeded in stabilizing its position after the devastation of the war : this made it essential to " increase the rate of our economic development ". But how was such an increase to be achieved ? Under war communism it had been hoped to " establish a planned economy at once and so reach socialism ". Under NEP it was recognized that the way to socialism lay through " a whole series of transitional forms ". What must, above all, be kept in mind

[1] The speech was published in two versions — in *Pravda*, April 24, 1925 (a week's delay in the publication even of important speeches was not unusual at this time, and does not suggest hesitation or textual revision), and in *Bol'shevik*, No. 8, April 30, 1925, pp. 3-14 ; No. 9-10, June 1, 1925, pp. 3-15. Though the *Pravda* version is described as " abbreviated ", the *Bol'shevik* version is of about the same length. But it shows substantial verbal differences : the style has been revised throughout and, in particular, many controversial phrases, including most of the references to the *kulak*, have been toned down. The less discreet *Pravda* version clearly has better claims to be regarded as an authentic record of what Bukharin said, and has been followed in the text ; some significant variants in the *Bol'shevik* version are quoted in footnotes. *Bednota*, April 26, 1925, printed a shortened version of the *Pravda* text. Rykov, at the fourteenth party congress eight months later (*XIV S"ezd Vsesoyuznoi Kommunisticheskoi Partii (B)* (1926), pp. 415-416), quoted the speech in the *Bol'shevik* version, which had by this time evidently been accepted as official.

was the indispensability of " a process of accumulation in the peasant economy ". Bukharin distinguished between two stages in Lenin's attitude to the positive function of NEP. In the first stage, of which the article of May 1921 *On the Tax in Kind* was the typical expression, Lenin had argued that it was possible to attain to socialism through an alliance with capitalism under a system of " state capitalism ". Concessions had to be made to the petty bourgeois peasantry which still, however, was regarded as " our chief enemy ". At this period Lenin treated the co-operatives as " the most important link in the chain of state capitalism ", something which " aids the *kulak* element in the countryside ". In the final stage, represented by Lenin's last article of January 1923 *On Cooperation*, Lenin had abandoned the conception of an alliance with capitalism and adopted a new view of the rôle of the cooperatives. According to this view, "the peasantry, organized in cooperatives, plus our (socialist) state industry, take the offensive against big capital and against the remains of private capital in general ". Relying on this somewhat strained interpretation of Lenin's last utterance on the subject,[1] Bukharin was able to depict the dead leader as an advocate of the unhampered development of a well-to-do peasant economy.

Bukharin then offered his audience a coloured, but not wholly unfair, outline of conditions in some parts of the Soviet countryside :

> The well-to-do top layer of the peasantry — the *kulak* and in part the middle peasant — *is at present afraid to accumulate.*[2] . . . If the peasant wants to put up an iron roof, tomorrow

---

[1] For conflicting views of the cooperatives see pp. 276-277 below ; Bukharin's later elaboration ot this argument, and Krupskaya's refutation of it, will be discussed in Part III in the following volume.

[2] In the version in *Bol'shevik* this passage began : " The well-to-do top layer of the peasantry and the middle peasant who is striving to become well-to-do *is at present afraid to accumulate* ". There were numerous variants in the following paragraphs. The *Bol'shevik* version avoided the terms *kulak* and *batrak* ; on the other hand, it introduced an argument absent from the *Pravda* version : " We have in the countryside a mass of peasants who are in fact working nowhere, but must eat. . . . This unused surplus population, concealed or unconcealed, presses fearfully on the towns, increasing the volume of unemployment." The phrase, " If a peasant puts on a new roof, they say he is a *kulak* ", occurred in Stalin's interview with the *sel'kor* delegation (see p. 247 above). This does not necessarily prove that Bukharin borrowed it from Stalin : it may have been a catch-phrase of the period.

he will be denounced as a *kulak* and that will be an end of him.
If the peasant buys a machine, he does it " so that the communists may not see ".   The technical improvement of agriculture is enveloped in a kind of conspiracy.

If we look at the different strata in the countryside, we shall
see that the *kulak* is displeased with us because we *prevent him
from accumulating*.  On the other hand, the poor peasants sometimes grumble at us for preventing them from hiring themselves out as *batraks* to this same *kulak*. . . . The poor peasant
who has no horse and no implements of production, and who
sits on his land, is displeased with us because we prevent him
" earning his bread " with the *kulak*.

Bukharin then proceeded to his recommendation :

Our policy in relation to the countryside should develop in
the direction *of removing, and in part abolishing, many restrictions
which put the brake on the growth of the well-to-do and* kulak
*farm*.  To the peasants, to all the peasants, we must say :
*Enrich yourselves*, develop your farms, and do not fear that
constraint will be put on you.

However paradoxical it may appear, *we must develop the
well-to-do farm in order to help the poor peasant and the middle
peasant.*[1]

Bukharin attempted the tactical manœuvre, later more successfully pursued by Stalin, of depicting himself in a middle position
between two opposite deviations, of which the first, at any rate
in the present instance, was purely hypothetical :

Some comrades think that it is necessary to develop *kulak*
farms — and *that is all*.  They do not understand the other half
of the problem : the question of compensation, of aid to the
middle and poor peasant.

But there was an opposite and more dangerous deviation :

Others say that capitalism in the countryside is developing,
that the *kulak* will establish large-scale farming, that renting
will grow, that the *kulaks* will turn into new landlords, and

---

[1] This passage was also toned down in the *Bol'shevik* version by minor
verbal variants, and by inserting, before the injunction to the peasants to enrich
themselves, an entirely new sentence : " The struggle against the *kulaks* must
be conducted by *other* means, on *other* lines ; it must be conducted by *new*
means, and conducted energetically so that the result of the change may not
be, so to speak, a ' wager on the *kulak* ' ".   The phrase " a wager on the *kulak* "
also occurred in the *Pravda* version, but in another context (see p. 261 below).

that then we shall have to carry through a second revolution
on the rural front.

This view Bukharin denounced as " theoretically incorrect and
practically senseless ". He ended on a consoling note. Just as
in capitalist countries cooperatives and credit had been used to
force the peasant to " grow into " a bourgeois society, so similar
means would be used " through the cooperatives to lead the
peasant to socialism ". The class war in the countryside would
not " die away all at once " ; it might even temporarily become
more acute. But there was no excuse for talking of " expropriat-
ing the capitalists in a second revolution ".[1]  Bukharin concluded :

> Is this a " wager on the kulak " ?  No.  Is it a declaration of
> a sharpening of the class war in the countryside ?  Also not.  I
> am not at all for sharpening the class war in the countryside.

Bukharin's speech provoked no immediate reaction. Even the
phrase, " Enrich yourselves ", addressed to the peasants, which
Bukharin was never afterwards allowed to live down, passed for
the moment unchallenged.

The fourteenth party conference assembled ten days later, on
April 27, 1925. The debate on the single agricultural tax, which
was taken first, yielded no surprises. Tsyurupa introduced it in
a long and rambling speech. He detailed the tax reductions proposed, and claimed that the annual tax falling on the peasantry
was now only 4 rubles a head against a pre-war annual charge,
covering taxes, rent and other obligatory payments of 10 rubles.[2]
He spoke of the need for greater mildness in the measures taken
to enforce payments of the tax. On the future basis of the tax,
he guardedly expressed the view that, however unsatisfactory its
present form, an income-tax was technically unworkable, and
expressed his own preference for a tax on rental value. But he
concluded by reading the decision of the Politburo to defer the
matter to the next party congress, which had the effect of closing
the discussion.[3]  The broader issues of agrarian policy were lightly
touched on in Rykov's general report. The main task of the
period was the revival and re-equipment of industry. In the

[1] In the Bol'shevik version Bukharin refused to " proclaim a St. Bartholomew's
night for the peasant bourgeoisie " (Bol'shevik, No. 9-10, June 1, 1924, p. 6).
[2] Chetyrnadtsataya Konferentsiya Rossiiskoi Kommunisticheskoi Partii (Bol'-
shevikov) (1925), pp. 60-62.              [3] Ibid. pp. 66-68, 75.

country its main characteristics were " relative agricultural over-population " and the development of production for the market. These factors necessarily led to differentiation in all its forms :

> In conditions of free trade and a petty bourgeois peasant economy, it is quite unavoidable that there should be more rapid growth for some peasant households, slower growth for others.

Rykov cautiously advocated the abandonment of " administrative obstacles " to the hiring of labour and the leasing of land : these facilities were necessary to " the development of the productive forces of the countryside ". He took the vexed question of the *kulak* by the horns. Kalinin and others had tried to get round it by drawing a fine distinction between " the well-to-do rich peasant " and the *kulak* : this, Rykov thought, was unreal and pointless. It was better frankly to recognize " the inevitability, during the present phase of restoration, of the growth in the countryside of relations of a bourgeois type, and the indispensability of fixing a clear political line in regard to this bourgeois peasant stratum ". This led up to a broad declaration of policy :

> By ensuring conditions for free accumulation in *kulak* households the tempo of accumulation in the whole economy is raised, the national income grows more rapidly, the material possibilities of real economic support for weak and poor households are increased, the possibilities of absorbing surplus population are broadened, and, finally, a more favourable atmosphere is created for the growth of co-operatives and the guiding of peasant savings into the co-operative channel.

Having thus staked on the *kulak* the prospects of revival throughout the economy, Rykov concluded with a pious hope that, " while developing capitalism in agriculture we may be able, more fully than hitherto, to turn our face to the poor peasant and the middle peasant ".[1]

Rykov's speech seemed at first likely to pass off without exciting serious controversy. This was evidently the hope and intention of the party leaders. Kamenev was in the chair ; and neither Stalin nor Zinoviev spoke throughout the discussion. Trotsky was not present and was not a delegate. Several minor orators entered the lists. One delegate, with evident reference

[1] *Chetyrnadtsataya Konferentsiya Rossiiskoi Kommunisticheskoi Partii (Bol'-shevikov)* (1925), pp. 83-86.

to Bukharin's admonition, complained that it had been left obscure up to what point the peasant had " the right to grow rich ", and thought that Rykov tended to call every expanding peasant household a bourgeois household. Another wanted to go even further in removing restraints on the *kulaks*, and attacked the draft resolution of the conference on the cooperatives for proposing to exclude " openly *kulak* elements " from the administrative organs of cooperative societies.[1] Then Larin rose. Kamenev evidently scented trouble, tried to declare the debate closed, and, when this was voted down, grudgingly gave Larin ten minutes, which were later extended by a show of hands.[2] Larin began by supporting Rykov's argument in favour of legitimizing the hiring of labour and the leasing of land : this was the first time, he declared with his usual perversity, that NEP had been extended to the countryside. He followed Rykov in attacking Kalinin's attempt to deny the existence of differentiation and to pretend that the *kulaks* were only middle peasants : Kalinin and his friends were like " Catholic priests who baptise meat ' fish ' for use on fast days ". Having thus cleared the ground, Larin launched into the main theme of the speech, a slashing attack on Bukharin's recent pronouncement. Bukharin had not merely given his approval to temporary expedients of policy in the spirit of NEP, but was opposed in principle to the intensification of the class struggle in the countryside, and wanted to give an undertaking against any " second revolution ", i.e. against a possible expropriation of the *kulaks* 15 or 20 years hence. Larin, on the contrary, while he applauded as indispensable the temporary measures of conciliation now proposed, hoped to see the ending of NEP and the expropriation of the *kulaks* in 15 or 20 years' time.[3] Larin throughout his speech skilfully pretended to accept wholeheartedly the line of Rykov and the party central committee, from which Bukharin had diverged.

[1] *Ibid.* pp. 109-110, 122.
[2] *Ibid.* pp. 135, 138 ; according to a later statement in Stalin, *Sochineniya*, vii, 382, Zinoviev " demanded that the attack on Bukharin should not be allowed ". Larin, a former Menshevik, was never treated very seriously in the party ; as a Menshevik at odds with official Menshevism, he had been described by Lenin in 1911 as " the *enfant terrible* of opportunism " (Lenin, *Sochineniya*, xv, 126).
[3] *Chetyrnadtsataya Konferentsiya Rossiiskoi Kommunisticheskoi Partii (Bol'-shevikov)* (1925), pp. 135-142.

This assault demanded an official reply. Rykov protested, appositely enough, that nothing could have been more inopportune than Larin's threat to the *kulaks* at a moment when it was desired to conciliate them,[1] but did not attempt to answer the awkward question about the intended duration of the tactics of conciliation and the ultimate goal of agrarian policy. On the following day Bukharin, who had not intended to speak and had not been present when Larin attacked him, denied that there was any difference of opinion between the central committee and himself, or that he was guilty of a " *kulak* deviation ". He accused Larin of simplifying the picture by ignoring the middle peasant and talking as if there was nothing between the *kulak* at one end of the scale and the kolkhoz at the other. Cooperatives for the middle peasant and kolkhozy for the poor peasant could go hand in hand with freedom of opportunity for the bourgeois peasant. Progress depended on uniting all these methods : " the kolkhoz is a powerful instrument, but it is not the high road to socialism ".[2] Bukharin could always attract personal sympathy where Larin merely aroused antagonism. The conference appears to have listened to his explanation with sympathy, or without overt dissent.

Since none of the other leaders was prepared to speak, the debate died away ; and the thorny issue of the *kulak* was reflected only indirectly in the resolutions of the conference on other matters. An anodyne passage in the introductory section of its general resolution on party work declared that the link between proletariat and peasantry was as indispensable as ever, that differences of interest between them were " not irreconcilable ", that capitalist elements were bound to grow in the countryside, but that they could be overcome only " by methods of economic struggle " and by developing the cooperatives.[3] The introductory section of the resolution on the cooperatives [4] went further in recognizing the problem. It began by insisting on increased production in agriculture as the paramount aim : this required

[1] *Chetyrnadtsataya Konferentsiya Rossiiskoi Kommunisticheskoi Partii (Bol'-shevikov)* (1925), pp. 143-144.

[2] *Ibid.* pp. 181-189 ; for Bukharin's previous remark that the kolkhozy were " not the high road to socialism " see p. 221 above.

[3] *VKP(B) v Rezolyutsiyakh* (1941), ii, 4-5.

[4] For the operative parts of this resolution see pp. 281-282 below.

" the most rapid replacement of the three-field system by the many-field system and the transition to efficient methods of agricultural production ". But it foresaw, as an inevitable concomitant of this process, " a sharpening of competition between the socialist and capitalist elements of the economy ", and " a differentiation in the countryside " which, beginning with the introduction of NEP, must be expected to find expression in, " on the one hand, a further increase and strengthening for a certain time of the new peasant bourgeoisie emerging from the well-to-do strata of the peasantry, and, on the other hand, the proletarianization of the poor peasant elements ".[1] But the frank diagnosis of a rising rural bourgeoisie, put in this context, led up to the practical conclusion of the need to develop and strengthen the cooperatives, and carried the implication that this would prove a sufficient remedy. The effect was to suggest that alarm was unnecessary, and the situation well in hand. The resolution of the conference on the agricultural tax made the firm recommendation for a reduction of the total for 1925-1926 to 280 million rubles. But, on the crucial issue of the incidence of the tax, it restated the contradictions inherent in the problem without resolving them. The grading of the tax, while " guaranteeing the interests of the poor and middle peasantry ", was also to " open the possibility for the further development of the peasant economy ". The new proposals which the central committee was to prepare for the next congress were to guarantee the " untrammelled development " of agriculture and at the same time to " protect the interests of those strata in the countryside which are the bulwark of the Soviet power ".[2] Even the last phrase, which was intended to refer to the poor peasant, seemed ambiguous and ambivalent in the light of the new policy.

A party conference, unlike a party congress, was not a sovereign body, and any resolutions passed by it became binding only when they had received the formal endorsement of the party central committee.[3] The central committee remained in session throughout the conference for this purpose. It did not, however, confine

[1] *VKP(B) v Rezolyutsiyakh* (1941), ii, 14-15.      [2] *Ibid.* ii, 23-24.
[3] This had been true of the important resolutions passed by the thirteenth party conference in January 1924 : see *The Interregnum, 1923-1924*, p. 351.

itself to confirmation of the conference decisions. The issues of policy which had been glossed over by the conference were taken up, once the conference was safely over, in the more restricted forum of the central committee. The fourteenth party conference ended on April 29, 1925 ; and on the following day the central committee adopted a considered resolution, which was published a few days later in the press, and remained the major party pronouncement of this period on agrarian policy. The resolution was submitted to the committee by Molotov, whose accompanying report drew a careful distinction between the *kulaks* and the evil practices of which they were guilty : " in tolerating the *kulaks*, we shall struggle against oppressive bargains, against the dishonest exploitation of the poor peasantry ".[1] The author of the resolution, as was later divulged, was Bukharin.[2]

The preamble to the resolution, which bore the title, " Current Tasks of Party Economic Policy in Connexion with the Economic Needs of the Countryside ", noted that the result of the civil war had been " a significant decline in agricultural production and impoverishment of the countryside (an acute shortage of livestock and implements), which is now finding expression in a pronounced relative *over-population of the countryside* (so-called ' rural unemployment ') ". It then proceeded to an enunciation of principles :

> The interests of the real expansion of agriculture, which is developing at the present period predominantly in the form of small, individual peasant farming, demand an increase in the marketability of the production of peasant farms and, in connexion with this, a decisive elimination of the remnants of " war communism " in the countryside (for example, a cessation of the struggle by administrative methods against private trade, the *kulaks*, etc.), which are incompatible with the development of market relations in the country as sanctioned in NEP conditions. The development of market relations in the countryside itself, and the strengthening of the commercial link of the country with the town and the external market, is, and must increasingly be, accompanied by a strengthening of the basic mass of middle peasant farms, together with the simultaneous growth (at any rate for the next few years), on the one hand, of

---

[1] The resolution is in *VKP(B) v Rezolyutsiyakh* (1941), i, 642-649 ; Molotov's report was published in full in *Pravda*, May 9, 1925.
[2] *XIV S"ezd Vsesoyuznoi Kommunisticheskoi Partii (B)* (1926), pp. 149, 415.

the well-to-do strata in the countryside, among which capitalist elements can be distinguished (the *kulaks*), and, on the other hand, of farm labourers and the country poor.

The existence in the countryside of a substantial amount of unemployed labour power, which is not being absorbed by the development of agriculture and by industry, throws into particular relief at the present time the necessity of an effective removal of the administrative obstacles which slow down the growth and strengthening of peasant households (including the well-to-do strata), together with the necessary carrying out of legal (and especially economic) measures of struggle against *kulaks* who are connected with rural usury and the dishonest exploitation of the poor peasantry.

The formula seemed, like Molotov's report, to admit that there might be *kulaks* who were not exploiters.

Having established the principle of a carefully guarded encouragement for *kulaks*, the resolution went on to pay tribute in conventional terms to the cooperatives and, with qualifications, to the Sovkhozy and kolkhozy. It announced the support of the party for every attempt to " unite working peasant households by way of the development of cooperatives ", and declared that only this method could " transform the slow growth of a much enfeebled mass of small peasant households into a powerful development of the productive forces of the countryside ". Cooperative societies for the use of machines and societies for the common cultivation of land should both be encouraged. This policy would " create the preliminary conditions for a gradual (voluntary) transition to large-scale collective agriculture, combined with the application of the latest methods of mechanization and with electrification ". Approval of the Sovkhozy was qualified by a recommendation for " a partial liquidation of Sovkhozy which cannot really be restored in the next two or three years ". The kolkhoz movement was to be supported only " in so far as it is growing through the completely voluntary participation of peasant and poor peasant households uniting together "; and kolkhozy, too, must " demonstrate their economic viability ".

This long and diffuse preamble served as an introduction to three sections on " practical measures " which were the operative part of the resolution. The first section covered agrarian policy properly so called, and was consistently designed to favour the growth of relatively large individual holdings in the hands of efficient and prosperous peasants. It pronounced against frequent redistributions of land " in contravention of the agrarian code ",[1] offered credits to facilitate the grouping together of isolated strips and dispersed holdings, and encouraged the formation of separate units whether in the form of the *otrub* or of the *khutor*, as well as all forms of cooperation and mechanization. It sanctioned " the broader utilization of the right of leasing land by peasants " up to a maximum period of two rotations or, where the three-field or four-field system was still in operation, of 12 years. Even this limit might be exceeded in the case of state lands leased to peasants. Material support was to be given to the Sovkhozy. The resolution endorsed the decree of the RSFSR of March 6, 1925, on the strengthening of the Sovkhozy,[2] and proposed that it should be extended to the other republics. But doubt was expressed whether " the existence of the present number " of Sovkhozy justified itself. This section of the resolution quoted the Politburo resolution of April 16 and the Sovnarkom decree of April 18,[3] confirmed the authority given by it for the unrestricted hiring of labour, specifically adding (the point had presumably been contested) that it applied to leased lands as well as to those held by other forms of tenure. The second section of the resolution recommended the abandonment of " the recently existing practice of limiting prices of grain and agricultural products ", and the adoption of the practice of agreements through " state and cooperative purchasers " without " compulsory prices for peasant-sellers ". This registered the victory of the well-to-do peasants who had broken the attempt to impose fixed prices for grain after the 1924 harvest. The third section briefly confirmed the reduction in the total of the agricultural tax, and proposed public works and resettlement to relieve over-population in rural areas. For this purpose state land should be placed at the disposal of the peasant ; the number of Sovkhozy in the densely populated areas should be reduced, and their land distributed.[4] The resolution

---

[1] The agrarian code of 1922 had, in fact, been vague and neutral on questions of land tenure, and nearly everything had been left to the discretion of the local authorities (see *The Bolshevik Revolution, 1917–1923*, Vol. 2, p. 296) ; the new pronouncement was intended as a guide to interpretation.

[2] See p. 218 above.     [3] See pp. 256-257 above.

[4] *VKP(B) v Rezolyutsiyakh* (1941), i, 642-649.

was noteworthy as containing the first recognition in party literature of a problem which was to become a dominant theme in the next few years — the problem of rural over-population.[1] But, though it contained the conventional genuflexions to such familiar symbols of party doctrine as the Sovkhozy and the kolkhozy, it was a clear-cut announcement of party acceptance of the wager on the *kulak* as the mainstay of agricultural recovery. While Bukharin's slogan, " Enrich yourselves ", was not formally endorsed, the resolution represented Bukharin's policy.

The annual decree on the rates of assessment for the agricultural tax was issued a few days after the party conference and the central committee had ended their labours.[2] It was a document of immense complexity, containing more than 100 tables of rates of assessment and conversion for different regions. Apart from these additional refinements, it differed from its predecessors in three important respects. It reduced the total amount of the levy, as already decided ; it reduced the rate of conversion for animals ; and it contained elaborate tables fixing statutory minimum holdings for different regions below which no tax was levied, so that an increased number of the poorest peasants secured exemption.[3] The first two of these changes were substantial concessions to the well-to-do peasant ; the third was a sop to those party consciences which were troubled by the new turn towards the *kulak* and demanded that something should be done to uphold the poor peasant. In both respects the new line accurately reflected the mood of the party leadership. The official economic journal hailed it as " a transition to intensified

---

[1] Kamenev, in a speech earlier in April 1925, had referred to " the growth of population in the countryside " which " goes on at such a rate that the excess of hands cannot be used either in the towns or in the country ", and put the natural annual increase at 2·5 per cent (L. Kamenev, *Stat'i i Rechi*, xii (1926), 131-132) ; F. Lorimer, *The Population of the Soviet Union* (League of Nations, Geneva, 1946), p. 89, on the basis of the 1926 census figures, estimates the increase at " slightly less than 2 per cent a year ".

[2] *Sobranie Zakonov, 1925*, No. 31, art. 209 ; it appeared on this occasion for the first time as a decree of the USSR. Its importance is indicated by the fact that it was printed in full (though without the accompanying tables) in *Izvestiya*, May 8, 1925 (the day after its promulgation) ; it was printed in full with the tables in *Bednota*, May 12, 1925, occupying practically the whole number.

[3] In practice, according to *Finansy i Kredit SSSR*, ed. V. Dyachenko and G. Kozlov (1938), p. 120, the proportion of exemptions rose only from 20 per cent in 1924–1925 to 25 per cent in 1925–1926.

forms of NEP ", which corresponded to " the present stage of development of the national economy ".[1]

Once the hurdles of the party conference and the party central committee had been surmounted, legislative formalities presented less difficulty.  The TsIK of the RSFSR met at the beginning of May 1925 and confirmed, apparently without debate, the amendment of the agrarian code on the leasing of land adopted by its presidium a fortnight earlier.[2]  The ninth All-Ukrainian Congress of Soviets which met at the same time in Kharkov was an occasion of greater importance, and Kamenev, as president of STO, came down from Moscow to expound and defend official economic policy.  He spoke of the need for measures " which will take the shackles off the peasant economy " : this meant to extend the period for which security of tenure of land was given by the existing law (in the Ukraine, nine years), and to remove restrictions on the leasing of land and the hiring of labour.[3]  The congress resolution on agriculture ranged far and wide.  Recognizing over-population as the central problem of the Ukrainian countryside, it dwelt at length on such remedies and palliatives as the extension of agricultural credit, the expansion of sugar-beet cultivation and of the sugar industry, as well as of technical crops and of animal husbandry, " the strengthening and deepening of the experiment of collectivization of poor and middle peasant farms " through cooperatives and other forms of joint cultivation, and the organization of migration to less crowded regions in other parts of the Soviet Union.  These generalities helped to mask the controversial character of the new policy.  Of the three decisions taken in Moscow, the reduction of the agricultural tax was not mentioned at all in the agricultural resolution, though the general resolution noted the reduction in the amount of the tax with " satisfaction ", and recommended " a more

[1] *Ekonomicheskaya Zhizn'*, May 6, 1925.

[2] *Vserossiiskii Tsentral'nyi Ispolnitel'nyi Komitet XI Sozyva : Tret'ya Sessiya* (1925), p. 25 ; *Sobranie Uzakonenii, 1925*, No. 29, art. 207 ; for the amendment see p. 257 above.

[3] L. Kamenev, *Stat'i i Rechi*, xii (1926), 189-190 ; the speech was printed in *Izvestiya*, May 6, 1925.  The full proceedings of the congress have not been available, so that it has been impossible to trace the course of the debate.  The relevant resolutions are in *Resolyutsii Vseukrain'skikh Z"izdiv Rad* (1932), pp. 197-206, or in Russian in *Puti Ukrepleniya Raboche-Krest'yanskogo Bloka* (1925), pp. 71-82.

flexible approach to the assessment of different regions and groups of households, having in mind the greatest possible encouragement of an increase in livestock " ; [1] in the Ukraine, where the proportion of " horseless " peasants was high, a low assessment on animals was, even more than elsewhere, a particular interest of the well-to-do peasant. The agricultural resolution dealt with the other two decisions in its two brief concluding paragraphs, which registered approval of " the extension of the leasing of land to 12 years ", and of " the law promulgated by the union government on the use of hired labour in the peasant economy ".

Kamenev's most delicate task was, however, to pilot the new proposals through the third Union Congress of Soviets which followed immediately on the Ukrainian congress. He devoted the main part of his speech to the reading of a long draft resolution " On Measures to Raise and Strengthen the Peasant Economy " which he submitted to the congress, interspersed with a running commentary of his own.[2] The preamble of the resolution took care to hold the balance between agriculture and industry and between different strata of the peasantry. Support from agriculture must go " side by side with the further development of industry " ; and, while " the application of any kind of administrative measures against the bourgeois (kulak) upper stratum in the countryside which is growing on the basis of freedom of trade " was dismissed as " inappropriate ", the policy of the Soviet Government must be " directed to real assistance for the poor and middle elements of the peasantry ". Freedom for the peasant — meaning freedom for the strong — was, however, the keynote of the resolution and of the speech. " Free choice by the peasantry of forms of land utilization " must be preserved. Kamenev admitted that " the

[1] *Resolyutsii Vseukrain'skikh Z"izdiv Rad* (1932), p. 199 ; the agricultura resolution is *ibid.* pp. 201-206. A year earlier the Ukrainian TsIK demanded " the firm adoption of a class policy in tax legislation " (*Byulleten' II⁰ⁱ Sesii Vseukrain'skogo Tsentral'nogo Vikonavchogo Komitetu VIII Sklikaniya*, No. 7, April 17, 1924, p. 7). Little satisfaction was given to this demand in 1925, when the tax was assessed in the Ukraine in such a way as to give the maximum relief to the well-to-do peasant ; a statement by Krupskaya to this effect at the fourteenth party congress in December 1925 was qualified, but not denied, by Kaganovich (*XIV S"ezd Vsesoyuznoi Kommunisticheskoi Partii (B)* (1926) pp. 160, 234).

[2] *Tretii S"ezd Sovetov SSSR* (1925), pp. 323-358 ; Kamenev's main speech and concluding remarks are also in L. Kamenev, *Stat'i i Rechi*, xii (1926), 202-244.

separation of the peasantry into individual *khutor* or *otrub* farms
would undoubtedly delay the process of the collectivization and
mechanization of agriculture ", and thought that propaganda
should be conducted " for forms of land utilization which facilitate
the transition from small, scattered, out-of-date peasant farming
to large-scale collective farming ". But the resolution firmly took
its stand on the " strict observance of the right of free choice ".
The right to lease land, and to hire labour under the " temporary
rules " of April 18, 1925, were reaffirmed. Price regulation by
the government must not be applied in such a way as to " establish
ruinous obligatory prices for the peasant seller ". Meanwhile the
peasant must receive financial help through the co-operatives to
increase the number of horses and tractors. The resolution was
more cautious in its approach to the still nebulous and half-
realized problem of rural over-population. Kamenev took up a
phrase which was just becoming current as a panacea for this
problem — the " industrialization of agriculture " — and challenged its usefulness. In the place of " this clever, foreign,
difficult word ", it would be better to tell the peasant that what
was really meant was that the products which he provided, instead
of being exported in a raw state for the benefit of the foreign
capitalist, should be worked up in factories at home.[1] One way
to relieve the pressure of " the surplus labour force in the countryside " was to extend the cultivable area through land improvement and reclamation : the government was granting 77 million
rubles for such work in the south-eastern regions affected by
periodical droughts. A second way was to encourage rural industries and handicrafts : the resolution expressed approval of a
decree of April 10, 1925, granting tax exemptions to rural workers
and artisans who did not employ hired labour.[2] A third way was

---

[1] Advocates of the " industrialization of agriculture " in the nineteen-twenties
meant mechanization of agricultural work. But it so happened that Lenin in
his early writings had applied the term to " the development of commercial-
industrial capitalist agriculture ", i.e. the transition from subsistence farming
to organized production for the market, bringing with it the growth of " rural
enterprises " and " rural hired labour " (Lenin, *Sochineniya*, iii, 460 ; cf.
*Leninskii Sbornik*, xix (1932), 62 : " the industrialization of agriculture separates
landed property from agriculture ") ; and Kamenev now interpreted, or professed to interpret, the phrase in this sense. Trotsky, *Sochineniya*, xxi, 435,
had distinguished between the use of the term in this sense and in the sense of
" the industrialization of cultivation " through mechanization and improved
technical methods.　　　　　　　　[2] For the decree see p. 361 below.

to promote migration of peasants from the over-populated areas to " free lands " in less congested regions.[1]

Having completed his reading of the resolution, Kamenev made a broader attempt to defend the principles on which it rested. Nothing short of a " cultural revolution " would really raise the standard of agricultural production : to liquidate illiteracy, to overcome the " old servile traditions ", the reluctance to work in common, the absence of initiative — this was the prime need. The implication was obvious that those who had advanced furthest on this path were the most deserving of support. Kamenev dwelt eloquently on " the need to remove certain restrictions which were left over in our countryside as a heritage of the old outworn epoch of war communism ", and to allow the peasant to " employ his labour more freely within the limits of the law and enjoy more freely the results of his labour ". Such measures were essential to increase production, though, of course, " it is first and foremost the well-to-do groups which are able to take advantage of these measures ".[2] He skirted delicately round the vexed question of the definition of a *kulak* :

> We refuse to consider as a *kulak* the peasant who, making use of the Soviet power, making use of Soviet credit, making use of the Soviet cooperatives, improves his farm, raises its technical level, acquires new machines on credit from Soviet industry.

Finally, Kamenev stoutly denied that the measures adopted since the autumn of 1924 represented any change of policy, and ended his speech with a galaxy of quotations from Lenin designed to prove that what was now being done was exactly what Lenin had prescribed several years in advance. This did not prevent Kamenev from describing these measures, in a later summing up, as " a definite political act ".[3]

---

[1] For this policy see Note A : " Migration and Colonization ", pp. 519-529 below.

[2] At the twelfth All-Russian Congress of Soviets, which met simultaneously with the third union congress, the People's Commissar for Agriculture of the RSFSR spoke still more frankly on this point : " If there are no other means of resolving this question, then there is only one way out : we must create conditions in which some can sell their superfluous labour power and others can buy it " (*XII Vserossiiskii S"ezd Sovetov* (1925), p. 201).

[3] *Tretii S"ezd Sovetov SSSR* (1925), p. 411.

In spite of Kamenev's plea, the party stalwarts in the congress showed little enthusiasm for the proposals. Behind the scenes Zinoviev addressed the party fraction with rather more frankness than was permissible on the floor of the congress. He referred openly to " concessions " to the well-to-do peasantry, and described the recognition of the right to lease land as " an absolutely indispensable concession ". He admitted that the *kulak* in the village was a greater danger than the nepman in the town, but used the specious argument that the way to curb the *kulak* was to help the poor peasant :

> For the present the evil resides not in the power of the *kulaks*, but in the fact that there are so many peasants in the villages who have no horse.

The conclusion was balanced and consoling :

> To a certain extent we free the hands of the *kulak*, but at the same time we prepare to circumvent him, to isolate him politically.[1]

Scepticism was, however, not altogether eliminated. The debate on Kamenev's report in the congress elicited some frank comments from peasants and others familiar with conditions in the countryside. A delegate from Tambov, where the 1924 harvest had been a failure, spoke of " 50 to 60 per cent of hungry people " and deaths from starvation. A delegate from White Russia, referring to the proffered remedy of migration, objected that only the well-to-do peasants could afford the costs of migration and resettlement, and they had no incentive to move ; another delegate dwelt wistfully on the land hunger in the Ukraine and White Russia, and on the " empty lands " in Siberia. One speaker remarked that he could not imagine " how they will receive us at home when we come and tell them that the renting of land is permitted ". Another complained that in Samara, in spite of official disapproval of frequent redistributions of land, such redistributions occurred every year.[2] But no serious opposition was offered. No alternative policy was, or could be, propounded.

---

[1] The speech was published in full in *Pravda* and *Izvestiya*, May 26, 27, 1925.

[2] *Tretii S"ezd Sovetov SSSR* (1925), pp. 369, 391, 407-408.

The resolution was carried unanimously with a few minor amendments.[1]   The wager on the *kulak* must be tried.

The attempt made at this time to revive the agricultural co-operatives was part of the campaign to assist the peasant and to raise the level of agricultural production.   These once powerful organizations had lost their *raison d'être* under war communism, and had virtually ceased to exist.   After the introduction of NEP they had been re-established under the terms of a decree of August 16, 1921, which encouraged " the working population of rural localities " to form " agricultural cooperative societies or *artels* for carrying out agricultural production in common, for organizing the labour of their members, for providing them with the necessary agricultural implements, seeds, fertilizers and other means of production, for working up and selling the output of agricultural production and, finally, for taking other measures designed to increase the quantity or improve the quality of the agricultural products of their members ".[2]   But, in spite of this comprehensive description of their functions, agricultural co-operatives, like other producers' cooperatives, seem at this time to have found little support ; and in the autumn of 1923, when the turn-over of all cooperatives did not exceed 6 million rubles a month, producers' cooperatives of all kinds accounted for only one-sixth of the total.[3]   The troubles of the scissors crisis drew the attention of the party to the importance of an adequate system for the marketing of agricultural products ; and the year 1924 marked an important stage in the rehabilitation of the agricultural cooperatives.   The establishment in February 1924 of a Central Agricultural Bank made possible the provision of credit facilities to be dispensed through the agricultural co-operatives.[4]   The thirteenth party congress in May 1924, in the course of a long general resolution on the cooperatives, referred especially to " the weakness of the cooperatives in the countryside ", and declared that the peasant must be organized not only

---

[1] The text is in *Tretii S"ezd Sovetov SSSR : Postanovleniya* (1925), pp. 21-29, and in *Sobranie Zakonov, 1925*, No. 35, art. 248.

[2] *Sobranie Uzakonenii, 1921*, No. 61, art. 434.

[3] *Vtoraya Sessiya Tsentral'nogo Ispolnitel'nogo Komiteta Soyuza Sovetskikh Sotsialisticheskikh Respublik* (1924), p. 39.          [4] See p. 471 below.

as a consumer, but as a producer : mention was made of collective marketing organizations already in existence for dairy products, flax and potatoes.[1] The summer of 1924 following the thirteenth party congress witnessed a general revival of interest in the cooperatives and the adoption of measures to assist them.[2] From this the agricultural cooperatives also benefited ; and a decree of the USSR of August 22, 1924, issued jointly by VTsIK and Sovnarkom, redefined the functions of agricultural cooperatives in the same broad terms as the decree of three years earlier, but with more attention to details of organization. Agricultural cooperatives were to be placed under the " general supervision " of the Narkomzem of the republic in which they worked, and were also subject, where they engaged in credit operations, to the " special supervision " of the appropriate financial authorities.[3] This decree, and the new credit facilities now available to cooperatives, marked a substantial advance. At the beginning of 1923 the capital resources of the agricultural cooperatives of the RSFSR had amounted to no more than 57 million rubles ; by October 1, 1924, they were said to have risen to 470 millions (or 560 millions including the Ukraine). The number of local societies in the RSFSR for granting agricultural credit rose from 1600 in October 1923 to 11,500 in April 1925.[4] Nevertheless, only 3 million peasants belonged to agricultural co-operatives as against 12 millions before 1914.[5]

The return to official favour of the agricultural cooperatives brought to light two different views of these organizations which prevailed in the party. Both schools of thought could fortify themselves with quotations from Lenin. On the one view, agricultural cooperatives were — as they had been before 1914 — organizations designed to help the individual peasant to produce and to market his crops efficiently and profitably : they were fundamentally capitalist institutions appropriate to the market

---

[1] *VKP(B) v Rezolyutsiyakh* (1941), i, 586. These three organizations — Maslotsentr, L'notsentr and Soyuzkartofel' — together with a similar organization for fruit and wine growers, Plodovinsoyuz, apparently remained independent of Sel'skosoyuz, the general organ which controlled all other agricultural cooperatives (*Planovoe Khozyaistvo*, No. 4, 1925, p. 64).

[2] See pp. 429-430 below.    [3] *Sobranie Zakonov, 1924*, No. 5, art. 11.

[4] *XII Vserossiiskii S"ezd Sovetov* (1925), p. 78.

[5] *Chetyrnadtsataya Konferentsiya Rossiiskoi Kommunisticheskoi Partii (Bol'-shevikov)* (1925), pp. 92, 132.

economy of NEP. This was exactly what Lenin had said when, after the introduction of NEP, he had accepted " cooperative " capitalism as a temporarily useful form of state capitalism :

> The cooperation of small producers of commodities . . . inevitably begets petty bourgeois capitalist relations and contributes to their development, brings capitalists into prominence and gives them the greatest advantage.[1]

And Lenin's diagnosis was borne out by a warning of the party conference of August 1922 against those anti-Soviet forces which " systematically attempt to convert the agricultural cooperatives into a weapon of *kulak* counter-revolution ".[2] But in two articles, or two drafts of one article, written in the last weeks of his active life, Lenin modified this criticism, and appeared to accept the view that agricultural cooperatives might serve as stepping-stones to the eventual collectivization of agriculture which was the goal and condition of a socialist economy. " With us ", he had written, " the cooperatives, thanks to the peculiarities of our governmental order, acquire an altogether exceptional significance ", and " in our conditions again and again perfectly coincide with socialism ".[3] These differences reflected two different basic attitudes towards NEP, both of which could claim support from Lenin's writings.[4] Those who regarded NEP primarily as a retreat into capitalism which would one day have to be retrieved were mistrustful of the cooperatives as capitalist survivals which must be temporarily and reluctantly tolerated. Those who treated NEP as the high road of advance towards socialism believed in the cooperatives as one of the major instruments of the transition.

The agricultural cooperatives thus quickly became involved in the controversies of 1924 and 1925 about the attitude to be adopted towards the *kulak*. The argument that the cooperatives were the bulwark of socialism in the countryside could be rendered plausible only if they served the interests of the poor and middle peasants. The thirteenth party congress in May 1924 had declared it indispensable " in every way to preserve the lower

---

[1] Lenin, *Sochineniya*, xxvi, 336 ; for Lenin's sceptical attitude towards producer cooperatives before the revolution see *The Bolshevik Revolution, 1917–1923*, Vol. 2, p. 120.
[2] *VKP(B) v Rezolyutsiyakh* (1941), i, 465.
[3] Lenin, *Sochineniya*, xxvii, 396.
[4] See *The Bolshevik Revolution, 1917–1923*, Vol. 2, pp. 276-279.

278 THE ECONOMIC REVIVAL PT. II

organs of the cooperatives from being captured and influenced by *kulak*-speculative elements which make, and will continue to make, efforts to use them as a springboard for their own advantage, thus discrediting the idea of cooperation among the broad peasant masses ".[1] The decree of August 22, 1924, confined the right of membership of agricultural cooperatives to those " enjoying the right to participate in elections to Soviets " — a limitation which theoretically excluded persons employing hired labour and therefore classified as members of the bourgeoisie. But the application of this provision, at any rate in the RSFSR, was modified by an instruction of the Narkomzem of the RSFSR of November 12, 1924, which laid it down that, whereas the limitation in question should be strictly applied to persons founding new agricultural cooperatives, it was not necessary, pending further orders, to expel persons not enjoying electoral rights from existing cooperatives.[2] Whatever party doctrine might prescribe, no regulation could do away with the hard fact that the membership of agricultural cooperatives consisted mainly of well-to-do peasants. The agricultural cooperatives were primarily concerned to provide an organization for the common working up and marketing of agricultural products ; and it was the well-to-do peasant who produced for the market. These facilities were useless to the poor peasant engaged in small-scale subsistence farming. A modest beginning had also been made in the establishment of cooperatives for the joint ownership and hiring of agricultural machines.[3] But a Gosplan report of the period described these associations as " little share companies " working for a profit and not making a " rational " use of their machines ;[4] in other words, the poor peasant could not afford to hire them. It was well established that participation in agricultural cooperatives was greatest among " the higher groups in the village ".[5] Participation was also

---

[1] *VKP(B) v Rezolyutsiyakh* (1941), i, 586.

[2] Quoted in L. Povolotsky, *Kooperativnoe Zakonodatel'stvo* (3rd ed. 1925), p. 178.

[3] According to an article in *Ekonomicheskoe Obozrenie*, No. 12, 1929, there were, in 1925, 4500 cooperative machine depots lending out tractors and agricultural machinery.

[4] *Ekonomicheskaya Zhizn'*, February 21, 1926.

[5] *Sel'skoe Khozyaistvo na Putyakh Vosstanovleniya*, ed. L. Kritsman, P. Popov, Y. Yakovlev (1925), p. 717. Detailed figures appeared to show that the percentage of participation among different groups of the peasantry rose

greatest among producers of specialized and " technical " crops ;
and these were once again the well-to-do peasants working on a
large scale for the market.[1] If the well-to-do peasant applied for
credit, he easily got 100 rubles ; the poor peasant was lucky if he
got 5 or 10 rubles.[2] Examples were quoted of the hostile attitude
of poor peasants to the agricultural cooperatives, which were
regarded as the preserve of rich and " strong " peasants, and of
refusal to admit poor peasants to membership.[3] In the Vladimir
province, the few kulaks " know how to organize and creep into
the boards of the cooperatives ".[4] " Old speculators and traders
quickly worm their way into the cooperatives ", complained a
Siberian delegate to the TsIK of the RSFSR in October 1925 ; [5]
and it was alleged that, in the same year, a good half of " the
leading personnel of the cooperative branches in the Ukraine "
were counter-revolutionaries who had fought against the Soviet
régime in the civil war.[6] Abundant evidence could be found to
support the conclusion of a party investigator who reported that
" capitalist principles have secured most favourable conditions
for themselves under the cooperative flag ", and that the party
leadership had taken as " an example of a movement towards
socialism " what was really a movement towards capitalism.[7]

In these conditions, the campaign in favour of the conciliation
of the kulak, which gathered strength in the first months of 1925,
enormously strengthened the case for the agricultural cooperatives ;
and insistence on their socialist character, bolstered by the famous
quotation from Lenin's last article, became a popular item of
propaganda for the party line. The decision to subordinate the

progressively with the wealth of the group till the highest group of all was
reached, when it began to fall off : the richest peasants could afford to remain
outside the cooperatives (Na Agrarnom Fronte, No. 3, 1925, pp. 118-130 ;
see also ibid. No. 11-12, 1925, pp. 51-66 ; No. 4, 1926, pp. 95-96).
  [1] In 1925, 28 per cent of all peasant households were said to belong to
agricultural cooperatives ; but among producers of specialized products (milk,
potatoes, tobacco, sugar) proportions up to 70 or 80 per cent were registered
(A. Arutinyan and B. Markus, Razvitie Sovetskoi Ekonomiki (1940), p. 214).
  [2] Vserossiiskii Tsentral'nyi Ispolnitel'nyi Komitet XII Sozyva : 2 Sessiya
(1925), pp. 215-216.          [3] Na Agrarnom Fronte, No. 7-8, 1925, p. 36.
  [4] Pravda, November 21, 1924.
  [5] Vserossiiskii Tsentral'nyi Ispolnitel'nyi Komitet XII Sozyva : Vtoraya
Sessiya (1925), p. 213.
  [6] Kooperatsiya v SSSR za Desyat' Let, ed. V. P. Milyutin (1928), p. 243.
  [7] Y. Yakovlev, Nasha Derevnya (1924), p. 65.
  VOL. I

kolkhozy to the agricultural cooperatives [1] was, in effect, a victory for the well-to-do peasant.　But even this could be plausibly depicted as a step towards the spread of socialist principles in the cooperatives.　It was the fourteenth party conference of April 1925, with its growing emphasis on the development of agricultural production and on the conciliation of the well-to-do peasant, which attempted the first detailed discussion of the rôle of agricultural cooperatives.　Rykov in his speech at the conference was less concerned than Bukharin had been to dwell on their potentially socialist character.　Silently dismissing the kolkhozy from the picture, he claimed that under NEP the cooperatives had become " our chief, almost our sole, lever " over the peasantry. Since, however, the party and the government were committed by NEP to an agriculture organized on petty bourgeois principles, the chief aim of the cooperatives must be not " the socialization of the process of agricultural production ", but " the organization of the peasants as producers of commodities ".　Rykov condemned the illusions of collective agriculture, as a year earlier he had poured scorn on the illusions of planning :

> The collective organization of a few peasant households working the land with the wooden plough is not socialist construction.　We shall not build a socialist economy with the help of the wooden plough.

On the other hand, he eulogized the tractor which was " revolutionizing the productive process to a far greater degree than a thousand agitators " by demonstrating the necessity of cooperative methods in production as well as in marketing.[2]　A minor difficulty was caused by the attempt to combine the credit functions formerly exercised by separate credit cooperatives with the other work of

---

[1] See pp. 220-221 above.

[2] *Chetyrnadtsataya Konferentsiya Rossiiskoi Kommunisticheskoi Partii (Bol'-shevikov)* (1925), pp. 87-90 ; for Rykov's earlier attack on planning see *The Interregnum, 1923–1924*, p. 126.　The wooden plough (*sokha*) was a favourite rhetorical symbol of the backwardness of Russian agriculture.　According to G. T. Robinson, *Rural Russia under the Old Regime* (1932), p. 244, it is " probably quite safe to say " that more than one-half of the peasants were still using it in 1917.　On the other hand, though it survived in many regions, it had disappeared from the most advanced agricultural provinces well before 1914 ; Mackenzie Wallace (*Russia* (2nd ed. 1905), ii, 202-203), returning to the Smolensk province after 25 years in 1903, noted that the wooden plough had in the interval been everywhere replaced by the iron plough.

the agricultural cooperatives. This placed the local agricultural cooperatives in a position of dual responsibility to the central cooperative organization and to the Central Agricultural Bank ; and peasants were said to be unwilling to entrust savings to organizations which also engaged in trade. Rykov explained that the two functions would in future be kept entirely separate in order to give the peasant " full assurance that credit resources are guaranteed against all risks ".[1]

The resolution of the conference, more eclectic in character than Rykov's report, laid down the functions of the agricultural and credit cooperatives in carefully balanced terms which took account of all points of view. Their first function was to organize credit for the peasants through independent agricultural credit societies financed by the Central Agricultural Bank. The second was to organize the collective processing and marketing of agricultural produce. Both these functions, which represented a revival of activities which had been well developed before 1914, accorded fully with the views of those who supported the cause of the independent and well-to-do peasant. The third item, the development of " all forms of collective agriculture, kolkhozy and communes of every kind, etc.", was a concession to those party doctrinaires who saw in current policy an unwarrantable deviation in favour of the *kulak*, and believed that collective cultivation was the one and only key to the development of a socialist agriculture. The fourth item, which declared it to be a function of the cooperatives to supply the peasants with the means of production, reflected Rykov's eulogy of the tractor, and satisfied all schools of thought, though the possibilities of realising it in practice were still extremely limited.[2] Finally, the resolution returned to the vexed question of the social composition of the cooperatives by recommending that " openly *kulak* elements " should be excluded from their boards of administration.[3] But, since nobody was " openly " *kulak*, and the interpretation of the rule was in the

---

[1] *Chetyrnadtsataya Konferentsiya Rossiiskoi Kommunisticheskoi Partii (Bol'-shevikov)* (1925), pp. 90-91, 107 ; the Central Agricultural Bank was opposed to the mingling of the two functions, and wished to keep the credit cooperatives separate and subordinate to itself (*Vestnik Finansov*, No. 5, May 1925, pp. 26-34).      [2] *VKP(B) v Rezolyutsiyakh* (1941), ii, 17-19.

[3] *Ibid.* ii, 15 ; this recommendation applied to all cooperatives, not merely the agricultural cooperatives. A delegate at the conference had protested even against this restriction (see p. 263 above).

hands of bodies on which the well-to-do peasant predominated, this restriction would have little effect. Of the two approved purposes which agricultural cooperatives might theoretically serve — assistance to the independent farmer and the encouragement of collective farming — the first continued throughout the middle nineteen-twenties to eclipse the second ; and the corollary of this was that the agricultural cooperatives, which should, in the official party view, have served as a bulwark against the encroachment of capitalism in the countryside, began more and more to serve the interests of the small rural capitalist, the efficient and enterprising *kulak*. The contrast which had been drawn at the thirteenth party congress between the alternatives of cooperation and capitalist development in the countryside [1] was mainly wishful thinking. As a writer in *Pravda* admitted, " the cooperative movement as a rule springs from the depth of the peasant masses, and lies outside the sphere of our influence ".[2] Once the major policy had been adopted of encouraging the efficient and prosperous individual peasant as the best means of securing increased agricultural production, every subsidiary instrument was mobilized and directed to this end. The agricultural cooperatives, rooted in a pre-revolutionary tradition and representing the interests of the efficient and prosperous peasant, were among the most important of such instruments.

### (d) The Harvest of 1925

The resolution of the party central committee of April 30, 1925, carried into effect by the legislative acts of the following month, was the high-water mark of the campaign to support the efficient and well-to-do peasant farmer, to encourage him to increase and develop his individual holding, and to make him the fulcrum of a national economic revival. To its supporters the policy looked like the logical outcome of NEP : once it had been decided to allow the peasant to trade freely and to base his economy on the bourgeois practices of the market, it seemed necessary in the interests of efficiency to carry this licence to its conclusion. The introduction of NEP — also a one-sided measure designed to favour the peasant with surplus grain to dispose of —

[1] See *The Interregnum, 1923–1924*, p. 148.
[2] *Pravda*, October 10, 1925.

had paved the way for a striking economic revival. If this new extension of NEP had the immediate effect of increasing the prosperity and the productivity of the upper stratum of the peasantry and of drawing new recruits into this stratum — in popular language, of strengthening the *kulaks* and of turning the more prosperous middle peasants into *kulaks* — the general spread of prosperity which might be expected from the success of the policy would affect other strata of the peasantry and other sectors of the economy, and thus, in the long run, like NEP, make its contribution to the eventual realization of a socialist order. But the new policy had another close resemblance to its prototype. It revived the controversy between those who praised NEP as the only true path to the socialist goal and those who emphasized its character as a " retreat ", and accepted it as a temporary, though necessary, evil.[1] Precisely the same issue now arose in regard to the new policy. Was it to be described positively as " an extension of NEP " or negatively as a retreat ? Zinoviev, who was about to make a report to a party meeting in Moscow on the proceedings of the fourteenth party conference and of the third Union Congress of Soviets, inquired of his colleagues in the Politburo whether the concessions in regard to leasing and to the employment of labour were to be described as a " retreat ". One or two members of the Politburo demurred to the use of the word, but did not press their objections, and it was formally sanctioned.[2]

The lines of division in the party leadership on the interpretation of the decisions of the fourteenth party conference and on the attitude to be adopted towards the well-to-do peasant were slow to crystallize. The position of Stalin was still highly ambiguous. His relations with Zinoviev, once Trotsky had been defeated, were marked by growing mutual mistrust. But Stalin, with his usual astuteness, seems to have recognized at an early date that the sharp turn of policy in favour of the *kulak*, executed behind the back of the party conference, had outraged a large section of party opinion, and that the position had been dangerously compromised by Bukharin's disarming frankness. While the fourteenth party conference was still in progress, he told the presidium

---

[1] For this argument see *The Bolshevik Revolution, 1917–1923*, Vol. 2, pp. 274-277.
[2] *XIV S"ezd Vsesoyuznoi Kommunisticheskoi Partii (B)* (1926), pp. 113-114.

of the conference, " in the presence of Sokolnikov, Zinoviev, Kamenev and Kalinin ", that " the slogan ' Enrich yourselves ' is not our slogan ".¹ This disavowal received, however, no publicity, and no word of dissent among the party leaders was allowed to appear. A few weeks later, an article in the newly established Komsomol daily newspaper, *Komsomol'skaya Pravda*, cautiously endorsed the slogan " Enrich yourselves ". This provoked a letter of censure from the party secretariat, which described the slogan as " not ours " and " incorrect ".² But the rebuke, like Stalin's previous criticism of the slogan, remained unpublished and unknown to the party in general. It continued to be assumed, both in the Soviet Union and by observers abroad, that Bukharin had spoken, if not for a united party, at any rate for a united leadership.³

Meanwhile those who had been primarily responsible for the April concessions hastened to consolidate their victory on the theoretical plane. The issue of the party journal *Bol'shevik* dated June 1, 1925, carried not only the second half of Bukharin's speech of April 17 containing the incriminating slogan, but two other articles clearly designed to support the current policy. The first, by one Litvinov, argued that the alliance with the peasantry,

¹ The authority for this is Stalin's own statement at the fourteenth party congress in December 1925 (Stalin, *Sochineniya*, vii, 382) ; but the statement would hardly have been made if it was open to contradiction. In a speech of May 9, 1925, Stalin, in rejecting what he called " the capitalist path " for the development of agriculture, described this " path " as meaning " development through the impoverishment of the majority of the peasantry in the name of the enrichment of the upper strata of the urban and rural bourgeoisie " (*ibid.* vii, 111) — a veiled, but unmistakable, allusion to Bukharin's slogan.

² The letter purported to convey " first impressions " and was written in the singular, but was signed collectively by Stalin, Molotov and Andreev. It was quoted at the fourteenth party congress by Stalin, who added that a few days later the Orgburo decided, " with Bukharin's full agreement ", to dismiss the editor of the journal (*ibid.* vii, 383-384) ; the full text of the letter was not published till 1947, when it appeared in Stalin's collected works (*ibid.* vii, 153-155).

³ Kamenev at the fourteenth party congress alleged that Bukharin's slogan " travelled round our party for a whole half year " (*XIV S"ezd Vsesoyuznoi Kommunisticheskoi Partii (B)* (1926), p. 254) ; according to Krupskaya, it was " taught to the broad party masses as the slogan of the party, as the slogan of the central committee " (*ibid.* p. 160). The Menshevik journal in Berlin described the phrase " Enrich yourselves " as " an appeal of the Politburo to the bourgeoisie " (*Sotsialisticheskii Vestnik* (Berlin), No. 10 (104), May 29, 1925, p. 4) ; for the *smenovekh* reaction see p. 301, note 1 below.

being everywhere a necessary condition of the victory of the revolution, demanded " an *intelligent* sacrifice of the temporary interests of the proletariat for the sake of its permanent class interests ". The second, by a certain Bogushevsky, bearing the title *On the Rural Kulak and on the Rôle of Tradition in Terminology*, quoted A. P. Smirnov and Kalinin with approval, spoke of the *kulak* as " a type of pre-revolutionary Russia, . . . a bogy, a phantom of the old world ", and suggested that the word should be dropped from current Soviet usage.[1] The same issue of *Bol'shevik* significantly published for the first time Lenin's note to the Politburo on the eve of the eleventh party congress in March 1922 resisting and denouncing Preobrazhensky's theses on the rise of the *kulak* and of differentiation in the countryside, and his letter to Osinsky on the same occasion.[2] The opportune disclosure of unpublished material from the Lenin archives became a familiar manœuvre as party controversies sharpened.

At this point an unexpected intervention threatened to bring the issue to a head. Krupskaya, angered by what she regarded as a perversion of her late husband's views, wrote an article attacking the Bukharin line and the policy of indulgence for the *kulak*, and sent it to *Pravda* for publication. Bukharin, the editor of *Pravda*, wrote a counter-article defending himself, and submitted both articles to the Politburo. It was a delicate situation. To veto the publication in *Pravda* of an article by Lenin's widow still seemed invidious and shocking to party consciences. But the argument against a public airing of differences between leading party members on so explosive a subject was also strong, and eventually prevailed. By a majority, the Politburo decided that neither Krupskaya's article nor Bukharin's reply should be published : the minority consisted of Zinoviev and Kamenev.[3]

[1] *Bol'shevik*, No. 9-10, June 1, 1925, pp. 16-38, 59-64.
[2] For these documents see *The Bolshevik Revolution, 1917–1923*, Vol. 2, pp. 292-293 ; they were originally published in *Leninskii Sbornik*, iv (1925), 389-396, with a note by Preobrazhensky explaining that the theses were drafted by him on behalf of a committee set up by the party central committee to examine the question.
[3] *XIV S"ezd Vsesoyuznoi Kommunisticheskoi Partii (B)* (1926), p. 270. This must have happened in the first part of June 1925 ; Krupskaya's article is said to have been sent in " on the day after the appearance of Bukharin's article " (the issue of *Bol'shevik* containing it was dated June 1, 1925, but journals were not always published punctually). Neither Krupskaya's article nor Bukharin's

By way of counterpart to this snub to Krupskaya, the central committee requested Bukharin to write an article in the press renouncing the slogan "Enrich yourselves" as incorrect.[1] But the voice
of the central committee was less authoritative than the voice of
the Politburo. The request does not appear to have been treated
as urgent. Bukharin went off for his summer vacation with the
peccant slogan still unwithdrawn.

The unwonted experience of defeat in the Politburo piqued
Zinoviev's pride. The first public hint of dissent at the summit
of the party seems to have come in a speech delivered by Zinoviev
in Leningrad on June 21, 1925, at a conference of party workers
in the Red Army. The occasion did not invite a pronouncement
on agrarian policy; that Zinoviev should have chosen to make
one, and to secure prominence for it in the press,[2] increased its
significance. Zinoviev began by announcing that he proposed
" to dwell a little on the peasant question ". " Face to the countryside " meant " Face to the middle and poor peasant " ; some
peasants had apparently interpreted it as " a turning towards the
well-to-do strata in the countryside ", as a proof of the determination of the leadership to rely, not on " the wretched peasant nag ",
but on " the fat *kulak* horse ". The decisions on leasing and on
hired labour had, in fact, been a " serious concession to the rich
top stratum in the countryside " : to pretend otherwise was to
offer the party a dose of " sugared water ". Zinoviev proceeded,
evidently with reference to Bogushevsky's article, though he did
not mention it, to poke fun at those who called the *kulak* " an

counter-article was ever published, though both were printed and circulated
privately by the opposition at the fourteenth party congress. According to
Petrovsky, to whom Krupskaya showed the article at the time, it " pointed
directly to the incorrect policy of the central committee, and sought to turn
party policy towards the smashing of the *kulak* " (*XIV S"ezd Vsesoyuznoi
Kommunisticheskoi Partii (B)* (1926), p. 168). Bukharin in his reply was evidently quite unrepentant. One passage from it was quoted at the fourteenth party
congress : " When we demand from the recipient of a concession an increased
volume of production, we are in effect addressing to him the slogan : Make
profits. Not only the *kulak*, but the concessionnaire enters into the system
of collaboration on the broadest lines " (*ibid.* p. 383).

[1] Stalin, *Sochineniya*, vii, 382-383.

[2] The main report of the conference appeared in *Leningradskaya Pravda*,
June 24, 1925 : Zinoviev's speech on the agrarian question (described as an extract — the commonplaces appropriate to the particular occasion were obviously
omitted) was printed nearly a week later (*Pravda* and *Leningradskaya Pravda*,
June 30, July 1, 1925).

obsolete category ", " a ghost of the old world ", " an unreal creature ". On the contrary, " the *kulak* in the countryside is *more dangerous*, far more dangerous, than the nepman in the town ". To deny the existence of differentiation was the symptom of a " *kulak* ideology ". The economic régime which had existed in the Soviet Union since the currency stabilization was not socialism, but, in Lenin's phrase, " state capitalism in a proletarian state " ; it was " a NEP Russia with growing elements of socialism ". The very existence of the *kulak* was enough to prove that socialism had not yet been attained.

Opinion in the party was now on the alert, and any hint of further concessions provoked a sharp reaction. In July 1925 it was discovered that the People's Commissariat of Agriculture of the Georgian SSR had drafted a project for legalizing the purchase and sale of land. It was explained that nearly half of the officials of the Georgian Narkomzem were drawn from families of " former princes, gentry or clergy ", and that the scheme had obtained the support of " some communists ". When news of it reached party headquarters, those concerned were heavily censured and nothing more was heard of the plan.[1] Some time during the year the Narkomzem of the White Russian SSR drew up theses on the restoration of agriculture which appeared to encourage the multiplication of individual *khutors* and *kulak* farms : this project was also vetoed by the party central committee.[2] In Uzbekistan " the new course " was interpreted as " a surrender to the bey and the manap " ; in the Urals, the slogan " Face to the countryside " seemed equivalent to " Face to the *kulak* ", and caused the poor peasants to lose all confidence in the party.[3]

The most serious heart-searchings of all were aroused in the Ukraine. It was in the Ukraine that every agrarian issue assumed its sharpest and acutest form, that the pressure of population on

[1] The fullest available record of this incident is an account of it given to the fourth congress of the Georgian Communist Party in the autumn of 1925 and quoted in *Leningradskaya Pravda*, December 23, 1925 : it was referred to at the fourteenth party congress by Zinoviev and by Orjonikidze, who described it as a " fool's project " which had met with " proper resistance " in the party *XIV S"ezd Vsesoyuznoi Kommunisticheskoi Partii (B)* (1926), pp. 118, 223). The allegation in L. Trotsky, *Stalin* (N.Y. 1946), p. 397, that it emanated from Stalin is improbable.

[2] *Istoricheskie Zapiski*, xlvi (1954), 302 ; no precise date or other details are given. [3] *Na Agrarnom Fronte*, No. 10, 1925, pp. 10-11.

land was fiercest, and differentiation between well-to-do and poor peasants most marked. Parts of the Ukraine had suffered badly from the partial harvest failure of 1924. In Volhynia, according to a statement made at the Ukrainian TsIK in February 1925, the situation was " extremely grievous ". Some 20 to 30 per cent of the peasants were eating only potatoes ; 10 per cent had no potatoes, and could exist only " by making oppressive bargains with the *kulak* or by begging or in some other way ".[1] This no doubt fanned the flames of animosity against the *kulaks*. In this atmosphere the concessions to the well-to-do peasant decided on in Moscow at the time of the fourteenth party conference were received with mixed feelings by the party leadership in the Ukraine ; and the unanimous endorsement of this policy at the ninth All-Ukrainian Congress of Soviets [2] did not prevent a crisis breaking out in the summer of 1925. It occurred on the specifically Ukrainian issue of the committees of poor peasants (komnezamozhi or KNS). These committees were created in the Ukraine at a time when their counterparts in the RSFSR had already been disbanded.[3] But, like the Russian kombedy in 1918, the Ukrainian komnezamozhi inevitably became rivals of the village Soviets ; and the struggle between poor peasants and well-to-do peasants assumed in this period the form of a struggle between komnezamozhi and Soviets for political power in the countryside — a struggle further complicated by the issue of Ukrainian nationalism, whose strongest adherents were commonly to be found among the supporters of the well-to-do peasant and the enemies of the komnezamozhi. The campaign for " the revitalization of the Soviets " in the spring of 1925 was an implicit challenge to their political ascendancy ; and the new encouragement given to Ukrainian nationalism was a blow to institutions which had never associated themselves with the nationalist cause.[4] But the major adverse factor was the decision of the fourteenth party conference, endorsed by the ninth Ukrainian congress of Soviets in May 1925, in favour of the conciliation of the *kulak*.

---

[1] *Byulleten' 4 Sesii Vseukrains'kogo Tsentral'nogo Vikonavchogo Komitetu*, No. 2, February 16, 1925, p. 23.    [2] See pp. 270-271 above.

[3] See *The Bolshevik Revolution, 1917–1923*, Vol. 2, p. 159.

[4] The campaign for the revitalization of the Soviets will be discussed in Part IV in the following volume, the rise of Ukrainian nationalism in a later volume.

A few days after the congress the Kharkov party journal, *Proletarskaya Pravda*, was vigorously echoing the argument of Bukharin :

> The peasant is afraid to accumulate, is afraid to purchase a new seed-drill, reaper, harrow or plough, since he may at once be dubbed a *kulak*, deprived of the right to vote, and in one way or another simply *dekulakized*.

The new line was necessary in order to stabilize the peasant economy and to increase production. It was regrettable that " some party workers in the KNS, and many members of them, interpret the new course as a wager on the middle peasant and as treason to the komnezamozhi ".[1]

The issue came to a head at the session of the central committee of the Ukrainian party in July 1925, when a debate on the future of the komnezamozhi was made the occasion for a radical dispute on agrarian policy. Petrovsky, at this time president of the TsIK of the Ukrainian SSR, seized the occasion to deliver a sweeping indictment of the growth of bourgeois tendencies in party policy :

> I am reminded a little of the days when the bourgeoisie prepared for the great French revolution on the basis of the doctrine of the class struggle, and then, when it had got power into its own hands, and the working class began to organize and to carry on a struggle against it, the bourgeoisie made a stand against the class struggle. Are we not here, comrades, making a sharp turn ?
> . . . Today in some villages I am met mainly by middle peasants and well-to-do inhabitants, who thrust aside the poor peasant at such meetings. Can we wink at such a phenomenon ? This tendency, comrades, is now declaring itself very strongly. NEP is, so to speak, getting a move on ; and, if we do not fear it in the town, do not forget that in the villages the poor peasants and the still weak communist organizations will find it difficult to cope with these NEP tendencies.[2]

[1] *Proletarskaya Pravda* (Kharkov), May 12, 1925.
[2] The speech, though not apparently published, was quoted by Zinoviev at the fourteenth party congress five months later (*XIV S"ezd Vsesoyuznoi Kommunisticheskoi Partii (B)* (1926), p. 119) ; Kaganovich accused Zinoviev of " tearing a single quotation " out of its context (*ibid.* p. 234), but quoted no other passages of Petrovsky's speech. It may be significant that Petrovsky had had a conversation with Krupskaya, who was visiting Kharkov at the time (*ibid.* p. 168).

But Popov, another Ukrainian party leader, the ıght that too many favours had been shown to those " *uneconomic elements among the village poor* " who like to lead a parasitical existence at government expense ", and wished to dissolve the remaining komnezamozhi as " an anachronism of the epoch of war communism ".[1] The demand for the liquidation of the committees was staved off by a decision to transform them into " social and voluntary organizations " to uphold the interests of the poor peasantry, and thus deprive them of their political authority, which would be automatically transferred to the Soviets.[2] This looked like a compromise which enabled the institution to survive while destroying its efficacy. The wings of the komnezamozhi had been effectively clipped. Popov, who defended the compromise, afterwards admitted that " remains of old anti-Soviet feelings among the mass of middle peasants had not been eradicated ", that " anything but confidence was felt between the middle peasant and the poor peasant ", that the reorganization of the komnezamozhi had been " painful ", and that, at the next Soviet elections, " while the activity of the middle peasant conspicuously increased, the activity of the poor peasants and of the KNS was relatively small ".[3] In the Ukraine, as elsewhere in the Soviet Union, the well-to-do peasant was in the ascendant in the summer and autumn of 1925. A report received at this time from the Ukraine by the party central committee showed, in the words of Krupskaya, " how bold the *kulak* has become ", and how the poor peasants " feel themselves abandoned ".[4]

By this time the party struggle was approaching its decisive phase. As in 1923 and 1924, it was the harvest, the decisive event of the Russian year, which gave the signal for battle. Early fears notwithstanding, the harvest of 1925 was excellent. The sown area showed a further increase over the previous year — once

[1] Also quoted by Zinoviev, *XIV S"ezd Vsesoyuznoi Kommunisticheskoi Partii (B)* (1926), p. 122.

[2] This resolution passed by the central committee of the Ukrainian party at its session on July 23-25, 1925, was endorsed by the central committee in Moscow in October 1925 (*VKP(B) v Rezolyutsiyakh* (1941), ii, 40).

[3] M. Popov, *Naris' Istorii Kommunistichnoi Partii (Bil'shovikiv) Ukraini* 2nd ed. 1929), p. 287.

[4] *XIV S"ezd Vsesoyuznoi Kommunisticheskoi Partii (B)* (1926), p. 160.

more a slight increase in the area under rye, and a much more substantial increase in the area under wheat and the " technical " crops.[1] In 1924 the harvest had yielded some 2800 million puds of grain, of which 450 millions had been marketed. When the first prognostications for 1925 were made, the grain harvest appeared likely to reach a total of from 4000 to 4200 million puds, of which, 1200 millions might be made available for the market, and from 350 to 400 millions exported.[2] Working on this basis, Sokolnikov estimated that it should be possible for the state purchasing organs to collect 800 or 900 million puds of grain.[3] The official estimate made in July and formally approved by STO more cautiously put the collection at 780 million puds ; [4] even this prospect was sufficiently rosy. In August these hopes were damped by unusually early rains, which caused losses amounting to 230 million puds : and this led to a hasty cut in export estimates from 380 million to 235-265 million puds, with a corresponding revision of import programmes.[5] Even after this adjustment, it was still a splendid harvest. The figures of preceding years were left far behind. When the final accounts were made up, the value of the 1925 harvest at pre-war prices was estimated at more than 10 million rubles against slightly less than 8 millions for the 1924 harvest. The grain harvest reached 4400 million puds or 80 per cent of the 1913 figure (which allowed for large exports). Other agricultural products registered higher percentages ; potatoes, milk, fruit and vegetables, and tobacco had outstripped pre-war levels of production.[6] It was the harvest of the recovery.

The troubles of 1925 began not, like those of 1924, from a partial failure of the harvest, but from unexpected difficulties

---

[1] *Kontrol'nye Tsifry Narodnogo Khozyaistva na 1926-1927 god* (1926), p. 337 ; *Itogi Desyatiletiya Sovetskoi Vlasti v Tsifrakh, 1917-1927* (n.d.), pp. 168-171.

[2] *Na Agrarnom Fronte*, No. 7-8, 1925, p. 52 ; *Planovoe Khozyaistvo*, No. 9, 1925, pp. 8-9 ; No. 10, 1925, pp. 47-48. This estimate was repeated by Kamenev in his speech of September 4, 1925 (see p. 292, note 3 below).

[3] *Sotsialisticheskoe Khozyaistvo*, No. 4, 1925, p. 5.

[4] *XIV S"ezd Vsesoyuznoi Kommunisticheskoi Partii (B)* (1926), pp. 263-264, 266.

[5] *Planovoe Khozyaistvo*, No. 10, 1925, p. 54 ; No. 1, 1926, pp. 41-42.

[6] See the tables in *Kontrol'nyi Tsifry Narodnogo Khozyaistvo na 1926-1927 god* (1926), pp. 339-344. Grain accounted for 85 per cent of all agricultural production in terms of value.

in marketing it. Under the new policy laid down in April 1925, the attempt to apply maximum fixed prices to the 1925 harvest was officially abandoned. The state purchasing organs now worked on what were called " directive " prices. These prices, which were to be maintained, not by administrative action, but by the economic manipulation of supply and demand, were intended rather to protect the peasant against a fall in prices (such as might be expected in a good harvest year) than to protect the state or the consumer against a rise. The Soviet leaders were no more immune than capitalist economists of the same period from fears of over-production and falling agricultural prices. At the third Union Congress of Soviets, in May 1925, Rykov had announced that " this year the government will be strong enough, and will not allow prices to fall so far as to be injurious to the peasant ".[1] Throughout the summer anxiety was constantly felt lest the expected good harvest should bring grain prices down to a ruinous level. On August 28, 1925, *Pravda* published, over the signature of a high official of STO and with an introductory note by Kamenev,[2] an elaborate estimate of the results of the harvest, which concluded by confidently predicting " *a universal fall in grain prices* " and diagnosing the presence of " the preconditions for a new opening of the scissors ". The moral was drawn by Kamenev at a party meeting in Moscow exactly a week later :

> The task of price regulation in 1925 . . . has consisted in not allowing the price of grain to fall below a given level. In connexion with this, the policy of so-called *directive prices* has been adopted in 1925, i.e. a system of mass state purchasing, which should guarantee the peasantry a definite equitable price consistent with its interests and with the interests of the consumer of grain — of the worker and of the peasant who buys grain. If we see that prices are beginning to fall, we must increase the demand on the spot and thus raise prices. If prices soar too high, we must call off our purchasers.[3]

[1] *Tretii S"ezd Sovetov SSSR* (1925), p. 155.
[2] Kamenev's note (reprinted in L. Kamenev, *Stat'i i Rechi*, xii (1926), 299-300) for the first time mentioned the " most serious *social* consequences " of the concentration of grain surpluses in the hands of the well-to-do peasant, and quoted the figures which he used in his speech of September 4, 1925 (see p. 299 below).
[3] Kamenev's speech of September 4 was published in *Pravda*, September 17, 18, 1925, and reprinted in L. Kamenev, *Stat'i i Rechi*, xii (1926), 303-337.

Early in September 1925 the price paid to the grower for a pud of rye fell steeply, under the influence of bumper harvest prospects, to 101 kopeks, or less than half the price four months earlier. On the assumption that the fall would continue, the " directive " price was fixed at 75-80 kopeks,[1] at which level it was to be supported, if necessary, by extensive state purchases. The " directive " prices for other grains were fixed in proportion on the same principle.

These apparently plausible calculations went hopelessly astray, and brought unmerited discredit on those responsible for them. The largest harvest since the revolution was paradoxically followed not by abundance, but by stringency, on the internal grain market, and by a strong upward pressure on prices. In the previous year the fixed prices of the state purchasing organs had held their own throughout the autumn in spite of competition from higher prices on the free market. In 1925 the " directive " prices of the state purchasing organs failed to bring out buyers and were almost at once forced up in an unequal struggle to compete with the free prices. The first note of alarm was sounded in an article in *Ekonomicheskaya Zhizn'* of September 24, 1925, which admitted that the grain collections were falling behind the plan and prices rising. A few days later an article in *Pravda* deplored the falling off in offers of grain, " especially in the districts of the Ukraine ", which was attributed to the shortage of industrial goods and the lowering of the agricultural tax.[2] In August 1925 only 10 per cent of all grain purchases were taken by the free market ; in succeeding months the percentage doubled or more than doubled.[3] The official directive prices for wheat, which was the commodity subject to the keenest free market competition, were immediately raised. But in spite of these elastic prices grain was slow to come forward. Stocks in the " consuming "

[1] *Planovoe Khozyaistvo*, No. 9, 1925, p. 12.
[2] *Pravda*, September 25, 1925 ; the writer was Milyutin. The discrepancy between rising agricultural prices in the autumn of 1925 and the prognostications of Gosplan is shown in a table in *Planovoe Khozyaistvo*, No. 2, 1926, p. 74.
[3] *Na Agrarnom Fronte*, No. 5-6, 1926, p. 18 ; this figure is confirmed in *Ekonomicheskoe Obozrenie*, No. 3, 1926, pp. 43-44, where it is added that the percentage of the grain harvest in the central provinces taken by the " main state organizations " (meaning, apparently, Khleboprodukt and Gosbank, but excluding the cooperatives) fell from 73 in August to 49·5 in September (for detailed figures see *ibid.* p. 48).

provinces actually fell between October and December 1925.[1] Nor
did the increases in the directive prices narrow the margin between
official and free prices, or enable the state organs to regain control
of the market.  In December 1925 the free market price for rye
was 50 per cent above the directive price, and for wheat 33 per
cent.[2]

Many explanations were given of this unwelcome and unexpected turn of events.  The substitution of " directive " for
" fixed " prices had meant the abandonment of attempts to
coerce the peasant, and the fruits of this change of policy were
now apparent ;  poor organization and competition between
different state purchasing agencies[3] were a source of weakness ;
growing prosperity and the expansion of credit had increased
the amount of capital in the hands of private purchasing concerns, and enabled them to trade on a larger scale.  " Private
capital," wrote Smilga in *Pravda* on January 1, 1926, " driven
out of trade in industrial goods, has in greater part transferred
itself to the purchase of grain."  But the major reason was,
beyond doubt, the increased independence and bargaining power
of the well-to-do peasant.  The peasant enjoyed " a much greater
freedom " than ever before " in the choice of the time and the
terms for disposal of his surpluses owing to the decrease in
' forced sales ' ".[4]  It was the first time, as Stalin later said, that
" the peasant and the agents of the government met face to face
on the market as equals ".[5]  The " tax pressure " exercised by
the agricultural tax had been substantially reduced ;  compulsory
price-fixing had been shelved ;  measures of " administrative
pressure " on the well-to-do peasant had been abandoned, and he
had even been allowed to rise to positions of influence in local
administration.  Moreover, the peasant had learned the lesson
of the harvest of 1924, when prices had been kept low in the
autumn and winter and had soared in the following spring.  A
year earlier the purchase and hoarding of grain by *kulaks* had
been noted for the first time.[6]  After the 1925 harvest it became a
regular symptom :

[1] *Ekonomicheskoe Obozrenie*, March 1926, p. 46.
[2] *Na Agrarnom Fronte*, No. 5-6, 1926, p. 121.
[3] See pp. 295-297 below.
[4] Preobrazhensky in *Pravda*, December 15, 1925.
[5] Stalin, *Sochineniya*, vii, 320.          [6] See p. 193 above.

The less prosperous peasants bring in their grain in the autumn, the more prosperous in the spring. The more prosperous peasants and the middle peasants sometimes buy grain in the autumn and keep it till the spring in the hope of making money on it.[1]

From the Urals and from Siberia, from the Ukraine and from the north Caucasus, reports came in of a deliberate holding back of grain by the well-to-do peasants.[2] A writer in *Leningradskaya Pravda* of November 13, 1925, could still express the pious hope that " the *kulaks* who now hoard grain in the hope of re-selling it in the winter and the spring at a still higher price will pay dearly for their irresponsible speculation ". But, in fact, no such retribution was in sight. The well-to-do peasant, no longer pressed for money, and with little in the way of available supplies of industrial goods on which to spend it, found himself in the position of being able to hold the state to ransom. In November 1925 Sokolnikov was still talking of a reduced export figure of 200 million puds of grain.[3] But in the following month the full gravity of the situation became apparent. The grain collection for the year 1925–1926 was likely to fall short by 200 million puds of the estimated 780 millions ; and a decision of the Politburo suspended all exports.[4] The vision of industrial expansion on a broad front financed on the proceeds of ample grain surpluses faded away. The *kulak* had shown himself master of the situation.

One minor factor, which helped the *kulak* to force up grain prices in the autumn of 1925, was a glaring defect in the machinery of collection. The essence of NEP had been the removal of restrictions on trade in the peasant's grain surpluses. Narkomprod

[1] *Vserossiiskii Tsentral'nyi Ispolnitel'nyi Komitet XII Sozyva : Vtoraya Sessiya* (1925), p. 470.
[2] *Ekonomicheskoe Obozrenie*, No. 2, 1926, pp. 128-129 ; *Na Agrarnom Fronte*, No. 5-6, 1926, pp. 20-21. According to the latter source, " the *kulak* was more favourable and friendly to the Soviet power " (i.e. sold his grain to the state purchasing organs at lower prices) in the north Caucasus than in the Ukraine. The Ukrainian *kulak* profited by greater concentration due to a more intensive economy, and by greater proximity to markets ; Rykov confirmed that difficulties had been greatest in the Ukraine (*ibid.* No. 10, 1925, p. 5). Sharp variations of price occurred between different regions (*Vestnik Finansov*, No. 9, September 1925, p. 3).
[3] G. Sokolnikov, *Finansovaya Politika Revolyutsii*, iii (1928), 231.
[4] *XIV S"ezd Vsesoyuznoi Kommunisticheskoi Partii (B)* (1926), pp. 263-264, 416.

with its requisitions had been wound up ; and in the new conditions of trade no organ to replace it seemed necessary or appropriate. In 1924 Narkomvnutorg began to establish its grain-purchasing machinery in the form of a share company working on commercial lines under the name of Khleboprodukt. But a decree of July 1925, which purported to give the Narkomvnutorg of the USSR, acting through the Narkomvnutorgs of the republics, unlimited powers over all official collections of grain,[1] evidently came too late. By this time other organizations were already active in the grain market [2] — the cooperatives, the gostorgi, state and provincial milling trusts, and, above all, Gosbank, which, retaining the tradition of the inflation period when grain was the most stable, and at the same time most liquid, of all assets, continued to carry substantial quantities as part of its reserve.[3] All these organizations were independently engaged in the autumn of 1925 in purchasing grain from the peasant, and often competed with one another in the same localities, where the mutual animosities and public recriminations of their respective agents were a frequent source of confusion and scandal.[4] Kamenev spoke angrily of the

---

[1] *Sobranie Zakonov, 1925*, No. 60, art. 444.

[2] A decree of the White Russian SSR of September 1, 1924, named the following agencies as entitled to participate in the purchase of grain in the territory of the republic : the White Russian branch of Khleboprodukt ; the White Russian office of Gosbank ; the food industries department of the White Russian Vesenkha ; and agricultural and consumer cooperatives (*Zbor Zakonau i Zahadau BSSR, 1924*, No. 21, art. 197). Attention had been drawn in 1924 to the inconvenience of this proliferation of agencies ; but, before the crisis of the autumn of 1925, nothing was done (L. Kamenev, *Stat'i i Rechi*, xii (1926), 348).

[3] Throughout eastern Europe before 1914 the financing of the harvest was an important function of the banks, which commonly held substantial stocks of grain. In the summer of 1925 Gosbank defended its interest in the grain market on the specious ground that " the conditions of realization of the harvest have the most powerful influence on the volume of the bank note issue " (*Ekonomicheskaya Zhizn'*, July 1, 1925). The following table shows the respective shares of the three largest purchasing organizations, expressed in millions of puds of grain purchased :

| | 1923–1924 | 1924–1925 | 1925–1926 |
|---|---|---|---|
| Khleboprodukt | 85·6 | 128·9 | 204·9 |
| Gosbank | 50·9 | 77·5 | 107·5 |
| Tsentrosoyuz | 37·3 | 28·5 | 60·0 |

(*Na Agrarnom Fronte*, No. 5-6, 1926, p. 22.)

[4] The abuse was a burning issue at the TsIK of the RSFSR in October 1925 ; the spokesman of the Narkomvnutorg of the RSFSR blamed the inexperience of the Narkomvnutorg of the USSR (*Vserossiiskii Tsentral'nyi Ispolnitel'nyi*

" bacchanalia of competition " over the collection of the grain.[1] The issue stirred up all those interdepartmental rivalries which were from time to time added to the other strains and stresses of the Soviet economy ; and the peasant with grain to sell had every reason to encourage the perpetuation of a system which confronted him with a choice of competing buyers.[2]

The development of these untoward symptoms was accompanied throughout the autumn of 1925 by a progressive growth of tension and dissent within the party. The disappointment of the 1925 harvest was in no way comparable to the stringency of 1924 or the calamity of 1921. But it raised, in a far more direct and dramatic way than the troubles of the earlier years, the issue of political power, and brought about a new alignment of forces. Since 1923 the opposition, divided and unorganized though it was, had consistently objected to the policy of further concessions to the peasantry and of extended freedom of trade on the ground that this was incompatible with the necessary expansion of industry and the development of planning. This view was shared by Trotsky, Preobrazhensky, Pyatakov and the group which had formed the nucleus of " the 46 ", and was the antithesis of the official view that the expansion of industry required the development of a prosperous peasant economy through concessions to the peasant. The sequel to the harvest of 1925 seemed to justify the apprehensions of this group, and to show up the unwarranted optimism of the official line. But the events of this year had also created a new focus of opposition in the form of a group which, while it had originally stood for concessions to the peasantry on a wide front as the necessary consequence of NEP and of the policy of the " link ", was now shocked by the growing differentiation in the countryside and by a policy which favoured a small

---

*Komitet XII Sozyva : Vtoraya Sessiya* (1925), pp. 445, 448, 452, 485). A lively account of the situation at Rostov appeared in *Ekonomicheskaya Zhizn'*, September 30, 1925 ; a good general description is in *Planovoe Khozyaistvo*, No. 1, 1926, pp. 54-58.  [1] L. Kamenev, *Stat'i i Rechi*, xii (1926), 526.

[2] Krasin rather surprisingly defended this return to free market practices : " Even if, as a result of this parallelism, a certain rise occurs in the price of the commodities purchased, it benefits the peasant or the small producer, and the USSR as a whole does not lose, but rather gains, from this competition " (L. B. Krasin, *Voprosy Vneshnei Torgovli* (1928), p. 127).

section of *kulaks* and would-be *kulaks* at the expense of the great masses of poor and middle peasants. In the autumn of 1925 the new opposition found leaders in Kamenev and Zinoviev. Both had personal reasons for resenting the rise of Stalin's authority, and differences on official policy were no doubt sharpened by their dislike of Stalin's increasing predominance in the party. But the sincerity of the apprehension which they felt at the recent turn in policy need not be questioned. On this point Kamenev's subsequent *apologia* rings true :

> We felt a genuine profound alarm at a definite policy of the party at the moment in question : it seemed to us that a number of comrades, and the press in particular, were underestimating all those processes which go on not in the socialist, but in the capitalist, sector of our economy. It seemed to us that the party was not taking into account the difficulties created by the increasing accumulation in the hands of the *kulak* and the nepman, the increasing estrangement from us of the poor peasant. Whether we were right or wrong, is another matter ; but we thought, comrades, that it was our legitimate obligation, of course within the limits of the party statute, to say this to the party.[1]

This " new " opposition had for the present little in common with the " industrial " opposition, though a certain solidarity could be established between them on an anti-*kulak* platform and in a shared resistance to the official party line. The weakness of both opposition groups lay in their lack of a definite positive policy. The introduction of NEP had been dictated by the impossibility of continuing to coerce the peasant. Everything that had happened since had followed from recognition of the same overriding factor. To withdraw support from the peasants who had surpluses to dispose of, to deny them the right to use their bargaining power to dispose of those surpluses on the best terms, meant a resumption of coercion, in however mild and attenuated a form. Neither opposition group frankly faced this issue, or showed any sign of having considered how far it was prepared to go. The defenders of the official line could always meet their attack by quoting Lenin on the need to maintain the " link " with the peasantry and by invoking the bogy of a return to the methods of war communism.

[1] *XV Konferentsiya Vsesoyuznoi Kommunisticheskoi Partii (B)* (1927), p. 483.

Kamenev's speech of September 4, 1925, in which he had laid down the official price policy for the harvest [1] also contained critical passages which foreshadowed the platform of the new opposition. In spite of the excellent harvest there were " black spots " which could not be ignored :

> We should be bad Marxists if, for example, we simply rejoiced that we have a good harvest and did not put to ourselves the question : What is the social content of the harvest ?

Kamenev estimated that 14 per cent of all the peasants had harvested 33 per cent of the grain, and held 61 per cent of the grain surpluses available for the market : out of an estimated total of 1200 million puds of marketable grain, 700 millions were in the hands of this 14 per cent of well-to-do peasants. Kamenev accused some party members of seeking to conceal or gloss over these facts. One comrade had asked : " Is the stressing of such figures compatible with the policy of developing productive forces in the countryside ? " Kamenev went on to criticize the view of the cooperatives as the agents of socialism :

> Whom do the cooperatives help most ? It is impossible to deny, and it would be an ostrich policy to deny, that the cooperatives as at present organized, being inevitably and spontaneously drawn into commercial exchange, give more help to the stronger strata.

He quoted Trotsky — irrelevantly, but in order to emphasize dissent from his heresies ; Kamenev was at this very moment lending his support to the sceptics who were attacking the first " control figures " of Gosplan. [2] He equally dissociated himself from those who still thought it necessary " to save the revolution by an agreement with the capitalist west ". He concluded with a reassuring declaration of policy. " We are not in the least powerless in face of the growth of differentiation which can be observed in the countryside." It was necessary to " set certain limits on the emergence of a top layer of *kulaks* " ; but this must be done " *by way of helping the middle and poor peasant to rise* ". The speech kept within the formal framework of the party line. But the change of emphasis was a challenge which, coming from the president of STO, could not be indefinitely ignored. It was

---

[1] See p. 292 above.          [2] See p. 504 below.

apparently about this time that the Lenin Institute, which was
under the direction of Kamenev, published an article by Lenin
which had somehow escaped inclusion in the first edition of
Lenin's collected works. The article had been written in August
1918 at the height of the campaign in support of the committees
of poor peasants, and carried the title, *Comrade Workers, We are
Marching to the Last Decisive Battle.* It contained some of Lenin's
most savage denunciations of the *kulak*, who " hates the Soviet
power and is ready to strangle, and cut the throat of, hundreds
of thousands of workers " ; and it demanded " merciless war "
against the *kulaks*, described as " these blood-drinkers, vampires,
robbers of the people, speculators making their profit out of
hunger ".[1] It was a reminder that quotations from Lenin, when
invoked in current disputes, did not at all point in one direction.

At the same moment when Kamenev was launching his
guarded criticisms, Zinoviev attempted a far more direct and
provocative attack. Early in September 1925 he submitted for
publication in *Pravda* a long article entitled *The Philosophy of an
Epoch.* The philosophy discussed was that of the *smenovekhovtsy*,
and the article was a critique of a volume of essays by Ustryalov
recently published in Harbin with the title *Under the Sign of
Revolution.* The volume was treated by Zinoviev as significant
of a " turning-point " in the history of NEP. Ustryalov had
written that " behind the nepman the socialist bourgeoisie is
bound to arrive, . . . and, first and foremost, of course, the
' strong peasant ' without whom no recovery of the health of our
agriculture is conceivable ". He looked forward to a recovery
which would " follow in the wake of peaceful economic evolution "
on these lines, and observed with satisfaction that " the peasan
is becoming the only real master of the Soviet earth ". This,
declared Zinoviev, was the same " canonization of the *kulak* "
which could also be found in *émigré* publications. It was an
illustration of the " danger of degeneration " which Lenin had
always regarded as inherent in the prolongation of NEP and the
continued delay in world revolution. When Ustryalov said that
" the country is ready for normal life ", he meant bourgeois life.

[1] Lenin, *Sochineniya*, xxiii, 205-208, where it is said to have been published
in " a special edition of the Lenin Institute ", presumably a pamphlet, in
1925 ; the exact date of publication has not been ascertained, but Zinoviev
quoted it in October 1925 (see p. 304 below).

He was " the ideologue of the bourgeoisie (though, of course, of the
' new ' bourgeoisie) ". But Zinoviev was not really interested in
Ustryalov. As Kaganovich afterwards said, the article " professed to be about Ustryalov, but the real target was comrade
Bukharin ".[1] Zinoviev was primarily concerned to draw attention
to the dangers of the *kulak* policy :

> *Yes* [he wrote in italics], *the development of NEP together
> with the delay in world revolution is really pregnant, among other
> dangers, with the danger of degeneration. Lenin pointed this out
> a dozen times.*[2]

This was followed by " sallies against Bukharin " — whether by
name or not, can no longer be established.[3] Then, by way of
counterblast to the alleged encouragement of differentiation under
the party line, Zinoviev launched into a rhetorical hymn of praise
to " equality " :

> Do you want to know of what the mass of the people really
> dreams in our days ? To express this dream in one word, the
> word is " equality ". There is the key to an understanding of
> the philosophy of our epoch. . . .
> In the name of what in the great days of October did the
> proletariat rise, and after it the great masses of the whole people ?
> *In the name of the idea of equality, of the idea of a new life on
> principles that are not bourgeois.*

The article concluded by predicting " a serious struggle for the
*interpretation* of the revolutionary line " as laid down in the
decisions of the fourteenth party conference, and by calling on
the proletariat to " help " the party to interpret this line " in a
strictly Leninist spirit ".[4]

---

[1] *XIV S"ezd Vsesoyuznoi Kommunisticheskoi Partii (B)* (1926), p. 238.
Zinoviev's use of parable was facilitated by the fact that Ustryalov had just
published, in the Harbin *smenovekh* journal, *Novosti Zhizni*, an article entitled
*Now Lettest Thou Thy Servant* . . . in which Bukharin's " Enrich yourselves "
speech was extravagantly applauded (N. Ustryalov, *Pod Znakom Revolyutsii*
(2nd ed. 1927), pp. 209-211).

[2] The passages so far quoted appeared in the final version published in
*Pravda*, September 19, 20, 1925, and separately as a pamphlet, G. Zinoviev,
*Filosofiya Epochi* (1925).     [3] Stalin, *Sochineniya*, vii, 375.

[4] These passages were quoted by Uglanov at the fourteenth party congress
(*XIV S"ezd Vsesoyuznoi Kommunisticheskoi Partii (B)* (1926), pp. 195-196) ;
the original text has never been published in full.

When this explosive document was received, Kalinin and Molotov were the only full members of the Politburo in Moscow.[1] Molotov, in charge of the secretariat, sent it to Stalin, who, on September 12, 1925, from his vacation retreat, indited a " rude and cutting criticism ". Zinoviev, he wrote, had been guilty of " a distortion of the party line in the spirit of Larin ". In calling in question the interpretation of the resolutions of the fourteenth party conference, he had launched an attack on them. He had passed over altogether " the central theme " of the conference — the middle peasant and the cooperatives — in order to compare Bukharin with Stolypin and raise the slogan of " equality ", which was nothing but a piece of SR demagogy. He had even referred to the 1917 revolution as " non-classic ", which smacked of Menshevism. As a result of this broadside, Zinoviev's article was subjected to " amendments and additions " by " members of the central committee ". In the revised version, all allusions to Bukharin and his slogan had been expunged, though a faint reference to Bogushevsky's indiscretion was retained ; and a passage was inserted insisting on the importance of the middle peasant and on the dangers of the Left deviation of ignoring him.[2] The rhetorical passage about equality was radically expurgated and rephrased, and the saving epithet " socialist " inserted before " equality ". In this form the article was published in *Pravda* on September 19 and 20, 1925, and also appeared as a pamphlet. Though Stalin remained formally in the background, his guiding hand in the affair must have been apparent to Zinoviev, and made the breach between the two men irretrievable. How far the issue of policy still remained open is, however, shown by the appearance in the party journal at the end of September of an article attacking not only Bogushevsky, but Kalinin and A. P. Smirnov, for their " attempt to legalize the *kulak* by denying his existence ".[3]

The stir caused in the inner councils of the party by the

---

[1] *XIV S"ezd Vsesoyuznoi Kommunisticheskoi Partii (B)* (1926), p. 318 ; Kuibyshev, a candidate member of the Politburo and, as president of the central control commission, an influential figure, was also in Moscow, and was concerned in this episode (*ibid.* pp. 441-442).

[2] The episode is related, and Stalin's letter quoted, in Stalin, *Sochineniya*, vii, 375 ; a reference to the cooperatives was also inserted (*XIV S"ezd Vsesoyuznoi Kommunisticheskoi Partii (B)* (1926), p. 194).

[3] *Bol'shevik*, No. 17-18, September 30, 1925, pp. 51-59.

incident over *The Philosophy of an Epoch* had scarcely died away when, in October 1925, a substantial volume of 400 pages was published from Zinoviev's pen with the title *Leninism* and the sub-title *An Introduction to the Study of Leninism*. The work was evidently designed to re-establish Zinoviev's claim, after the inroads made on it by Stalin, to be the authoritative exponent of Leninist doctrine. The preface was dated September 17, 1925 — the very moment when *The Philosophy of an Epoch* was being revised in the secretariat. It contained an expression of "warm gratitude" to Krupskaya, who had " twice read it in manuscript and proof and given me much valuable advice ". The book was said to have had its " beginning " in lectures delivered at the end of 1924 at the Communist Academy and the Institute of Red Professors. Indeed, the first half of it entirely reflected the atmosphere of the previous winter and of the campaign against Trotsky, dealing with such safe and well-worn topics as the relation between the bourgeois-democratic and socialist revolutions, the supreme importance of the link with the peasantry (" Lenin ' discovered ' the peasantry as the ally of the working class in the proletarian revolution "), permanent revolution, and Trotsky's neglect of the rôle of the peasant.

Half-way through the book, however, the theme changed, and Zinoviev turned to a more topical, though still muffled, attack on the present party leadership. At the end of a long chapter on " Leninism and the Dictatorship of the Proletariat " he suddenly introduced Ustryalov, whose collection of articles had " recently come into my hands ". This section closely followed the argument of *The Philosophy of an Epoch*, though in rather more explicit terms. Zinoviev quoted at length the passage in which Ustryalov had welcomed the slogan " Enrich yourselves ", though he omitted the mention of Bukharin's name ; and he noted that, according to Ustryalov, NEP was " not tactics, but evolution ". Two dangers existed : the danger that the reins of the proletarian dictatorship might be drawn too tight, and the danger that " petty bourgeois forces will ' put water into the wine of ' the proletarian dictatorship, unscrew the essential ' nuts and bolts ', and thus offer to the bourgeoisie (and to the *kulak*) . . . the possibility of striking a direct blow at the basic pillars of the proletarian dictatorship ". No doubt could be felt that the second danger

was the real one in present conditions in the USSR.[1]  This was
followed by two chapters devoted to " Leninism and NEP ".
Zinoviev proved by copious quotation that, in Lenin's conception,
NEP was a retreat.  Moreover, the retreat was still in progress :

> We cannot celebrate the supposed victory of non-capitalist
> evolution in agriculture at the very moment when we are having
> to make supplementary concessions precisely to the capitalist
> elements in agriculture.[2]

Zinoviev went on to attack " those who seriously maintain that
there is no such thing as a *kulak* in the contemporary Russian
countryside and who hold that ' accumulation ' will take place in
the countryside almost by way of some immaculate conception ".[3]
The economic order established by NEP was " state capitalism in
a proletarian state ", state capitalism being a step on the road to
socialism.  At this point Zinoviev quoted Lenin's recently published article of 1918 in denunciation of the *kulaks*.[4]  The moral
was plain.  The class struggle continued and must continue ; this
reality must not be obscured by talk about cooperation between
classes.  If in 1923 the greatest enemy had been " irresponsible
grumbling, pessimism and whining about ' the ruin of the
country ' ", in 1925 the danger was " complacency, when it turns
into a glossing over of the class struggle in the countryside and a
playing down of the danger from the *kulak* ".[5]

So far Bukharin, though nowhere named, had been the
principal target.  But next Zinoviev turned his guns on " socialism
in one country ", massing quotations from Lenin to prove the
impossibility of creating a socialist economy in a single backward
country.  The controversy, like all discussions of this subject,
had an air of unreality.  On the one hand, Zinoviev was committed by the resolution of the fourteenth party conference of the
previous April, which he had himself moved, to the principle of
building socialism ; and he now reiterated with the emphasis of

[1] G. Zinoviev, *Leninizm* (1925), pp. 215-220 ; the greater part of this
chapter was published as an article in *Leningradskaya Pravda*, September 26,
1925.               [2] G. Zinoviev, *Leninizm* (1925), p. 255.
[3] *Ibid.* p. 260.          [4] See p. 300 above.
[5] G. Zinoviev, *Leninizm* (1925), p. 281 ; the NEP chapters appeared as an
article running through six issues of *Leningradskaya Pravda*, September 29–
October 4, 1925.

italics that " *we must, can, dare to and are obliged to* " build it.[1]
On the other hand, Stalin did not deny that the completion of the
building depended on the victory of the socialist revolution in other
countries.  At times the argument seemed to turn on the differences between two moods of the same verb, the one denoting the
process, the other the completed act, of building.  The difference
could be reduced to one of emphasis.  Yet nobody who read this
chapter could fail to understand that the president of the Communist International was opening fire, in the name of the international character of the revolution, on the Stalinist conception
of socialism in one country.

> Lenin was from head to foot an *international* revolutionary.
> His teaching was applicable not only to Russia, but to the
> whole world.  We, disciples of Lenin, must banish as a hal-
> lucination the mere thought that we can remain Leninists if we
> weaken by a single jot the international factor in Leninism.[2]

It was the first time that socialism in one country had been openly
and publicly assailed.  Stalin can hardly have failed to regard it
as a declaration of war.  Having delivered his broadside, Zinoviev
returned in the next chapter to the relatively safe ground of the
rôle of the party.  But here, too, he registered a controversial
point by insisting on the predominance of workers over peasants
in the leadership of the Komsomol, and returned a belated
answer (though still without mentioning Stalin's name) to Stalin's
strictures on his assertion of the " dictatorship of the party ".[3]
After these alarms and excursions, the book ended quietly with
a commonplace and harmless chapter on " Leninism and the
Dialectic ".

The situation was already tense when the party central committee met at the beginning of October 1925 to prepare the
ground for the forthcoming fourteenth party congress in December.
This meeting offered the opportunity for a trial of strength, for
which, as the sequel showed, neither side was yet ready.  Kamenev
submitted a report on the economic situation which was lengthily
discussed.  It distinguished between three recent periods in

[1] G. Zinoviev, *Leninizm* (1925), p. 326 ; the resolution of the party conference and the controversy about socialism in one country will be discussed in
Part III in the following volume.                                [2] *Ibid.* p. 318.
[3] *Ibid.* pp. 358, 360 ; for the " dictatorship of the party " see p. 104,
note 3 above.

economic policy : the period May-June 1925 when caution about
the rate of expansion still prevailed ; the period of optimism in
July-August, when massive grain surpluses were expected and
ambitious export and import plans drawn up ; and the new
period of doubt in August-September, when the losses caused by
the early rains proved the portent of more serious difficulties in
collecting the grain.   The report contained, however, some contentious passages :

> The good harvest itself, though an enormous asset in
> strengthening and facilitating the cause of further socialist con-
> struction, is distributed among different groups of the peasantry
> in such a way as to intensify differentiation.

While, therefore, the conclusion of the report that grain prices
must be kept in hand " by way of purely administrative measures "
was unexceptionable, the committee preferred merely to take note
of Kamenev's report, and " instructed the Politburo, on the basis
of the exchange of opinions which took place, to discuss practical
measures and take a final decision on comrade Kamenev's report ".[1]
The puzzling situation in regard to the collection and marketing
of the harvest provided a valid reason to postpone any important
decision of policy ; and everybody was relieved.[2]   The second

---

[1] Kamenev's report and draft resolution are in L. Kamenev, *Stat'i i Rechi*,
xii (1926), 347-371 ; an abbreviated text of the report appeared in *Pravda*,
October 24, 1925.  The decision of the central committee was published in
*Pravda*, October 15, 1925, and is in *VKP(B) v Rezolyutsiyakh* (1941), ii, 32.
Kamenev's report contained a table which showed the distribution of the
surpluses :

| Area | % of Population | % of Total Production | % of Surpluses |
|---|---|---|---|
| No sown land | 3 | — | — |
| Up to 1 desyatin | 12 | 3 | — |
| 1-2 desyatins | 22 | 12 | — |
| 2-3 ,, | 20 | 16 | 3 |
| 3-4 ,, | 14 | 15 | 11 |
| 4-6 ,, | 15 | 21 | 23 |
| 6-8 ,, | 7 | 12 | 19 |
| 8-10 ,, | 3 | 7 | 12 |
| Over 10 desyatins | 4 | 14 | 30 |

(L. Kamenev, *Stat'i i-Rechi*, xii (1926), 355-356 ; the last column leaves 2 per
cent unaccounted for.)
[2] The relief was plainly expressed in a leading article in *Pravda*, October
15, 1925.

major item on the agenda consisted of a report and a resolution
submitted by Molotov on " party work among the village poor ".
The resolution, which apparently embodied some proposals put
forward by Zinoviev [1] and could therefore be regarded as a compromise between the two wings, had been approved " in its
fundamentals " by the Politburo on October 1, 1925.[2] The
report which introduced the resolution was carefully balanced.
The report admitted that differentiation was growing, but claimed,
in opposition to Kamenev's figures, that " the mass of grain is
produced and thrown on the market by the mass of middle
peasants ".  Molotov repeated the now familiar diagnosis of the
two potential deviations for and against the *kulak*, using Bogushevsky as an example of the first and Larin [3] of the second ; as a
further example of the latter, he now added an article in a Ukrainian
fournal protesting against the altered status of the komnezamozhi.[4]
The resolution underwent minor amendments in the committee.
A passage recommending that the organizations of poor peasants
in Central Asia should follow the example set by the komnezamozhi
was omitted and referred back to the Politburo.  A recommendation for the creation of a fund to provide credits for poor peasants
was inserted.[5]  In its final form the resolution could be said to
mark a slight shift towards the Left.  It did not revert to the
concessions to the *kulak* sponsored in the central committee's
resolution of April 30 : these were now past history.  It offered a
slightly different definition of the two deviations "in the direction
of an underestimate of the negative sides of NEP, and in the
direction of failure to understand the significance of NEP as an
indispensable stage in the advance to socialism ".  The first
deviation implied "neglect of the interests of the village poor and

---

[1] *XIV S"ezd Vsesoyuznoi Kommunisticheskoi Partii (B)* 1926), pp. 457-
458.
[2] *Pravda*, October 2, 1925 ; *Leningradskaya Pravda*, October 6, 1925.
[3] Larin, who had already challenged Bukharin at the fourteenth party conference (see p. 263 above), published in the summer of 1925 a book entitled
*Sovetskaya Derevnya*, which drew attention to growing " differentiation " in
the countryside and denounced the current policy of favouring the *kulak* ; it
was attacked by Maretsky, one of Bukharin's disciples, in a long article in
*Bol'shevik*, No. 19-20, October 30, 1925, pp. 26-46, and defended by Larin
himself in a " discussion article " in *Pravda*, December 16, 1925.
[4] See p. 290 above.
[5] *Leningradskaya Pravda*, October 15, 1925 ; the amendments can be
verified by comparing the original draft with the final form of the resolution.

an under-estimate of the *kulak* danger ", the second neglect of the middle peasant and a breach of the " link " between proletariat and peasantry. The resolution endorsed the decision of the Ukrainian central committee on the reorganization of the komnezamozhi. But it dwelt with greater emphasis than before on the work of Vserabotzemles and on the need to protect " the interests of the poorest strata in the countryside " ; and it recommended " special meetings of poor peasants " to defend their interests at Soviet elections.[1] The third item before the committee was the time-honoured project to transform the single agricultural tax into an income-tax. But nobody except Sokolnikov, now the staunchest defender of the interests of the small peasant against the *kulak*, thought that the time was ripe to bring up this proposal to the party congress ; and it was allowed to drop.[2] The session ended without any burning of boats. All decisions had been taken unanimously.

The embarrassments of the grain collection were now becoming more apparent every day. A week after the central committee dispersed, Kamenev addressed the Moscow party organization on the results of the session. Significantly, he now for the first time reversed the order of the two deviations, dismissing first the deviation of Larin, which was contrary to Lenin's view that " NEP is the unavoidable path to socialism ", and then dwelling more heavily on the deviation of Bogushevsky, who thought that the *kulak* was a figure of the past and could be ignored in the present ; and, just as Molotov had reinforced the Larin deviation with an additional up-to-date quotation, so Kamenev, by way of showing that the pro-*kulak* deviation was by no means dead, quoted from the current issue of the journal *Pechat' i Revolyutsiya* a dictum that " agrarian capitalism . . . will serve as the lever

---

[1] Molotov's resolution in its final form is in *VKP(B) v Rezolyutsiyakh* (1941), ii, 38-39 ; the question of the " special meetings of poor peasants " will be dealt with in Part IV in the following volume.

[2] Stalin, *Sochineniya*, vii, 361. Sokolnikov's statement was as follows : " The agricultural tax should be reconstructed in such a way as to make it a real force in the countryside organizing the poor peasants and middle peasants on the side of the Soviet power against the rich and the rural *kulaks*. This task should be put on the agenda " (G. Sokolnikov, *Finansovaya Politika Revolyutsii*, iii (1928), 22-23 ; this article was originally published in *Vestnik Finansov*, No. 10, October 1925, pp. 3-22, and evidently represented the substance of his report to the party central committee).

for the growth of the socialist elements in the Soviet economy ",[1] and cited Molotov's own resolution in proof of the untenability of such a view. He repeated, without referring to Molotov's contradiction, and with a slight variation of his original figures, the statement that 12 per cent of the peasantry held 60 per cent of the marketable grain, adding, somewhat mysteriously, that the figures did not correspond to a " class division of the peasantry ". Bogushevsky, whose name had become an indispensable shuttlecock in these debates, made his only recorded personal appearance at this meeting to defend his views, and was mildly answered by Kamenev. Nobody had anything new to suggest.[2]

The magnitude of the problem seemed likely for the moment to still the voice of controversy in the party; and throughout November preparations went forward for the fourteenth party congress on the assumption that a compromise would once more be reached. If the issue had turned solely on divergent opinions about economic policy, this hope would probably have been realized. In the middle of November, Bukharin at last published his belated and somewhat grudging recantation of the " Enrich yourselves " slogan. It was embodied in an article in *Pravda* on the same theme as Zinoviev's *Philosophy of an Epoch* : the *smenovekh* movement. Writing under the title *Caesarism under the Mask of Revolution*, Bukharin attacked Ustryalov for rejecting democracy not in the name of Bolshevism, but " *in the name of Fascist Caesarism and bourgeois* dictatorship". Bukharin's article, in opposition to that of Zinoviev, was in essence a defence of NEP : it was Ustryalov who derived the " degeneration of Bolshevism " from the introduction of NEP, and was the author of the view which Zinoviev was now trying to popularize. The *clou* of the article was, however, Bukharin's reply to the passage in which Ustryalov, describing Bukharin as " the most orthodox and most pure-blooded " of the Bolsheviks, had praised him for

[1] The passage occurred in a book review and, read in its context, scarcely justified the pro-*kulak* interpretation placed on it by Kamenev (*Pechat' i Revolyutsiya*, No. 5-6, July-September, 1925, p. 335 : this issue had gone to press on June 14).
[2] The meeting was reported in *Pravda*, October 20, 24, 1925, and *Leningradskaya Pravda*, October 21, 22, 1925 ; Kamenev's speech and concluding remarks are in L. Kamenev, *Stat'i i Rechi*, xii (1926), 372-408.

addressing the injunction " Enrich yourselves " to the peasants on behalf of the party. Bukharin disengaged himself from this embarrassing compliment. He briefly expressed regret for the phrase, and explained that " this formulation was an undoubtedly erroneous formulation of the correct proposition that the party should set its course for raising the prosperity of the countryside ".[1] Towards the end of November the Politburo approved the economic report to be made by Kamenev to the congress ; and it was duly published in *Pravda* on November 27, 1925. Its main emphasis was on the expansion of industry. The aim of agrarian policy was " to draw the peasantry into socialist construction with the help of a broad development of the cooperatives on the basis of the industrialization of the countryside ". It noted the reduced prospects of the grain collection, but drew no practical conclusion from the failure. The resolution of the fourteenth party conference on " the strengthening and development of the peasant economy " was reaffirmed, but only the middle and poor peasants were specifically mentioned. As the conflict between the leaders sharpened and entirely absorbed the attention of the party, the question of agricultural policy faded temporarily into the background. As president of STO, Kamenev bore the chief responsibility for prognostications which had been falsified and for policies which had admittedly been found wanting. Zinoviev had raised the slogan " Face to the countryside " in the course of the campaign against Trotsky, and thus became, as others besides Trotsky doubtless recalled, " one of the initiators of the peasant deviation ".[2] The new opposition was ill placed to give battle on the agrarian front. Kamenev's figures of the distribution of surpluses between different categories of peasant became the object of fierce attack ; and the censure extended to the central statistical administration which had originally supplied them. A report of the central control commission estimated that even poor peasants sold two-fifths of what they grew, and middle peasants from one-third to one-half, and that poor, middle and well-to-do peasants produced respectively 21·7 per cent, 48·6 per cent and

---

[1] *Pravda*, November 13, 14, 15, 1925 ; the article also appeared in pamphlet form.

[2] The phrase occurs in an unpublished note written by Trotsky at the time of the fourteenth congress ; this note, which is in the Trotsky archives, T 2975, will be discussed in Part III in the following volume.

29·7 per cent of the marketable grain.[1] In the absence of any
agreed definition of the categories of peasant, the statistics could
clearly be manipulated to suit any purpose.

The defeat of the opposition at the fourteenth party congress,
resulting among other things in the cancellation of Kamenev's
report and the dismissal of Kamenev and Sokolnikov from their
official posts, did not therefore turn on differences of agrarian
policy. Both at the congress itself, and at the Leningrad and
Moscow provincial conferences which preceded it, the theme of
the two deviations was repeated *ad nauseam*, each side concentrating in turn on its chosen targets. But the discussion of this issue
was now stale and perfunctory. Stalin comforted the congress
with the assurance that " agriculture, unlike industry, can for a
certain time advance rapidly even on its present technical base ",
and that " the further development of agriculture does not for the
present encounter the same difficulties as industry ".[2] Bukharin,
having purged himself of his error, could be unreservedly defended;
and even Bogushevsky's " deviation " was leniently treated.
Stalin continued to insist that the party needed to " concentrate
its fire " on the second deviation, i.e. on " the inflation of the rôle
of the *kulak* ", and specifically dissociated himself from Larin's
conception of a " second revolution " against *kulak* dominance.[3]
Bukharin, now apparently at the height of his authority and influence, devoted more time than any other delegate to the vindication
of the party's agrarian policy and of concessions to the well-to-do
peasant. Yet signs were not wanting that the shift in emphasis,
faintly adumbrated in the October resolution of the central committee, was being carried a step further, and that the defeat of the
opposition did not necessarily mean the rejection of all their
criticisms. Stalin admitted the existence in the party of a tendency
to suppose " that the policy of a firm alliance with the middle
peasant might imply an ignoring of the poor peasantry ", and
that " some elements of the poor peasantry and even some communists thought that the abandonment of dekulakization and of
administrative repression mean an abandonment of the poor
peasant, a neglect of his interests " ; and he stressed the October

[1] *Pravda*, December 11, 1925 ; Kamenev's figures were violently attacked
by Yakovlev in a series of articles (*ibid.* December 9, 10, 16), and by Stalin at
the fourteenth party congress (Stalin, *Sochineniya*, vii. 329-330).
[2] *Ibid.* vii, 315-316.     [3] *Ibid.* vii, 336-337, 373.
VOL. I

resolution as designed to counteract these errors.[1]  Molotov in
his speech at the congress was still more explicit :

> At the present time we do not really have the middle
> peasant behind us.  This task that we have set ourselves — the
> task of rallying the poor and middle peasants round our party
> — we are now beginning to realize ; but this task we are still
> performing weakly, and for this reason it is the most important,
> as well as the most difficult, central task of our party in the
> countryside. . . . We cannot really regard the poor peasants
> as yet organized round our party.

He went on to urge the party to " struggle against forgetfulness
of the interests of the poor peasants ", and to " struggle against
the *kulak* danger, for the isolation of the *kulak*, for the expulsion
of the *kulak* from those economic and political positions which he
still holds in the countryside ".[2]  This was far nearer to the
language of the opposition than to the language of Bukharin.

Officially nothing in the agrarian policy of the party was
changed or affected by the congress.  It passed no special resolution on the agrarian question, and its main resolution, which was
as usual a general survey of the party line in foreign and domestic
affairs, was wholly non-committal on the subject.  It approved
the decisions taken by the central committee to rectify errors
committed in regard to the grain collections and to foreign trade.
It cautiously established the doctrine that " the building of a full
socialist society " was possible in the Soviet Union, and proclaimed " state socialist industry " as " the advance guard of the
national economy ".[3]  This, however, led to " contradictions "
and to " dangers and difficulties ", including the " growth of
*kulak* farms in the countryside and the growth of differentiation ".
Among measures designed to " ensure the victory of socialist
forms of economy over private capital " in the agricultural sphere,
the resolution recommended " the raising of agricultural  technique
(introduction of tractors), the industrialization of agriculture, the
development of land reclamation and the support by all possible
means of the various forms of collectivization of agriculture ".
Having travelled so far on the socialist path, the authors of the

---

[1]  Stalin, *Sochineniya*, vii, 331-332.
[2]  *XIV S"ezd Vsesoyuznoi Kommunisticheskoi Partii (B)* (1926), pp. 476-477.
[3]  For the rôle of the congress in industrial policy see pp. 352-353 below.

resolution evidently felt the need to redress the balance. A later passage repeated the diagnosis of the two opposite deviations made by the party central committee in October, which it inaccurately attributed to the fourteenth party conference. It endorsed the decisions of the fourteenth party conference on agrarian policy, erroneously including among them the decisions on hired labour and the leasing of land, which were recorded not by the conference, but by the central committee after the conference had dispersed,[1] and observed that this " turn of policy " had " radically improved the position in the countryside " and " raised the authority of the proletariat and of its party among the peasantry ". This emphatic approval of the policy of support for the *kulak* was balanced by a promise of party sympathy and support for the poor peasant. But, once again, the qualification quickly followed : " there can be no question here either of a return to the committees of poor peasants or of a return to the pressure system of the period of war communism, to the practice of dekulakization etc." [2] In this field, the resolution of the fourteenth party congress reflected, not the victory of one opinion over another, but the unresolved dilemma of those responsible for the conduct of Soviet agrarian policy. Nevertheless, the *obiter dicta* of Stalin and Molotov were straws in the wind, revealing the direction in which the party line would inevitably be driven under the impulse of intensive industrialization.

### (e) The Uncertain Prospect

The storms of the fourteenth party congress were followed by a period of reaction in which nobody was for the moment eager to raise controversial issues. The panic over the grain collection in the autumn of 1925 might even seem, on a short view, to have been exaggerated. Disaster had been averted. The cities and factories were being fed, though at a higher cost than had originally been contemplated : only the intended exports of grain had had to be abandoned. Progress might be slower or faster, but things would go on much as before. Yet, while this mood prevailed for some time, it presently became apparent that the fourteenth party

---

[1] See p. 268 above.
[2] *VKP(B) v Rezolyutsiyakh* (1941), ii, 50-52.

congress had, in fact, marked a change of outlook on economic policy which left no part of that policy unaffected. The demands of industrialization henceforth had priority and would set the pace for other sectors of the economy. The defeat of the opposition facilitated the emergence of new alignments, which did not preclude the adoption by the party leadership of some of the arguments and points of view of the former opposition. On the agrarian front, mistrust of the *kulak* had, in the autumn of 1925, been a main plank in the platform of the opposition. Now that the opposition had been routed, the leadership had less difficulty in recognizing the solid grounds for this mistrust. In January 1926 *Pravda* carried a report from Kharkov indicating that the reorganization of the Ukrainian komnezamozhi on a voluntary basis by no means indicated a weakening of their effectiveness as spearheads against the *kulak*.[1] Early in February *Pravda* gave belated prominence to a long speech of Mikoyan, said to have been delivered " not long before the fourteenth party congress " at a North Caucasian regional conference of Vserabotzemles, in which " class warfare " against the *kulak* was unashamedly preached.[2] These indications showed that the campaign against the *kulak* had not been forgotten ; and the need for the better organization of the poor peasants became a constant theme of official publicists.[3]

Part of the new programme was an attempt to raise the prestige and effectiveness of Vserabotzemles, the trade union of agricultural workers.[4] The party central committee at its session in October 1925 gave its blessing to Vserabotzemles as " the organization of the broad proletarian and semi-proletarian masses, of the agricultural workers, *batraks* and semi-proletarian elements of the countryside, whose basic occupation is working for a wage ", though it was significant that the passage occurred in the resolution " On Party Work among the Poor Peasants " and not in the

---

[1] *Pravda*, January 12, 1926.     [2] *Ibid.* February 13, 1926.

[3] See, for example, a leading article in *Izvestiya*, February 27, 1926 ; the *kulak* had not, however, forfeited the sympathy of Bukharin, who in a speech in Leningrad in February 1926 described the poor peasant as seeing in the *kulak* " a father-benefactor who, though he flays the skin off him, none the less gives him something, whereas we feed him with excellent decrees and fine speeches about Chamberlain, but give him practically nothing " (N. Bukharin, *Doklad na XXIII Chrezvychainoi Leningradskoi Gubernskoi Konferentsii VKP(B)* (1926), p. 30).     [4] See pp. 231, 234-235 above.

resolution on the trade unions.[1]  In December 1925 the fourteenth party congress for the first time included Vserabotzemles in a general resolution on the trade unions ;[2] and at the end of January 1926 Vserabotzemles held its fifth congress in Moscow (apparently the first to attract any publicity).  It mustered 537 delegates from all parts of the USSR, of whom 103 were described as *batraks*, and passed a number of resolutions, including one on " the struggle with the *kulak* in the countryside " and another containing such far-reaching demands on behalf of the *batrak* as wages above the state minimum, the payment of wages in cash and not in kind, and better lodging, sleeping accommodation and food.[3]  Meanwhile the membership of Vserabotzemles had risen from 250,000 (of whom less than 5000 were *batraks*) on January 1, 1923, to 770,000 (including 260,000 *batraks*) on October 1, 1925.[4]  But, even if (which is more than doubtful) this membership was effective, it covered only a small fraction of the mass of semi-proletarianized rural workers ; it also included a substantial number of agronomists, land surveyors and other non-manual workers employed in work connected with agriculture.[5]

The pronouncements of the fourteenth party congress on industrialization and on the building of a socialist society added topicality and urgency to the question on which Bukharin had crossed swords with Preobrazhensky at the end of 1924.  How could the necessary capital for the development of industry be extracted from the only available or potential source within the country — the surpluses of Soviet agriculture ?  The issue had been put in the foreground by Gosplan in the introduction to its first set of " control figures " :

> Inasmuch as the whole system of control figures, the system of economic equilibrium and prices, is based on the presupposition of the full extraction of the commercial surpluses of peasant production, and inasmuch, therefore, as non-fulfilment of this task threatens to destroy the equilibrium, the directive must be accepted that the conquest of the peasant market, the extraction

---

[1] *VKP(B) v Rezolyutsiyakh* (1941), ii, 40.       [2] *Ibid.* ii, 68.
[3] The fullest available account of the congress is in *Na Agrarnom Fronte*, No. 1, 1926, pp. 3-9 ; it was intermittently reported in *Pravda* between January 27 and February 2, 1926.
[4] *Na Agrarnom Fronte*, No. 11-12, 1925, p. 18.
[5] This was stated by Kalinin in a speech reported in *Pravda*, February 2, 1926.

of the whole commercial production of agriculture, . . . is the
first and most important task of our economic policy.[1]

The attempt to turn this dilemma by the appeasement of the
peasant had been carried to its limit.  The well-to-do peasant had
received every encouragement to rent land, to hire labour, to grow
abundant crops, to become prosperous.  In 1925 he had complied
with all these requirements and had reaped an excellent harvest.
The one thing he had been unwilling to do was to bring his crops
to a market ill stocked with the cheap consumer goods which he
presumably wanted to buy.  The vital surpluses had not been
forthcoming to promote the long-term expansion of industry.  The
situation was described by Preobrazhensky with his usual clear-
headedness in January 1926 :

> Socialist accumulation may threaten a break with the
> peasant, the liquidation of the link, etc.  On the other hand we
> have at the present time a dissolution of the link (rassmychka)
> with the peasantry thanks to the goods famine, and the goods
> famine results from insufficient accumulation.[2]

Appeasement of the peasant could be carried no further except by
expanding exports of grain in payment for imports of consumer
goods ; and this counsel of despair, involving an abandonment or
indefinite postponement of industrialization, had been decisively
rejected by the fourteenth party congress.[3]  It was not surprising
that, at the beginning of 1926, a mood of reaction should have

---

[1] *Kontrol'nye Tsifry Narodnogo Khozyaistva na 1925–26 god* (1925), p. 43.

[2] *Vestnik Kommunisticheskoi Akademii*, xv (1926), 251.

[3] It is significant that Shanin, a " Narkomfin professor " who had been a
conspicuous advocate of the Sokolnikov line (for his attack on industrialization
in November 1925 see p. 351 below), and had been denounced by Stalin at
the fourteenth party congress (Stalin, *Sochineniya*, vii, 298), was allowed once
more to ventilate this view in an article in *Bol'shevik*, No. 2, January 30, 1926,
pp. 65-87 ; Shanin argued that it was " absolutely false to suppose that the
development of our industry can in the near future keep in step with that of
agriculture ", that industry was already developing too fast for the balance of
the economy, and that a large importation of consumer goods was the only
remedy.  Shanin was a straightforward *laissez-faire* economist who wanted
capital to flow into that branch of production which would yield the highest
immediate profits at the lowest rate of capital investment : this was, indisput-
ably, agriculture.  A perfunctory attempt was made to reconcile these views
with the resolution of the party congress.  This was probably the last time that
they found expression in the party press ; Sokolnikov, their only influential
exponent, was dismissed from his post as People's Commissar for Finance
about the time the article appeared.  Shanin continued to write articles in
defence of financial orthodoxy in *Planovoe Khozyaistvo* (see p. 486 below).

set in against the favour shown to the *kulak*, a growing inclination to take up the wrongs of the poor peasant, a renewed search for other ways and means of inducing the rich peasant to part with his surpluses.

Two attempts were made to pursue this line of thought. The first was a proposal to bring the price weapon again into play ; and, since the present crisis was due to inability to keep agricultural prices down, this could only mean a rise in industrial prices. One of the apparent causes of the unwillingness of the *kulak* to market his stocks was the shortage of manufactured goods available for him to purchase. An article in the journal of Gosplan in February 1926 pointed out that this shortage was relative, and a result of price policy : had prices of industrial goods been set higher, the peasant would have had to sell more grain in order to acquire them. A rise in industrial prices was the only way to restore " the equilibrium of supply and demand ", and represented a " transition from administrative to economic planning ".[1] A far-reaching " discussion article " in the party journal in April by a party member named Ossovsky, a former worker, called in question the whole course of price policy since the closing of the scissors in the autumn of 1923, which was attacked as the fundamental " fall from grace ". As a result of the policy of favouring the peasant, the peasant surpluses, instead of being drained off for the benefit of industry, " now weigh upon us, shattering our planned economy ".[2] The weakness of this argument was that it assumed a high degree of elasticity in peasant demand for manufactured goods, and ignored the danger that a steep rise in prices might be met by a strike of consumers. What was in effect a proposal to reopen the scissors seemed quite unrealistic at a time when the general rise in prices was being everywhere treated as the main menace to economic security. It provoked a sharp reaction. The article was at once denounced as a plea for Preobrazhensky's " primitive socialist accumulation ", and was rather oddly described as a " return to capitalism " ; [3] and Ossovsky became for some months a favourite target for upholders of the official line.

[1] *Ibid.* No. 2, 1926, pp. 107-120 : this was printed as a " discussion article ", i.e. without editorial responsibility.
[2] *Bol'shevik*, No. 7-8, April 30, 1926, pp. 86-100.
[3] *Planovoe Khozyaistvo*, No. 5, 1926, pp. 10-11.

The second proposal was to ply the tax weapon more vigorously. Since the drastic tax reductions of 1925, especially in favour of the well-to-do peasant, had failed to produce the desired results, and had, by common consent, been one of the factors which enabled the peasant to hold back his surpluses, a reversal of that policy seemed a logical deduction. Strong criticism began to be directed, not at first against the reduction of the total amount of the tax (to demand higher taxation is rarely a popular course), but against the manner in which the burden had been distributed. The facts were difficult to establish. It was certain that, in a year when the harvest had surpassed its predecessor in terms of value by more than 25 per cent, the effective total of the agricultural tax had been reduced by some 40 per cent.[1] A cautious analysis which appeared in the party journal at the end of October 1925 showed that, whereas, in 1924–1925, 17 per cent of the total income of the peasantry had been taken in taxation, the proportion had declined in the current year to 10·8 per cent. Out of 22 million peasant households, the 6 million poorest were said to have been altogether exempt from the agricultural tax : the number may be slightly overstated. The claim that, as between those liable, the tax had become more progressive, was open to more serious doubt. It was said that the lowest categories of tax-payers had profited most, some of them getting a reduction of 50 per cent ; but it was admitted that the highest categories had profited as much, or almost as much, as the middle categories.[2] Later official statistics were quoted to show that the poor peasant in 1924–1925 paid 6·2 per cent of the total tax, the middle peasant 76·9 per cent and the *kulak* 16·9 per cent, and that in 1925–1926 the corresponding percentages were 4, 74·8 and 21·2.[3] But such calculations were open to the same uncertainties as others based on the three categories of peasant. Other authorities suggest that, while some attempt was made to place a larger share of the much diminished burden on the shoulders of the rich peasant, the

---

[1] The amount collected in 1924–1925 was 332 million rubles, exclusive of 100 millions collected for local budgets, and in 1925–1926 252 million rubles, inclusive of the allocation to local budgets (*Kontrol'nye Tsifry Narodnogo Khozyaistva SSSR na 1927–1928 g.* (1928), p. 553).

[2] *Bol'shevik*, No. 19-20, October 30, 1925, pp. 52-59.

[3] *Finansy SSSR sa XXX Let, 1917–1947* (1947), p. 248 ; this is not a particularly reliable work, and the sources of the figures are not stated.

attempt was only partly successful, and that not in all parts of the country.  The complaint made by Krupskaya at the fourteenth party congress that the tax for 1925–1926 had been assessed in some parts of the Ukraine in such a way as to give the maximum relief to the well-to-do peasant and place the heaviest burden on the poor had been implicitly admitted by the official spokesman;[1] and another observer from the Ukraine wrote that " rejection of progressive norms of calculation for animals employed in production, and the general lowering of norms, has led in the present year to a great reduction in the tax for households with many animals, and therefore better off, while the poor households, which have no animals or are less well supplied with animals, have encountered a smaller reduction, or sometimes an increase, of tax ".[2]  A few facts emerge clearly.  In the first place the total amount of the tax had been sharply reduced ;  and any such reduction, unless accompanied by a steep increase in the progressiveness of the tax, necessarily benefits the large tax-payer most.  Secondly, the lower conversion rate for animals, added to the exemption of subsidiary earnings and of technical crops, increased the advantage already enjoyed by the rich peasant.  Thirdly, though a larger number of poor peasants secured total exemption,[3] not much was done to temper the burden for the lower grades of tax-payers, i.e. the middle peasants.  Lastly, under the current system of assessment, much depended on the local authorities ;  and, at a time when complaints were frequently heard of the growing influence of *kulaks* in the local Soviets,[4] it is not unlikely that this influence was sometimes brought to bear on the assessments.  This was particularly liable to happen in the Ukraine and in the North Caucasian region, where the *kulak* was strongest.

After the defeat of the opposition at the fourteenth party congress, the case for a drastic revision of the agricultural tax was no longer contested.  In a report to STO on the shortcomings of

[1] See p. 271, note 1 above.

[2] *Bol'shevik*, No. 6, March 31, 1926, p. 74 ;  the statement is general in form, but seems to be based on observations in the Ukraine.

[3] By a belated concession, assessments amounting to less than one ruble were cancelled, presumably as not being worth the trouble of collection :  the sum involved was estimated at not more than 300,000–400,000 rubles (Stalin, *Sochineniya*, vii, 361–362).

[4] This will be discussed in Part IV in the following volume.

the control figures for the current economic year, Gosplan attributed the troubles which had arisen to a new " disproportion " between agriculture and industry, and proposed that the single agricultural tax should be amended in three ways, by increasing the amount of the tax, by making it more steeply progressive, and by curtailing the period of payment.[1] At the same time Narkomfin was pressing for the inclusion of earnings from subsidiary agricultural occupations (wine-growing, bee-keeping, poultry-keeping, etc.) in the assessment of the tax, and the computation of assets for purposes of the assessment in terms of money (not, as hitherto, by conversion at conventional rates into arable land).[2] At the beginning of March 1926 all these proposals were approved by Sovnarkom as the basis for the assessment of the tax in the forthcoming budget.[3] It was still necessary to proceed with caution. When a writer blurted out in *Pravda* that what was needed was to double the receipts of the agricultural tax by raising them to at least 400 million rubles, Rykov protested against the " tactlessness " of the proposal.[4] But, though party leaders still hesitated to speak openly in such terms, the logic of necessity was soon to drive party policy in this direction.

The winter of 1925–1926 saw a revival of the slogan which Kamenev had deprecated at the third Union Congress of Soviets earlier in the year — the " industrialization of agriculture ".[5] The primitive character of the technical equipment of Soviet agriculture was notorious. The crux of current embarrassments in agrarian policy was the dependence of the state and of industry, in the present phrase of development, on the grain surpluses of the *kulak*, who alone had the necessary tools for the job. Could not this dependence be overcome if the efficiency of peasant agriculture as a whole were raised by measures of mechanization, cooperation and the encouragement of large-scale cultivation ? In November 1925 the journal of Gosplan published, in the form of a discussion article by one of its Ukrainian workers, a project for the industrialization of agriculture, i.e. the introduction of large-scale cultivation with machines on an American pattern. Three circumstances, in the view of the

[1] *Planovoe Khozyaistvo*, No. 2, 1926, pp. 41-42, 60.
[2] *Na Agrarnom Fronte*, February 1926, pp. 11-19.
[3] *Pravda*, March 6, 1926.
[4] *Izvestiya*, March 3, 1926.    [5] See p. 272 above.

author, made the project urgent : the transition in the Soviet economy from the period of recovery to a period of reconstruction ; the difficulties experienced by the planners after the harvest of 1925 in controlling the production of a backward and fragmented peasant economy ; and the growth of the *kulak* element in the countryside. An investment of 45 million rubles to purchase machinery would, it was calculated, make possible " the complete industrialization of the Ukraine " in 14 years.[1]

This far-reaching and utopian excursion into the future seemed too remote from current realities to be taken very seriously. Nobody had ever doubted the virtues of mechanization as a solution on paper of the agrarian problem. Asked by a British correspondent in 1919 what was agricultural Russia's greatest need, the then People's Commissar for Agriculture had replied in one word : " Tractors ".[2] About the same time, at the eighth party congress, Lenin had exclaimed that, " if we could tomorrow provide 100,000 first-class tractors ", the peasants would all be for communism.[3] When NEP was introduced, a decree of TsIK instructed state economic organs to " treat the manufacture of agricultural machines as a matter of extreme state importance ".[4] But this was a vision of the future. In 1922 the whole country is said to have possessed only 1500 tractors, of which 25 per cent were in working order. Imports began on a small scale in 1923 and 1924, when a few hundred American tractors were ordered : most of these went to the North Caucasian region, where collective farming was strongest and man-power least abundant. At the beginning of August 1924 the prospect of a bumper harvest drew attention to the need for tractors, and STO authorized an emergency order for 1000 tractors from the United States. These were actually delivered in Novorossiisk before the end of September.[5] Even the cautious and conservative People's Commissar for Agriculture of the RSFSR, A. P. Smirnov, expressed enthusiasm :

> The tractor is of extreme importance for collectivization. 80 per cent of our whole number of tractors have provided an

[1] *Planovoe Khozyaistvo*, No. 11, 1925, pp. 33-50.
[2] A. Ransome, *6 Weeks in Russia in 1919* (1919), p. 100.
[3] Lenin, *Sochineniya*, xxiv, 170.
[4] *Sobranie Uzakonenii, 1921*, No. 28, art. 157.
[5] *Na Agrarnom Fronte*, No. 11-12, 1925, p. 225 ; L. B. Krasin, *Voprosy Vneshnei Torgovli* (1928), pp. 222-224.

occasion for the formation of various types of cooperatives and collectives for their utilization. The tractor is undoubtedly one of the greatest factors in eliminating the habits of individual peasant production. If the tractor is linked with a cooperative and correctly utilized, it is not only an agricultural machine, but a new factor in the growth of the socialist element in the village. The tractor unites the poor peasants, in particular by raising their production and really preventing their exploitation by the *kulak*.[1]

In the same year the first attempts were made to produce tractors in Soviet factories. At the beginning of 1925 Gossel'sindikat was said to have repaired 50 old tractors for the use of Sovkhozy, to have obtained an unspecified number of new tractors from the Kolomensky factory, and to have ordered a further 1000 from the Gomza works. But these experiments were still on an insignificant scale, and seem to have had little success.[2]

It was only by slow degrees that this question began to attract attention in influential party circles. A decree of STO of August 5, 1925, attempted to eliminate the large discrepancy between the cost of foreign and Soviet tractors by laying down standard prices. A Fordson tractor from the United States and a comparable product from the Putilov works were both to sell at 1800 rubles — the lowest figure quoted. Sales were to be financed by the Central Agricultural Bank : payments for foreign tractors might be spread over two harvests, and for Soviet tractors over three.[3] On October 15, 1925, *Pravda* published an article on *The Mechanization of the Countryside* headed by the quotation from Lenin at the eighth party congress. Larin pointed out that, whereas the poor peasant could be " driven into the collectives by need ", the middle peasant would have to be attracted by offers of tractors and

[1] *Ekonomicheskaya Zhizn'*, November 7, 1924.

[2] Kamenev claimed in May 1925 that " in the last few months, after a series of failures ", the Kolomna factory had begun to produce a tractor " which works quite satisfactorily ", though the number produced was still totally inadequate (L. Kamenev, *Stat'i i Rechi*, xii (1926), 292) ; according to *Itogi Desyatiletiya Sovetskoi Vlasti v Tsifrakh, 1917–1927* (n.d.), p. 245, 481 tractors were produced in the Soviet Union in 1924–1925, and 815 in 1925–1926. At the end of 1926 the Putilov works in Leningrad were producing an " extremely insignificant " number of tractors at five or six times the cost of the foreign article (*XV Konferentsiya Vsesoyuznoi Kommunisticheskoi Partii (B)* (1927), pp. 122-123).

[3] *Sbornik Dekretov, Postanovlenii, Rasporyazhenii i Prikazov po Narodnomu Khozyaistvu*, No. 23 (44), August 1925, pp. 14-15.

electrification, " so that his mouth may water and he may be speedily converted into a collectivist ".[1] On the eve of the fourteenth party congress in December 1925, *Pravda* published two more detailed articles by Mikoyan entitled *On the Way to the Tractorization of Agriculture.* Mikoyan claimed that 2000 imported tractors were at work in the Soviet Union, mainly in the North Caucasian region. In the current year 15,000 more tractors were to be imported, and funds had been assigned for the setting up of a Soviet tractor factory at Stalingrad, once it had been decided which of three types was best suited to Soviet conditions.[2] Thus prompted, the congress included " the raising of agricultural technique (introduction of tractors) " in its list of desiderata for the countryside ; and this was the signal for a broad and rather desultory discussion. Krzhizhanovsky recorded in January 1926 that " the demand for tractors is growing from month to month ".[3] Preobrazhensky recommended " the provision of state tractors on a mass scale for the poor peasant " as a step towards " the divorce from small-scale production of those operations which lend themselves most easily to socialization ".[4] *Pravda* published two articles on the electrification of the countryside. This would favour collective enterprise, since threshing, milling and rural industry generally could be carried on collectively with electric power. On this ground electrification was said to be opposed by the *kulak*, and cases of sabotage were reported.[5] A congress of agricultural organizations of the RSFSR met in Moscow at the end of February 1926 and discussed such matters as the provision of tractors and of technical assistance.[6]

Meanwhile effective progress was so slow as to seem wholly

[1] Yu. Larin, *Rost Krest'yanskoi Obshchestvennosti* (1925), p. 37.

[2] *Pravda*, December 17, 18, 1925. 10,000 tractors were imported from the United States in the spring of 1926 (*Vlast' Sovetov*, No. 23, June 6, 1926, pp. 12-13) ; preparatory work on the Stalingrad factory began in 1926, but it was not expected to come into production before 1929 (M. Latsis, *Sel'skokhozyaistvennye Kontsessii* (1926), p. 15).

[3] G. Krzhizhanovsky, *Sochineniya*, iii (1936), 117.

[4] E. Preobrazhensky, *Novaya Ekonomika* (1926), p. 208.

[5] *Pravda*, January 12, 13, 1926 ; the prospect of the availability of electricity in the countryside on any significant scale was so remote as to make the argument rather unreal.

[6] *Ibid.* March 3, 1926, which reported the congress, also carried an article stressing the importance of the supply of tractors and machinery to the agricultural cooperatives.

illusory ; and those who regarded such projects with scepticism
or disfavour clung to Lenin's interpretation of " the industrialization of agriculture ".[1]   They pointed out that the development of
an intensive agriculture, supported by an abundant drying, processing and refrigerating plant, was required to meet the conditions of an over-populated countryside in the Ukraine and other
parts of the Soviet Union, rather than an extensive mechanized
agriculture of the American type ;  that the Sovkhozy, as proto-
types of " socialist grain factories ", had enjoyed little success,
and least of all in densely populated areas ;  and that the " industrialization of agriculture " in the sense of mechanization was
in the long run inseparably bound up with the industrialization
of the whole national economy, which could alone provide the
necessary machines and equipment, and could alone offer an
outlet for the labour rendered superfluous by the mechanization
of work in the countryside.[2]   Trotsky in his article " Towards
Socialism or Capitalism ? " in the autumn of 1925 had spoken of
" scientific methods of cultivation, electrification and technical
improvements generally " as conditions of a socialist agriculture,
and added that " the technical and socialist progress of agriculture
is inseparable from an increasing preponderance of industry in
the economic life of the country ".[3]   Stalin, in his speech to the
Leningrad party organization in April 1926, made his contribution
to the current theme :

> Agriculture itself cannot advance unless you give it in good
> time agricultural machines, tractors, products of industry
> etc. . . . It relies, and has already come to rely, on the direct
> development of industry.[4]

Later statistics put the production of agricultural machinery in the
Soviet Union in the year 1925–1926 at 70 million rubles at pre-
war prices as compared with a pre-war figure of 67 millions.[5]
This indicated no startling development.  The discussion petered

[1] See p. 272, note 1 above.
[2] Articles on the subject appeared in *Planovoe Khozyaistvo*, No. 4, 1926,
pp. 120-127 ; No. 5, 1926, pp. 107-127 ; No. 6, 1926, pp. 112-121 ; and in
*Ekonomicheskoe Obozrenie*, No. 3, 1926.
[3] *Pravda*, September 20, 1925 ; for this article see p. 505 below.
[4] Stalin, *Sochineniya*, viii, 119.
[5] *Kontrol'nye Tsifry Narodnogo Khozyaistva SSSR na 1929–1930 god* (1930),
p. 437.

out in an atmosphere of scepticism about the prospect of any early advance along these lines.    But it revealed, even at this period, some acute appreciations of the far-reaching character of the only effective alternative to the current policy of the wager on the *kulak*, and the revolutionary social changes which would be required to give effect to it.

The meeting of the party central committee of April 6-9, 1926, was the first important party occasion since the fourteenth congress in the previous December ;   and both sides uneasily awaited a renewal of the struggle.    The tide was now setting strongly away from the policy of indulgence to the *kulak*.    A few days before the meeting, Yaroslavsky, reporting to the central control commission on party work in the countryside, had referred indignantly to the party workers who pretended that " there were no poor peasants among whom they could work ".[1]    Stalin, whose spokesman Yaroslavsky was, had made it clear that he would not give battle on that issue ;   and this rather took the wind out of opposition charges of a *kulak* deviation.    The main resolution proposed by Rykov was the work of a committee appointed three months earlier by the party central committee on a proposal of Trotsky.[2] It was eclectic in character, and clearly represented an attempt to conciliate divergent points of view.    Trotsky, taking an active part in the proceedings of a party organ for the first time for nearly two years, proposed a series of amendments which, in fact, constituted an alternative draft resolution.    The debate proceeded on the two documents.[3]    Trotsky argued that a lack of balance in the economy fatal to the " link " between the proletariat and the peasantry might arise in two ways :   either through the extraction,

---

[1] *Pravda*, April 6, 1926.

[2] This was stated by Trotsky in his speech at the central committee in April 1926 (see following note).

[3] The proceedings of this session were not published, and Rykov's resolution is not available in its original form ; Rykov's report on the session to a party meeting appeared in *Pravda*, April 23, 1926 ; see Trotsky's speech and his amendments in the Trotsky archives,T 2982, 2983 ; salient passages from the amendments were quoted in *XV Konferentsiya Vsesoyuznoi Kommunisticheskoi Partii (B)* (1927), pp. 122, 126, 138.   For parts of the debate bearing on industry see pp. 354-356 below ; its significance for relations between Trotsky and the new opposition will be discussed in Part III in the following volume.

for investment in industry, of too large a proportion of the national surplus or through the extraction of too small a proportion of the surplus. The symptom of the former error would be a supply of industrial goods in excess of demand, of the latter error an excess of demand over supply. Clearly it was the second error which was being committed : " state industry is lagging behind agricultural development ". In such circumstances even a good harvest " may become a factor not hastening the tempo of economic development, but on the contrary disorganizing the economy ". Trotsky's main recommendation in agricultural policy was a stiffening and steeper grading of the agricultural tax in order to secure " the correct redistribution of accumulation in the national economy ". He also desired to raise wholesale, though not retail, prices for industrial goods.[1] Kamenev supported Trotsky, particularly in his prediction of the potential adverse consequences of a good harvest and in his advocacy of increased taxation for agriculture. When Kamenev was once more blamed for his mistakes of the previous autumn, somebody — presumably Trotsky — retorted that " this is not according to Marx, and it is obligatory to seek a class content for our economic difficulties ".[2] The identity of views between Trotsky and Kamenev was so close that Stalin at one point called out ironically : " What's this ? A bloc ? "[3]

But, whatever the importance of the debate from the point of view of relations between the leaders, the discussion of issues of substance was largely shadow-boxing. Stalin was not prepared either to fight for the *kulak* or to resist the growing pressure of industrialization. While Trotsky's amendments were formally

[1] An article of April 2, 1926, preserved in the Trotsky archives, T 874 (evidently written by someone closely connected with Gosplan, possibly Smilga), carried to its logical conclusion the argument of Ossovsky's article (see p. 317 above) : " By guaranteeing through a change in price policy a flow of resources into industry in a measure sufficient for the normal expansion of production, we shall at the same time free ourselves from the need to feed it artificially. . . . At the foundation of our price policy we should . . . place the principles of a correspondence of price with the tempo of development which we consider necessary in this or that branch of industry, and not the principle of correspondence of price and cost at a given moment." Judging from his attitude, Trotsky was not fully convinced of the practicability of such a course.

[2] This exchange was recorded by Rykov (*Pravda*, April 23, 1926).

[3] *Sotsialisticheskii Vestnik* (Berlin), No. 10 (128), May 22, 1926, p. 15 ; since this was reported before the " bloc " actually materialized, it is probably true.

rejected, the resolution unanimously adopted at the end of the session bore the stamp of a compromise not unfavourable to the opposition. Its agricultural section referred to " the struggle between different social groups in the peasantry ", to " the inevitable strengthening of the *kulaks* in the present period of NEP ", and to " the struggle of *kulak* elements to control the countryside ". It drew the conclusion that the party should give material aid to the poor peasant and seek " a strengthening of the link of the proletariat and poor peasantry with the middle peasant in order to isolate the *kulak*". More significantly, it enjoined on the party " the task of carefully studying the experiment of the application of hired labour in the countryside and of the development of leasing ". The most important passage in the agricultural section dealt with the reform of the agricultural tax. It approved the assessment of assets in terms of money, the inclusion of revenues from subsidiary agricultural occupations as well as from rural industry, and a more progressive system of assessment, meaning " complete exemption from tax of the poorest groups and a heavier assessment on the well-to-do and *kulak* strata of the peasantry ". The section of the resolution dealing with the co-operatives called for " a complete guarantee within the cooperatives of the interests of the poor and middle strata of the peasantry, and a struggle against attempts to utilize the cooperatives by *kulak* elements ".[1] By no means all these projects were destined to be carried out in the near future. The resistance of the *kulaks*, and of the section of the party which supported them, was tenacious and effective. But, by the spring of 1926, industrial ambitions were in the ascendant ; and the incompatibility of the policy of intensive industrialization with the policy of conciliating the well-to-do peasant was already becoming apparent.

In addition to the general resolution on economic policy, the central committee, on a report by Kamenev, passed a special resolution " On the Organization of the Grain-purchasing Apparatus for the Campaign of 1926–27 ". This was an attempt to get rid of the scandal of competing organizations which had complicated the marketing of the harvest of 1925, and inflated the costs of collection.[2] It was now laid down that in principle all

---

[1] *VKP(B) v Rezolyutsiyakh* (1941), ii, 95-96.
[2] See pp. 295-297 above.

state purchases of grain should be made either by Khleboprodukt or by the cooperative organizations. Mills and milling trusts might purchase only for their own requirements and only in the areas where they operated : milling was to be brought under the control of Narkomtorg. Gosbank might facilitate purchases outside the plan by credits or by transactions on commission. But its activity as a purchaser of grain on behalf of the state would be confined to " minimal proportions " ; and the remainder of its grain-purchasing organs would be liquidated in agreement with Narkomtorg.[1] By these means, it was hoped " to simplify and cheapen our grain-purchasing apparatus ".[2]

[1] *VKP(B) v Rezolyutsiyakh* (1941), ii, 97-100 ; earlier editions attribute the report to Kamenev, whose name was suppressed after 1936. The reform was reflected in the balance-sheet of Gosbank, where " investments " (i.e. investments in commodities, primarily grain) fell from 89 million rubles on October 1, 1925, to 52 millions on October 1, 1926, and 5·9 millions a year later (*The State Bank of the USSR, 1921-1926* (Moscow, 1927), pp. 31-32).

[2] Stalin, *Sochineniya*, viii, 134.

CHAPTER 6

# INDUSTRY

THE resolution of the scissors crisis in the winter of 1923–
1924 opened the way for a fresh advance of industry all
along the line, thus defeating the gloomy prognostications
of the opposition and illustrating once more the validity of the
underlying principle of NEP that the revival of industry, like
everything else in the Soviet economy, depended on a revival of
agricultural production. In the period of steady economic recovery
which ran from 1923 to 1926, agriculture and industry moved
forward in step with each other. The efforts of the champions
on both sides to create an antithesis between them proved unnecessary and futile. This miscalculation discredited the opposition of 1923. The official line, which maintained that increased
prosperity in agriculture was the first condition of increased
prosperity in industry, seemed to have been triumphantly vindi-
cated. But, as the advance proceeded, fresh problems began to
confront the makers both of agricultural and of industrial policy ;
and these problems sprang from the same source. The development both of agriculture and of industry stimulated by NEP followed capitalist rather than socialist lines. In agriculture it
meant the encouragement of the *kulak*. In industry, it favoured
the growth of light industries working with limited capital for the
consumer market and earning quick profits rather than of the heavy
industries which were, by common consent, the basis of a future
socialist order, but required an initial volume of long-term capital
investment ; for this contingency the principles and practices of
NEP made no provision.[1] Hence the struggle in agricultural
policy against the predominance of the *kulak*, which began in 1924
and remained acute throughout 1925, was matched at the same

[1] See *The Bolshevik Revolution, 1917–1923*, Vol. 2, pp. 315–316.

period by a similar struggle in industrial policy centring on the
requirements of heavy industry.  The history of industrial progress between 1923 and 1926 falls into three stages.  In the first
stage, approximately corresponding to the economic year 1923–
1924, the " spontaneous " forces of recovery stimulated by
NEP were still in the ascendant, and light industries continued
to advance more rapidly than heavy industry.  In the second
stage, running from the autumn of 1924 to the end of 1925,
a confused battle was fought with varying fortunes between
conflicting policies and interests.  In the third stage, which
began with the fourteenth party congress in December 1925,
the expansion of heavy industry became the predominant aim of
economic policy.

The revival of industry in the year 1923–1924 was due to
two immediately favourable factors.  In the first place, the
rationalization of industry through concentration, which was first
undertaken on a serious scale in the summer of 1923,[1] lowered the
costs of production ;  and the reduction of prices of industrial
products, enforced or stimulated by official action as a result of
the scissors crisis,[2] combined with the good harvest to expand
the market at a rapid rate, so that the volume of production, as
well as the number of those employed in industry, increased as
prices fell.  This expansion of production required little fresh
investment of fixed capital.  It was achieved primarily by bringing
back into production labour power, premises and plant which had
lain idle during the years of stagnation.

The second favourable factor was the easing of credit.  The
shortage of working capital had been acutely felt in the crisis of
1923, when the banks had forced down industrial prices by with-
holding credits from industry.[3]  The first consequence of the
accomplishment of financial reform had been another constriction
of credit, due to the excessive caution of Narkomfin and the
bankers, and fears for the stability of the new currency ;[4] and
this accounted for the hostility to the reform displayed by most

[1] See *The Interregnum, 1923–1924*, p. 118, note 4.
[2] See *ibid*. pp. 110-112.        [3] See *ibid*. pp. 96-98.
[4] See p. 476, note 1 below.

industrialists.[1]  But by the autumn of 1924, these fears had been dissipated, and the problem of working capital was no longer acute. The stability of the currency had been established. Short-term credit flowed freely for concerns which could find a ready market for their goods. The total indebtedness of industry to the five principal banks rose from 161 million rubles on October 1, 1923, to 448 millions on October 1, 1924, and 953 millions on October 1, 1925.[2]  Moreover, the rate of circulation of capital had also increased, so that a given volume of credit was being more economically utilized.[3]  The only complaint was, once more, that a credit system based on market conditions favoured the consumer industries, which yielded quick profits, and starved the heavy industries producing capital goods for long-term projects of reconstruction.[4]

These conditions had made the year 1923–1924 a year of steady progress in all branches of Soviet industry. Industrial production (excluding rural and artisan industries, which formed only a small proportion of the whole) had stood in 1922–1923 at 1,950,000 pre-war rubles, representing about double the low level of 1920 and 34 per cent of the production of 1913. In 1923–1924 it reached 2,570,000 pre-war rubles, or more than two-and-a-half times the total of 1920 and 40 per cent of pre-war production.[5] But the gains were unevenly distributed. In the major industries

---

[1] *Sotsialisticheskoe Khozyaistvo*, No. 3, 1924, p. 12, quotes a number of articles of the first months of 1924 expressing mistrust of the reform on these grounds : a year later Bronsky recalled " the serious and obstinate campaign against the punctual carrying out of the reform " conducted by " our comrade-managers ", who had, however, since " learned by experience " (*ibid*. No. 5, 1925, p. 11).

[2] *Promyshlennost' SSSR v 1925–26 godu* (Vesenkha, 1927), p. 8.

[3] Y. S. Rozenfeld, *Promyshlennaya Politika SSSR* (1926), p. 408, quotes the testimony of Pyatakov on this point.

[4] *Shestoi S"ezd Professional'nykh Soyuzov SSSR* (1925), pp. 260-261 ; it was stated that at this time 60 per cent of all bank credits went to light industry and less than 30 per cent to heavy industry (*ibid*. p. 271).

[5] These totals are taken from *Kontrol'nye Tsifry Narodnogo Khozyaistva na 1925–1926 god* (1925), p. 82. Slightly higher figures of industrial production for 1923–1924 are found in *Kontrol'nye Tsifry Narodnogo Khozyaistva na 1926–1927 god* (1926), p. 321, slightly lower figures in *Promyshlennost' SSSR v 1925–26 godu* (Vesenkha, 1927), pp. 20-21 ; the latter are apparently net figures, excluding half-finished products transferred from one industry to another. Such variants are characteristic of all statistics of the period ; it is seldom clear whether the later figures represent corrections of the earlier ones, or have been reached on a different basis. But they are rarely important enough to affect the general picture.

which came under the control of Vesenkha, an average of only 36 per cent of pre-war production had been reached ; and this average represented a wide variation in achievement. The metal industry, which accounted for the major part of the output of capital goods, registered only 28·7 per cent and the textile industry, the largest consumer goods industry, 35 per cent.[1] According to another calculation made at this time heavy industry, which in 1913 accounted for 22·6 per cent of all industrial production, produced only 17·7 per cent in 1922–1923 and 17·4 per cent in 1923–1924.[2] In the period of general recovery they continued to lag behind. The output of the basic iron and steel industries was lowest of all. In 1923–1924 only 660,000 tons of pig iron, 990,000 tons of steel and 690,000 tons of rolled metal were produced as against 1913 figures of 4 millions, 4 millions and 3½ millions respectively.[3]

The industrial recovery of 1923–1924 meant that industry as a whole was for the first time earning profits. The conception of profit in Soviet industry was, and remained, to some extent arbitrary and uncertain. But, with the adoption of *khozraschet*, it became an essential feature of Soviet accounting ; and the results, though subject to criticism in detail, provided a sound general barometer of progress. A calculation of the profit-and-loss account of state industry for 1922 showed a global loss for that period of 11 million rubles. The heaviest losses were sustained by the metallurgical, chemical and paper industries ; the food, salt and leather industries earned a substantial margin, the textile industries a bare margin, of profit.[4] The period of losses

---

[1] *Kontrol'nye Tsifry Narodnogo Khozyaistva na 1925–1926 god* (1925), p. 79.

[2] *Sotsialisticheskoe Khozyaistvo*, No. 1, 1925, pp. 82-83 ; the term " heavy industry ", which normally comprised the fuel industries (coal and oil), and the chemical and electrical industries as well as iron, steel and engineering, is here evidently confined to the metal industries. For the desperate condition of these in 1920 see *The Bolshevik Revolution, 1917–1923*, Vol. 2, pp. 118-119 ; for their slow recovery in 1923 see *The Interregnum, 1923–1924*, pp. 118-119. In December 1924 Dzerzhinsky claimed that heavy industry as a whole had reached 46 per cent of pre-war production (*Pravda*, December 4, 1924).

[3] *Ekonomicheskoe Obozrenie*, October 1927, p. 110 : the 1913 figures are from *Kontrol'nye Tsifry Narodnogo Khozyaistva na 1926–1927 god* (1926), p. 320. (The writer in *Ekonomicheskoe Obozrenie*, to judge from his percentage calculations, used somewhat lower figures for 1913, probably owing to a different adjustment for territorial changes.)

[4] *Sotsialisticheskoe Khozyaistvo*, No. 2, 1924, pp. 165-182.

had now been finally left behind. The financial year 1922–1923 was the first for which state industry purported to show a small profit, though this was probably achieved by making inadequate allowances for amortization — a problem whose importance was still unrecognized. Everyone agreed that industry earned profits in 1923–1924, though the total figure varied widely with the authority making the return, the estimate of the industrial trusts themselves being 55 million rubles, while the Vesenkha figure rose to 83 millions and the figure of Narkomfin to 102 millions. It was not difficult to detect an interest of the trusts in understating their rate of profit, or of Narkomfin in overstating it. But all authorities agreed that the textile industries had been the most profitable, followed by coal and oil, and that the metal and timber industries were the only ones which had actually worked at a loss.[1] A later Vesenkha return put the total profits of industry for the year at 105 million rubles.[2]

The " spontaneous " and unplanned industrial recovery of 1923–1924 served to mask for the time being the basic problem of the capital goods industries, to which Trotsky and the opposition had begun to draw attention in 1923. The revival of heavy industry was given a place in every important party pronouncement. The thirteenth party conference of January 1924, which condemned the economic theses of the opposition, none the less adopted the resolution drafted by the scissors committee which described " socialist accumulation " as " the fundamental and decisive factor in the fate of the proletarian dictatorship under NEP ", and firmly recommended that the metal industry should " be advanced to the front rank and receive support of all kinds from the state ".[3] The appointment of Dzerzhinsky in February 1924 as president of Vesenkha was a move to make this recommendation effective. At the thirteenth party congress in May

[1] *Planovoe Khozyaistvo*, No. 9, 1925, p. 301. Dzerzhinsky, in a speech to Vesenkha on December 2, 1924 (*Pravda*, December 4, 1924), put the profits of industry at 4 million rubles for 1922–1923, and 45 millions for 1923–1924.
[2] *XV Konferentsiya Vsesoyuznoi Kommunisticheskoi Partii (B)* (1927), p. 109. According to S. G. Strumilin, *Ocherki Sovetskoi Ekonomiki* (1928), pp. 166–167, trusts directly responsible to the Vesenkha of the USSR showed a profit of 100 million rubles in 1923–1924, after setting aside 130 millions for amortization ; but trusts responsible to organs of the republics still failed to show profits. For the question of amortization see p. 343, note 2 below.
[3] See *The Interregnum, 1923–1924*, p. 115.

1924 Zinoviev proclaimed that the moment had come for the revival of heavy industry and for the production of the means of production.[1] But any practical step towards the achievement of these ends continued to encounter powerful opposition. Sokolnikov, speaking a few days after the end of the congress, put in a caveat about " the tempo of the development of state industry ", and thought that the congress had called a halt to " excessive forms of enthusiasm for so-called ' socialist accumulation ' ".[2] A little later a spokesman of Narkomfin argued that it was the development of light rather than of heavy industry which was required to feed the peasant market, and that capital available for investment should go into industries processing agricultural produce for export.[3] The view that the Soviet Union, instead of developing its own heavy industry, should expand its agricultural production for export, and import machinery and capital goods from abroad, was by no means dead. As the economic year 1923–1924 came to an end, few signs pointed to radical changes in industrial policy in 1924–1925.

The situation in the autumn of 1924 offered little encouragement to the advocates of intensive industrialization. The harvest had partly failed. The party was seriously alarmed by symptoms of peasant unrest, and Zinoviev had proclaimed the slogan, "Face to the countryside ".[4] Trotsky had resumed his offensive against the party leadership ; and the campaign against him was being built up around the charge of underestimating the peasant.[5] Financial restraint and the conciliation of the peasant were the two main watchwords of current policy. It was in this atmosphere that the Chief Administration of the Metal Industry (Glavmetall

[1] See *The Interregnum, 1923–1924*, pp. 143-144.

[2] *Ekonomicheskaya Zhizn'*, June 5, 1924. When a few days later Sokolnikov spoke in the same sense at the Business Club, Smilga asked in reply whether " the industrialists exist for the benefit of Narkomfin, or Narkomfin for the benefit of the industrialists " ; Smilga was at this time editor of the newly founded *Vestnik Promyshlennosti, Transporta i Torgovli*, the journal of the council of congresses, the organ of the industrial managers (see *The Interregnum, 1923–1924*, pp. 42-45).

[3] *Sotsialisticheskoe Khozyaistvo*, No. 5, 1924, pp. 102-103.

[4] See p. 195 above.

[5] The struggle in the party will be discussed in Part III in the following volume.

or GUMP) [1] put forward in the autumn of 1924 a plan for the industry for 1924–1925 which provided for output of a total value of 306 million rubles — an expansion of 55 per cent over the total of the preceding year.[2] This proposal provoked an immediate crisis. During the first years of NEP industry had been financed by advances from the budget ; these took the form partly of direct subsidies for the restoration of fixed capital, and partly of advances under the head of working capital, mainly for the purchase of raw materials. From 1923 onwards the policy had been to restrict these budgetary sources (in 1923–1924 state industry received 82 million rubles in subventions from the budget), and to finance industry primarily through long-term and short-term credit from the banks.[3] The difference was one of substance as well as of method. Banking credit was made available on the basis of tangible security and potential profits. It met the needs of consumer industries whose capital requirements were comparatively low, and whose products found a ready market and a quick turn-over, but not of the metal industries which were still in course of construction or reconstruction, whose capital requirements were extensive, and whose profits lay far ahead in the future. If the programme of Glavmetall for 1924–1925, or anything like it, were to be approved, substantial advances from budgetary sources were indispensable.[4]

The clash of policies was sharp. Sokolnikov, in submitting his budget to TsIK in October 1924, reiterated his old thesis that the expansion of industry was dependent on the expansion of agriculture : industry was " fettered to the condition of the

[1] Glavmetall and Glavelektro were the only *glavki* which survived from the period of war communism into NEP (*Voprosy Istorii*, No. 7, 1953, p. 42) ; this was scarcely accidental, since they controlled two industries where private enterprise had no footing.

[2] *Sotsialisticheskoe Khozyaistvo*, No. 1, 1925, p. 81.

[3] See *The Interregnum, 1923–1924*, p. 8 ; Bukharin in his polemic against Trotsky in January 1925 complained that the opposition wished to use state subsidies as the " centre of gravity " for financing industry, whereas the party wished industry to rely on bank credit (N. Bukharin, *Kritika Ekonomicheskoi Platformy Oppozitsii* (1926), p. 83).

[4] Besides receiving advances industry was, of course, also paying taxes : a table in R. W. Davies, *The Development of the Soviet Budgetary System* (1958), p. 103, purports to show that in both the years 1924–1925 and 1925–1926 industry paid more than it received. Even so, the process was significant, since the tax on profits came mainly from light industry and the advances were made to heavy industry.

peasant economy ".[1]   This principle dominated the budget.
Owing to the bad harvest, state assistance to agriculture was to
be raised from 59 million rubles in the previous year to 88 millions
in 1924–1925.   Transport, which had received 50 million rubles
to cover its deficit in the previous year, would be expected to
balance its accounts in 1924–1925.   Industry would have to
economize, its subvention of 82 millions in the preceding year
now being reduced to 59 millions ; expenditure on electrification
was cut from 46 millions to 37 millions.[2]   At the end of October
1924 Sokolnikov told the party central committee that " we have
rather overstrained the line of support for heavy industry ", and
stressed the alleged dangers of over-developing the metal industry.[3]
At the sixth congress of trade unions in the middle of November
Rykov propounded the view, congenial to a trade union audience,
that " the growth of our metal industry is the criterion of the
recovery of the productive backbone of our whole economy, both
in industry and in agriculture ".   But he excused himself from
giving " definite data on the question of the prospects of the metal
industry for the coming year " on the ground that the commission which was dealing with the question had not yet reported.[4]

The result was a compromise.   The decision taken by STO a
few days later was for a 10 per cent cut in the programme put
forward by Glavmetal : production for the year was reduced from
the planned total of 306 million rubles to 270 millions.[5]   At the
beginning of December 1924 Kamenev addressed the Moscow
Soviet in terms which revealed the prevailing confusion of thought.
He pointed out that in the past " our metal industry rested not

---

[1] *SSSR : Tsentral'nyi Ispolnitel'nyi Komitet 2 Sozyva : 2 Sessiya* (1924),
p. 136 ; for Sokolnikov's enunciation of the same thesis in 1922 see *The Bolshevik Revolution, 1917–1923,* Vol. 2, p. 317.

[2] *SSSR : Tsentral'nyi Ispolnitel'nyi Komitet 2 Sozyva : 2 Sessiya* (1924),
pp. 150-151.

[3] *Ekonomicheskaya Zhizn'*, November 2, 1924.

[4] *Shestoi S"ezd Professional'nykh Soyuzov SSSR* (1925), p. 239.  Delegates
of the metal workers and the railway workers spoke of the growing demand for
the products of heavy industry :  the whole production of Yugostal' had been
sold for a year ahead, and the repair of 3000 locomotives was being held up for
lack of material (*ibid.* pp. 258, 273).  The congress passed a resolution stressing
the significance of the metal industry " which at the present time still lags
behind the development of other branches of the economy " (*ibid.* p. 485).

[5] *Ekonomicheskaya Zhizn'*, November 21, 22, 1924 ; *Pravda*, December 2,
1924 ; the decision of STO is dated November 24 in *VKP(B) v Rezolyutsiyakh*
(1941), i, 634.

on rural demand, but on the orders of the railways, of shipping companies, on the military requirements of the Tsarist govern- ment ". But he drew no moral for present conditions from this correct and pertinent observation ; and, having quoted the figures of the decision of STO, he proceeded to invoke the same criterion as Bukharin and Sokolnikov : " Can our basic con- sumer, the peasant, consume what our industry produces ? " [1] At the same moment Dzerzhinsky warned an all-union conference of Vesenkha that shortage of capital resources placed a limit on the industrial expansion ; the metal industries, in particular, were chronically in debt to the banks, and unable to meet their wages bills without further credit. The resolution of the conference damped down the recent emphasis on heavy industry, observing that the development of industry " must go hand in hand with the development of the peasant economy, and proceed from the require- ments of that economy ".[2] The moral was pointed in a leading article in *Ekonomicheskaya Zhizn'* of December 11, 1924, which applauded the resolution as a contribution to a " Bolshevistically restrained economic policy " and condemned " the ' permanentist ' underestimate of the peasant economy in our country ". The check to the hopes of the industrialists was subtly linked with the new crisis in relations between Trotsky and the party leadership. The fight with Trotsky was on ; and every official pronouncement on policy at this time must add its quota to the weight of denuncia- tion. No cause associated with Trotsky's name or with Trotsky's supporters could be allowed to receive any semblance of official corroboration.

It was, therefore, both significant and characteristic when the party central committee, having in January 1925 administered a resounding reproof and defeat to Trotsky and his followers, and having relieved Trotsky of his functions as president of the Revolutionary Military Council, should once more have reversed the trend of economic policy. Now that there was no longer any immediate risk of strengthening Trotsky's hand by seeming to prove that he had not been entirely wrong, it was possible to resume the forward march of heavy industry, and his supporters

[1] L. Kamenev, *Stat'i i Rechi*, xi (1929), 265-268.
[2] *Ekonomicheskaya Zhizn'*, December 2, 4, 1924 ; the resolution of the conference was reported in full *ibid.* December 7, 1924.

could be made to feel that they had been tilting at windmills. The objections of Narkomfin were brushed aside. The central committee authorized Vesenkha to increase by 15 per cent the programme of production for heavy industry approved by STO in November 1924 (thus more than restoring the cuts then made), and decreed " a corresponding increase in budget allocations and an expansion of credit for industry ". In more general terms the central committee demanded " the working out of a plan to restore fixed capital, to re-equip the factories and to erect new factories to meet the needs of the whole economy "; special attention was to be given to " the supply of metal goods to the countryside "; and the production of lead and zinc, as well as the output of locomotives and wagons, were to be expedited.[1] An explicit programme for the expansion of heavy industry had for the first time received the endorsement of the highest party authority.

While, however, the zigzags of Soviet economic policy in the winter of 1924–1925 may be attributed in part to the exigencies of the struggle against Trotsky, deeper explanations are required of the ultimate victory of the industrializers. The most obvious cause was the extent and volume of the economic recovery, reflecting itself in the unexpected resilience of the public revenue. In spite of the poor harvest, receipts continued to flow into the treasury at a rate which reversed all the cautious prognostications of the previous summer, and called in January 1925 for a substantial upward revision of the budget.[2] Since funds were available to finance industrial expansion on a more generous scale, Narkomfin was deprived of its most convincing argument. Yet this evidently does not answer the question why these surpluses should not have been used merely to relieve the burden of taxation, or, if retained, should not have been directed to consumer industries rather than to the production of the means of production. The tenacity of purpose which in these years insisted on the maintenance and intensification, in face of every obstacle and opposition, of the programme of development of heavy industry, requires explanation on a different level, and would appear to have derived both from the dynamic of national self-sufficiency

[1] *VKP(B) v Rezolyutsiyakh* (1941), i, 634-635.
[2] For this see p. 460 below.

which had been set in motion before the revolution and from the dynamic of Marxist socialism set in motion by the revolution itself. The Soviet economy advanced in the nineteen-twenties and nineteen-thirties under the inspiration of these two powerful forces.

One particular factor brought the issue of industrialization to a head in the year 1925 and helped to shape the character of the decisions taken. Since the introduction of NEP economic advance had taken the spontaneous and relatively uncontroversial form of economic recovery, of regaining lost ground. No new decisions, no influx of fresh capital, had been called for. Hitherto it had been possible to promote a partial revival of industry by bringing back disused factories and plant into operation and by processes of rationalization : this did not involve any large invest- ment of new capital. By the end of 1924, however, what could be achieved by this method had been achieved ; it was officially estimated that existing factories and plant were being utilized up to 85 per cent of capacity.[1] Makeshift expedients were no longer available. The next requirement, if industry was to expand further, or even to maintain its existing level of produc- tion, was the accumulation of capital resources, not merely to repair the damages and wear-and-tear of war and civil war and to make up for a long period of neglect, but to transform and modernize obsolescent factories, plant and equipment. A large programme of capital investment, such as had been envisaged by Dzerzhinsky at the thirteenth party congress in May 1924,[2] had become indispensable.

The party resolution of January 1925, while giving its blessing to " budget allocations " as well as to " an expansion of credit ", had ignored the details of the financial problem ; and it was round this that the battle raged throughout 1925. The situation was analysed in a set of theses on the question of the restoration of fixed capital, now described as " the central problem of in- dustrial policy ", which were adopted by Vesenkha at the end of February 1925.[3] What was at stake was no longer " simple repro- duction " of capital, but an expansion of production through an

[1] *Ekonomicheskaya Zhizn'*, February 25, 1925.
[2] See *The Interregnum, 1923–1924*, pp. 143-144.
[3] *Ekonomicheskaya Zhizn'*, February 25, 1925.

increase of fixed capital. The obsolescent equipment of many enterprises must be renewed, and new industries must be created. The problem of the renewal of fixed capital must be approached " with a plan worked out on an all-state scale ". The first step was to ensure a planned allocation of amortization allowances " on the basis of general state interests and not of particular economic interests of separate state undertakings ". But these resources were not large enough for a policy of expansion. After everything possible had been done by way of rationalization, increased rate of circulation and increased productivity of labour, and by the extraction of the savings of the population through taxation, it would be necessary to create an " industrial fund ", to be built up out of amortization allowances, state subsidies and loans, for capital investment in industry. The fund was to be administered by the appropriate banks (Prombank and Elektrobank) under the direction of Vesenkha.

Following the adoption of these theses, Vesenkha convened a " special conference on the restoration of fixed capital in industry " (Osvok), which remained active for the next eighteen months [1] and marked an important stage in the history of industrial planning. Its initial recommendation for an " industrial fund for long-term credit ", to be administered by Prombank under the general supervision of Vesenkha, contemplated a total investment for the year 1925–1926 of " not less than 300 million rubles " : it was estimated that 80 millions could be provided by industry, and that 250 to 300 millions might be furnished from the budget.[2] At the beginning of May 1925 Gosplan set up a " special commission on the question of fixed capital " to coordinate the investigations of Vesenkha and of other departments working on the same question.[3] Meanwhile, STO established a further " special conference on the improvement of the quality of production ", which continued to work throughout the year and with which Trotsky actively associated himself.[4]

---

[1] Among other functions, it " published a series of works on different sectors of industry " (G. Krzhizhanovsky, *Sochineniya*, ii (1934), 296).

[2] A report on the conference is in *Planovoe Khozyaistvo*, No. 6, 1925, pp. 269-272.          [3] *Ibid.* No. 6, 1926, p. 253.

[4] Decision of STO of April 2, 1925 quoted in M. Saveliev, *Direktivy VKP(B) v Oblasti Khozyaistvennoi Politiki* (1928), p. 171 ; an address by Trotsky to the plenum of the conference was reported in *Ekonomicheskaya Zhizn'*, August 18, 1925.

The spring of 1925, when these ambitious projects were under active discussion, and when short-term credit flowed freely from the banks,[1] was a period of intense optimism in industry, and especially in heavy industry. The fourteenth party conference in April, which wrestled uneasily with the *kulak* problem in agriculture, could console itself with the favourable picture drawn by Dzerzhinsky of the situation on the industrial front. It was, said Dzerzhinsky, " the metal industry, its condition and its level ", which was decisive for " the level, the dynamic and the line of development of all other branches ". Much remained to be done. While industry as a whole had attained in 1923–1924 from 42 to 45 per cent of pre-war production and was expected to reach 65 to 70 per cent in the current year, the metal industry had attained only 30 per cent in the previous year, and was estimated to reach only 47 per cent in 1924–1925. But the blast furnaces of the Ukraine were now being restored ; automobiles and tractors were being manufactured for the first time in the current year ; and the foundations of an aviation industry had been laid in the creation of an Aviotrust. In the current year 70 million rubles were being invested in the metal industry. Dzerzhinsky submitted a three-year plan for industry providing for an 80 per cent increase in present levels of production. He himself was more optimistic, and believed that this increase could be achieved in one and a half or two years.[2] A special resolution of the conference " On the Metal Industry " sanctioned a further increase in the programme of 270 million pre-war rubles approved by STO in November 1924, and raised to 310 millions by the central committee in January 1925, to a grand total of 350 millions. Greater efficiency and greater economy must be practised everywhere, and those trusts which were still working at a loss must be set the " rigorous task " of making ends meet. Finally, a distinction was drawn between the long-term " perspective " plan for the metal industry and the new three-year " directive " plan. The latter was specifically approved, and the construction of new factories was declared a " priority task ".[3]

The same atmosphere of enthusiasm dominated the third

[1] For credit policy at this time see pp. 474-475 below.
[2] *Chetyrnadtsataya Konferentsiya Rossiiskoi Kommunisticheskoi Partii (Bol'-shevikov)* (1925), pp. 151-180, 208.
[3] *VKP(B) v Rezolyutsiyakh* (1941), ii, 24-25.

Union Congress of Soviets in the following month. This, declared Rykov in his opening speech, was " a turning-point in the sense that we do not intend to stop at what we have already achieved, or at what was achieved by the Russian national economy before the October revolution ".[1]  Hitherto recovery had been the watchword : lost ground had still to be made up.  Now the prospect of fresh advances began to appear.  Dzerzhinsky gave a somewhat broader and more popular view of industrial achievement than he had offered to the party conference.  The congress in its resolution called particular attention to the burning question of the moment — " the organization of long-term credit " for industry. This must be drawn, in the first instance, from the amortization reserves accumulated by industry itself : but industry was also encouraged to look both to the state budget and to the banks as sources of credit.  " The task of restoring the fixed capital of industry acquires the greater significance the more fully enterprises are loaded." [2]  The spring of 1925 marked the turning-point, first contemplated by Zinoviev a year earlier [3] and often spoken of by party leaders at this time, when recovery and restoration could be left behind and a programme of advance and new development taken in hand.  But this automatically implied not only a measure of planning, but of planning for consciously chosen ends.  It was time, an industrial leader told the Leningrad Soviet, " to introduce considerations not of capitalist utility, but of our socialist utility ".[4]

The problem of amortization was now seriously tackled for the first time, and began to enter into all calculations of the profits of industry.  Before 1921 destruction and wear-and-tear of plant and machinery had gone unchecked.  Nothing but the most essential repairs could be thought of ;  and, though no detailed statistics are apparently available before 1923, it seems clear that funds set aside for amortization during the first two years of NEP were insufficient to keep pace with the continuing deterioration, so that the value of fixed capital in industry continued to decline. The first serious attempt to deal with the problem of depreciation was made in the year 1923–1924, when the trusts working under

[1] *Tretii S"ezd Sovetov SSSR* (1925), p. 52.
[2] *Id. : Postanovleniya* (1925), p. 36.
[3] See *The Interregnum, 1923–1924*, p. 144.
[4] *Leningradskaya Pravda*, June 2, 1925.

the control of Vesenkha of the USSR showed a profit of 100 million rubles after setting aside an allocation of 130 millions for amortization.   In 1924–1925, when complete figures were available for the first time, profits of all trusts reached 436 million rubles, in addition to an allocation of 271 millions to amortization,[1] though it is doubtful whether even this sufficed to balance the long-standing process of deterioration.[2]   It was not till 1925–1926 that investment in industry effectively overtook depreciation, and that proper accounting processes, with adequate amortization allowances, became effective in state industry.

The revival of industry in 1924–1925 was bound up with certain local conditions and influenced by local pressures.   The disintegration of heavy industry, and especially the metal industries, during and after the civil war had affected the two areas where they had been mainly centred :  the Ukraine and Leningrad. The Ukraine, in particular, had been a seat of war, foreign occupation and anarchy for a prolonged period, during which its former place in iron and steel production had been partly usurped by the more strategically located industry of the Urals.   The first condition of a revival of Ukrainian heavy industry was a resumption of the largely disabled coal production of the Donbass region.   This was a constant preoccupation from 1921 onwards, and by 1924 serious — even spectacular — results had been achieved.   At the beginning of the economic year 1923–1924, an ambitious plan had been drawn up to raise 412 million puds of coal from Donbass during the year ;  by January 1924 it had been possible to raise the target to a minimum of 450 or a maximum of 500 millions ;

---

[1] See the table in S. G. Strumilin, *Ocherki Sovetskoi Ekonomiki* (1928), pp. 166-167 ; the corresponding figures for 1925–1926 were 614 million and 365 million rubles.

[2] According to an article in *Planovoe Khozyaistvo*, No. 4, 1926, pp. 146-156, amortization and fresh investment in industry did not overtake depreciation before 1925–1926, though S. G. Strumilin, *Ocherki Sovetskoi Ekonomiki* (1928), pp. 164-165, gives reasons for contesting this calculation.  Failure to adjust pre-war values to replacement values vitiated amortization figures before 1925–1926 :  " The profit of industry for the past period is largely fictitious, since in calculating amortization in 1923, 1924 and 1925 no account was taken of the change in the purchasing power of the chernovets ruble in comparison with the pre-war ruble " (*Torgovo-Promyshlennaya Gazeta*, March 18, 1926, quoted in *Sotsialisticheskii Vestnik* (Berlin), No. 6 (124), March 31, 1926, p. 8).   In the year 1925–1926 fixed assets acquired before October 1, 1923, were revalued at replacement cost on October 1, 1925, and carried in the books at the new figure (*Planovoe Khozyaistvo*, No. 8, 1939, p. 39).

by the end of the economic year the total amount mined had risen to 540 million puds — a figure which for the first time exceeded 50 per cent of the pre-war output of the region.[1]   Once, however, coal was once more freely available, the revival of Ukrainian iron and steel, organized in the mammoth trust Yugostal,[2] followed almost automatically ; for the rival industry in the Urals, before the opening of the coal-mines of the Karaganda basin in the nineteen-thirties, had no accessible supply of coking coal.[3]   Rivalry between Yugostal and the Urals was a burning topic at the sixth trade union congress in November 1924.   Some delegates thought the rivalry "unhealthy" and "menacing", and drew from it an argument in favour of planning.   Rykov, on the other hand, regarded it with official satisfaction :

> One of the positive merits of our new economic policy con-
> sists in testing out our work in market conditions. . . . We
> suffer not from an excessive amount of competition, but from
> too little competition, in our state industries.[4]

Since, however, a committee had been set up in October 1923 to regulate the proper distribution of state orders to the metal industries,[5] and since these industries were entirely dependent on such orders, it is clear that the ultimate decision was one of public policy, and that the revived heavy industry of the Ukraine was a powerful factor in the crucial year 1925 in promoting an expansion of state support for heavy industry.

The other former centre of heavy industry which had suffered from the prolonged industrial eclipse was Leningrad.   Here heavy industry had incurred less physical devastation than in the Ukraine, but had been subject to the same processes of decay : political influences had protected it from the full consequences of rationalization, but at the cost of efficiency.[6]   Since Leningrad had been especially associated with the metal industries, the low level to which they had sunk had been a blow to its prosperity

---

[1] L. Kamenev, *Stat'i i Rechi*, xi (1929), 139-140.
[2] See *The Bolshevik Revolution, 1917–1923*, Vol. 2, p. 307.
[3] For the difficulties of iron and steel production in the Urals in this period and the rapid revival of production in the Ukraine see *Sotsialisticheskoe Khozyaistvo*, No. 5, 1924, p. 80 ; *Promyshlennost' SSSR v 1925–1926 goda* (Vesenkha, 1927), p. 12).
[4] *Shestoi S"ezd Professional'nykh Soyuzov SSSR* (1925), pp. 263-264, 275, 287.    [5] *Ibid*. p. 288
[6] See *The Interregnum, 1923–1924*, p. 10, note 2.

and prestige scarcely less severe than the removal of the capital to Moscow : no city had so much at stake in their revival. With the decision of January 1925 for an intensified industrial expansion, Vesenkha set up a special commission to draw up a five-year plan for the development of industry in Leningrad.[1] Recovery had already set in. At one time, Zinoviev told the Leningrad Soviet in April 1925, Leningrad had been " looked on as a city with a glorious past, but without present or future " ; now, however, " we have once more assembled in our factories the workers who had gone away ".[2] In June 1925 particulars were published of the reports of the Vesenkha commission. At one time it had apparently been suggested that light engineering should be developed in Leningrad and heavy engineering removed elsewhere. But all thought of this was now abandoned. The commission recommended an investment of 290 million rubles spread over the five years, of which 150 millions would be drawn from local and internal resources and 140 millions from the union budget. On this basis it was estimated that industrial production in Leningrad at the end of the five-year period would rise to 130 per cent of its pre-war level, and that costs of production would fall from a present 175 per cent to 122 per cent of the pre-war level, allowing for a limited rise in real wages. At the same time Dzerzhinsky made a speech in Leningrad in which he declared that " the tasks laid down by the third congress of Soviets in connexion with the problem of the restoration of fixed capital relate first and foremost to Leningrad ", and attacked the " financial fetishism " of dependence on finance as " some kind of force lying outside ourselves ".[3] A week later the presidium of Vesenkha gave its approval to the plan drawn up by the commission, drawing particular attention to the need to expand the engineering and electrical industries, and raising the proposed total capital investment for the five years to 465 million rubles, of which, however, only 125 millions would come from the union budget.[4]

[1] *Leningradskaya Pravda*, January 27, 1925.
[2] *Ibid.* April 12, 1925 : for a further speech in the same vein see *ibid.* April 15, 1925.
[3] The recommendations of the commission and Dzerzhinsky's speech are both in *Leningradskaya Pravda*, June 18, 1925.
[4] *Ibid.* June 27, July 5, 1925 ; discussions of the plan filled the columns of *Leningradskaya Pravda* through the first half of July 1925.

The plan was a landmark in the revival of Soviet heavy industry and in the resuscitation of Leningrad.

Throughout the summer of 1925 the question of long-term credit for industry was a burning issue. The project of an " industrial fund " elaborated by Vesenkha in the previous March had apparently undergone modification, and now appeared in two alternative forms. The first was for a fund to be administered by Vesenkha under the supervision of STO, and to be built up by allocations from the profits of industry, by grants from the budget, and by state loans and loans from the banks. This proposal encountered strong opposition, being branded as a return to *glavkism*, i.e. to the financing of industry directly by Vesenkha instead of through the banks.[1] The alternative proposal, endorsed by Gosplan, was to entrust the task of financing industrial expansion to Prombank. This raised the objection that the same institution should not be asked to handle both long-term and short-term credit.[2] By way of compromise, the issue of a state loan of 300 million rubles for the explicit purpose of financing industrial reconstruction was announced in the middle of August.[3] This device was less helpful than it seemed. In spite of the official abandonment of the practice of forced subscriptions to state loans,[4] no funds were forthcoming for voluntary subscriptions, and state enterprises were required to take up their quota of the new industrial loan. Since subscribers were allowed to pay for the bonds by instalments, and since they could only meet the instalments by mortgaging the bonds with the State Bank, the transaction merely made the state a partner in the indebtedness of industry to the banks without increasing its total volume,[5] and illustrated the fundamental dilemma of industrial expansion — inability to find in any part of the national economy savings to

[1] *Planovoe Khozyaistvo*, No. 8, 1925, pp. 23-25    for *glavkism* see *The Bolshevik Revolution, 1917–1923*, Vol. 2, pp. 177-182.

[2] *Planovoe Khozyaistvo*, No. 6, 1925, p. 254 ; *Ekonomicheskoe Obozrenie*, September 1925, p. 18.

[3] *Sobranie Zakonov, 1925*, No. 53, art. 398.        [4] See p. 470 below.

[5] In Sokolnikov's words, the loan " rests on deposits accumulated in the banks, and represents a form of state guarantee of the long-term debts of enterprises " (G. Sokolnikov, *Finansovaya Politika Revolyutsii*, iii (1928), 211) ; Rudzutak at the fourteenth party congress taunted Sokolnikov with having supposed that ." by re-naming short-term credits long-term credits we can really raise a loan for economic recovery " (*XIV S"ezd Vsesoyuznoi Kommunisticheskoi Partii (B)* (1926), p. 330).

finance capital development.  But, while it solved no problem, it
was a signal of continued encouragement to industry to expand
and to the banks to pursue a generous credit policy in aid of the
expansion.  The warning light of inflation was still ignored.[1]

Under the impetus of these powerful driving forces, the year
1924–1925 was one of record achievement for Soviet industry.  It
consolidated the rapid recovery of consumer industries in the two
preceding years ;  it laid the foundations of a correspondingly
spectacular recovery in the capital goods industries ;  and it prepared the way for fresh development beyond the limits and levels
of pre-war Russian industry.  According to Gosplan calculations,
the total value of the production of " census " industry in terms
of pre-war rubles rose from 2627 millions in 1923–1924 to 4000
millions in 1924–1925 — an increase of 54 per cent.[2]  According
to the slightly lower returns of Vesenkha, the production of consumer industries rose in terms of value from 788 million pre-war
rubles in 1923–1924 to 1318 millions in 1924–1925, and the production of heavy industry from 1620 millions to 2642 millions ;
within this category the production of capital goods rose from 820
millions to 1312 millions.  The production of iron and steel in
terms of quantity almost doubled during the year.[3]  While, however, the increase in the production of capital goods had, for the

[1] For the financial aspects of the situation see pp. 475-481 below.  Trotsky
afterwards claimed to have given warnings on June 12, and again on June 24,
1925 (in what form is not clear), that the drive for expansion in certain sectors
of industry threatened to produce a financial and credit crisis, and that " industry,
or at any rate certain branches of it, is expanding beyond our means " (see his
speech at the party central committee of April 1926 preserved in the Trotsky
archives).  Dezen, an official of Narkomfin, who regarded the loan with mis-
givings, wrote that " the character of the loan and the possibility of placing it
do not in themselves exclude the inflationary danger ", but that " in view of the
impossibility of enlarging the fixed capital of industry in the normal way " the
risk must be taken (*Ekonomicheskoe Obozrenie*, September 1925, pp. 16-17) ;
like most officials of Narkomfin, Dezen believed in " normal " methods of
finance and regarded large-scale industrial expansion as impossible without
foreign capital (*ibid.* p. 22).

[2] *Kontrol'nye Tsifry Narodnogo Khozyaistva na 1926–1927 god* (1926),
p. 321 ; virtually identical totals are given by the central statistical administration in *Itogi Desyatiletiya Sovetskoi Vlasti v Tsifrakh, 1917–1927* (n.d.), p. 230.

[3] These figures are in *Promyshlennost SSSR v 1925–26 godu* (Vesenkha,
1927), pp. 15, 20-21.  Other figures give similar results : a table in *Sotsialis-
tichiskoe Khozyaistvo*, No. 5, 1925, p. 27, shows a rise in the production of
state industry from 1553 million pre-war rubles in 1923–1924 to 2524 millions
in 1924–1925 and in sales of state industry from 1278 millions to 2290 millions

first time, kept pace with the increases in the production of the consumer industries, this did not mean that the legacy of past neglect had been overcome. The basic metallurgical industries still lagged far behind. In 1924–1925 the volume of production of textiles had reached 66 per cent of the 1913 total; of salt, 57 per cent; of matches, 85 per cent; of cigarettes, 102 per cent; of coal, 55 per cent; of oil, 76 per cent. In the same year the production of iron ore was no more than 23·8 per cent of the 1913 total ; of pig iron, 31 per cent; of steel, 43·8 per cent ; of rolled metal, 38 per cent.[1]

It was in these conditions that Gosplan produced, in August 1925, its first " control figures of the national economy " for the ensuing economic year.[2] Restrained optimism prevailed. A repetition of the phenomenal increase of total industrial production by 48 per cent in 1924–1925 could not be expected. The total increase for 1925–1926 was set at 33 per cent. But the increase was not uniformly spread. A more rapid rate of increase was predicted for large-scale (46 per cent) than for small-scale (26 per cent) or artisan (8 per cent) industry, and the highest rates of all for heavy industry. Thus, while the production of textiles was to rise in 1925–1926 by 42 per cent, of salt by 15 per cent, and of matches by 10 per cent, the corresponding rates of increase for the metal and electrical industries were 63 per cent and 73 per cent. This would raise the total of industrial production for the year at pre-war prices to 89 per cent of the 1913 level, the metal industries reaching 90 per cent and the textile industries 92 per cent.[3] In the event the global increase of industrial production forecast by

— increases of 62·5 per cent and 77·5 per cent respectively. According to another calculation in calendar years and chervonets rubles, the production of heavy industry rose from 4660 million rubles in 1924 to 7739 millions in 1925, the production of capital goods rising from 2109 millions to 3356 millions, and of consumer goods from 2551 millions to 4383 millions (*Sotsialisticheskoe Stroitel'stvo SSSR* (1936), p. 2).

[1] *Kontrol'nye Tsifry Narodnogo Khozyaistva na 1926–1927 god* (1926), p. 320. Similar, though not identical, figures for pig iron, steel and rolled metal are given in *Ekonomicheskoe Obozrenie*, October 1927, p. 110 ; lower figures are apparently used for 1913, giving a less unfavourable percentage for the later years (see p. 332, note 3 above).

[2] For the production of these figures and the reception accorded to them see pp. 500, 503-505 below.

[3] *Kontrol'nye Tsifry Narodnogo Khozyaistva na 1925–1926 god* (1925), pp. 17-18, 79.

Gosplan was attained, but the share of large-scale industry, and of heavy industry in particular, in that increase proved to have been over-estimated. The " spontaneous " forces were still more powerful than the efforts of the planners.

The production of the control figures was the high-water mark of the wave of optimism which had swept over the whole Soviet economy in the first half of 1925. In the autumn the results of the harvest called for the usual reappraisal of the economic situation, and led to a strong reaction. The hesitations of the previous autumn were renewed, though the causes and symptoms were radically different. The baffling difficulties of the grain collection, the unmistakable signs of a credit and currency crisis, and the widespread scepticism and hostility provoked by the publication of so great a novelty as the control figures combined to create an atmosphere unpropitious to the claims of an expanding industry. Expedients continued to be canvassed. A scheme for a special bank to finance state industry was mooted ; [1] and Narkomfin produced a plan for a special department for long-term credit to be set up in Prombank.[2] Zinoviev, swinging rapidly over to the side of industry, insisted that without a re-tooling of existing factories or the construction of new factories no further expansion was possible : " every usable bench is fully loaded ".[3] Kamenev echoed that " at the end of this year there will not be in Soviet territory a single factory, a single workshop, not working to full capacity, not a single enterprise not carrying its full pre-war load ".[4] But it had now become impossible to evade the basic issue of the sources from which capital for investment in industry could be drawn and, therefore, of the rate of investment in industry. Sokolnikov represented the natural reaction of Narkomfin to a policy which was placing an intolerable strain on financial resources and threatened the stability of the currency. In the autumn of 1925 the advocates of caution were not short of arguments. The cancellation of the export programme for grain

[1] *Ekonomicheskaya Zhizn'*, October 3, 4, 1925 ; the scheme originally emanated from an official of Narkomfin (*Ekonomicheskoe Obozrenie*, September 1925, p. 18).                [2] *Ekonomicheskaya Zhizn'* October 27, 1925.
[3] *Leningradskaya Pravda*, November 11, 1925.
[4] L. Kamenev, *Stat'i i Rechi*, xii (1926), 571.

meant, as Sokolnikov explained at the beginning of October, a
reduction in imports of raw materials, semi-manufactured goods
and consumer goods ; and " in connexion with this we shall have
to revise to some extent a whole series of our plans for the development of industry and proceed at a more cautious pace ". Sokolnikov propounded a philosophy of restraint :

> The difficulties that have arisen on this path lead up to the
> general question of methods of planning. They show that, after
> all our measures, we should none the less avoid excessive
> enthusiasm. If the Soviet state disposed of large reserves, if
> we really already had far more solid ground beneath our feet,
> we could more boldly pursue a policy of economic development.
> But, as it is, we must appeal for greater caution.[1]

A few weeks later he attacked with still greater vehemence the
demand for a further expansion of credit to meet the crisis. The
proposal to cure the goods hunger by a further deterioration in
our currency position " could, of course, yield no positive results " ;
and " a whole series of our plans for the development of industry
must be subjected to revision ". A studiously vague note of
warning was sounded on the prospects of the loan for industrial
reconstruction :

> Even if there is no reason to doubt that it will be possible
> to realize it in its entirety in the current [financial] year, a
> certain elasticity will have to be introduced . . . in the dates
> of fulfilment of the loan, in connexion with which a partial
> postponement for a month or two of financing on the basis of
> the loan for economic reconstruction appears indispensable.[2]

In the period of uneasy uncertainty which preceded the
fourteenth party congress, and so long as compromise between
the leaders still seemed probable or possible, nobody was in a
hurry to grasp this nettle. The seventh all-union congress of the
trade union of metal workers, which met in Moscow on November
17, 1925, had evidently been planned as a demonstration of the

---

[1] G. Sokolnikov, *Finansovaya Politika Revolyutsii*, iii (1928), 19 ; this article
was first published in *Vestnik Finansov*, No. 10, October 1925, pp. 3-22.

[2] G. Sokolnikov, *Finansovaya Politika Revolyutsii*, iii (1928), 40-43 ; this
article, in its original form apparently a speech at a Narkomfin conference,
was published in *Vestnik Finansov*, No. 11-12, November-December 1925,
pp. 3-15.

new drive for the supremacy of heavy industry, and received much publicity. But Dzerzhinsky, who made the major governmental speech, confined himself to eloquent platitudes ; and Zinoviev, the only top-ranking leader to appear at the congress, devoted the major part of his speech to the international situation.[1] The opponents of industrialization seemed momentarily in the ascendant. Shanin, the " Narkomfin professor ",[2] wrote an article which stated the case against industrialization in its extreme form. Industry, in Shanin's view, was " developing too fast and also developing incorrectly ". The current crisis arose from attempting to force capital development at a time when the demand for consumer goods still exceeded supply. The first essential of economic policy was to increase agricultural exports ; and this end could be promoted by fostering " those branches [of production] which serve agricultural export ", i.e. processing industries and industries producing consumer goods.[3] But, though Shanin's article appeared in the official economic journal as a " discussion article " without editorial endorsement, a speech delivered by Sokolnikov to officials of Gosbank revealed him as a convert to similar views :

> The more rapid development of agriculture in comparison with industry can in no way be a handicap to the economic development of the country ; on the contrary, it is a fundamental condition of its more rapid economic development. Contradictions between the levels attained by industry and agriculture must be resolved by going to the foreign market, and realizing the surplus of agricultural raw materials on the foreign market in order to organize the import of capital.[4]

Preobrazhensky broke silence with a pessimistic article in which he expressed the fear that the goods shortage would increase, and that, instead of seeking the fundamental cure of increasing capital accumulation in industry " at the expense of the whole economy of the country ", many would be found to advocate " *the line of least resistance* " and intensify imports of consumer goods.[5]

---

[1] The opening of the congress was reported in *Pravda*, November 18, 1925, Dzerzhinsky's speech on November 22, 24, Zinoviev's on December 1.
[2] See p. 316, note 3 above.
[3] *Ekonomicheskoe Obozrenie*, November 1925, pp. 25-40.
[4] *Pravda*, December 1, 1925.      [5] *Ibid.* December 15 1925.

As the fourteenth party congress approached, therefore, it seemed scarcely likely that the congress would prove a decisive landmark in the process of industrialization. Stalin, now firmly in command of the powerful levers of the party machine, rallied to the support of the peasant against the attacks of Zinoviev and Kamenev, the new-found champions of industry ; and the advocacy of Trotsky continued to be compromising to policies of out-and-out industrialization and comprehensive planning. Yet, in spite of these omens, the congress earned its place in party history under the name of " the congress of industrialization ". The logic of socialism in one country compelled Stalin to call a halt to the policy of concessions to the *kulak* and, almost in spite of himself, carried him over to the camp of the industrialists. In effect, this issue cut across the lines of division which had established themselves in the personal struggle. On one side Stalin found allies in Bukharin, Rykov and Kalinin, all of them supporters of the peasant cause. Bukharin's embarrassing attempt at the congress to reconcile socialism in one country with his defence of the peasant by a theory of " snail's pace " industrialization was long remembered, and exposed him to much contumely and derision in later years :

> We came to the conclusion that we could build socialism even on this wretched technological level . . ., that we shall move forward at a snail's pace, but that all the same we shall be building socialism, and shall build it.[1]

But on the other side the opposition was involved in still graver inconsistency. For it consisted not only of Zinoviev and Kamenev, who were moving towards an industrialist policy, even at the price of a *rapprochement* with Trotsky, but also Sokolnikov, the only prominent figure in the party who openly advocated grain exports and imports of consumer goods, and relegated industrialization to an indefinite future. It was noticeable that Stalin, in his reply to the debate at the congress, refuted the heresies of Sokolnikov with far greater vigour and conviction than those of Kamenev and Zinoviev, reiterating his determination not to allow the USSR to be " converted into an agrarian country for the benefit of any other country whatever ", and to put it in a position to " produce

[1] *XIV S"ezd Vsesoyuznoi Kommunisticheskoi Partii (B)* (1926), p. 135.

machines and other instruments of production ".[1] The main resolution of the congress announced the decision, as one of " the fundamental propositions " by which the policy of the central committee was to be guided, " to conduct economic construction from the standpoint of converting the USSR from a country that imports machines and equipment into a country that produces machines and equipment, in order that the USSR may not, in the circumstances of a capitalist environment, be converted into an economic adjunct of the capitalist world economy, and may constitute an independent economic unit in course of construction on socialist lines ". The resolution went on to record, as its first " directives in the field of economic policy ", the determination of the party and the government to " give first place to the task of securing by every means the victory of socialist forms of economy over private capital ", to " ensure the economic independence of the USSR in such a way as to protect the USSR from becoming an adjunct of a capitalist world economy ", and to " pursue a policy aimed at the industrialization of the country, the development of the production of means of production and the formation of reserves for economic manœuvre ".[2] Everything in the execution of the policy remained vague and subject to controversy. But, after the fourteenth party congress, certain principles were fairly established in party doctrine and could no longer be openly contested. In the first place, the remedy for the ," disproportion " in the Soviet economy was to be sought not through the development of agriculture, but through the expansion of industry. Secondly, industrialization was to proceed first and foremost through the development of the means of production, i.e. of capital goods industries as the broad basis on which consumer goods industries could be ultimately expanded. Thirdly, industrialization was to be financed out of internal resources, and had as its overriding purpose the conversion of the USSR into an economically powerful and self-sufficient unit. Industrialization was the economic corollary of socialism in one country.

If, however, the opponents of industrialization could no longer give battle on the issue of principle, its supporters showed no

---

[1] Stalin, *Sochineniya*, vii, 355 ; the debates of the congress will be described in Part III in the following volume.

[2] *VKP(B) v Rezolyutsiyakh* (1941), ii, 48-50.

immediate eagerness to convert principles into practice. It was perhaps embarrassing that the only leading party figure to speak with enthusiasm of the resolution of the congress on " the growth of socialist state industry " was Trotsky, who repeated that " to raise the peasant without a rise in industry is impossible ".[1] Controversy now turned, not on the question whether to industrialize, but on what rate of industrialization was practicable and desirable ; tempo became the catchword of Soviet economic vocabulary. Narkomfin proposed to reduce the total of the 300 million ruble loan, which had run into serious difficulties, to 225 millions : a compromise figure of 240 millions was eventually approved by STO.[2] On March 24, 1926, STO fixed the plan of industrial production for the current year at a total of 5050 million chervonets rubles or 3020 million pre-war rubles. This was higher than the control figures of the previous summer, but apparently lower than had been foreseen in some intermediate projects.[3] The writer of a leading article in the current number of the party journal *Bol'shevik* found it difficult to reconcile such adverse phenomena as rising prices and symptoms of inflation with the evident fact that " our economy is in a state of expansion, not of decline ", but concluded that the remedy lay not in Shanin's " agrarianization " but in further industrialization.[4] The growth of this somewhat bewildered faith in the virtues of industrial expansion was characteristic of the new mood.

The session of the party central committee of April 6-9, 1926, while it appeared to be concerned primarily with agricultural policy, was strongly marked by the growing emphasis on industrialization. Trotsky's reappearance on the party scene was significant of the trend. To finance the expansion of industry was the motive of Trotsky's demand for a stiffening of the agricultural tax, and he reproached Kamenev for not seeing that the problem of differentiation in the countryside could not be solved

[1] Trotsky's speech to the Moscow provincial congress of the textile workers' trade union was published in *Pravda*, January 31, February 2, 1926 ; it was uncontroversial in tone.        [2] *Ekonomicheskaya Zhizn'*, March 21, 1926.

[3] *Ibid*. March 25, 1926. This was said to represent a 40 per cent increase on the previous year ; but the different classification of industries makes precise comparison with the Gosplan figures of August 1925 (see pp. 347-348 above) impossible.

[4] *Bol'shevik*, No. 5, March 15, 1926, pp. 3-8 ; for Shanin's article see p. 351 above.

except through progressive industrialization.[1] There was some discussion of the grandiose project to construct a dam on the Dnieper (afterwards famous under the name Dnieprostroi) for the generation of electrical power. As president of the commission on electrification, Trotsky was an enthusiastic promoter of this project. Stalin came out on the side of caution, and made some remarks which Trotsky later published and held up to opprobrium :

> The means required here are enormous, some hundred millions. We should be falling into the position of a peasant who had saved up a few kopeks and, instead of repairing his plough or renewing his stock, bought a gramophone and ruined himself. Can we fail to take account of the decision of the congress that our industrial plans must correspond with our resources ? Comrade Trotsky, however, evidently does not take account of this decision of the congress.[2]

Trotsky's main speech and his " amendments " to Rykov's resolution [3] revolved round the central theme that " state industry is lagging behind agricultural development " and that " our fundamental economic difficulties arise from the fact that *the volume of industry is too small* ". His recommendations were familiar — to make the agricultural tax more progressive and wholesale prices " more flexible ", to curtail unnecessary expenditure, remembering that " we have not yet emerged from the stage of primitive socialist accumulation ", to increase long-term credits to industry and capital investment in industry, to press forward with electrification, and to intensify the application of planning. The resolution

---

[1] Trotsky had expressed the same view in an introduction written in November 1925 for the English translation of his pamphlet, *Towards Socialism or Capitalism ?* : " We have every reason to anticipate that, with proper guidance, the growth of industry will keep ahead of the process of stratification among the peasantry and thus neutralize it, creating a technical base and economic possibilities for a gradual transition to collective farming " (L. Trotsky, *Towards Socialism or Capitalism ?* (1926), p. 11). The argument evidently attracted attention ; Trotsky recalled it in his letter of July 12, 1928, to the sixth congress of Comintern, a copy of which is in the Trotsky archives (translation in L. Trotsky, *The Third International after Lenin* (N.Y., 1936), p. 281).
[2] Quoted from the unpublished records of the committee in *Byulleten' Oppozitsii* (Paris), No. 29-30, September 1932, p. 34 (the speech is incorrectly dated April 1927, *ibid.* No. 19, March 1931, p. 17, and No. 27, March 1932, p. 3).      [3] See p. 325, note 3 above.

in its final form went far enough in its support of industry to secure unanimous acceptance. It contrasted the first years of NEP, when the revival of agriculture had been the main concern, with the present period of " disproportion " in the growth of the economy, when " the development of industry and, in general, the industrialization of the country is the decisive task, the fulfilment of which determines the further progress of the economy as a whole on the road to the victory of socialism ". In spite of difficulties, a rosy picture was painted of the prospects of industry :

> Industry is growing far more vigorously than other branches of the economy of the USSR. The general production of industry in comparison with that of the previous year is being once more increased, approximately by 30 to 40 per cent. For the first time in the present year substantial investments are being made in industry for re-equipment and for new con- struction, which determines the possibility of the expansion of industry in succeeding years.

The only specific decision on aid for industry was the endorse- ment of an unpublished and apparently otherwise unrecorded decision of the Politburo of February 25, 1926, on " capital ex- penditure in industry, the state budget and the creation of a special reserve fund in the state budget ".[1]

The moral was promptly driven home. Stalin, speaking on the results of the session to a Leningrad party meeting (an audi- ence which would be predisposed to welcome emphasis on industry), referred to the " fundamental slogan " of industrializa- tion proclaimed by the fourteenth party congress. It was not enough, however, to speak of developing industry : even colonial countries had industries. What was essential was the development of heavy industry. The conclusions were " to advance the industry of our country as the foundation of socialism and the driving force which carries forward the whole national economy ", " to increase the tempo of our socialist accumulation ", and " to ensure the correct utilization of the reserves which are being accumulated and establish the strictest régime of economy ".[2]

---

[1] *VKP(B) v Rezolyutsiyakh* (1941), ii, 91-97 ; for the sections of the resolu- tion on agriculture and on planning see p. 327 above and p. 512 below.
[2] Stalin, *Sochineniya*, viii, 119-122, 147.

Those who listened to such phrases must have assumed that the battle of industrialization had been won ; and this was in part, though only in part, true. The problem of the rate of capital accumulation, and of the sources from which it was to be drawn, had become all-important. But the leaders were no more eager than of old to face the issue.[1] The struggle between agriculture and industry, and between different agrarian policies, which had marked the earlier stages of the controversy was soon to be resumed in the new setting with unabated acrimony. Progress had, however, in fact been made. Industry continued to advance, while the same protagonists on both sides exchanged the same arguments about the tempo of industrialization.

While the major problems of industrial recovery turned on large-scale industry working under direct state management, two other forms of industrial production remained important. In the first place, state enterprises not directly operated by the state were leased to private individuals or cooperatives ; these were included in the category of census industry. Secondly, outside the category of census industry, small private industrial concerns continued to operate in the towns,[2] and rural industries played an extensive part in the trade of the countryside, though this must have been extraordinarily difficult to evaluate with any approach to accuracy. According to Gosplan statistics, the production of both these forms of industry continued to increase during the period of recovery, though the percentage of both in total production somewhat declined. A table prepared in Gosplan early in 1926 gave a

[1] The superficial optimism of the official line was well displayed in an address to the Communist Academy on April 27, 1926, by Milyutin, People's Commissar for Finance of the RSFSR, who re-hashed the old themes, branding Preobrazhensky and Sokolnikov respectively as spokesmen of the two deviations (*Vestnik Kommunisticheskoi Akademii*, xvi (1926), 216-227).

[2] The original limitation of private enterprise to concerns employing not more than five workers with mechanical power or ten workers without (see *The Bolshevik Revolution, 1917–1923*, Vol. 2, p. 174) had been raised in 1921 to 20 workers (see *ibid.* Vol. 2, p. 300) ; according to a Vesenkha report quoted in *Chastnyi Kapital v Narodnom Khozyaistve SSSR*, ed. A. M. Ginsburg (1927), p. 33, " hundreds " of private enterprises exceeded this limit. By a decree of the RSFSR of May 1925 private concerns were allowed in certain conditions to employ up to 100 workers (Y. S. Rozenfeld, *Promyshlennaya Politika SSSR* (1926), p. 494).

comprehensive picture of industrial production in millions of chervonets rubles : [1]

|  | Census Industry | | | Rural and Artisan Industry | Total |
|  | State | Cooperative | Private | | |
| --- | --- | --- | --- | --- | --- |
| 1923–1924 | 3346 | 204 | 195 | 1668 | 5,414 |
| 1924–1925 | 4985 | 394 | 252 | 1935 | 7,567 |
| 1925–1926 (estimates) | 7100 | 500 | 291 | 2322 | 10,214 |

Thus, while the production of the cooperative and private sectors of census industry, and of rural and artisan industry, continued to rise, it declined relatively to total production ; and this decline was more marked in the private than in the cooperative sector. The official attitudes varied from grudging toleration to active support.  A commission of Vesenkha set up in February 1926 to study the position of private capital in industry and trade reported that " in the present position of the national economy the existence of a certain body of private industrialists and traders is unavoidable, and the task of the state is to utilize productive elements in a practical way in the interest of a broadening and cheapening of production ".[2]

The leasing of industrial enterprises by the state to private individuals or cooperatives willing and able to exploit them had been practised on a modest scale since the early days of NEP.[3] According to the latest regulation issued by Vesenkha on September 10, 1924, enterprises could be leased up to a maximum of 12 years — the same limit adopted for the leasing of land.[4]  The leased enterprises were almost invariably small, so that their total

---

[1] *Planovoe Khozyaistvo*, No. 2, 1926, p. 122.  A table in *Itogi Desyatiletiya Sovetskoi Vlasti v Tsifrakh, 1917–1927* (n.d.), p. 284, relating to census industry alone, gives somewhat higher totals (presumably owing to differences of classification) : according to these figures, the proportion of state industry in terms of value produced remained constant in the years 1923–1924 to 1925–1926 at about 90 per cent of all census industry, while the proportion of the cooperatives rose slightly at the expense of the private sector.

[2] *Chastnyi Kapital v Narodnom Khozyaistve SSSR*, ed. A. M. Ginsburg (1927), p. 35.

[3] See *The Bolshevik Revolution, 1917–1923*, Vol. 2, pp. 301-302.

[4] Y. S. Rozenfeld, *Promyshlennaya Politika SSSR* (1926), pp. 499-500.

weight in the economy was never significant.[1]  Out of a total of
6500 leased enterprises on March 1, 1924 (it is not clear that all
were in effective operation), 30 per cent were in the food industry,
24 per cent in the leather industry ; half of them were leased to
private individuals (half of these being the former owners), the
remainder to *artels*, cooperatives or institutions.[2]  The origin of
one such enterprise is recorded from the environs of Moscow.
In August 1924 a new factory producing screws, bolts and nails
was started under the name *Proletarskii Trud* by combining two
former factories in the hands of lessees, possibly the previous
owners of one of them.  The project was approved by the Moscow
provincial Vesenkha, by the metal workers' trade union, and
finally by the provincial planning department (Gubplan), before
the necessary credits, amounting in all to 100,000 rubles, were
granted.  The lessees had to demonstrate that they were capable
of restoring the factory to production and could do it more quickly
and better than a state concern.  The factory started in August
1924 with 192 workers, and in October 1925 was employing 653.[3]
This account is probably typical of the empirical and somewhat
haphazard methods by which industrial production in small units
was built up again after the collapse of 1918–1921.  Private
industry had its share in the industrial revival of the middle
nineteen-twenties ; in 1925 it was employing 36 per cent more
workers in the industrial region of Moscow than in the previous
year, though this still accounted for only 12–13 per cent of all
industrial workers in the region.[4]  In the long run, however,
leased industries faced an awkward dilemma.  If they were

[1] According to S. G. Strumilin, *Ocherki Sovetskoi Ekonomiki* (1928), p. 179,
they accounted, in terms of capital invested, for only 1·1 per cent of all capital
invested in industry in 1925 ; according to Y. S. Rozenfeld, *Promyshlennaya
Politika SSSR* (1926), p. 504, they employed less than 5 per cent of all labour
employed in state industry, but produced 8 per cent of the output.  These
figures show that, as was to be expected, the proportion of fixed capital employed was far less in leased enterprises than in state industry.

[2] *Ibid.* pp. 502-503.

[3] Supplementary bulletin issued with *Planovoe Khozyaistvo*, No. 12, 1925,
pp. 6-7.

[4] *Bol'shevik*, No. 14, July 30, 1926, pp. 36, 43.  Workers in private factories
received higher wages than in state industry, but were politically more backward ; " patriarchal relations " were said to exist between employers and
workers, religious holidays were observed, and the workers were " often stubbornly convinced that state industry would inevitably be beaten by private
capital if the latter were ' given its head ' " (*ibid.* pp. 40-43).

unsuccessful, they failed to earn profits and collapsed; if they were successful, they were sooner or later likely to be taken over by the state or the region.[1]

Rural and artisan industries, ranging from the part-time output of peasant households to small enterprises employing hired labour or organized as *artels* or industrial cooperatives,[2] made a larger contribution than the leased enterprises to the sum of industrial production. Such industries, though they almost entirely escaped central control, had enjoyed official approval and encouragement since the early days of NEP, and had regained a measure of prosperity more rapidly than factory industries.[3] Industrial cooperatives, which represented the highest form of organization of artisans and rural workers, but covered only a fraction of them,[4] were exempted by article 57 of the civil code from the limitation on the number of hired workers imposed on " private " industrial concerns,[5] but never rivalled the agricultural cooperatives (and still less the consumer cooperatives) in importance. In 1925 they were organized in the RSFSR in four large cooperative unions, a general All-Russian Union of Industrial Cooperatives, which included leather-workers, textile workers, metal workers, woodworkers and builders, and special unions for forestry, fishing and hunting. The cooperatives organized in these unions were said to have about 600,000 members on October 1, 1925. Industrial cooperatives in the Ukraine had 65,000 members, and there were incipient industrial cooperatives with a few thousand members in White Russia, Transcaucasia and Uzbekistan.[6] These enterprises had from the first been regarded with jealous hostility both by the trade unions and by the spokesmen of large-scale industry.

---

[1] *Chastnyi Kapital v Narodnom Khozyaistvo SSSR*, ed. A. M. Ginsburg (1927), pp. 31-32.

[2] According to an article in *Pravda*, March 24, 1925, from 80 to 90 per cent of rural industrial enterprises were either smithies or village mills.

[3] See *The Bolshevik Revolution, 1917-1923*, Vol. 2, pp. 297-300, 310.

[4] According to figures in *Kontrol'nye Tsifry Narodnogo Khozyaistva na 1928-29 god* (1928), pp. 424-425, private enterprises in 1925-1926 accounted for 19·9 per cent of all industrial production and industrial cooperatives for 8·2 per cent ; since a considerable proportion of industrial cooperatives worked in census industry (see table on p. 358 above), the predominance of private enterprise in rural industries must have been very great.

[5] See p. 357, note 2 above.

[6] These particulars are taken from a detailed, and obviously official, account of industrial cooperatives in *Ekonomicheskoe Obozrenie*, April 1926, pp. 134-146.

Like the agricultural cooperatives, the industrial cooperatives operating in rural industry were frequently accused of being merely a cloak for the revival of the type of petty capitalism especially associated with the *kulak*.[1]

It was logical that, at a moment when agricultural policy was moving in favour of the well-to-do peasant, these predominantly rural forms of industrial production should also have attracted renewed attention. On April 10, 1925, a decree was issued granting tax exemptions to workers in rural industries and to craftsmen and individual artisans in towns.[2] A leading article in *Pravda* on the following day described the falling off in rural industries as one of the causes of agrarian over-population, and hence of unemployment. At the fourteenth party conference in April 1925, with the defence of the *kulak* usurping a central place in party preoccupations, Rykov launched a campaign on behalf of industrial cooperatives and rural industries. Industrial cooperatives had been neglected because " our economic organs have. often regarded, and even now still regard, industrial cooperation as a rival ". It was important to develop them if only to employ the surplus population which would otherwise " press on the cities and increase the reserve industrial army ". Rykov denounced trade union jealousies of rural industries :

Some of our workers in the trade union movement do not understand that the rural craftsman, whose whole budget consists of his earnings in rural industry, is certainly no further removed from the working class than the peasants who flock to the city for temporary work and are organized by the trade union.[3]

The resolution of the conference demanded greater attention than hitherto to " questions of rural industry and cooperatives of rural workers and craftsmen ", who must not be treated as non-workers and deprived of electoral rights. State industry was to work in conjunction with rural industries, and not conclude " oppressive

[1] See *ibid.* June 1927, pp. 109-110, for a development of this thesis ; according to a later source, 47 per cent of all workers engaged in industrial cooperatives in 1926 were hired workers (A. Arutinyan and B. L. Markus, *Razvitie Sovetskoi Ekonomiki* (1940), p. 209).
[2] *Sobranie Zakonov, 1925*, No. 25, art. 168.
[3] *Chetyrnadtsataya Konferentsiya Rossiiskoi Kommunisticheskoi Partii (Bol'-shevikov)* (1925), pp. 93, 149.

contracts " with rural workers.[1]  A few days later the third Union
Congress of Soviets passed a resolution in favour of assistance
for small-scale and rural industry in the form of " the granting
of credit, the provision of raw materials, half-finished products
and material, the marketing of products of rural industry, the
encouragement of workers' cooperatives etc.".[2]  With such
encouragement small-scale private industry continued to prosper
and participated in the general recovery, though an absolute
increase in production represented a relative decline in the proportion of private production to that of the state and cooperative
sectors.

[1]  *VKP(B) v Rezolyutsiyakh* (1941), ii, 19-21.
[2]  *Tretii S"ezd Sovetov SSSR : Postanovleniya* (1925), p. 26.

CHAPTER 7

# LABOUR

THE labour situation of the middle nineteen-twenties showed in most respects a slow and modest, but undeniable, improvement. The worst wages scandals of 1923 finally disappeared with the stabilization of the currency. The drive for higher productivity, steadily and remorselessly pursued, achieved a measure of success and laid the foundation for a progressive restoration and expansion of industry, especially heavy industry; better labour discipline as well as technical improvements contributed to this result. On the other hand, the wage structure, now no longer controlled by the state, remained chaotic and encouraged jealousies between different branches of industry and different regions ; the machinery for the settlement of conflicts was overburdened, and cumbrous in operation ; and the trade unions found it to be more and more difficult to combine their rôles as loyal instruments of state and party and as representatives of the group interests of the workers. Above all, no solution was found, or appeared to be in sight, for the problem of mass unemployment, which continued without abatement through this period, and dominated the labour situation. Fear of dismissal still largely replaced other forms of discipline ; and the abundance of manpower had the usual effect of concentrating attention on the increase of individual output rather than on improvement of technical means of production.

The number of unemployed rose steadily during the first half of 1924, those registered at 70 labour exchanges increasing from 754,000 on January 1 to 822,000 on July 1, corresponding to an estimated total of 1,240,000 registered unemployed on January 1 and 1,340,000 on July 1.[1] At the sixth trade union congress in

[1] *Sotsialisticheskoe Khozyaistvo*, No. 4, 1925, p. 413.

November 1924 Shmidt repeated the familiar explanations which, however true, did little to mitigate the harsh realities of the problem.  Of the numbers registered on July 1, 25 per cent had never worked for wages at all ; 17 per cent had worked for less than three years ; others were office workers or peasants recently arrived from the country ; only 300,000–400,000 were genuine unemployed industrial workers.[1]  In July 1924 the labour exchanges repeated the operation, already several times attempted,[2] of purging the sham registrations.  This, combined with the seasonal outflow for the harvest, produced a startling diminution in the figures for October 1, which showed only 473,000 unemployed registered at the same 70 exchanges, with an estimated total of 775,000 : the Moscow and Leningrad exchanges were particularly zealous, reducing the numbers of unemployed on their books from 140,000 to 70,000 and from 170,000 to 13,000 respectively.[3]  But these manipulations, which a trade union delegate compared with the practice adopted by Narkomfin in the inflation period of revaluing the currency by striking off noughts from the sum of rubles,[4] failed to keep down the numbers for any length of time ; and, after the drastic reorganization of January 1925, which transformed the labour exchanges into voluntary organizations for the recruitment of labour,[5] the purging process was set in motion once more.  In the first three months of 1925 the number of registered unemployed was reduced by 60 per cent, most of whom are said to have failed to present themselves for re-registration.  On the other hand, the percentage of trade unionists among the registered unemployed rose, as a result of the operation, from 38 to 61 ; and this was taken as a sign that the lists were now more nearly representative of the *bona fide* unemployed.[6]  In October 1925 *Pravda* quoted the official figure of unemployed on September 1 as 1,100,000 with the remark that,

[1] *Shestoi S"ezd Professional'nykh Soyuzov SSSR* (1925), p. 186.
[2] See *The Interregnum, 1923–1924*, pp. 54-55.
[3] *Sotsialisticheskoe Khozyaistvo*, No. 5, 1925, p. 413 ; *Shestoi S"ezd Professional'nykh Soyuzov SSSR* (1925), p. 138 (even Shmidt admitted that the Leningrad figure was " fictitious ").  [4] *Ibid.* p. 199.
[5] For this reform see *The Interregnum, 1923–1924*, p. 64 ; an instruction was issued by the party central committee that party members who were unemployed should be registered at the labour exchanges (*Spravochnik Partiinogo Rabotnika*, v, 1925 (1926), 251).
[6] Y. S. Rozenfeld, *Promyshlennaya Politika SSSR* (1926), p. 321.

since registration at the labour exchanges was no longer obligatory, the real total might be nearer 1,300,000.[1]

It was now becoming apparent that the problem of unemployment in the Soviet Union differed in one fundamental respect from the similar problem in the west. The observation was justly made that, whereas in western countries the curve of unemployment varied inversely to the curve of employment, in the Soviet Union between 1924 and 1926 the numbers of unemployed and of employed workers were both rising. The number of workers employed in industry, which stood in 1913 at about 2,600,000, had fallen in 1921–1922 below 1,250,000, but thereafter had risen steadily, reaching 1,620,000 in 1923–1924, and estimated totals of 1,900,000 and 2,300,000 in the two following years.[2] As early as 1924 a critic noted that the country " is once more throwing on to the town a ' reserve army of labour ', and at the present time our industry is not in a position to digest all the labour power offered ".[3] The rapid expansion of heavy industry in the following year brought to light a new problem in the form of a shortage of skilled labour without diminishing the incidence of unemployment as a whole. " Thus ", wrote Smilga at this time, " we have on one flank a shortage of labour power, and on the other an over-production of labour power." [4]

Industrial unemployment in the Soviet Union could, therefore, be clearly diagnosed as a reflexion of the phenomenon of rural over-population to which attention had been drawn in the resolution of the party central committee in April 1925.[5] The rapid natural increase of population combined with the growing

[1] *Pravda*, October 14, 1925.

[2] Y. S. Rozenfeld, *Promyshlennaya Politika SSSR* (1926), pp. 317, 319 : the figures quoted were those of Gosplan. Other calculations were current : the number of workers in census industry was given as 1,190,775 (including 1,110,539 in state industry) for 1923–1924, 1,429,515 (including 1,319,973 in state industry) for 1924–1925, and 1,728,364 (including 1,592,750 in state industry) for 1925–1926 (*Planovoe Khozyaistvo*, No. 2, 1926, p. 132). A later calculation put the total of workers in large-scale industry for the three years in question at 1,795,000, 2,107,000 and 2,678,000 respectively (*Sotsialisticheskoe Stroitel'stvo SSSR* (1934), p. 306 ; *id.* (1936), p. 508).

[3] *Sotsialisticheskoe Khozyaistvo*, No. 3, 1924, p. 218.

[4] *Planovoe Khozyaistvo*, No. 9, 1925, p. 24. From the autumn of 1925 onwards constant complaints are heard of lack of skilled labour ; in February 1926 Gosplan reported to STO that " the supplies of skilled labour on the labour market have been exhausted " (*ibid.* No. 2, 1926, p. 53).

[5] See pp. 266-267 above.

process of " differentiation " in the countryside to promote a continuous exodus of unskilled and unwanted peasants seeking employment in towns and factories.

> If we gave work to 2000 [Zinoviev told the Leningrad Soviet at this time], very often at the same moment 10,000 new unemployed would arrive from the country seeking work and imagining that it was possible to find a tolerable livelihood in the town. . . . You cannot empty this sea of unemployment with a teaspoon.[1]

Since the productivity of the worker in industry was many times greater than that of the worker in agriculture,[2] this movement could only be regarded as in the long run potentially desirable ; and, since the rationalization and mechanization of agriculture would tend to diminish the demand for rural labour, the ultimate solution of the problem of unemployment in the Soviet Union could only be to expand industry rapidly enough to absorb a growing rural population. Preobrazhensky was the first to insist that all attempts to solve the unemployment problem in a radical way led to the problem of accumulation.

> Hundreds of millions in terms of value [he wrote in 1925], in the form of the unused labour of unemployed workers, in the form of idle factories with unused machinery or of uneconomic utilization of the factories that are at work — these hundreds of millions are being lost, and people are starving, simply because we are only just beginning to accumulate and are still achieving only minor successes in this field.[3]

But this analysis did not make the immediate issue any less embarrassing or any less painful. The trade unions were frankly unsympathetic to the mass of unemployed who weighed on the labour market :

> The unions [wrote the official trade union newspaper] have never made it their business to defend the interests of persons not earning wages. The mere intention to get a job and join a union is not enough. The unions fight for organized labour, for the organized securing of jobs, and, as a matter of course,

[1] *Leningradskaya Pravda*, April 15, 1925.
[2] According to a contemporary calculation the net product of the industrial worker was nearly five times greater (*Ekonomicheskoe Obozrenie*, October 1925, p. 80).    [3] E. Preobrazhensky, *Novaya Ekonomika* (1926), p. 240.

for the employment of those already organised.  If the unions
acted otherwise, they would lose their class character and deny
their essence — the defence of their members' interests. . . .

Unemployed who are not union members will not agree
with us on this, because they want to get jobs. . . . Yet it
must be stressed once again that the unions will not and cannot
protect all those who are not wage-earners.[1]

Nor were the unemployed outside the trade unions entitled to
unemployment benefit ;  as Bukharin said in a sympathetic
analysis of the question at the fourteenth party congress, " since
we do not now work on social security principles, the position of
the unemployed is at present very grave ".[2]

The complaint was constantly heard at the time that unemployment was falling more heavily on women than on men,
and even that women were being driven out of industry.  Here,
too, the crisis displayed the same anomalous feature of a simultaneous increase both of employment and of unemployment.  The
number of women employed in industry rose steadily throughout
this period from 414,000 in January 1, 1923, to 679,000 in January
1, 1926 ;  the percentage of women employed fell from 29·5 in
January 1, 1923, to 27·5 a year later, thereafter recovering slightly
and remaining stationary at between 28 and 29.[3]  In spite of these
figures, the evidence makes it clear that the turnover in female
labour was disproportionately high,[4] and that the prevalence of

---

[1] *Trud*, July 1, 1925.

[2] *XIV S"ezd Vsesoyuznoi Kommunisticheskoi Partii (B)* (1926), p. 815 ;
for previous discussions of this question see *The Interregnum, 1923–1924*,
pp. 56-58.  In the Ukraine an unemployed person was allowed to exercise
electoral rights only if he was registered at a labour exchange, or had a certificate
from a regional executive committee, a village Soviet or the militia, or was a
member of a trade union :  other unemployed were disfranchised (*Zbirnik
Uzakonen' ta Rosporyadzhen', 1924*, No. 34, art. 235).  This provision seems to
have been peculiar to the Ukraine.

[3] Official figures cited in *International Labour Review* (Geneva), xx (1929),
No. 4, p. 518 ;  similar percentages are given in *Itogi Desyatiletiya Sovetskoi
Vlasti v Tsifrakh, 1917–1927* (n.d.), p. 337.

[4] A point to be noted in this context is the unusually high excess of females
over males at this period in the Soviet population, especially in the age group
from which the labour force would be mainly drawn, the disproportion being,
of course, due to the war and the civil war.  The population of the Soviet Union
shown by the 1926 census consisted of approximately 71 million males and
76 million females.  In the age group 25-29, the ratio of males to females was
only 83 to 100 ;  in the population above the age of 30, the ratio was 87·9 to
100 (F. Lorimer, *The Population of the Soviet Union* (Geneva, 1946), pp. 41-42).

unemployment produced in some trade unions (especially among the metal workers) a movement to drive out women. The incompatibility of this state of affairs with orthodox party doctrine had been remarked by the thirteenth party congress in May 1924 :

> In connexion with the continual exclusion of women workers from production, the congress insists that the maintenance of a female labour force in production has a political significance, and sets the party the task of intensifying the development of skilled female labour and of drawing women, where this is possible, into branches of production in which female labour has hitherto not been utilized at all or insufficiently utilized.[1]

An awkward situation was created by the reluctance of managements to employ women owing to the limitations imposed by protective legislation and, perhaps also, to the objections of male workers to the special privileges and exemptions accorded to them.[2] The whole issue was discussed at length, and a firm stand taken by the trade union leadership, at the sixth trade union congress in November 1924. Shmidt, the People's Commissar of Labour, dilated on the attempt to exclude women from industry (" women have always been the first to be discharged ") and went on :

> A great number of our laws prohibiting the employment of women at night, or barring unhealthy occupations to them, must be revised. Where working conditions are onerous, legislation must be amended so as to facilitate the admission of women.

Another delegate complained that " the dismissal of women from enterprises has assumed a mass character ". A woman delegate explained that the disinclination of managements to employ women was the result of protective legislation applying to them ; for example, the prohibition of the employment of women on nightwork " disorganizes production ". Hence the clue was to relax these restrictions : it was better for a woman to " have the possibility to earn a crust of bread as a worker than to go and sell herself on the boulevard ". The only reason for not repealing the

---

[1] *VKP(B) v Rezolyutsiyakh* (1941), i, 619.

[2] An early case of collusion between management and workers to ignore the decree prohibiting night work for women is recorded in *The Bolshevik Revolution, 1917–1923*, Vol. 2, p. 70.

restrictions was " fear of what they will say in the west ".[1] The
congress took the hint, recommended the removal of the prohibition on the employment of women in certain unhealthy occupations and on night work as leading to " the exclusion of women
(especially skilled women) from production ", and instructed trade
unions to " combat the present tendency to replace women in
production by men ".[2] No formal change in the law appears to
have been made. But in April 1925 the Narkomtrud of the
USSR issued a circular to the Narkomtruds of the republics
confirming " the necessity in future of allowing night work for
women in all branches of production with the exception of those
specially unhealthy industries where female labour is prohibited
altogether " ; pregnant women were to be put on day shifts.[3]
Women's wages at this time amounted to from 60 to 65 per cent
of those of men, the difference being due to the lower grading of
women workers.[4]

A much more vexed question was the widespread unemployment among juveniles.[5] This not only threatened to demoralize
the workers of the future at the outset of their career, but deprived
them of the practical training required to make them into skilled
workers. Since May 1922 a decree had been in force obliging

[1] *Shestoi S"ezd Professional'nykh Soyuzov SSSR* (1925), pp. 185, 208, 222-223.                                                                              [2] *Ibid.* p. 488.

[3] *Sbornik Dekretov, Postanovlenii, Rasporyazhenii i Prikazov po Narodnomu Khozyaistvu*, No. 19 (40), April 1925, p. 58. Article 129 of the labour code of November 1922, instructing Narkomtrud to draw up a list of " specially heavy and unhealthy occupations " from which women and young people were to be excluded, had not at this time been carried out in regard to work for women. A list of occupations closed to women was belatedly issued on October 30, 1925 (*Byulleten' Finansovogo i Khozyaistvennogo Zakonodatel'stva*, No. 1, January 8, 1926, pp. 34-35). But exceptions, where they were justified by the " local conditions of the industry ", could be authorized by the Narkomtrud of the republic concerned. The list of occupations closed to women included " all underground work " in mining. A German labour delegation which visited the Soviet Union in 1925 was shocked to find women working underground in mines, and questioned Tomsky about this : he did not attempt to deny the facts and specifically defended night work for women, objections to which rested on " old bourgeois prejudices " (*Trud*, August 20, 1925).

[4] A. Rashin, *Zarabotnaya Plata za Vosstanovitel'nyi Period Khozyaistva SSSR* (1928), pp. 140-141.

[5] Lists of occupations from which juveniles were excluded had been published from time to time in accordance with art. 129 of the labour code, the latest on February 24, 1925 (*Izvestiya Narodnogo Komissariata Truda*, No. 11-12, 1925, pp. 16-17).

industrial enterprises to employ a certain percentage (the so-called " ironclad minimum ") of youths between the ages of fifteen and seventeen to give them training.[1]  But with the growth of unemployment, and with adult labour available in abundance, managements and workers conspired together to evade this proviso.  Cases occurred in which unemployed youths were given work in factories, ostensibly for training, without regular wages ; and an order was issued to prohibit this abuse.[2]  At the thirteenth party congress in May 1924, 48 per cent of juveniles wanting work in factories were said to be unemployed.[3]  At the sixth Komsomol congress of July 1924, which debated the question of juvenile unemployment at length, the official spokesman admitted the existence of " contradictions between today's rigid *khozraschet* and the training of skilled workers in expectation of tomorrow's expansion of industry ", and could only express the hope that the " ironclad minimum ", which had been " almost fulfilled " in 1923, would be strictly observed in the current year.  Another delegate complained that all employers, except a few of the most successful trusts, regarded the employment of juveniles as " an overhead charge", and were waging a campaign against juvenile labour:  the prohibition on the drafting of juveniles into unhealthy work also stood in the way of their employment and training.[4]  At the sixth trade union congress in November 1924 it was stated that juveniles formed only 2·9 per cent of the total employed labour force, as against 3·7 per cent two years earlier ; Shmidt complained that as many juveniles were unemployed as were employed, and that, since those who had never worked were not entitled to insurance benefit, their plight was desperate.[5]  The congress resolution demanded strict observance of the " ironclad minimum " of juveniles in view of " the necessity of training skilled workers ".[6]

---

[1]  *Sobranie Uzakonenii, 1922*, No. 39, art. 447 ; the obligatory percentage varied from industry to industry, and from a minimum of 2·5 to a maximum of 13.

[2]  *Sbornik Dekretov, Postanovlenii, Rasporyazhenii i Prikazov po Narodnomu Khozyaistvu*, No. 12, September 1924, p. 59.

[3]  *Trinadtsatyi S"ezd Rossiiskoi Kommunisticheskoi Partii (Bol'shevikov)* (1924), pp. 549-550.

[4]  *Shestoi S"ezd Rossiiskogo Leninskogo Kommunisticheskogo Soyuza Molodezhi* (1924), pp. 201, 206-208.

[5]  *Shestoi S"ezd Professional'nykh Soyuzov SSSR* (1925), pp. 88, 184, 201-202.    [6]  *Ibid.* p. 464.

But the disappearance of the obligation to recruit labour through the labour exchanges made the stipulation difficult to enforce,[1] though a party circular of February 1925 came out against the abolition or reduction of the wages of juveniles in industrial employment, stressing the need to maintain the " ironclad minimum " and to concentrate on training.[2]

The issue was still acute at the time of the fourteenth party congress in December 1925, when a Komsomol delegate denounced " proposals to lower the wages of juvenile workers, to introduce unpaid apprenticeship, especially among artisans, to convert factory schools from schools for the masses into schools for the training of highly skilled workers, craftsmen etc., and to deprive juvenile workers in unhealthy occupations of extra holidays ". Bukharin replied in some embarrassment to this tirade that he personally thought that " the old norms should be retained ", but that the whole question was being considered by the party authorities.[3] The congress resolution pronounced it " indispensable to maintain all fundamental legislative provisions about the labour and training of young workers ".[4] How far this victory in principle was translated into practice remains uncertain. In February 1926 the central committee of Komsomol called for measures " to regulate and utilize the labour of juveniles on night work, to which they are not at present admitted, and to which they could be admitted on certain conditions ".[5] The official *rapporteur* at the seventh Komsomol congress in the following month bravely enunciated the official doctrine :

> *The socialist organization of the labour of youth means that the labour of juveniles in production is subject not only to economic tasks, not only to considerations of economic advantage, but also to tasks of instruction.*

But it was admitted at the congress that " not in a single branch of industry has the percentage fixed by government legislation [for

---

[1] *Ibid.* p. 202 ; for the by-passing of the exchanges see *The Interregnum, 1923–1924*, p. 64.
[2] *Izvestiya Tsentral'nogo Komiteta Rossiiskoi Kommunisticheskoi Partii (Bol'-shevikov)*, No. 13-14 (88-89), April 6, 1925, p. 14.
[3] *XIV S"ezd Vsesoyuznoi Kommunisticheskoi Partii (B)* (1926), pp. 829, 852.
[4] *VKP(B) v Rezolyutsiyakh* (1941), ii, 78.
[5] *Trud*, February 27, 1926.

the employment of juveniles] been fully maintained ".[1] Complaints of chronic unemployment among juveniles were still being repeated at the end of the year.[2]

The question of training both for skilled factory work and for administrative posts continued to present an enormous problem. From 1921 onwards stipends had been granted on the nomination of the trade unions to students in higher educational institutions, technical schools and factory schools.[3] From 1923 onwards, when unemployment first declared itself, industrial enterprises and Soviet institutions were required by decree to fill 1 and 2 per cent respectively of their vacancies from graduates of universities or technical schools.[4] But the campaign for economy in the budget reduced available funds. In 1923–1924, 67,000 students were in receipt of state stipends of from 15 to 20 rubles a month (a rough approximation to the current wage of an industrial worker). In 1924–1925 the number had fallen to 47,000, and the People's Commissariat of Education was threatening to charge tuition fees in institutions under its control. At the sixth trade union congress of November 1924, where these facts were stated, the plight of the students was painted in lurid colours : many of them, unable to afford lodgings, lived in underground cellars or slept at the railway stations.[5] A year later at the fourteenth party congress Bukharin, in a gloomy picture of the plight of the young throughout the Soviet Union, spoke once more of the " desperate material position " even of " those strata of our youth which stand on the top rungs of the educational ladder (students in higher institutions, factory schools etc.) ".[6] Throughout this period a series of decrees attempted to ensure that those graduating from technical schools and colleges should find occupation suited to their talents

[1] VII S"ezd Vsesoyuznogo Leninskogo Kommunisticheskogo Soyuza Molodezhi (1926), pp. 32-33, 341.

[2] Sed'moi S"ezd Professional'nykh Soyuzov SSSR (1927), pp. 181-182, 217, 352. The following percentages of juveniles among workers in census industry were given for the years named : for 1923, 6·6 ; for 1924, 5·5 ; for 1925, 5·2 ; for 1926, 5·7 (Statisticheskii Spravochnik SSSR za 1928 g. (1929), pp. 532-533).

[3] Sobranie Uzakonenii, 1921, No. 56, art. 353 ; No. 76, art. 621 ; Sobranie Uzakonenii, 1922, No. 24, art. 270 ; No. 35, art. 413.

[4] Sobranie Uzakonenii, 1923, No. 49, art. 484 ; Sobranie Zakonov, 1925, No. 34, arts. 236, 237. The percentages were later raised to 1·25 and 2·5 (Sobranie Zakonov, 1926, No. 44, art. 320).

[5] Shestoi S"ezd Professional'nykh Soyuzov SSSR (1925), pp. 112-113.

[6] XIV S"ezd Vsesoyuznoi Kommunisticheskoi Partii (B) (1926), p. 814.

and training.  They were to be employed, in general, as specialists,
in posts appropriate to their qualifications ; Narkomtrud was given
powers to direct them to such posts.[1]  But the very multiplicity
of these decrees throws doubt on their effectiveness.  In an age of
chronic unemployment man-power was still a cheap commodity,
and flagrant waste of it at every stage of development a neglected
evil.

Throughout this period the theory was maintained that the
conditions of employment were freely determined by individual
agreement or by collective agreements between trade unions and
employers.  State intervention was limited to the now almost
entirely nugatory prescription of a statutory minimum wage, which
was announced month by month.[2]  Wage questions, though still
acute, occupied a less prominent place in the labour policies of
the period from 1924 to 1926 than in the preceding years.  In
1924, thanks to the financial reform, to the general rise in prosperity, and to increased industrial production, the position of the
worker had substantially improved.  The acute labour discontent
of the autumn of 1923 had been appeased, and the remuneration of
the worker reached a level which at any rate enabled him to live
and work.  His wages were paid in a stable currency, which pre-
cluded a repetition of past juggling with the price-index ; [3] and
the crying abuse of delayed wage payment, prompted in part by
desire to take advantage of a falling exchange, had largely disappeared, though complaints of wages paid a month or six weeks
in arrears were still heard, especially from the Ukraine and the
Urals.[4]  In some places, only 40 per cent of wages were paid in

---

[1] *Sobranie Uzakonenii, 1924*, No. 80, art. 801 ;  No. 90, art. 915 ;  *Sobranie Zakonov, 1925*, No. 34, arts. 236, 237.
[2] See *The Interregnum, 1923–1924*, p. 61, note 3.  In the first half of 1924
the statutory monthly minimum varied from 6 rubles to four and a half rubles
according to zone (*Sbornik Dekretov, Postanovlenii; Razporyazhenii i Prikazov
po Narodnomu Khozyaistvu*, No. 8, 1924, ii, 66) ;  similar decrees recording
slightly higher minimum rates appeared in the same publication at intervals
during 1925.          [3] See *The Interregnum, 1923–1924*, pp. 75-77.
[4] *Shestoi S"ezd Professional'nykh Soyuzov SSSR* (1925), pp. 120, 255, 282.
The position was summarized in an article in *Voprosy Truda*, No. 11, 1924,
pp. 56-60, on *Unpunctuality in Payment of Wages and the Struggle against it.*
Employees carried on state budgets, and employees and workers in state enterprises (e.g. railways), were now, as a rule, paid punctually ; employees carried on
local (i.e. provincial or county) budgets were less well placed, teachers, agrono-

cash and the balance in credit notes on the cooperatives : cases
were said to have occurred where the whole wage had been paid
in credit notes. The sixth trade union congress in November
1924 refused to deprive the cooperatives of this factitious support
by condemning the practice.[1] But it seems to have fallen into
disrepute, and gradually declined.[2] Protests, moreover, began to
be made against increasingly large deductions made from wages
in the form of subscription to MOPR (the International Association for Aid to Revolutionaries),[3] to ODVF (the society for the
encouragement of aviation) and Dobrokhim (the organization for
the development of chemical warfare): these were said to amount
in some cases to 8 or 9 per cent of wages. The sixth trade union
congress passed a resolution that such deductions should not
exceed 4 per cent of total wages.[4]

Apart from these alleviations, it seems clear from the confused
and admittedly unreliable statistics that wages rose throughout
the greater part of 1924. Average monthly wages in industry in
the quarter January–March 1924 stood at 36·2 chervonets rubles
or 20·39 conventional rubles. In October 1924 they had risen to
42·25 chervonets rubles or 25·58 conventional rubles.[5] The

mists and doctors being mentioned as the worst sufferers. Workers in light
industry were punctually paid, but there were still black spots in heavy industry,
notably in the mining and metallurgical industries. In February 1926 delays
of three or four months still occurred in payment of wages from local budgets
(*Ekonomicheskaya Zhizn'*, February 13, 1926). As late as April 1926 complaints
occurred of delays in wage payments by " private persons or firms working on
government contracts ", and a decree required such firms, before receiving the
sums due to them, to submit evidence that their wages bills had been paid
(*Sobranie Zakonov*, *1926*, No. 25, art. 158).

[1] *Shestoi S"ezd Professional'nykh Soyuzov SSSR* (1925), pp. 255, 629-630.
[2] In the following year the trade union central council issued an order
limiting the amount that might be deducted from wages in the form of credits
at the cooperatives to a maximum of 15 per cent (*Trud*, July 18, 1925) ; and
the average of such deductions fell from 12·7 per cent in March 1925 to 4·8
per cent in September 1926 (A. Rashin, *Zarabotnaya Plata za Vosstanovitel'nyi
Period Khozyaistva SSSR* (1928), p. 60).
[3] For MOPR see *The Bolshevik Revolution*, *1917–1923*, Vol. 3, p. 405.
[4] *Shestoi S"ezd Professional'nykh Soyuzov SSSR* (1925), p. 520.
[5] Average monthly wages for every quarter from October–December 1922
in conventional rubles and from January-March 1924 in chervonets rubles
are given in A. Rashin, *Zarabotnaya Plata za Vosstanovitel'nyi Period Khozyaistva SSSR* (1928), pp. 6, 11 ; monthly figures from October 1924 in conventional and in chervonets rubles are in *Planovoe Khozyaistvo*, No. 2, 1926,
p. 54 ; annual figures (in chervonets rubles only) are in *Kontrol'nye Tsifry
Narodnogo Khozyaistva na 1926–1927 god* (1926), pp. 376-377. For the basis

official trade union spokesman at the congress of November 1924 claimed that wages had now reached 109 per cent of their pre-war level in Moscow, 90 per cent in Leningrad and 75 per cent in the country as a whole.[1] Another review of wages at this time, using the same general figure of 75 per cent, recorded that in the food industry wages had reached 129·2 per cent of the pre-war figure, in the textile industry 92·8 per cent, in the metal industries 62·5 per cent and in mining only 48·6 per cent.[2] What is clear is that all statistics purporting to record average wage levels concealed great divergences between different industries, different regions

on which the " conventional " ruble was calculated see S. Zagorsky, *Wages and Regulation of Conditions of Labour in the USSR* (Geneva, 1930), pp. 191-193. Some uncertainty prevails about the precise definition of " wages " in this period for the purpose of wages statistics ; and practice may not have been uniform. When NEP was introduced, the remuneration of labour lost the character of social maintenance which it had had under war communism, and became payment for value received (see *The Bolshevik Revolution, 1917–1923*, Vol. 2, p. 320). The term " wages " was officially defined in 1925 to cover all forms of direct payment for labour, including overtime and bonus payments and payments in kind, but excluding social insurance benefits, free housing, and other free services provided by the state, by local authorities or by employer institutions (*Sobranie Zakonov, 1925*, No. 14, art. 107). On the other hand, these indirect components of the worker's income were commonly referred to as " the socialized part of wages " by those who, for purposes of domestic or foreign propaganda, desired to prove that the worker was better off than he appeared to be ; and industrialists, for a similar reason, treated the cost of social insurance and other benefits for the worker as part of their wages bill, " combining both wages and expenditure on social insurance in their calculations under the same general rubric of costs of labour power " (*Planovoe Khozyaistvo*, No. 8, 1925, p. 36, where Strumilin argues at length the case for keeping them apart). An example of this practice was a table in *Ekonomicheskoe Obozrenie*, April 1926, p. 115, where it was calculated that of all " expenditures on labour " in state industry in 1924–1925, 81 per cent represented wages (approximately 75 per cent in cash and 6 per cent in kind), and 19 per cent social expenditures, including 12 per cent for social insurance and minor items covering the cost of special clothing, cultural activities, hospitals, etc. Lodging for workers, commonly provided in the mining and oil extraction industries, and in other cases where factories were remote from urban centres, appears to have been generally treated as wages in kind.

[1] *Shestoi S"ezd Professional'nykh Soyuzov SSSR* (1925), p. 98. A critic at the congress, however, attacked these figures, alleging that they had been drawn exclusively from heavy industry and from factories employing at least 250 workers (whereas the average number of workers in a single enterprise was 14) : he also attacked the pre-war statistics on which the percentages were based as unrepresentative. The official spokesman claimed in reply that the statistics covered the wages of 1,300,000 workers (*ibid.* pp. 138-140, 157).

[2] *Sotsialisticheskoe Khozyaistvo*, No. 5, 1925, p. 28 ; similar figures with slight variations are quoted from a Gosplan publication in Y. S. Rozenfeld, *Promyshlennaya Politika SSSR* (1926), p. 348.

and different categories of workers : these, as Tomsky pointed out
at the sixth trade union congress, were inevitable unless it was
proposed to return to a system of state regulation.[1]  The highest
wages were always paid in Moscow, followed closely by Leningrad ;  the Ukraine generally came next, and the Urals rated
lowest among the great industrial regions.[2]  Broadly speaking, the
more highly industrialized the region, the higher the wages, the
difference being presumably offset in part by differences in the
cost of living.  The metal workers of the Ukraine and the Urals
were said in 1924 to be receiving only 8·50 rubles a month (in
" conventional " rubles) as against 18 rubles paid to their colleagues in Moscow :  in the Donbass, wages were still at 50 per
cent of the pre-war level.  In general, an official spokesman
admitted that, while " a very few workers " were drawing as much
as 50 chervonets rubles a month, there was " a fairly wide area "
of " black spots ", where workers received less than 10 rubles.[3]

More important was the vexed question of the discrepancy
in remuneration between different grades of worker.  After
September 1921, when the egalitarian principle of levelling out
wages had been officially abandoned [4] in the interests of efficiency,
the spread between the wages of the skilled and unskilled worker
increased rapidly.  The wage scale recognized by the trade unions
was divided into 17 grades of which the first 9 represented manual
workers and the remainder clerical and administrative staff.  In
the original scale of 1921 the spread between the grades was in
the ratio of 1 for the lowest, 2·7 for the ninth (the highest grade
of manual worker) and 5 for the seventeenth :  by the end of 1923
the spread had increased to the ratio 1 : 3·5 : 8.[5]  But such
differentiation, where it arose from differential wages for different
grades of work, continued to shock party members and trade
unionists reared in an egalitarian tradition.  In the wages crisis of

---

[1] *Shestoi S"ezd Professional'nykh Soyuzov SSSR* (1925), pp. 170-171.

[2] Detailed figures for different industries are quoted in A. Rashin, *Zarabotnaya
Plata za Vosstanovitel'nyi Period Khozyaistva SSSR* (1928), pp. 86-88, 95.

[3] *Shestoi S"ezd Professional'nykh Soyuzov SSSR* (1925), pp. 117, 143,
294 ; in the spring of 1925, wages in the Gomza engineering works were less
than 50 per cent of the pre-war level, though this level had been exceeded in
corresponding works in Moscow and Leningrad (*Chetyrnadtsataya Konferentsiya Rossiiskoi Kommunisticheskoi Partii (Bol'shevikov)* (1925), p. 203).

[4] See *The Bolshevik Revolution, 1917-1923*, Vol. 2, p. 320.

[5] *Trud*, October 7, 1923.

1923 the plight of lower paid workers was so desperate that it became difficult to resist pressures to allow them a somewhat larger share in a depleted wages fund.[1] The sixth trade union congress confirmed the existing wages scale with its 17 grades in face of some opposition from those who demanded a still higher number of grades with a wider spread.[2] But the most striking differences in wages received were henceforth due less to different categories in the wages scale than to differences in work done or in time worked. The further extension of piece-rates in 1924, which, being primarily applicable to skilled workers, threatened to widen still further the gap between skilled and unskilled, was counteracted by a decision to extend a system of bonuses to unskilled workers as well as to administrative staffs.[3] Between 1924 and 1926 the gap between the wages of skilled and unskilled workers continued to widen, but less rapidly and regularly than in the previous period.[4] Egalitarian principles were the foundation of party precept, and still served as some brake on contrary practice. The fifteenth party conference in November 1926 gave its blessing to recent increases in the wages among " low paid groups of workers " as " a further step towards overcoming the abnormal discrepancy in the wages of different categories of workers ".[5] It was still seriously expected at this time that, with the spread of general education and with the progressive elimination of the remaining bourgeois elements from responsible positions, differences of remuneration between different categories of workers would gradually be wiped out. This was a necessary part of the realization of socialism.[6]

[1] See *The Interregnum, 1923-1924*, pp. 72-79 ; the reference in the resolution of the scissors committee (*ibid.* p. 115) to the necessity of bringing up low wages to the " average level " related, however, to the levelling up of low-paid industries, not of low-paid grades within the same industry.

[2] *Shestoi S"ezd Professional'nykh Soyuzov SSSR* (1925), pp. 463, 613-616.

[3] *Trud*, October 16, 1924 ; further pronouncements of the trade union central council in 1925 and 1926 in favour of " removing the gap in wages between time-workers and piece-workers, skilled and unskilled workers " are quoted in A. Rashin, *Zarabotnaya Plata za Vosstanovitel'nyi Period Khozyaistva SSSR* (1928), p. 38.

[4] Figures for a number of industries are quoted in *ibid.* pp. 67-78 ; skilled workers in 1925 earned on an average about twice as much as unskilled workers.

[5] *VKP(B) v Rezolyutsiyakh* (1941), ii, 135.

[6] Preobrazhensky, who believed that wages in the Soviet economy had been in large measure emancipated from the law of value, none the less admitted that the system of differential wages " has nothing, and can have nothing, in

It was a symptom of the strength of egalitarian sentiment that, whenever the wages issue came up for discussion during this period, the malcontents seized the opportunity to invoke the salaries of specialists as a standing grievance. Though the situation had ostensibly been regularized by the decree of November 1923 fixing limits for " individual " or " personal " salaries falling outside the seventeen-grade scale of wage rates determined by collective agreements,[1] the wide divergence between these salaries and the wages even of the most highly paid worker continued to attract attention ; and in the wages crisis of the winter of 1923–1924 individual salaries were cut by 10 or, in the case of the highest salaries, by 20 per cent.[2] But the agitation did not abate, and in May 1924 a new ceiling of 250 rubles a month was fixed for individual salaries, though once more with the saving proviso that salaries above this rate required the special sanction of Vesenkha.[3] Another attempt was made to deal with the scandal of party members in receipt of high salaries. A decree was issued prohibiting party members from receiving salaries in excess of the ordinary wage rates of workers, except such as were required by their work to have dealings with representatives of bourgeois countries.[4] But the new drive for productivity increased the influence of the " Red industrialists " and made further attempts to " squeeze " the specialists inopportune. Not much was said on the matter at the sixth trade union congress, though one delegate repeated the old complaint that specialists were unwilling to work outside the great cities, and the official spokesman, who claimed that the trade union central council had taken " a very stiff line " on the subject, resisted a demand to assimilate specialists' salaries to the ordinary wage scales.[5] The congress, in its resolution on incentives to increased productivity, specifically recommended " bonuses to auxiliary and administrative staff in

---

common with socialism " (E. Preobrazhensky, *Novaya Ekonomika* (1926), p. 176). The argument that differences of remuneration will disappear with the growth of socialism is also developed in A. Rashin, *Zarabotnaya Plata za Vosstanovitel'nyi Period Khozyaistva SSSR* (1928), pp. 136-137.

[1] *Sobranie Uzakonenii, 1924*, No. 11, art. 90.

[2] *Ibid.* No. 53, art. 525 ; No. 64, art. 646. For these cuts see *The Interregnum, 1923–1924*, p. 115, note 2.

[3] *Sbornik Dekretov, Postanovlenii, Rasporyazhenii i Prikazov po Narodnomu Khozyaistvu*, No. 8, 1924, ii, 66.        [4] *Ibid.* No. 10, 1924, ii, 86-87.

[5] *Shestoi S"ezd Professional'nykh Soyuzov SSSR* (1925), pp. 103, 178.

proportion to their productive achievements ", and instructed
factory committees to take special care to establish " normal
business and comradely relations " between specialists and
workers.[1] But all this did little to allay the bitter feelings that
still persisted in trade union circles.[2]

In the summer of 1925 the rising tide of prosperity, and the
increasing indulgence to bourgeois elements in the economy
shown by the new attitude to the *kulak*, led to renewed pressure
to improve the position of the specialists. A decree of July 10,
1925, once more commended the principle of bonuses (*tantièmes*)
to members of the administrative and technical staffs in state
enterprises or enterprises working with state capital, who had
been responsible for raising profits or lowering production costs :
Vesenkha was instructed to make detailed proposals for the
extension of this principle in heavy industry in view of the particular importance of this sector of production.[3] Four days later
an order of Vesenkha explained that, with the new campaign for
increased productivity, qualified higher staff was more than ever
in demand, and that its remuneration still bore no relation to the
value of its work in increasing output. Salaries should conform to
the quality and results of work done, and bonuses should be paid
for special services on a basis of percentage of salary. The
percentage rates should be agreed with the trade unions.[4] A
further decree of August 1925 offered financial and other benefits
to professional or administrative workers appointed to posts in
the Archangel or Murmansk provinces and provinces or autonomous regions of the RSFSR in Asia.[5] These concessions did
nothing to appease the critics ; and in the same month the question was sufficiently acute to engage the attention of the party
central committee, which explained that it was necessary to

[1] *Ibid.* pp. 460-461.

[2] *Ekonomicheskya Zhizn'*, November 29, 1924, replied to what was evidently
a general attack on the specialists in *Trud* ; a protest by a speaker at a Vesenkha
conference against bonuses for specialists is recorded *ibid.* December 6, 1924.

[3] *Sobranie Zakonov, 1925*, No. 43, arts. 324, 325 ; according to *Ekonomicheskaya Zhizn'*, July 4, 1925, an agreement between Vesenkha and the trade
union central council preceded the issue of this decree.

[4] *Sbornik Dekretov, Postanovlenii, Rasporyazhenii i Prikazov po Narodnomu
Khozyaistvu*, No. 22 (43), July 1925, ii, 105-106.

[5] *Byulleten' Finansovogo i Khozyaistvennogo Zakonodatel'stva*, No. 16, October
2, 1925, pp. 24-25 ; for a further definition of these special posts see *ibid.*
No. 1, January 8, 1926, pp. 32-33.

" guarantee normal conditions of work for specialists in industry, agriculture, transport and other branches of economic and state life ". Recommendations were made to curb " wholesale criticism " of the specialists in the party and trade union press, and give them further " material incentives " in the form of " individual and collective bonuses for achievements in improving production ". The resolution also pronounced in favour of a special salary scale for specialists " in order to avoid the system of personal contracts ".[1] But this last proposal apparently proved too far-reaching, and was not carried out.

However great the resentment of the workers at the exceptional status enjoyed by the specialists, the element of high and un-regulated remuneration for outstanding managers and organizers was still a necessary incentive to production — an ineradicable remnant of capitalism in the mixed economy created by NEP. *Pravda* sensibly argued that the only fundamental remedy was for the régime to improve technical education and so build up its own cadres of " Red specialists ".[2] Rykov, in a speech at the Moscow provincial party conference, tried to discredit the campaign against the specialists by treating it as part of the opposition attack on the so-called " bourgeois degeneration of our state " ; [3] and Tomsky explained to the fourteenth party congress that there was only one way, short of the militarization of labour, to induce scarce specialists to put their services at the disposal of the state : to offer them higher salaries.[4] The agitation against the privileges of the specialists none the less continued. A decree of March 15, 1926, drawing attention to the variety of systems followed in the remuneration of specialists and to the need to establish " uniform principles for the payment of *tantièmes* ", ordered a suspension of all fresh agreements for such payments.[5]

---

[1] *Pravda*, August 23, 1925 : the text appeared in *Izvestiya Tsentral'nogo Komiteta Rossiiskoi Kommunisticheskoi Partii (Bol'shevikov)*, No. 41 (116), October 26, 1925, p. 5, in *Spravochnik Partiinogo Rabotnika*, v, 1925 (1926), 306-307, and in *VKP(B) o Profsoyuzakh* (1940), pp. 241-242, under the date September 18, 1925, which may be the date on which it was circulated as an instruction to party organizations.

[2] *Pravda*, November 29, 1925.        [3] *Ibid.* December 9, 1925.

[4] *XIV S"ezd Vsesoyuznoi Kommunisticheskoi Partii (B)* (1926), p. 199.

[5] *Sobranie Zakonov, 1926*, No. 22, art. 147. According to tables in A. Rashin, *Zarabotnaya Plata za Vosstanovitel'nyi Period Khozyaistva SSSR* (1928), pp. 126-127, 547 directors who were " responsible party workers " were being paid

It may be doubted whether this attempt at restraint had much effect. For, while the resolution of the party central committee of April 1926 avoided the question, Stalin in his subsequent report on the proceedings spoke severely of those who had nothing better to do than " to ' slate ' the managers, accusing them of every mortal sin ". Industry needed new recruits to management, and this required " not the castigation of the managers, but, on the contrary, support for them in every way in the task of building industry ".[1]

The prejudice against any state fixing of wages inculcated by NEP remained so strong in this period that it applied even to direct employees of the state. It was not till 1925 that a beginning was made in the creation of a civil service with graded staffs and uniform salaries. Before the budget of 1925–1926, no real control was exercised by Narkomfin over the way in which sums allocated to the departments were expended, and no uniformity imposed.[2] Up to this time the People's Commissariats concluded collective agreements with trade unions covering subordinate staffs, and engaged responsible officials, like specialists in industry, on personal contracts. This naturally produced inequalities and anomalies. According to a table published at the time, the lowest salaries were paid by state departments ; then, in ascending order, came economic administrations, state trading organs, cooperatives and finally banks. A typist who received 13 goods rubles a month in a state department received 32 goods rubles in a bank ; the corresponding figures for cashiers were 24 and 42, for secretaries 30 and 56, and so on.[3] On January 2, 1925, Sovnarkom passed a resolution declaring it " indispensable to establish a uniform

in March 1926 at an average monthly rate of 187·9 chervonets rubles, and 282 (presumably non-party) directors at an average rate of 309·5 rubles ; of the latter, 17·4 per cent received salaries of more than 400 rubles. Directors also generally received free and privileged living accommodation and transport, and sometimes other facilities.                    [1] Stalin, *Sochineniya*, viii, 139.

[2] *Ekonomicheskoe Obozrenie*, November 1925, p. 14 ; the system is described in *Vestnik Finansov*, No. 4, April 1925, pp. 130-131.

[3] *Planovoe Khozyaistvo*, No. 6, 1925, p. 266 ; a speaker at the sixth trade union congress in November 1924 had complained that salaries in Gosbank and Prombank were many times higher than in Narkomfin or in any other People's Commissariat (*Shestoi S"ezd Professional'nykh Soyuzov SSSR* (1925), p. 143). An article in *Voprosy Truda*, No. 2, 1925, pp. 44-47, gave instances of variations of 400 to 500 per cent in rates of pay for the same job in different institutions.

nomenclature for posts in state organs and institutions carried on the state budget (central or local), as well as in institutions working on *khozraschet*, fixed salaries being assigned to these posts ".[1] In the following month a scale of salaries was fixed by decree for the seven lowest grades of Soviet workers : these included chiefs of district police, people's judges, teachers, and presidents and secretaries of rural district Soviets.[2]  In June the foundations of a regular system were laid.  A list of posts in every state institution and enterprise was to be drawn up with an authorized salary attached to each post.  The numbers of staff were to be fixed by Rabkrin and the trade union central council, salaries by Narkomfin, Narkomtrud and Rabkrin.  A loophole was left for supplementary payments to individuals of particularly high qualifications and for the engagement of temporary staff.[3]  In July the opposition of the trade unions to the abandonment of collective contracts for Soviet workers was finally overruled, and a circular letter from the trade union central council gave its blessing to the reform.[4]  The new system seems to have become effective, at any rate in Moscow, by the autumn of 1925, and was later extended by decree of September 1926 throughout the Soviet Union,[5] though the trade unions continued to lament the passing of this aspect of wage control out of their hands.[6]

It was significant of the changing outlook that disputes about wages and salaries were no longer conducted, as they had been in the previous period, in terms of social justice or of the living standards of the workers.  A new factor of overwhelming importance was now decisively invoked in every discussion : the need to raise the productivity of labour.  The problem of stimulating the

[1] *Planovoe Khozyaistvo*, No. 6, 1925, p. 267.
[2] *Sobranie Zakonov, 1925*, No. 9, art. 86 ; about one-quarter of " Soviet workers " were carried on the budget of the USSR, one-half on local budgets, and one-quarter on the budgets of state economic enterprises (S. Zagorsky, *Wages and Regulation of Conditions of Labour in the USSR* (Geneva, 1930), p. 138).          [3] *Sobranie Zakonov, 1925*, No. 42, art. 321.
[4] *Trud*, July 18, 1925.
[5] *Sobranie Zakonov, 1926*, No. 67, art. 514 ; the official date for the introduction of the new system was October 1, 1926.
[6] *Sed'moi S"ezd Professional'nykh Soyuzov SSSR* (1927), pp. 131-133, 152-153, 184, 242, 788.

productivity of labour and the controversies to which it gave rise
— the charge of " Taylorism ", opposition to piece-rates, and
resistance to the creation of a " workers' aristocracy " through
differential rates of pay — had been familiar since the first weeks
of the régime.[1] Lenin had emphatically asserted the principle
in his famous article of 1919 on " communist Saturdays " :

> The productivity of labour is in the last resort the most
> important, the chief, factor in the victory of a new social order.
> Capitalism created a productivity of labour unknown under
> serfdom. Capitalism can be finally conquered, and will be
> finally conquered, through the creation by socialism of a new
> and far higher productivity of labour.[2]

The foundation in 1922 of a Central Institute of Labour under
the direction of Goltsman and Gastev to study the " scientific
organisation of labour " (Nauchnaya Organizatsiya Truda, or
NOT) had been followed by the formation of a counter-group led
by Kerzhentsev which also used the mystic initials NOT, and
denounced the institute for seeking to improve productivity not
by better organization, but by increasing the pressure on the
individual worker.[3] The controversy came to a head in February
1924 when an all-Russian conference was summoned to discuss
the " scientific organization of labour ". The directors of the
Central Institute of Labour issued in preparation for the conference a manifesto in which they quoted Lenin's qualified approval
of Taylorism, declared that it was " incorrect and unprofitable to
put the teaching of NOT on the basis of a polemic against Taylor
and others ", and proposed to send experts abroad to study
" foreign techniques of the management and organization of
labour ".[4] This manifesto provoked a strong counter-blast from
the Kerzhentsev group, whose " platform of 17 " appeared a

---

[1] See *The Bolshevik Revolution, 1917–1923*, Vol. 2, pp. 111, 114-115.
[2] Lenin, *Sochineniya*, xxiv, 342 ; for the article see *The Bolshevik Revolution, 1917–1923*, Vol. 2, p. 208.
[3] See *The Interregnum, 1923–1924*, p. 84, note 4. Both Gastev and Kerzhentsev were former workers in Proletkult (see pp. 48-50 above) ; Kerzhentsev
had been trade representative in Stockholm from 1921 to 1923, and in 1923,
when he concerned himself with NOT, founded a " League of Time " (Liga
Vremeni), with a journal *Vremya*, to inculcate the rationalization of work by
measurements in terms of time occupied, which won Trotsky's approval
(Trotsky, *Sochineniya*, xxi, 70, 471-472).
[4] *Trud*, February 5, 6, 1924.

week later in *Pravda* : they denounced the open appeal to capitalist methods, objected to " piece-rates for the individual worker for the individual job ", and advocated what they called an " objectively correct " system of wages.[1] Zinoviev came to the rescue of the central institute, and, reversing his attitude in the previous controversy with Goltsman in 1920, wanted to train party members among the workers as leaders and organizers, " who by their personal example will raise production ".[2] Then Kerzhentsev wound up with a broadside in the trade union newspaper which had published the original manifesto. He accused the central institute of lack of faith in the workers, and of desire to " civilize " them " from above " ; of an attempt to create " an aristocracy of the working class, high priests of NOT " ; and of failure to recognize NOT as " a class problem ".[3]

The conference, which met on March 10, 1924, elected members of both groups to its presidium, and was clearly designed to bring about a compromise between them. It ended by approving a long set of theses put forward by Kuibyshev. Attempts to treat NOT as " a complete system of the organization of labour " were condemned as non-Marxist. On the other hand, NOT could be welcomed as a means of improving the existing organization of labour by introducing improved means of production (mechanization, electrification, etc.), by rationalizing conditions of production, and by " raising the productivity of living human labour (raising of skills, intensification of labour, qualitative improvement of labour, etc.) ". The " first and fundamental task in the field of NOT in the USSR " was the plan of electrification : all schemes which evaded this initial requirement were " without substance ". On the other hand, " in the present economic situation, it is pure childishness, or lack of understanding of the tasks of the working class, or a disguised struggle against the proletarian dictatorship, to neglect questions of the raising of the productivity of living human labour in a country where, in virtue of weak technological development, this living labour plays a colossal rôle ".[4] But the fullest review of the problem of productivity at this time was

---

[1] *Pravda*, February 13, 14, 1924.     [2] *Ibid.* February 17, 1924.
[3] *Trud*, February 20, 22, 1924.
[4] The conference was reported at some length in *Trud*, March 11, 12, 1924 ; Kuibyshev's theses were printed in *Byulleten' 2ᵢ Vsesoyuznoi Konferentsii po NOT : 15 Marta 1924* (1924), pp. 27-36.

contained in a report read by an expert of Narkomtrud to a conference of industrial managers in May 1924.  The speaker started with the striking calculation that it took three Russian workers to produce as much as one American worker or one-and-a-half British workers.  This could not be taken unreservedly as a proof of the relative inefficiency of the Russian worker or as a justification for his lower rate of wages.  Other factors also helped to account for the catastrophic decline in the productivity of labour : deterioration in plant and equipment ; deterioration in the quality of raw materials ; failure to keep factories working at full capacity ; the maintenance of unprofitable enterprises for political reasons ; the excessive number of employees and auxiliary workers ; the eight-hour working day ; and shortage of capital leading to inefficient organization of work.  But, when all these factors had been taken into account, there remained two " subjective reasons " for the decline : the diminished skill of the workers, and the lower intensity of effort of the individual worker.[1]  The contrast between " objective " and " subjective " factors of productivity became a favourite *cliché* of the period.  While nobody denied the need to remedy the objective shortcomings which were independent of the worker, practical obstacles continued in many cases to make the remedy difficult ; and it was therefore natural to concentrate on the subjective reasons residing in the worker himself, which might be eliminated by suitable policies of exhortation, instruction, incentives and penalties.[2]  After all, Trotsky had insisted at the twelfth party congress on the imperative need for

---

[1] *Sotsialisticheskoe Khozyaistvo*, No. 4, 1924, pp. 62-106.

[2] A trade union writer in *Planovoe Khozyaistvo*, No. 5, 1925, p. 91, attempted to counteract this tendency by quoting a passage from Marx's *Wages, Price and Profit* : " The productivity of labour depends on progressive improvements in the sphere of social productive forces : improvements resulting from a broadening of production, concentration of capital, combination and division of labour, introduction of machinery, rationalization of methods, utilization of chemical and other natural agents, contraction of time and space through means of communication and transport, and all those special contrivances by means of which science compels the forces of nature to serve labour, and with the help of which the social or cooperative character of labour attains its full development ". But the argument cut both ways ; Rykov on one occasion countered a claim that the rise in productivity had not been adequately reflected in a rise in wages with the argument that part of the rise in productivity was due not to any merit of the worker, but to improvements in organization and equipment (*Trinadtsataya Konferentsiya Rossiiskoi Kommunisticheskoi Partii (Bol'shevikov)* (1924), p. 84).

increased productivity as the workers' contribution to " socialist accumulation ".[1]

Throughout the summer of 1924, the campaign for increased productivity of labour gathered momentum with Dzerzhinsky, as president of Vesenkha, leading the attack.[2] In June, Dzerzhinsky addressed an open letter " to boards of syndicates and trusts and Red directors " instructing them to work out with the trade unions ways and means of drawing the "masses of workers" into consultation about the revival of industry and the need for increased productivity.[3] But it was in the long run impossible to keep the question of productivity separate from the question of wages. Zinoviev, at the thirteenth party congress in May 1924, quoted some optimistic figures to show that the productivity of labour was rising *pari passu* with the rise in wages and had reached from 70 to 75 per cent of its pre-war level.[4] But this favourable estimate was flatly contradicted by Dzerzhinsky in an address to the trade union central council.[5] At the beginning of July, Vesenkha announced that it had set on foot, in agreement with the trade union council, an inquiry into levels of productivity and wages. In a speech at the Kolomensky works Dzerzhinsky alleged that in this factory 100 units of production which before the war had cost 27 rubles in wages now cost 108 rubles, and that the productivity of labour was only 39 per cent of the pre-war level ; at the Sormovo works it was still lower.[6] One com-

---

[1] See *The Interregnum, 1923–1924*, p. 84.

[2] *Sotsialisticheskoe Khozyaistvo*, No. 2, 1925, pp. 451-453, printed a lengthy bibliography of pamphlets, speeches and articles on the productivity of labour published in 1924 and the first months of 1925.

[3] *Ekonomicheskaya Zhizn'*, June 18, 1924 ; on the following day *Ekonomicheskaya Zhizn'* published a leading article advocating " the organization of mutual emulation (sorevnovanie) " among workers for this purpose.

[4] *Trinadtsatyi S"ezd Rossiiskoi Kommunisticheskoi Partii (Bol'shevikov)* (1924), p. 83.

[5] *Ekonomicheskaya Zhizn'*, June 24, 1924. At the sixth trade union congress in November 1924, Dogadov confessed that statistics of productivity issued by different authorities differed " as widely as heaven from earth " (*Shestoi S"ezd Professional'nykh Soyuzov SSSR* (1925), p. 99) ; about the same time Dzerzhinsky admitted that the fall in industrial prices had tended to depress figures of productivity (*Pravda*, December 4, 1924).

[6] *Ekonomicheskaya Zhizn'*, July 3, 1924. In a speech of December 2, 1924, Dzerzhinsky stated that, in the factories now controlled by Glavmetall, the number of workers employed was 99,000 in 1913 and 80,000 in 1923–1924 ; the value of production for the former year was 173 million gold rubles, for the latter 62 millions (*Pravda*, December 4, 1924).

mentator described the discrepancy between the curve of productivity and the curve of wages as " the new ' scissors ' " which
must be closed at all costs.[1] From this time onwards, wages
policy was unequivocally geared to production — a logical corollary of the adoption of the principle of *khozraschet*. It became
accepted doctrine that wages could not rise faster than the productivity of labour (or as fast, if capital accumulation was to take
place) ; and wages policy was scrutinized more and more exclusively from the standpoint of its capacity to raise production.

The campaign culminated in a resolution of the party central
committee of August 19, 1924, on wages and productivity. The
past rise in wages, declared the resolution, had been " unavoidable and in general legitimate " : it had been necessary to attain
" a certain level of satisfaction of the primary and indispensable
needs of the workers ". But the time had now come to revise this
attitude. Between October 1922 and January 1924 wages were
said to have increased by 90 per cent, productivity only by 23 per
cent : while it was admitted that these figures had " a conditional
character ", they " correctly indicate the tendency of a rise in
productivity to lag behind the rise in wages ". The conclusion
was that any further rise in real wages must be sought not by
raising wage rates, but by reducing prices through increased
efficiency in production. The duty of the trade unions was to make
common cause with the state economic organs in the campaign to
raise the productivity of the worker :

> The general opposition fairly often encountered between
> economic organs and trade unions must be overcome. Under
> the dictatorship of the proletariat, both Vesenkha and the trade
> unions should look on the increase of productivity as their
> business.

The resolution ended with a general injunction to increase productivity, not to allow any further increase in salaries, and to
settle industrial disputes by established peaceful procedures.[2]

---

[1] Y. S. Rozenfeld, *Promyshlennaya Politika SSSR* (1926), p. 360.

[2] *VKP(B) v Rezolyutsiyakh* (1941), i, 626-629. Varying estimates of the
relation between productivity and wages current at this time for the most part
reflected the prejudices of those who propounded them. A more careful calculation than most was made in *Sotsialisticheskoe Khozyaistvo*, No. 4, 1925,
pp. 419-420 : this showed that from October 1922 to October 1923 wages rose
much more rapidly than productivity : that from October 1923 to October

Kuibyshev, now president of the central control commission of the party, commented on this resolution at a conference of Red directors. Among the measures required to increase productivity he named " electrification, mechanization and the introduction of new technical processes " and " the more rational utilization of the installations at our disposal ". But he also pointed out that, in a country where production was still technically at a low level, the labour factor was predominant ; and he reminded his audience that " even Lenin declared that the practical application of piece-work and of everything which is scientific and progressive in the Taylor system is necessary ".[1] Kamenev devoted the major part of a speech in the Moscow Soviet to the theme of the productivity of labour, asserting that in the metal industry productivity in 1924 was still only 30 or 40 per cent of the 1913 figure, and ending with the quotation from Lenin to the effect that capitalism could be conquered by socialism only through a higher productivity of labour.[2] In October 1924 the party journal carried a leading article by Molotov, which concluded :

> Our path to socialism lies through the increased productivity of labour on the basis of electrification.[3]

A cautious plea from Leningrad for an increase of wages in the metal industry, supported by the argument that the mass entry of new workers into industry was bound to affect productivity adversely,[4] was a voice crying in the wilderness. When Larin complained that " our managers " had grasped the importance of raising the productivity of labour, but that the raising of wages had " recently often receded further into the background of their

1924 both increased more slowly, but *pari passu* ; and that after October 1924, as the result of the campaign and the decision of the central committee, wages remained stationary and productivity rose steeply. These conclusions may be approximately correct for the period after October 1923, if not for the earlier period. A table in *Ekonomicheskoe Obozrenie*, April 1926, pp. 109-110, showed that the proportion of expenditure on labour (i.e. wages plus social service expenditure) to the total costs of production in all state industry fell from 27 per cent in October 1924 to 23 per cent in April 1925.

[1] *Internationale Presse-Korrespondenz*, No. 135, October 17, 1924, pp. 1787-1789.

[2] L. Kamenev, *Stati'i i Rechi*, xi (1929), 134-145 ; the first sentence of the quotation from Lenin (for which see p. 383 above) appeared as a banner headline on the front page of *Ekonomicheskaya Zhizn'*, November 7, 1924.

[3] *Bol'shevik*, No. 12-13, October 20, 1924, p. 9.

[4] *Leningradskaya Pravda*, October 24, 1924.

consciousness than it should have done ",[1] it was party policy which he was covertly attacking.

The campaign dominated the sixth trade union congress in November 1924. Zinoviev opened the proceedings in a conciliatory mood by admitting that " *the productivity of labour in no case results exclusively from personal intensity of work* " and that " 50 per cent depends on the state, on the managers, on all-of us ".[2] To increase output merely by intensifying labour without fresh capital investment was to increase the ratio of labour to the capital engaged, and to encourage the retention of obsolete and uneconomic plant. But Rykov expressed what was plainly the general view : while " the restoration of the fixed capital of industry " was an important desideratum, the essential fact was that " out of this whole chain of questions the one most susceptible of solution with our present resources is an increase in the productivity of the individual worker " ; and two delegates complained with some reason that the object of the campaign was to increase production at the expense of " the muscle-power of the worker ".[3] Some managers were said to be aiming at the abolition of the eight-hour day, at an increase of 20 to 25 per cent in norms of output and at the dissolution of the factory committees, while another delegate quoted collective contracts recently renewed in which norms of output had been raised by 20 or 30 per cent with the consent of the workers.[4] But, though there was some grumbling, nobody except the permanent dissenter Ryazanov, who, under provocation from Tomsky, declared himself opposed altogether to raising the productivity of labour,[5] resisted the conclusion that the worker must produce more for the same money ; and the congress in its resolution formally admitted that " a further improvement in the material position of the workers will depend on the development of our industry and agriculture ", and accepted on behalf of the trade unions the obligation to assist the economic organs of the state " in the practical work of raising the productivity of labour ".[6]

---

[1] *Vserossiiskii Tsentral'nyi Ispolnitel'nyi Komitet XI Sozyva : Vtoraya Sessiya* (1924), p. 108.

[2] *Shestoi S"ezd Professional'nykh Soyuzov SSSR* (1925), p. 35.

[3] *Ibid.* pp. 115, 252, 257.  [4] *Ibid.* pp. 256-257, 281.  [5] *Ibid.* p. 166.

[6] *Ibid.* p. 438 ; a circular of the trade union central council emphasized that any further rise in wages was dependent on a rise in productivity (*Trud*, January 28, 1925).

Two well-tried expedients were discussed at the congress. The first was an extension of piece-rates. From the earliest days of the revolution, piece-rates had been reluctantly recognized by the trade unions and had been widely applied.[1] The formal limitation placed on piece-work by the labour code could always be over-ridden in " exceptional cases " and proved ineffective. The resolution of the party central committee of August 1924 on increased productivity reverted to this topic, explicitly demanding " the removal of limitations on extra pay for piece-work ", and " the periodical revision of norms of output and of piece-rates " to take account of improvements in the organization of production ; [2] and the trade union newspaper embarked on a campaign in favour of piece-work as an incentive to higher output.[3] The official spokesmen came out strongly in favour of piece-rates ; and the resolution of the congress recommended " an extensive use of incentive forms of wage payments by way of the introduction of direct and unrestricted remuneration of labour by piece-rates ". This recommendation, however, apparently excited more resistance than any other decision of the congress, and a qualifying amendment moved by Ryazanov in the drafting commission was rejected by the narrow majority of 143 votes to 132.[4] Secondly, while the resolution on wages policy enjoined in general terms a continuance of " the struggle against the use of overtime ", Shmidt successfully resisted a specific amendment to his report as People's Commissar of Labour condemning abuses of overtime : examples quoted by supporters of the amendment showed that in some factories as much as 60 hours' overtime a month was worked, and that the legal requirement of a continuous 42-hour break at the week-end was being ignored.[5] A year later, 20 per cent of industrial workers were said to be on overtime work.[6]

[1] See The Bolshevik Revolution, 1917–1923, Vol. 2, pp. 110, 199.
[2] VKP(B) v Rezolyutsiyakh (1941), i, 628.
[3] Trud, August 23, September 17, 1924.
[4] Shestoi S"ezd Professional'nykh Soyuzov SSSR (1925), pp. 296, 460, 596, 600-601 ; a statement in Voprosy Truda, No. 11, 1924, p. 7, that the congress displayed " striking unanimity " in favour of piece-work, only Ryazanov opposing it, was far from the truth. The campaign culminated in a detailed order of the trade unions for the application of an unrestricted system of piece-rates (Trud, February 4, 1925).
[5] Shestoi S"ezd Professional'nykh Soyuzov SSSR (1925), pp. 464, 639-640.
[6] XIV S"ezd Vsesoyuznoi Kommunisticheskoi Partii (B) (1926), p. 785.

Here, too, as in the case of restrictions on work for women, the urgent needs of the country were likely to over-ride the concern for the individual worker which inspired the ideals and professions of international socialism. One result of the increased prevalence of piece-work and of overtime was that standard wage rates came to play a less important part in the determination of wages actually paid. Wages actually paid were frequently double, or more than double, the rate laid down in the collective agreements under the 17-grade scale, especially in the higher grades of manual workers, where piece-work and overtime were most prevalent.[1]

The sixth trade union congress of November 1924 marked a decisive stage in the gearing of wages to productivity. Since the August resolution of the party central committee, what was virtually a wages stop had been in force. From the high point of October 1924 wages fell back sharply in November, then recovered a little and remained merely stationary for the rest of the winter.[2] Meanwhile productivity increased rapidly;[3] and this increase was due mainly, though not wholly, to increased intensity of labour.[4]

[1] See the detailed table in A. Rashin, *Zarabotnaya Plata za Vosstanovitel'nyi Period Khozyaistva SSSR* (1928), p. 53. The proportion of actual wages to wage rates was highest among the metal workers, followed closely by the chemical and paper industries.

[2] The monthly averages were as follows :

|  | In Conventional Rubles | In Chervonets Rubles |
|---|---|---|
| **1924** | | |
| October | 25·58 | 42·25 |
| November | 22·92 | 38·54 |
| December | 23·54 | 39·71 |
| **1925** | | |
| January | 23·56 | 40·07 |
| February | 22·72 | 39·77 |
| March | 23·02 | 41·74 |

(*Planovoe Khozyaistvo*, No. 2, 1926, p. 54.)

[3] According to a calculation in Y. S. Rozenfeld, *Promyshlennaya Politika SSSR* (1926), p. 361, the value of industrial output in terms of man-days had increased in April 1925 by 32 per cent over the same period in the previous year and for the whole year 1924–1925 by 46 per cent.

[4] The presidium of the Leningrad provincial trade union council in July 1925 held that increased productivity had been achieved " through heightened intensity of the labour of the workers, and only in a small degree through an improvement in organizational techniques of production or an increase of technical improvements in the enterprise " (*Leningradskaya Pravda*, July 26, 1925).

The relation between wages and productivity continued to be a burning issue. In February 1925 STO set up a commission to study it [1] — probably a device to appease criticism rather than a prelude to any change of policy; and at the fourteenth party conference two months later a trade union representative complained of the way in which norms of output were being constantly increased.[2] In 1925, 53·4 per cent of hours worked in large-scale industry were paid at piece-rates, and in 1926 the percentage had risen to 55·6. In the metal-working and textile industries nearly two-thirds of the hours worked were paid at piece-rates, in coal-mining about a half.[3] Figures for March 1926 showed that of all wages paid in large-scale industry 4·2 per cent were for overtime; and this percentage which was said to compare with a corresponding percentage of 6·4 in 1914, rose to 5·8 in the metallurgical industry and to 7·7 in mining.[4]

In the winter of 1924–1925 attention began to be drawn to a steeply rising accident rate in industry. The trade union newspaper reported that a circular had been sent out from Narkomtrud instructing labour inspectors " categorically to resist the tendency in some quarters to diminish attention to questions of labour protection in connexion with the campaign to raise the productivity of labour ".[5] But complaints continued to be heard. The head of the labour division in the Moscow trade union organization cautiously pronounced that " causes connected with the rise in the productivity of labour have a definite effect on the increase in accidents ". In Leningrad the number of industrial accidents had risen from 4000 to 10,000 (proportional figures to the number employed were not given), the main causes being " ignorance and pressure on the workers for increased quantity of production ". In the Donbass the number of accidents had doubled in the past year, and the number of fatal accidents increased by 40 per cent.[6]

---

[1] *Sobranie Zakonov, 1925*, No. 6, art. 35.

[2] *Chetyrnadtsataya Konferentsiya Rossiiskoi Kommunisticheskoi Partii (Bol'-shevikov)* (1925), pp. 193-194.

[3] *Sotsialisticheskoe Stroitel'stvo SSSR* (1934), p. 337. According to A. Rashin, *Zarabotnaya Plata za Vosstanovintel'nyi Period Khozyaistva SSSR* (1928), pp. 33-34, the percentages of workers on piece-rates (figures for September of each year) were 45·7 in 1923, 51·4 in 1924, 60·1 in 1925 and 61·3 in 1926 ; this confirms that 1924–1925 was the period of most rapid increase in piece-work.

[4] *Ibid.* p. 57 ; after March 1926 overtime began to decline.

[5] *Trud*, January 9, 1925.          [6] *Ibid.* February 6, March 11, 17, 1925.

Articles in the press described the prevalence of industrial accidents as " the scourge of the working class, a social calamity ",[1] and attributed it partly to increasingly obsolete and worn-out plant and equipment, but mainly to the productivity drive.[2] According to official figures, the accident rate in mining increased from 1095 per 10,000 workers in 1923–1924 to 1524 in 1924–1925.[3]

In the spring of 1925 this unremitting pressure for production led to a fresh wave of unrest among the workers. Large-scale strikes occurred, " without the knowledge of the trade unions, without the knowledge of the party organs, without the knowledge of the economic organs ", in the textile factories of the Moscow and Ivanovo-Vosnesensk regions ; [4] and these were only symptoms of a wider dissatisfaction. In the first half of 1925, one in every six trade unionists was involved in an industrial conflict, meaning a dispute which could not be settled by the local Assessment and Conflict Commission (RKK).[5] The precise causes of the trouble were difficult to discover. But the trade union newspaper did not hesitate to raise again the old complaint of collusion between unions and management :

Indeed what can be the relation of the workers to their union when their representatives in the Assessment and Conflict Commissions, instead of defending the interests of the workers, are occupied in dismissing and fining them ? What confidence can the worker have in his factory committee when, submitting to the influence of the economic organ, it confirms without any justification the lowering of an assessment or the raising of a norm ? [6]

The party central committee appeared to endorse this explanation. In July it announced that the trade union organization must not be allowed to become " an appendage of the administration ". With this end in view, " the practice of electing factory committees on prepared lists must at the present time be

[1] *Leningradskaya Pravda*, August 2, 1925.
[2] *Trud*, September 25, 1925.
[3] *Sed'moi S"ezd Professional'nykh Soyuzov SSSR* (1927), p. 382.
[4] *XIV S"ezd Vsesoyuznoi Kommunisticheskoi Partii (B)* (1926), p 722.
[5] For these commissions see *The Interregnum, 1923–1924*, p. 65.
[6] *Trud*, June 16, 1925.

abandoned " ; and new elections to factory committees were to
be held, " after careful preparation ", in the Moscow province
and in Ivanovo-Vosnesensk, where the strikes had been worst.[1]

Yet, when the balance was struck, there was no doubt that low
wages were the primary cause of the discontent of the workers.
Some alleviations of their plight could no longer be refused.
Under the impetus, partly of the strikes, and partly of the rapid
expansion of industry and easy credit conditions, the wages stop
imposed in the autumn of 1924 broke down. The fourteenth
party conference in April 1925 had concentrated on the productivity of labour. As regards wages, it had merely repeated
the old formula about bringing up wages " in backward branches
of industry and regions " to the general level — but " primarily
on the basis of increased productivity ".[2] But the third Union
Congress of Soviets in the following month, while still pressing
for increased productivity and " intensification of labour ",
cautiously conceded that, " as a result of higher productivity,
wages too must rise " [3] — the first official intimation for many
months that a wages increase might be once more on the agenda.
In fact, without any recorded formal decision, a general and substantial increase in wages, both nominal and real, took place
in the summer of 1925. From an average monthly total of 23·02
conventional rubles or 41·74 chervonets rubles in March 1925
wages rose to 25·24 conventional rubles or 45·05 chervonets
rubles in June, and to 30·6 conventional rubles or 51·14 chervonets rubles in September. On October 1, 1925, the real wages
of all industrial workers were from 5 to 10 per cent above the

[1] *Pravda*, July 17, 1925. The full text of the resolution does not seem to
have been published ; but the publication of extracts from it in *Pravda* was
accompanied by a leading article drawing attention to its importance. The
diagnosis of " inattentiveness of the trade unions to the interests of the workers "
and of " an unnatural bloc between trade unions, party and Red managers "
was confirmed by the fourteenth party congress in the following December
(*XIV S"ezd Vsesoyuznoi Kommunisticheskoi Partii* (B) (1926), pp. 722-724, 729,
785). According to Tomsky, the strike in one factory began owing to the
introduction of a new and unworkable method of feeding the yarn to the looms
(*ibid.* p. 734). Strikes were now a comparatively rare phenomenon : in 1924
there were 267 strikes (151 in state enterprises) involving 42,000 workers ; in
1925, 196 strikes (99 in state enterprises) involving 43,000 workers (*Sed'moi
S"ezd Professional'nykh Soyuzov SSSR* (1927), p. 90).
[2] *VKP(B) v Rezolyutsiyakh* (1941), ii, 25.
[3] *Tretii S"ezd Sovetov SSSR : Postanovleniya* (1925), p. 20 ; *Sobranie
Zakonov, 1925*, No. 35, art. 251.

levels of a year earlier ; those of transport workers, the lowest
paid category, had risen by as much as a third.[1] These sharp,
unplanned advances, conceded in response to industrial unrest,
seem to have taken everyone by surprise. In preparing its
control figures for 1925–1926 in the summer of 1925, Gosplan
had foreseen a rise of 16 per cent in wages for the forthcoming
year, which would have brought wages to an average of 48 chervonets rubles a month by September 1926. In fact, wages had
already reached an average of 51 rubles by September 1925.[2]

By the end of the summer this wages increase was causing
anxiety to the party leaders, both on grounds of substance and on
grounds of party tactics. In the first place, it had an inflationary
character ; and fears of inflation, and of a potential threat to the
currency, were now just beginning to be felt.[3] Secondly, the
effective pressure for an increase had come, not from the trade
unions, which, under the leadership of Tomsky and Andreev,
remained at this time unimpeachably faithful to the party leadership, but — incongruously enough — from Narkomfin and
STO ;[4] and Sokolnikov, the People's Commissar for Finance, and

[1] The global monthly totals in conventional and chervonets rubles are in
*Planovoe Khozyaistvo*, No. 2, 1926, p. 54 ; for more detailed particulars about
different industries see *ibid.* No. 5, 1926, pp. 280-281 ; *Ekonomicheskoe Obozrenie*, March 1926, pp. 112-113. According to figures in *Ekonomicheskaya
Zhizn'*, November 14, 1925, wages at the end of September 1925 were 14 per
cent above the level of the preceding year and had reached 95 per cent of the
level of 1913 : only in heavy industry and mining did wages still lag seriously
behind, having attained 74 and 66 per cent respectively of the pre-war level.
The table in *Ekonomicheskoe Obozrenie*, April 1926, pp. 109-110, 113, showed
a sharp rise in the proportion of expenditure on labour to total costs of production in state industry from 23 per cent in April 1925 to 32 per cent in July 1925 :
it fell back to 26 per cent in September 1925.

[2] *Planovoe Khozyaistvo*, No. 1, 1926, p. 40. The calculation was complicated by the fact that Gosplan had mistakenly expected a rise in the purchasing power of the ruble, and had estimated that the rise in wages would amount
to a 20 per cent increase in real wages ; on this basis a rise of nominal wages to
53 rubles would now have been required (*ibid.* No. 2, 1926, p. 54). Gosplan
had, moreover, recommended that workers' rents should be raised by 3 rubles
a month to provide a fund for repairs, and this recommendation had not been
accepted, so that the Gosplan wages estimates should theoretically have been
reduced by that amount.                          [3] See p. 479 below.

[4] An opposition delegate at the fourteenth party congress related that at one
factory meeting at this time a manager declared : " I have just arrived from
Moscow, I have been told there that a 10 per cent increase can be given ; and
I am giving it to you " (*XIV S"ezd Vsesoyuznoi Kommunisticheskoi Partii (B)*,
(1926), p. 785).

Kamenev, the president of STO, were already in the autumn of 1925 coming out in more and more open opposition to the policies not only of Bukharin, Rykov and the party " Right ", but of Stalin. It was Sokolnikov who had unexpectedly and ostentatiously raised the wages issue at a financial conference of June 15, 1925. Wages in the metallurgical industries, in the mines and on the railways were all, he declared, below their pre-war level, and it was time " in the eighth year of the Soviet power " to undertake the levelling up of these wages.[1] Kamenev in his speech to the Moscow party organization on September 4, 1925, admitted that the rise in wages had fallen behind the growth in productivity, blamed the trade unions for having failed to maintain the link between the party leadership and the masses on this question, and put forward a tentative suggestion that a scheme should be devised to enable industrial workers to share in the profits of enterprises.[2] That the " new opposition " of 1925, unlike Trotsky and the 46 in the autumn of 1923,[3] was ready to exploit the discontents of the workers was a tribute to the growing strength and importance of the proletariat as industry expanded ; in any case it could not fail to annoy the trade unions [4] and to alarm the party leadership, both of which had been caught unawares without any effective wages policy. That the strikers enjoyed a measure of sympathy in the party is shown by a speech of Uglanov, who deplored " the participation of communists in unorganized conflicts of workers with the management ", and threatened severe disciplinary measures, " including even expulsion from the party ", against those who offended in this way.[5]

In October 1925, with the fourteenth party congress looming ahead, and deep dissensions already threatened, the party central

---

[1] *Sotsialisticheskoe Khozyaistvo*, No. 4, 1925, pp. 18-19.

[2] For this speech see p. 292, note 3 above ; Kamenev reverted to the profit-sharing suggestion three months later in his speech at the Moscow party conference reported in *Pravda*, December 13, 1925.

[3] See *The Interregnum, 1923–1924*, pp. 326–328.

[4] At the fourteenth party congress in December 1925 Andreev denounced Sokolnikov's behaviour as " an attack on the authority of the trade unions " and proof of " an irresponsible attitude towards the workers " (*XIV S"ezd Vsesoyuznoi Kommunisticheskoi Parti (B)* (1926), pp. 795-796) ; other speakers attacked it as demagogy (*ibid.* pp. 242, 296, 339, 500). Sokolnikov had touched a sensitive spot. Kamenev's profit-sharing proposal also came under attack on the ground that it would unfairly favour workers in light industry (*ibid.* pp. 170, 196).     [5] *Pravda*, October 4, 1925.

committee skirted cautiously round the wages issue. The draft
resolution on economic policy submitted to the session by Kamenev
contained the following passage :

> The central committee, while placing wages policy in direct
> dependence on the level of labour productivity, thinks indis-
> pensable a further pursuit of the policy of raising wages and of
> bringing backward sections of industry up to standard.[1]

But the Kamenev resolution was shelved ; [2] and the committee, in
a carefully balanced resolution which reflected differences of
opinion within it,[3] showed itself mainly concerned to make the
trade unions the scapegoat for the labour unrest earlier in the
year. The unions were accused of having sometimes neglected
" their most important and chief task — the defence of the
economic interests of the masses organized by them ", though
they were at the same time praised for their work " in the matter
of the restoration of the economy and of industry ", and encouraged to pursue it. The injunction of August 1924 to cooperate with the " economic organs " [4] was passed over in
silence ; and " certain trade union organizations " were convicted
of " a one-sided, so-called ' managerial deviation ' ". The phrase
was cunningly contrived to cover two apparently opposite errors —
an unduly passive attitude towards managerial demands on the
workers, and unduly active interference in questions of management ; but it aptly characterized the growing tendency of the
unions to make common cause with managements against the
troublesome worker.[5] On the other hand, it was admitted that
there had sometimes been too much interference by the party

---

[1] L. Kamenev, *Stat'i i Rechi*, xii (1926), 371.    [2] See p. 306 above.
[3] The account of differences between the leaders on this question in *XIV
S"ezd Vsesoyuznoi Kommunisticheskoi Partii (B)* (1926), p. 201, is difficult to
follow but shows that differences did exist.    [4] See p. 387 above.
[5] For an earlier stage of this process see *The Interregnum, 1923–1924*, p. 94.
A leading article in *Ekonomicheskaya Zhizn'*, October 22, 1925, called for " a
clearer differentiation between the work of the trade unionist and of the
manager ", and accused some trade unionists of regarding their work as " subordinate and auxiliary to the work of the manager " ; *Bol'shevik*, No. 21-22,
November 30, 1925, p. 5, admitted that " the worker takes a particularly
negative attitude to the ' managerial ' deviation of the trade unions which
occurs here and there ", and that " the content of trade union work comes with
us to stand partially, in a certain limited sense, *in contradiction with the demands*
of the working masses ".

in trade union work and appointments. The resolution in effect registered a certain uneasiness, but left things where they were.[1]

Under cover of this resolution the wages question seemed to have been gently pushed out of sight. Almost by the way of an afterthought, the committee decided, on a report from the People's Commissar for Labour, to instruct the Politburo " to work out the question of the possibility of some increase of wages in the most backward sectors of industry and to review the declarations already made on questions of wages in particular front-line sectors of industry ".[2] An instruction so cautiously worded clearly did not invite drastic action. A leading article in *Pravda* on October 10, 1925, recommended " caution in the matter of a general rise in wages ", while admitting that wages in heavy industry and transport must be allowed to catch up with the rest. A leading article in the official economic journal explained that, while some rise in wages was inevitable, the increase " must not be either excessive in relation to the state of our industrial resources, or general " ;[3] and Shmidt himself in an article in *Pravda* was still less encouraging on the prospect of a further advance.[4] Whatever the intention of the resolution of the central committee, its effect was to call a sharp halt to the wage increases of the past six months. After a slight but general fall in October and November 1925, from which only the ill-paid transport workers were exempt, wages once more remained fairly stable throughout the winter of 1925–1926.[5] In the middle of November the theses on the trade unions, which Tomsky was to present to the party congress in December for endorsement, were approved by the party central committee and published. But they added little to

    [1] *VKP(B) v Rezolyutsiyakh* (1941), ii, 41-43.
    [2] *Ibid.* ii, 32 : the instruction does not seem to have been included among the formal resolutions of the committee.
    [3] *Ekonomicheskaya Zhizn'*, October 14, 1925.
    [4] *Pravda*, October 20, 1925.
    [5] *Planovoe Khozyaistvo*, No. 6, 1926, p. 258 ; since the purchasing power of the ruble declined during the winter, a fall in real wages was partly masked by stable nominal wages. The reference in the resolution of the party central committee of April 1926 to measures taken to bring about a general increase in wages " at the beginning of the economic year " (*VKP(B) v Rezolyutsiyakh* (1941), ii, 97), i.e. in October 1925, was presumably prompted by the desire to give the central committee credit for the wage increases of the previous summer ; but the misdating was flagrant.

the October resolution of the central committee, and, apart from a passing allusion to " the improvement in the material position of the workers ", passed over the wages question in silence.[1] Sokolnikov, nothing daunted, continued to harp on " the task of restoring not only the plant and buildings, but the living forces of the working class " and of " bringing back wages, where possible, to the pre-war level ".[2] But Dzerzhinsky, on the eve of the congress, roundly attacked the " higher wages " slogan as impracticable. All resources were needed for the indispensable purpose of building up basic industry ; and these must be drawn " from the only source of all wealth, namely, from the worker and the peasant ".[3] An intensive programme of industrialization, which concentrated on the production of the means of production, was not compatible with an inflated wages bill which would increase the pressure on the already over-strained market for consumer goods.

By the time the fourteenth party congress met in the second half of December 1925, all other issues had been overshadowed by the struggle between the party leadership and the Leningrad opposition. In the general debate the wages issue was raised only in order to taunt Sokolnikov with demagogy and irresponsibility, and not seriously discussed on either side. The debate on Tomsky's theses on the work of the trade unions took place at a late stage of the proceedings when the main battle had been fought and won, and excited no strong interest. Tomsky in his speech [4] merely embroidered the text of his theses, which in turn repeated most of the criticisms made in the resolution of the party central committee two months earlier. The strikes of the past spring had been evidence of a culpable inattention by the trade unions to the needs of the masses of workers. The correct desire of the unions " to support all reasonable measures . . . taken by the managers " had sometimes led to a threefold *bloc* (" a triple alliance ", interjected Ryazanov) between management, party and trade unions, which inevitably raised the question against whom this " united front " was directed. Contact with

---

[1] They were published in *Pravda*, November 21, 1925.
[2] G. Sokolnikov, *Finansovaya Politika Revolyutsii*, iii (1928), 230.
[3] *Ekonomicheskaya Zhizn'*, December 18, 1925.
[4] *XIV S"ezd Vsesoyuznoi Kommunisticheskoi Partii (B)* (1926), pp. 722-747.

the workers should be fostered by making the conclusion of collective agreements less " automatic ". Tomsky spent some time praising and advocating the system of " production conferences " and " production commissions " representing workers and management, which should be constituted not only for factories as a whole, but for separate shops within the factory.[1] But he added a warning "not to exaggerate the rôle of these conferences or to get it into the head of the participants that their decision is final ". Otherwise "our managers will not be able to work", and we shall be faced with the " worst form of collegiality ".[2] In a cursory general reference to strikes, Tomsky confined himself to the observation that in state enterprises strikes " are theoretically admissible, but do not occur in practice ", and that strikes were

---

[1] It was apparently the thirteenth party conference of January 1924 which, in its endeavour to allay current industrial unrest, first recommended the trade unions to convene production conferences in the trusts " with the participation of representatives of the factory committees and of the directors of the enterprises " (VKP(B) v Rezolyutsiyakh (1941), i, 539) ; and a system of production conferences and production commissions in all important industrial enterprises was prescribed in a joint circular of the central trade union council and of Rabkrin of May 21, 1924 (quoted in Y. S. Rozenfeld, Promyshlennaya Politika SSSR (1926), pp. 300-301). The project was once more commended by the sixth trade union congress in November 1924, being specifically associated with the campaign to raise the productivity of labour (Shestoi S"ezd Professional'nykh Soyuzov SSSR (1925), p. 462), and by the fourteenth party conference in April 1925 (VKP(B) v Rezolyutsiyakh (1941), ii, 5-9). But a detailed resolution of the party central committee of May 15, 1925 (VKP(B) o Profsoyuzakh (1940), pp. 236-240) admitted that the scheme had exhibited " a number of substantial defects " and had failed to cover " really broad strata of the workers ", and gave fresh instructions to bring it to life. It was not mentioned in the resolution of the party central committee of October 1925 criticizing the " managerial deviation " in the trade unions (see p. 397 above) ; and Tomsky's remarks at the fourteenth party congress show that the problem of reconciling effective production conferences with managerial powers and responsibility was as persistent in the Soviet Union as in capitalist countries. Molotov stated at the congress that 371 conferences covering 34,000 workers had been set up in Moscow, and 204 conferences covering 36,000 workers in Leningrad ; he put in a caveat against attempting to confine them to party members (XIV S"ezd Vsesoyuznoi Kommunisticheskoi Partii (B) (1926), p. 60). It may be doubted whether many existed outside the two capitals. According to an article in Bol'shevik, No. 13, July 15, 1926, pp. 45-58, they " grew weaker " in 1925, and were looked on with disfavour as encouraging the "managerial deviation " in the trade unions.

[2] For the opposition between " one-man management " and " collegiality " and the condemnation of the latter see The Bolshevik Revolution, 1917–1923, Vol. 2, pp. 187-191 : it was now necessary to make it clear that production conferences did not infringe the principle of one-man responsibility in industry.

evidence of " weakness, lack of preparedness, or abnormality in the work of the trade unions and the local party organs ".[1]

The debate on the report was perfunctory. Ryazanov mildly taunted Tomsky with having said nothing whatever about the two vital questions of the productivity campaign — wages and the protection of labour.[2] A spokesman of the opposition enumerated six features of the labour situation which pointed to a régime of state capitalism rather than of socialism : payment of wages at piece-rates ; relations between workers and management ; the army of unemployed ; methods of engagement and dismissal of workers ; the frequency of industrial disputes ; and the prevalence of overtime, amounting to an abandonment of the eight-hour day.[3] Andreev, defending the resolution, furnished a little further information. Having admitted that the trade union question had been placed on the agenda of the congress because " it has acquired particular acuteness in the recent period ", he repeated Tomsky's general criticisms on the inadequate defence by the unions of " the material interests of the workers ", on the " managerial deviation ", and on the weakening of " the link of the trade unions with the masses ". But on the wages question he was adamant. In apparent contradiction with the October resolution of the party central committee, he came out strongly against equalization of wages. Wage differences reflected " the specific weight of the industry and the degree of qualification of this or that section of the proletariat " ; they performed a necessary function and must continue. Piece-rates Andreev described as an old capitalist device which must, nevertheless, be retained for the present " owing to our weak technical equipment ".[4] The

---

[1] The resolution proposed by Tomsky and adopted at the end of the debate contained a separate paragraph on strikes in private enterprises. It quoted the statement from the resolution of the eleventh party congress of 1922 that " one of the chief tasks of the trade unions is an all-round and all-out defence of the class interests of the proletariat and struggle with capitalism ", and required only that the struggle should be " centralized to the maximum extent and conducted with the active support and leadership of the central trade union organs " (*VKP(B) v Rezolyutsiyakh* (1941), ii, 66-67). No strikes enjoying trade union support appear, however, to be recorded at this time, even in private enterprises.

[2] *XIV S"ezd Vsesoyuznoi Kommunisticheskoi Partii (B)* (1926), pp. 780-781.

[3] *Ibid.* pp. 784-785 ; the controversy about state capitalism and socialism will be discussed in Part III in the following volume.      [4] *Ibid.* pp. 793-796.

theses were then unanimously adopted without substantial amendment.

No doubt was allowed to remain as to the effect of this resolution on wage rates. A month later Shmidt, the People's Commissar for Labour, once again trounced the opposition for its demagogic pronouncements on the eve of the congress :

> Some very responsible comrades were then incautious enough to say that in the current year we should attain the pre-war wage level or perhaps even surpass it. That was very rashly said. The trade unions have had to explain to the workers that we cannot reach this level in the present year.[1]

In February 1926 Tomsky addressed the trade union central council in similar terms :

> This year we must openly and honestly recommend all trade unions to tell the workers that no increase in wages can at present be expected. Starting from what we have, we must cut our coat according to our cloth.[2]

In the same month the Politburo passed a resolution in favour of bringing wages in coal-mining and transport up to the general level.[3] But in March 1926 a circular letter from the trade union general council once more insisted that no further wage increases were practicable except on the basis of increased productivity.[4] Tomsky in a trade union meeting struck the patriotic note :

> If the Russian worker raises the question of the possibility of going more slowly (" we dont want sacrifices ", " we will advance more slowly "), then he must ask himself another question : " Where shall we get to by this method ? " The answer is clear : we shall gradually fall behind industrial Europe and become an agrarian colony.[5]

The party central committee at its meeting in April 1926 firmly refused to reopen the wages question. It demanded " decisive measures to raise the productivity of labour by means of the rationalization of production, in particular by the fuller use of

[1] *Trud*, February 4, 1926.     [2] *Ibid*. February 12, 1926.
[3] *VKP(B) v Rezolyutsiyakh* (1941), ii, 97.
[4] *Trud*, March 25, 1926.
[5] M. Tomsky, *Izbrannye Stat'i i Rechi, 1917–1927* (1928), pp. 327-328.

plant, by an increase in the skill of workers, by an improvement of the organization of factories and workshops, as well as by the intensification of the working day, the strengthening of labour discipline, and the struggle with absenteeism, etc.", and observed that the purpose of these measures was that " the wage level already attained should be guaranteed " ; further increase must depend on the expansion of industry and the rise of productivity.[1] Stalin, in his report on the proceedings of the committee, advocated " a campaign to get rid of absenteeism in the workshops and factories, to raise the productivity of labour, and to strengthen labour discipline in our enterprises ".[2] In May 1926 a lengthy decree of STO, which had a hortatory rather than a legislative character, demanded an all-round 10 per cent increase in productivity before the end of the year. This was to be achieved by technical and organizational improvements as well as by greater intensity of work and increased labour discipline.[3]

In spite of the stubborn resistance to further wage increases, the workers could not fail to share to some extent in the rising prosperity of industry or to be conscious of some mitigations of the hardships from which they suffered. Social insurance was for the first time becoming to some extent effective. At the sixth trade union congress in November 1924 Shmidt claimed that 90 per cent of those legally entitled to insurance, or five-and-a-half million workers in all, had been brought into the fund. Payments for temporary incapacity were now being punctually paid from the sickness fund. On the other hand, pensions for permanent disability were still being paid at half rates ; and the highest rate for permanent disability was only 15 rubles a month. Some 50,000 workers — a small number, but a beginning — had been sent to sanatoria and convalescent homes.[4] Difficulties in collecting contributions due from employers were still experienced : a large cancellation of arrears had taken place in April 1924.[5] But here, too, the situation was slowly improving. It was an

[1] *VKP(B) v Rezolyutsiyakh* (1941), ii, 97.
[2] Stalin, *Sochineniya*, viii, 137.
[3] *Sobranie Zakonov, 1926*, No. 35, art. 262.
[4] *Shestoi S"ezd Professional'nykh Soyuzov SSSR* (1925), pp. 189-192.
[5] *Ibid.* p. 204 ; see also *The Interregnum, 1923-1924*, pp. 55-56.

important step forward when, in February 1925, a Union Council of Social Insurance composed of representatives of the trade unions and of the departments concerned was established. This was not an independent body, and the representative of Narkomtrud, who presided over it, had a right of veto on its decisions subject to an appeal to Sovnarkom. But, subject to this control, it supervised the administration of the insurance funds and decided all questions of principle regarding them.[1] According to a report later in the same year, the number of insured persons on July 1, 1925, had risen to nearly seven and a half millions : during the first three-quarters of the financial year 1924–1925, contributions had amounted to 314 million rubles, out of which 196 millions had been paid out in benefits and 87 millions paid to the health authorities for medical services. Unemployment relief continued to be the least satisfactory part of the system. The total of those receiving unemployment benefit on July 1, 1925, was 304,000 ; of these 24 per cent were receiving 30 per cent of the standard wage for their job, and the remainder 20 per cent.[2] The niggardly treatment of the unemployed continued to be a matter of public policy, if not of public necessity.[3]

Another long-standing grievance of the workers began to receive attention at this time — the question of housing. The Russian industrial worker, whether left to fend for himself or lodged, as was the common practice outside the large cities, in barracks provided by the factories, had never enjoyed anything but the most primitive living conditions. The best that could be said of them was that they were slightly more advanced than those to which the Russian peasant was still accustomed. Like other material conditions, however, they had deteriorated since 1914. The building and repair of dwellings had virtually come to a standstill before the revolution, and had not been resumed. Industrial concerns had no funds, and little incentive, to improve or maintain the housing of their workers. Houses in the cities had been transferred to the municipal authorities. But the rents paid for them, even when rents were reinstated under NEP, were

---

[1] *Sobranie Zakonov, 1925*, No. 8, art. 74 ; an elaborate decree on the organization and control of the funds followed a year later (*Sobranie Zakonov, 1926*, No. 19, art. 124).

[2] The report was printed as a special supplement to *Planovoe Khozyaistvo*, No. 12, 1925, pp. 20-22.          [3] See p. 367 above.

for a long time nominal,[1] and upkeep was neglected. In these conditions nothing was done to mitigate the endemic evil of over-crowding, which became acute with the returning flow of popula-tion to the cities after 1921. A review of the situation in towns of the RSFSR in the autumn of 1925 showed that the urban population had increased in the past two years by 7·5 per cent, whereas living-space had increased by 0·6 per cent, and that the average space per head of population had declined by 6·8 per cent.[2] In 1924–1925 the capital value of dwelling-houses through-out the RSFSR was estimated to have depreciated in the past ten years by 48 million pre-war rubles. In May 1925 rents were raised, and out of 342 million chervonets rubles collected in rent in 1925–1926, 143 millions were spent on repairs, so that deprecia-tion for the year was reduced to a figure of 4 million rubles.[3] According to another estimate, it was not till 1934 that deteriora-tion of houses in Moscow was finally arrested.[4] If these were the conditions in the capital, they were probably even worse in other cities.

Where urgent repairs were neglected, new construction was scarcely to be looked for. The first recorded attempt at self-help seems to have come from the cooperative movement, which in

[1] See *The Interregnum, 1923–1924*, p. 68, note 2. In 1925 workers in Moscow were said to pay 5·6 per cent of their wages, and in the provinces 2·3 per cent, for rent and communal services which before the war took 20 to 25 per cent : non-workers paid very high rents, but the number of these was insignificant (*Planovoe Khozyaistvo*, No. 6, 1925, pp. 45-46). In some in-dustries a considerable proportion of workers (in coal-mining and oil as much as 72 per cent) were still lodged by the enterprises in which they worked (*Statistika Truda*, No. 7, 1927, pp. 12-13).

[2] *Sovetskoe Stroitel'stvo*, No. 5, December 1926, p. 113. In 1912 living-space in Moscow was at the rate of 7·4 sq. metres per inhabitant ; in 1920 owing to the exodus of population, and in spite of deterioration and destruction, it had risen to 9·3 sq. metres ; from 1921 onwards it fell rapidly to 5·2 sq. metres in 1925 (D. L. Broner, *Ocherki Ekonomiki Zhilishchnogo Khozyaistva Moskvy* (1946), p. 13). According to an estimate used by Gosplan at the beginning of 1925, the average living-space of the Soviet worker was 9 sq. arshins (approxi-mately equivalent to 5 sq. metres) as against an ideal minimum of 16 sq. arshins ; conditions in the Ukraine and elsewhere were worse (*Planovoe Khozyaistvo*, No. 6, 1925, pp. 36-37). For statistics showing wide variation between different towns see *Sovetskoe Stroitel'stvo : Sbornik*, ii-iii (1925), 28, and *Vestnik Finansov*, No. 6, June 1925, p. 226 : all these calculations date from 1925 when attention was first seriously directed to the question.

[3] *Sovetskoe Stroitel'stvo*, No. 5, December 1926, p. 117.

[4] D. Broner, *Ocherki Ekonomiki Zhilishchnogo Khozyaistva Moskvy* (1946), pp. 64-68.

December 1923 summoned a congress on housing with a view to the creation of building cooperatives, and brought into being an " all-union bureau " for the purpose.[1]  The issue was officially aired in the following month at the eleventh All-Russian Congress of Soviets, which, at the end of a long general resolution on labour conditions, instructed Narkomtrud, " in view of the critical housing situation ", to work out and submit to Sovnarkom a scheme for the building and repairing of houses with state credits.[2]  The authorities in the Ukraine, where the problem was most acute of all, seem to have been the first to move, and in March 1924 issued a decree sanctioning the creation of building cooperatives.[3]  In May 1924 a decree of the USSR prescribed the establishment of a committee " on aid to the cooperative building of workers' dwellings " to supervise the scheme, which was to be financed by a house-tax on well-to-do tenants.[4]

The new policy received the formal endorsement of the thirteenth party congress, which recognized that housing was " becoming the most important question in the material life of the workers ", and instructed all Soviet organs to support the housing cooperatives.[5]  In August 1924, as the result of the deliberations of the committee, a decree was issued providing for the establishment of three types of housing cooperative.  The first was to acquire houses from municipal authorities on twelve-year leases, maintain them and rent them to their members.  The second and third types were to undertake the building of houses, the former being confined to workers, the latter admitting all citizens, and also juridical persons, to membership.[6]  Shmidt

---

[1]  L. Povolotsky, *Kooperativnoe Zakonodatel'stvo* (3rd ed. 1926), p. 232.

[2]  *S"ezdy Sovetov RSFSR v Postanovleniyakh* (1939), p. 295 ; also in *Sobranie Uzakonenii, 1924*, No. 27, art. 262.

[3]  Quoted in L. Povolotsky, *Kooperativnoe Zakonodatel'stvo* (3rd ed. 1926), pp. 247-248.

[4]  *Sobranie Uzakonenii, 1924*, No. 63, art. 636 ; the committee was actually set up by STO on July 5, 1924 (*Sobranie Zakonov, 1924*, No. 1, art. 22), and included a representative of the " all-union organizing bureau for building cooperatives ".          [5]  *VKP(B) v Rezolyutsiyakh* (1941), i, 588.

[6]  *Sobranie Zakonov, 1924*, No. 5, art. 60.  This was a decree of the USSR ; decrees of November 1924 providing that house-leasing cooperatives should have precedence over other applicants for houses, and workers' building co-operatives precedence over other applicants for building land, and guaranteeing to builders' cooperatives rebates in the cost of building materials, were decrees of the RSFSR (*Sobranie Uzakonenii, 1924*, No. 89, arts. 893, 900).  As in many

assured the sixth trade union congress in November 1924 that the indirect subsidies offered to building cooperatives under existing decrees would be equivalent to 39 million rubles in the current financial year, and estimated that the workers' building cooperatives would be able during that time to build 16,500 dwellings for 95,000 workers. The resolution of the congress confined itself to describing " the increasingly acute housing crisis " as " one of the most important questions of the material life of the workers ", and advising trade unions to support the building cooperatives.[1]

By this time it was clear to all that the development of a serious housing programme depended on the availability of funds. Housing for the workers was now an urgent priority ; and specific measures came at last with the drive for industrial expansion which gathered momentum in the spring of 1925. In January 1925 a special bank, the Tsentral'nyi Bank Kommunal'nogo Khozyaistva i Zhilishchnogo Stroitel'stva (Tsekombank), had been set up to finance municipal and other housing.[2] It only remained to provide sources of credit. Decrees of February and March 1925 instructed state trading organizations, and then the state industrial trusts, to set aside 10 per cent of their profits as a welfare fund for the workers : 75 per cent of these funds was to be devoted to workers' housing.[3] At the beginning of April 1925 a municipal tax on dwellings was imposed for the specific purpose of financing the construction of workers' houses : 70 per cent of the receipts were to go to the local authorities and 30 per cent to a central building fund.[4] A trade union delegate at the fourteenth party conference in April 1925 described these measures as inadequate to meet a desperate situation. Some workers were sleeping at the railway stations : others were paying peasants 20 or 25 rubles a month for a bed.[5] The month following the conference was a period of great activity on the housing front. An all-union

other matters, uncertainty prevailed as to the respective competence of union and republican authorities ; decrees of the Ukrainian, White Russian and Transcaucasian republics are on record.
[1] *Shestoi S"ezd Professional'nykh Soyuzov SSSR* (1925), pp. 196-197, 490.
[2] A. Z. Arnold, *Banks, Credit and Money in Soviet Russia* (N.Y., 1938), p. 311. [3] *Sobranie Zakonov, 1925*, No. 26, arts, 176, 184.
[4] *Ibid.* No. 26, arts, 178, 179.
[5] *Chetyrnadtsataya Konferentsiya Rossiiskoi Kommunisticheskoi Partii (Bol'-shevikov)* (1925), pp. 204-205.

conference on housing and building was held at Gosplan; [1] the statutes of a union of housing cooperatives (Zhilsoyuz) were duly approved; [2] and the third Union Congress of Soviets approved of " substantial state aid for the building of workers' dwellings ", and voted a sum of 36 million rubles for the purpose in the budget.[3] In July 1925 the party central committee instituted a drive to support and develop the building cooperatives.[4]

Thanks to these efforts, the year 1924 witnessed a resumption of house-building — industrial, municipal, cooperative and private — on a substantial scale, which underwent a further spectacular increase in the following year. The sums expended on the building of dwelling houses by different authorities during these two years were shown in the following table (in millions of chervonets rubles) : [5]

|  | 1924–1925 | 1925–1926 |
|---|---|---|
| Industry and Transport | 73 | 110 |
| Municipal and Local | 26·6 | 63·2 |
| Cooperatives | 5·9 | 26·5 |
| Private | 51 | 87 |
| Total | 156·5 | 286·7 |

In Moscow five times as much living-space was built in 1925 as in 1924, and twice as much in 1926 as in 1925.[6] Gosplan in its control figures for 1925–1926 recommended a capital investment in housing for the year of 375 million rubles, of which 200 millions would be required for the repair, renovation or completion of existing houses and 175 millions for new construction (of which 70 millions would go for replacement and 105 millions for housing of new industrial population). Of the total of 375 millions, 135

[1] Its proceedings were published as *Voprosy Sovremennogo Zhilishchnogo i Promyshlennogo Stroitel'stva* (1926), reviewed in *Planovoe Khozyaistvo*, No. 4, 1926, p. 256.

[2] *Izvestiya Narodnogo Komissariata Truda SSSR*, No. 31-32, 1925, pp. 46-58.

[3] *Tretii S"ezd Sovetov SSSR : Postanovleniya* (1925), pp. 36-37.

[4] *Izvestiya Tsentrál'nogo Komiteta Rossiiskoi Kommunisticheskoi Partii (Bol'shevikov)*, No. 29-30 (104-105), August 10, 1925, p. 8.

[5] *Kontrol'nye Tsifry Narodnogo Khozyaistva SSSR na 1927-1928 god* (1928) p. 532 ; the total for 1924-1925 was said to represent a 7 per cent increase over 1923-1924, the figure for which was not given.

[6] D. Broner, *Ocherki Ekonomiki Zhilishchnogo Khozyaistva Moskvy* (1946) p 16.

millions would be furnished by industry, 100 millions from the budget, and the rest from the internal resources of the building industry and long-term credits. It was also proposed to cover part of the cost by raising workers' rents by 3 rubles a month.[1] These ambitious plans were not carried out in full. The proposal to raise rents was too unpopular to be adopted.[2] The allocation in the budget was restricted to 80 million rubles.[3] Even then the building programmes proved too large for the available materials ; and, with rising prices, the sums allocated did not go as far as was expected.[4] The party central committee in April 1926 once more demanded " intensive application " to building, since " the further growth of industry, the increase in the productivity of labour and the improvement of living conditions for the workers all come up against the housing crisis ".[5] While 1924–1925 was the year in which construction at last began to catch up with depreciation, the average floor space per head of population in all towns of the USSR is said to have actually declined from 6·1 sq. metres in 1924–1925 to 5·6 sq. metres in 1926–1927.[6] But, miserable though the conditions were, the quality as well as the quantity of houses slowly improved, and for the first time the prospect of further improvement was real. Among the material conditions which, in the middle and later nineteen-twenties, made the increasing pressure on the worker for higher productivity tolerable, and in part successful, the gradual amelioration in the housing situation probably occupied an important place.

While the standard of living of the industrial worker and the efficiency of his work were rising painfully and by slow degrees from the appallingly low level of the civil war period and the first two years of NEP, the character and status of the trade unions were also subjected to a gradual process of change. After

[1] *Kontrol'nye Tsifry Narodnogo Khozyaistva na 1925–1926 god* (1925), pp. 28-30, 33.
[2] For a discussion of this proposal see *Planovoe Khozyaistvo*, No. 9, 1925, p. 23.
[3] *SSSR : Tsentral'nyi Ispolnitel'nyi Komitet 3 Sozyva : 2 Sessiya* (1926), p. 44. [4] *Planovoe Khozyaistvo*, No. 2, 1926, p. 29.
[5] *VKP(B) v Rezolyutsiyakh* (1941), ii, 93.
[6] *Planovoe Khozyaistvo*, No. 1, 1928, pp. 123-136.

.the outbreak of strikes in the spring of 1925, both party and trade union leaders had been surprisingly ready to detect a radical disease in the movement, a breakdown of the link between the trade union leadership and the masses organized in the unions. The frank, almost ostentatious, avowal of this failure in the resolutions of the party committee in October 1925 and at the ensuing fourteenth party congress [1] was no doubt in part a matter of tactics : to make the trade union leaders responsible for what had occurred was a way of exonerating the party and party policy. But the diagnosis invited further discussion of the disease. The growing rift could be explained either from below or from above, from the standpoint of the changing character of trade union membership or of the changing status of the leadership. Both aspects were significant.

In numbers and organization the trade unions had advanced steadily during the past three years. Between the fifth trade union congress in September 1922 and the sixth in November 1924, the number of trade unionists rose from four and a half to six millions ; and on April 1, 1925, there were 6,900,000 and a year later 8,750,000.[2] But the quality of the growth required scrutiny. Though the increase was due in part to the expansion of industry, only about a third of trade union members were workers in industry.[3] Much of the large accretion of numbers was due to the success of the trade unions in organizing seasonal workers who had hitherto escaped the trade union net — notably the agricultural and building workers. Between January 1, 1923, and October 1, 1925, the membership of Vserabotzemles rose from 253,000 to 730,000 and of the builders' union from 107,000 to 575,000.[4] Moreover, not only the agricultural workers and the

[1] See pp. 397-399 above. Tomsky's own admissions were not less sweeping : " we often confuse our functions with the functions of the managers ", he said in a speech of October 1925 (M. Tomsky, *Izbrannye Stat'i i Rechi, 1917–1927* (1928), p. 298).

[2] *Shestoi S"ezd Professional'nykh Soyuzov SSSR* (1925), p. 87 ; *XV Konferentsiya Vsesoyuznoi Kommunisticheskoi Partii (B)* (1927), p. 267.

[3] The numbers of industrial workers covered by collective agreements were 1,727,200 on January 31, 1924, 2,198,600 on January 31, 1925, and 2,747,000 on January 31, 1926 : the proportion of industrial workers covered by collective agreements at these dates was 86·6 per cent, 94·1 per cent and 96·4 per cent respectively (A. Rashin, *Zarabotnaya Plata za Vosstanovitel'nyi Period Khozyaistva SSSR* (1928), p. 30).

[4] *XIV S"ezd Vsesoyuznoi Kommunisticheskoi Partii (B)* (1926), pp. 724-725.

builders, but a large majority of the new recruits to factory industry, were peasants fresh from the country who were totally strange to the demands and traditions of the trade unions, so that this rapid numerical growth represented, in the words of the resolution of the fourteenth party congress, " an undue inflation of the unions with non-proletarian elements and a conversion of union membership into a mere formality (the man with the trade union ticket) ".[1]  On the eve of the congress Bukharin spoke of " the new strata of population, the proletarianized peasant mass which is only just *becoming* working class ", and of the gigantic task of " educating these new strata of the working class ".[2]  The large influx into the unions of new and inexperienced members was used by Tomsky in his speech at the congress as an argument for an unusual emphasis on cultural work, which he described as " the most important branch of the work of the trade unions ". The worker could not be expected to " think only about the proletarian revolution and its problems ".  He needed also " healthy recreation and healthy laughter " ;  the workers' clubs must cater for these needs as well as for his " political development and education ".[3]  The advice was sensible.  But the conception of a trade union as an organization whose most important function was to care for the political and cultural education of untutored masses of half-proletarianized peasants was poles asunder from the conception of a trade union as an organization of class-conscious workers combining to voice their grievances and to enforce a remedy of their wrongs, and implied a totally different relation between leadership and members.  The character of a large part of the membership of Soviet trade unions in the middle nineteen-twenties made them very different from earlier organizations which had borne that name, whether in Russia or elsewhere.  It was an integral part of that legacy of backwardness in the Russian economic and social structure which made the building of socialism in these conditions an uphill task.

---

[1]  *VKP(B) v Rezolyutsiyakh* (1941), ii, 62.

[2]  *Pravda*, December 10, 1925 ; the speech was delivered at the Moscow provincial party conference.

[3]  *XIV S"ezd Vsesoyuznoi Kommunisticheskoi Partii (B)* (1926), pp. 736–738 ; in a previous speech of October 1925 Tomsky had pleaded that workers' clubs should be places of recreation as well as of instruction (M. Tomsky, *Izbrannye Stat'i i Rechi, 1917–1927* (1928), pp. 300-302).

More important, however, than the changing character of trade union membership was the continued evolution of the trade unions towards complete integration in the governmental machine — the process of which the so-called " managerial deviation " was a symptom rather than a cause. Even in capitalist countries the relation of the trade unions both to employers and to the state was soon to undergo a subtle change, of which the first signs were already apparent. In a socialist, or partly socialist, economy that relation could not be other than intimate. The formal independence conferred on the trade unions by NEP had in practice meant not real independence (for that was impossible), but a less influential standing in the hierarchy of power.[1] Once the proposition had been accepted — and it could not well be rejected — that the long-term interest of the worker was bound up with a total increase in production, trade union policy could not be divorced on any major issue from the policy of the government : the over-riding aim to build up the national economy on socialist lines. Zinoviev, in his opening speech at the sixth trade union congress in November 1924, reminded the unions that they were called on to pursue not " a narrow trade union policy ", but " *the policy of the working class in a peasant country* " : this was what distinguished " Leninism, Bolshevism " from " trade-unionism ".[2] It was no accident that Andreev, formerly a supporter of Trotsky's programme for the " statization " of the unions,[3] had risen to a commanding position in the trade union hierarchy, while Tomsky, having learned the lesson of his disgrace in 1921 for neglect of the party line,[4] was now a faithful advocate of conformity to state policy, and stood on all important issues on the Right wing of the party. The guiding principle was once more enunciated in the resolution of the fourteenth party congress :

The growth and development of Soviet industry and the increase of its production, which are the real pledge of the durability of the achievements of general socialist construction, demand in the future from the Communist Party, from the

---

[1] See *The Interregnum, 1923-1924*, pp. 61-62.

[2] *Shestoi S"ezd Professional'nykh Soyuzov SSSR* (1925), p. 29 ; the word " trade-unionism " in English was used by Zinoviev, as by Lenin (see *The Bolshevik Revolution, 1917-1923*, Vol. 2, p. 101), in a derogatory sense.

[3] See *ibid.* pp. 221-226.          [4] *Ibid.* pp. 324-325.

Soviet power and from the trade unions still greater attention and energy in strengthening and developing these achievements.[1]

In this triumphal progress towards the building of socialism, party, government, managers and trade unions marched hand in hand. Yet such an alliance, whatever the intentions and whatever the professions, could not fail to widen and deepen the gulf between trade union leaders responsible for giving effect to this policy and the inarticulate masses of workers who were the instruments of its application. The virtual abandonment of the strike weapon, the waning interest of the trade unions in social insurance and in the protection of labour, and the exclusion of the unions from any concern in the recruitment of labour,[2] were all symptoms of the new attitude. The trade unions were no longer organizations representative of the special interests of the working class (since no such special interests were recognized), but organs for the performance of certain specific functions within a governmental machine which identified the interests of the working class with those of the community as a whole.

Both the inertia of a large inactive membership and the identification of the leadership with governmental policy helped to further the process of concentration of authority in the trade union organization. The trade unions, like the party, had always stoutly rejected any taint of " federalism " in their organization.

The centralism of the trade union movement [declared Tomsky at the sixth congress in November 1924], the general leadership of the whole trade union movement from the centre, remains and must remain the unshakable guiding principle of our movement, since any dispersal in our work, any independent policy on the part of separate organizations, is harmful and serves as a first step towards the break-up of the working class.[3]

As time went on more decisions were taken by the central council or by its presidium ; less initiative was open to local branches or to individual unions ; more emphasis was laid on " trade union discipline " which required the unquestioning acceptance of the

[1] VKP(B) v Rezolyutsiyakh (1941), ii, 64.
[2] This was complete when the engagement of labour through the labour exchanges ceased to be obligatory in January 1925 : see The Interregnum, 1923–1924, p. 64.
[3] Shestoi S"ezd Professional'nykh Soyuzov SSSR (1925), pp. 172-173.

decisions of the leadership. At the fourteenth party congress in December 1925, Tomsky issued a sharp reminder to trade unionists that " all directives and decisions of all-union congresses and of the central council, as the directing organ between congresses, which works under the immediate guidance of the central committee of the party and under its untiring observation, are binding and must be carried out by local organs ". An appeal against what was thought a wrong decision of the central organ was permissible. But " democratic centralism makes it indispensable that the directives of the central trade union organs should be exactly carried out ".[1] The resolution of the congress, having paid lip service to the need for " independence and initiative in all domains of trade union work ", instructed the party fraction in the trade union central council " to maintain unswervingly the unity of the general line, the unity of action and unity of organizational structure of trade union organizations throughout the territory of the USSR ".[2]

The most conspicuous example of the process of concentration and centralization was the now common practice of concluding collective agreements covering wages and conditions of employment for a whole industry between the trade union central council and the central administration of the industry in Moscow. This had the advantage both of securing a reasonable uniformity of wages throughout the industry and of facilitating the adoption of a coherent wages policy, and was for this reason strongly favoured by industrial managers and by official economic organs. But the system sometimes led, as Dogadov, the spokesman of the central council at the sixth trade union congress, admitted, " to the most extreme bureaucratism ", agreements being concluded without any participation by local trade union organs and branches, and even without their knowledge ; and this complaint was echoed in harsher terms by other delegates.[3] The resolution of the congress spoke cautiously of the need for " a certain decentralization in the work of concluding agreements by way of limiting the system of all-union general agreements " and " an extension of the practice of local agreements supplementary to the general agreements ".[4]

[1] XIV S"ezd Vsesoyuznoi Kommunisticheskoi Partii (B) (1926), p. 743.
[2] VKP(B) v Rezolyutsiyakh (1941), ii, 70.
[3] Shestoi S"ezd Professional'nykh Soyuzov SSSR (1925), pp. 101-102, 121-122.        [4] Ibid. p. 462.

But little was achieved in face of official and managerial opposition. More than a year later at the fourteenth party congress, Tomsky reiterated the principle that a collective agreement ought to be open for " discussion by those in whose name the collective agreement is written ", but admitted that in practice " the workers on whose behalf the collective agreement is signed do not know what is being signed on their behalf ".[1] It is difficult to imagine a procedure more likely to create among the rank and file of trade union membership a sense of separation from a remote and bureaucratic leadership.

Relations between workers and trade union officials in the factories no doubt varied, and evidence about them is rarely available. The problem of the collection of trade union dues [2] was partly solved through the appointment of "collectors specially selected from the mass of ordinary members ", who had the duty of acting as " transmitters of the influence of the union over the mass of its members ", though a resolution of the sixth trade union congress in November 1924 repeated the demand for " a more rapid transition by the unions to the exclusively individual collection of members' dues ", and a delegate admitted during the debate that many unions were still " dependent on the office " for the collection of dues.[3] In the following year a decree was issued requiring state and private enterprises employing labour to contribute 1 per cent of their wages bill to trade union expenses.[4] A crucial link in relations between unions and industrial workers was provided by the factory committees, " the intermediary organ ", as Tomsky called them, " which unites us with the broadest working masses ".[5] One witness referred to the factory committee as the " collective father ", to whom the worker turned for advice on his personal problems, including such questions as " whether to divorce his wife or not ", and argued that it was the factory committee far more than the party which had " penetrated the daily life of the worker ".[6] But such model committees must have been very rare, and the status and character

---

[1] *XIV S"ezd Vsesoyuznoi Kommunisticheskoi Partii (B)* (1926), pp. 730-731.
[2] See *The Interregnum, 1923–1924*, pp. 60-61.
[3] *Shestoi S"ezd Professional'nykh Soyuzov SSSR* (1925), pp. 445, 449, 538.
[4] *Sobranie Zakonov, 1925*, No. 77, art. 585.
[5] *Shestoi S"ezd Professional'nykh Soyuzov SSSR* (1925), p. 90.
[6] L. Trotsky, *Voprosy Byta* (2nd ed. 1923), pp. 142-143.

of the committees varied from factory to factory. In some cases, the committees were completely dominated by trade union nominees who were inhibited by party policy from criticizing the management ; according to one delegate at the sixth trade union congress, the factory committees were being simply " transformed in the eyes of the workers . . . into a department of the economic organ ".[1]   In other cases, the committees seem to have asserted some measure of independence, so that they were regarded with as much mistrust by trade union authorities as by the managements.   But here, too, the authority of the trade unions and the power of the management tended to grow side by side at the expense of spontaneous and unorganized activities of the workers. " We notice ", declared the *rapporteur* on organization at the congress, " a tendency for practical union work to be concentrated in the hands of a narrow circle of professional workers." [2]

The same processes continued to operate in another important field of trade union practice — the settlement of disputes.   At the fourteenth party congress the complaint was heard that, where a dispute was not settled by the RKK, the managerial side, instead of resorting to the procedure of the conciliation court and the arbitral tribunal, carried it direct to the party county or provincial committee and there secured a decision.   This, as Tomsky pointed out, was resented by non-party workers, and often placed the responsibility for an unpopular decision on the party.[3]   The resolution of the congress deprecated this practice, and demanded a strengthening of the RKK, the conciliation courts and the arbitral tribunals as the normal channels for the settlement of disputes.   It proposed, however, only one substantive change. The unilateral right conferred by the labour code on the trade unions to demand reference to an arbitral tribunal without the assent of the other side, though apparently abandoned by a decree of March 1923,[4] was now reasserted and extended to the " economic organs ", which were placed on the same footing for this

---

[1] *Shestoi S"ezd Professional'nykh Soyuzov SSSR* (1925), p. 322.

[2] *Ibid.* p. 320.

[3] *XIV S"ezd Vsesoyuznoi Kommunisticheskoi Partii (B)* (1926), p. 735 ; an interesting and significant implication of this grievance was that the manager had more chance than the trade unions of obtaining a favourable decision from the party authorities.

[4] See *The Interregnum, 1923–1924*, p. 67.

purpose as the trade unions.[1] It is unlikely that this change made much difference in practice, since the right to refer a dispute to compulsory arbitration was already vested in the state authorities, TsIK, Sovnarkom or STO. But it became an item in the indictment of party policy drawn up by the opposition in the following year.[2]

In effect, whatever the point of contact, the essence of relations between trade unions and state, or between trade unions and managers, was determined by common dependence on the party and subordination to its will. This was the core of the so-called "managerial deviation", since a single party line was equally binding on the party representatives who controlled the trade unions and on those who controlled industrial policy. Dzerzhinsky, in reply to a trade union critic at the fourteenth party conference in the spring of 1925, tried once again to establish a reasonable balance between the functions of the trade unions and those of the economic organs under his direction :

> We have a single line, but in practice there is a certain amount of pulling and hauling. We press from one side, they press from the other. . . . Comrade Lepse insists that we should . . . expand the economy, cheapen production, and at the same time not strain things too far in respect of that element which determines everything, on which our politics and economics rest, and which is our base, i.e. the working class. We on the other side defend the interests of the economy against, perhaps, excessive pressure arising from the hard conditions in which our working class has to live.[3]

But the balance was artificial and unreal. Whatever might be said about the long-term interest of the workers in industrial expansion through the accumulation of capital, their short-term interests necessarily ran counter to the campaign for increased productivity without an increase in wages. It was their

---

[1] *VKP(B) v Rezolyutsiyakh* (1941), ii, 66.

[2] L. Trotsky, *The Real Situation in Russia* (n.d. [1928]), p. 49 ; the change is said, in rather exaggerated language, to have transformed the collective contract " from a bilateral agreement into an administrative instrument ". The allegation that most labour conflicts were now settled " by compulsory rather than conciliatory procedures " is not borne out by contemporary press reports ; but compulsory powers can be a decisive factor even if they are rarely used.

[3] *Chetyrnadtsataya Konferentsiya Rossiiskoi Kommunisticheskoi Partii (Bol'-shevikov)* (1925), pp. 207-208.

short-term interests of which they were conscious, and which
they looked to the trade unions in vain to defend. The relation
between the claims of increased production and the claims of the
workers was a major issue of policy. But the source of policy
was in the party, not in the unions ; and, as time went on, the trade
union leaders appear to have had less and less importance in the
framing of party policy.

The same dilemma arose over the method of election of the
trade union leaders and of important trade union officials. The
famous resolution of the tenth party congress in March 1921,
prescribing the dual principle of the choice of leading trade union
personnel " under the directing control of the party " and the
simultaneous application of " normal methods of proletarian
democracy ", had led from the first to embarrassments of interpretation, which were not removed by official recognition of
" contradictions between different tasks of the trade unions ".[1]
While party control grew in practice tighter and more effective,
party doctrine continued to insist on the rule of election in filling
all trade union posts. Andreev at the fourteenth party congress
touched a sore spot when he reproached the organization with
" a too one-sided responsibility, insufficient responsibility of trade
union officials to their electors, one-sided responsibility to higher
trade union and party organs ".[2] The resolution of the congress
once more asserted the principles of " election on a broad basis of
all trade union organs " and of " responsibility, public, simple and
comprehensible to every worker, of elected officials to their
electors ".[3] When Ryazanov said at the congress that not only
managers but trade union officials were " put there by the party ",
someone interjected : " they are elected ". But, when Ryazanov
rejoined, " Elected, removed, transferred — that is all one to me ",
everyone knew that he was telling the indiscreet truth.[4] Nobody
in the trade union movement was likely to forget that Tomsky's
removal and reinstatement had been the work, not of his trade
union constituents, but of the party. Nor could it be otherwise.
Formulae could be found which appeared to reconcile every
demand.

[1] See *The Bolshevik Revolution, 1917–1923*, Vol. 2, pp. 323-327.
[2] *XIV S"ezd Vsesoyuznoi Kommunisticheskoi Partii (B)* (1926), p. 794.
[3] *VKP(B) v Rezolyutsiyakh* (1941), ii, 63.
[4] *XIV S"ezd Vsesoyuznoi Kommunisticheskoi Partii (B)* (1926), p. 780.

Trade unions [declared Tomsky at the fourteenth party congress] strongly welded, strongly linked with the masses, centralized, independent, enjoying the absolute confidence and authority of the members united in them, with communists at the head working under the general leadership of their party — there is the firm bulwark and guarantee for the carrying out of the policy of the party, for the broadening of its authority, for the extension of its unlimited influence over the broadest strata of the workers.[1]

But the view of the trade unions as " the transmission shaft from the communist party to the masses "[2] carried with it the requirement that the party should have the last word in trade union policy and in the choice of those appointed to carry it out. The will of the party was the synthesis in which every conflict or contradiction between the trade unions and the political or economic organs of the state, between union officials and public administrators or managers, was ultimately resolved.

[1] *Ibid.* p. 741.
[2] The phrase was first used in the resolution of the party central committee of January 1922 (Lenin, *Sochineniya*, xxvii, 154), subsequently endorsed by the eleventh party congress (*VKP(B) v Rezolyutsiyakh* (1941), i, 423).

CHAPTER 8

# INTERNAL AND FOREIGN TRADE

## (a) Internal Trade

THE closing of the scissors in the winter of 1923–1924 ended a crisis which never recurred in any comparable form. Never again did the Soviet economy suffer from a " sales crisis " in which purchasers could not be found for manufactured goods. The " sales crisis " of 1923 was succeeded in the spring of 1924 by a " goods famine " which continued for many years, and could be accounted a natural symptom in a rapidly expanding economy. In 1923–1924 the total trade turnover was estimated by Gosplan at 9750 million chervonets rubles ; this compared with a total of 9938 million rubles for the same area in 1913, and, allowing for the rise in prices, represented something like half the pre-war volume of trade.[1] As trade increased it also spread more widely. Turnover on the Moscow exchange, which in 1922–1923 handled two-thirds of the total trade of the country, rose in 1923–1924 by 56 per cent ; but there was a corresponding rise of 137 per cent in the turnover of the provincial exchanges, which now accounted for one-half of the total. This process of decentralization continued in the following year.[2]

[1] *Kontrol'nye Tsifry Narodnogo Khozyaistva na 1926–1927 god* (1926), p. 373 ; for the 1913 figure see S. G. Strumilin, *Ocherki Sovetskoi Ekonomiki* (1928), p. 250. *Itogi Desyatiletiya Sovetskoi Vlasti v Tsifrakh, 1917–1927* (n.d.), pp. 368-369, gives a total of 8567 million rubles for 1923–1924 for the European provinces of the RSFSR and the Ukrainian and White Russian republics ; much higher figures emanating from Narkomvnutorg are quoted in Y. S. Rozenfeld, *Promyshlennaya Politika SSSR* (1926), p. 464. Figures for years before 1923–1924 must be regarded as mainly conjectural, though there is no reason to doubt that trade expanded progressively after 1921.

[2] *SSSR : Tsentral'nyi Ispolnitel'nyi Komitet 2 Sozyva : 2 Sessiya* (1924), p. 40 ; Y. S. Rozenfeld, *Promyshlennaya Politika SSSR* (1926), p. 466 ; slightly higher figures appear in *Sotsialisticheskoe Khozyaistvo*, No. 5, 1924, p. 65. By a decree of STO of April 11, 1924, government departments, institutions and agencies were required to register all commercial transactions at the appropriate exchange (*Izvestiya*, May 7, 1924).

Expanding trade was reflected in a fairly rapid rise in the incredibly low levels to which consumption even of the most primitive commodities had fallen in the period of the civil war. Official statistics of the annual consumption per head of population, in these years, of sugar, salt, matches and cotton textiles offer a revealing picture of misery and gradual recovery : [1]

| | Sugar (in pounds) | Salt (in pounds) | Matches (in boxes) | Cotton Cloth (in arshins) |
|---|---|---|---|---|
| 1913 | 20 | 33 | 25 | 25 |
| 1921–1922 | 2·3 | 13·1 | 5·7 | 3·8 |
| 1922–1923 | 4·4 | 17·7 | 11·5 | 5·3 |
| 1923–1924 | 7·4 | 21 | 14 | 9·5 |
| 1924–1925 | 11·4 | 22·9 | 17 | 15·6 |
| 1923–1924 (as percentage of 1913) | 35 | 64 | 56 | 40 |
| 1924–1925 (as percentage of 1913) | 57 | 69 | 68 | 62 |

But, in spite of the constant insistence of the party leaders at this time on the importance of the peasant market for industrial goods, the evidence appears to show that the expansion and rising wages payments of industry played a larger part in the increased demand for, and consumption of, industrial goods than the rising prosperity of the countryside. It was calculated that, whereas in 1913 the peasant consumed 66 per cent of consumer goods marketed, his share in 1923–1924 had fallen to a little over one-third.[2] According to a further calculation, the value at wholesale prices of the goods received by the country from the town amounted in 1922–1923 to 32 per cent in terms of pre-war prices, or 62 per cent in

[1] Y. S. Rozenfeld, *Promyshlennaya Politika SSSR* (1926), p. 518. The figures for the years down to 1923–1924 (which were also quoted by Dzerzhinsky in his speech reported in *Pravda*, December 4, 1924) were apparently taken from the records of Vesenkha, those for 1924–1925 from Narkomvnutorg, so that doubts may arise whether the basis of calculation has been identical; but they appear plausible.

[2] *Ekonomicheskaya Zhizn'*, August 8, 1924.

terms of current prices, of the value of commodities received from the country by the town ; in 1923–1924, when the scissors had been partly closed, the percentages were 44 in pre-war prices and 69 in current prices.[1]  The balance was made up by the large margin between wholesale and retail prices, especially for industrial goods sold in the countryside, and by the agricultural tax.  In so far as the standard of living of the peasant depended on buying and selling and could be expressed in monetary terms, the decline from the standard of 1913 was greater in the country than in the town.  For that larger part of peasant life which was still governed by the conditions of a natural economy, this comparison had little validity or relevance.  But the extreme poverty of the countryside was a barrier to every attempt of public policy to extract surplus resources from the peasant to develop and strengthen the economy.

However low the standard of living might still be, and whatever comparisons might be made between town and country, the immense and progressive improvement of conditions throughout 1924 was not in doubt.  The expansion of internal trade, the natural reflexion of the expansion of agricultural and industrial production, seemed the reward and the justification of NEP, which had been designed first and foremost to clear the channels for the exchange of products.  But, by the time the expansion was under way, the embarrassments of NEP had also become apparent, and policies of price regulation had been adopted which, if carefully scrutinized, were difficult to reconcile with NEP principles.[2]  From this time forward it became clear that the further development of trade depended not on the removal of obstacles and restrictions that stood in its way, but on positive measures taken to promote and regulate it.  The control and direction of trade became an important element in economic policy ;  and this required an extension of state activity in this sector of the economy, and a curbing of the rôle of the private trader.  The thirteenth party congress of May 1924, which coincided with the creation of the People's Commissariat of Internal Trade (Narkomvnutorg),[3] clearly indicated the two primary functions to be fulfilled by

[1] *Planovoe Khozyaistvo*, No. 2, 1925, pp. 126-127.
[2] See *The Interregnum, 1923–1924*, pp. 110-113, 145-146.
[3] See *ibid.* pp. 142-143.

the new commissariat. The first was " such organization of internal trade and such direction of the activity of state trade and of the co-operatives as will ensure their conquest of the market . . . and realize effective control of the state over the activity of private capital ". The second was " the regulation of all internal trade, and the establishment of firm prices ".[1] The internal trade policy of the ensuing years fell entirely under these two heads.

The rivalry between state, cooperative and private trade was a constant theme of discussion in this period, and official statistics carefully distinguished between the three " sectors ", state and cooperative trade being sometimes grouped together as the " socialized sector ". The figures of trade turnover showed a steady advance of the " socialized sector " over private trade : [2]

TRADE TURNOVER (IN MILLIONS OF RUBLES)

| | State Trade | | Cooperative Trade | | Private Trade | | Total |
|---|---|---|---|---|---|---|---|
| | Absolute | % | Absolute | % | Absolute | % | |
| 1923–1924 | 3025 | 29 | 2750 | 25 | 4833 | 46 | 10,608 |
| 1924–1925 | 4855 | 35 | 5137 | 38 | 3711 | 27 | 13,763 |
| 1925–1926 | 7760 | 35 | 8900 | 41 | 5351 | 24 | 22,011 |

[1] *VKP(B) v Rezolyutsiyakh* (1941), i, 582-584.
[2] S. G. Strumilin, *Ocherki Sovetskoi Ekonomiki* (1928), p. 273 ; these are Gosplan figures, slightly corrected for 1923–1924 from those in *Kontrol'nye Tsifry Narodnogo Khozyaistva na 1926–1927 god* (1926), p. 373, which are said to have rested on an underestimate of the volume of private trade based on tax returns. The figures in *Kontrol'nye Tsifry Narodnogo Khozyaistva SSSR na 1927–1928 god* (1928), pp. 484-485, have been further adjusted upwards and give a less favourable view of the proportion of state trade (see the following table). Strumilin also attempts to make a distinction, not followed in the control figures, between " private capitalist trade " and " simple goods exchange ", the latter representing direct peasant trade not handled by any licensed trader or trading organization. Any estimate of this peasant trade was highly speculative, though the trade was probably quite large and continued to swell the " private " sector ; figures given for it in *Planovoe Khozyaistvo*, No. 11, 1925, p. 10, were apparently derived from Narkomvnutorg (they are also quoted in Y. S. Rozenfeld, *Promyshlennaya Politika SSSR* (1926), p. 464).

A somewhat different picture emerged when the percentages of state, cooperative and private trade were shown separately for wholesale and retail trade : [1]

### PERCENTAGES OF WHOLESALE TRADE

|  | State | Cooperative | Private |
|---|---|---|---|
| 1923–1924 | 61 | 17 | 22 |
| 1924–1925 | 60 | 30 | 10 |
| 1925–1926 | 55 | 36 | 9 |

### PERCENTAGES OF RETAIL TRADE

|  | State | Cooperative | Private |
|---|---|---|---|
| 1923–1924 | 16 | 26 | 58 |
| 1924–1925 | 17 | 40 | 43 |
| 1925–1926 | 16 | 45 | 39 |

While these figures probably underestimate the volume of private trade, especially in the earlier years, the general conclusions are clear. Trade, both wholesale and retail, expanded rapidly in these years, though some of the increase in value in 1925–1926 merely reflects rising prices. In wholesale trade, especially in industrial products, state trading continued to predominate,[2] though it lost some ground to the cooperatives. In retail trade, the share of state trade remained consistently low, but the co-operatives gradually advanced at the expense of the private trader.

Ever since the introduction of NEP, and still more since the

[1] The percentages for the two last years are calculated from the table in *Kontrol'nye Tsifry Narodnogo Khozyaistva SSSR na 1927–1928 god* (1928), pp. 484-485 ; the percentages for 1923–1924 are quoted in Y. S. Rozenfeld, *Promyshlennaya Politika SSSR* (1926), p. 479, from Narkomvnutorg figures, which may not be strictly comparable, though they are so treated in *XV Konferentsiya Vsesoyuznoi Kommunisticheskoi Partii (B)* (1927), p. 129.

[2] Statistics purporting to show the proportion of private trade in the marketing of the products of state industries in 1923–1924 are in *Planovoe Khozyaistvo*, No. 4, 1925, pp. 81-83. The proportion ranged from nil in the oil industry to 40 per cent in the not very important starch industry ; the over-all proportion was 10 per cent. Complaints were heard that state industry sold its products to private traders, who resold them at a profit (*Vserossiiskii Tsentral'nyi Ispolnitel'nyi Komitet XII Sozyva : Vtoraya Sessiya* (1925), p. 497).

scissors crisis, it had been a commonplace of party and governmental policy that private trade, however necessary and useful in the transitional phase of the Soviet economy, must ultimately give way to state and cooperative trade. To develop and increase these forms of trade was a constant aim, pursued with varying degrees of intensity from time to time. But this long-term purpose was not inconsistent with toleration, and even intermittent encouragement, of the private trader. The much advertised arrests of nepmen in December 1923 [1] were not repeated, though the need to overcome the predominance of private capital in retail trade was a constant theme of propaganda. The thirteenth party congress noted " the inadmissibility in the sphere of private trade of any measures which would lead to a curtailment of, or interference with, the general process of exchange of goods ".[2] In the autumn of 1924, the private trader showed his power to disturb official plans for the marketing of the harvest and to break the fixed grain prices which the government had attempted to enforce ; [3] and the alarm caused by the difficulties of the grain collection produced a certain desire to appease so powerful an adversary. In November 1924, while credit was being tightened, and the programme of support for heavy industry hastily curtailed,[4] there was talk of a " new trade policy " directed to " a cessation of the further liquidation of private trade and its potential utilization for the more active sale of industrial goods in the countryside ".[5] This was a natural corollary of the policy of concessions to the peasant implied in the " Face to the countryside " slogan. Kamenev, in a considered pronouncement as president of STO, argued that the growth of trade had outstripped the

---

[1] See *The Interregnum, 1923–1924*, pp. 121-122.
[2] *VKP(B) v Rezolyutsiyakh* (1941), i, 583.
[3] See pp. 294-295 above.                    [4] See pp. 336-337 above.
[5] *Planovoe Khozyaistvo*, No. 4, 1925, pp. 77-92. The article from which the phrase is quoted contains the fullest available account of a brief and little publicized episode : the conclusion was that, once state and cooperative trading were fully established, there would be no room for the private trader, but that he was temporarily useful as " a supplement to the basic work of the state and cooperative chain ". A later writer in *Ekonomicheskoe Obozrenie*, October 1925, p. 161, applied the term " new trade policy " to the whole policy of pressure on the private trader inaugurated at the beginning of 1924, and described the interlude of November 1924–March 1925 as " a certain change of course which did not, however, have time to find expression in concrete measures ".

capacities of the state trading organs and the cooperatives, and concluded that a renewed tolerance must be extended to the private trader.[1] The private trader may have benefited indirectly from the restriction of credit to state industry and to the cooperatives ; but little positive action was taken to assist him. A decree of STO of March 31, 1925, which was primarily concerned to provide much needed capital for the cooperatives, made some attempt to hold the balance. It recognized the desirability of attracting private capital to play its part in financing the expansion of trade, " especially retail trade ". The practice hitherto followed by state industry of demanding cash payment in full from private traders, and of insisting on the so-called " obligatory selection " of goods (meaning that the purchaser had to take goods which he did not want in order to obtain those which he wanted),[2] was to be abandoned and tax concessions to private traders were to be considered.[3] The resolution of the fourteenth party conference in the following month noted that " cooperatives and state trade cannot handle the increasing turnover in the countryside ", and that " a significant place is left in the sphere of trade for the participation of private capital ".[4] An unusual feature of the campaign was a large public meeting in Moscow at which Sheinman, the People's Commissar for Internal Trade, and other Soviet notabilities addressed an audience of nepmen in the hope of attracting private capital into trade.[5] This was followed by a circular from Gosbank to its branches and agencies throughout the Soviet Union, which explained that, in view of " the observed growth in the purchasing power of the population and the significant expansion of our industrial production ", private as well as state trading institutions were needed to take part in the work of distribution, especially in areas where state and cooperative trading organs were weak. Credit should therefore be given for such purposes, though the interest charged on advances to private

---

[1] *Planovoe Khozyaistvo*, No. 1, 1925, pp. 10-12 ; this was an introductory article to the first issue of the monthly journal of Gosplan.

[2] For this apparently well-established practice see *ibid*. No. 3, 1925, pp. 49-50; it did not disappear in 1925.

[3] For this decree see p. 433 below.

[4] *VKP(B) v Rezolyutsiyakh* (1941), ii, 15.

[5] The meeting was reported in *Internationale Presse-Korrespondenz* (Wochenausgabe), No. 19, May 9, 1925, pp. 527-528.

merchants might properly be from 3 to 5 per cent per annum higher than the rate to state enterprises or cooperatives, a figure of 12 to 15 per cent being suggested.[1] But thereafter these official gestures of conciliation seem to have petered out, and the normal pressures in favour of state and cooperative trading were resumed.[2] Throughout the period of NEP the nepman was exposed to alternate bouts of official encouragement and official denunciation. But neither in his capacity as a private trader nor in his capacity as a private capitalist did he at any time constitute a serious menace. The essential problem of Soviet economic policy was, and remained, the backwardness and obduracy of the peasant who was required to bear the brunt of the burden of industrialization. The nepman was important only in so far as, by providing the peasant with a market which escaped official control, he reinforced peasant resistance to official policies.

The developments of 1923–1924 placed a new emphasis on the cooperatives as instruments for the " conquest of the market " and as antidotes to the menace of the private trader. The twelfth party congress in April 1923 described the cooperatives as " the trading organ which must increasingly link state industry with agriculture ", and " the basic intermediary between state industry and agricultural production ".[3] When the scissors crisis brought home to the party leaders the impossibility of giving free rein to the operations of the market, and intervention in one form or another was clearly inevitable, the cooperatives provided one of the most obvious and least invidious channels through which such intervention could make itself felt. The passage from Lenin's last article in which the growth of the cooperatives was treated as a symptom of the growth of socialism was quoted with wearisome reiteration in party literature and in the press. The resolution of the scissors committee of the autumn of 1923 adopted by the Politburo, and subsequently by the thirteenth party conference in January 1924, declared that every extension of the cooperatives and of state trade meant the extension of a " socialist economy ", and relied on this to " overcome the private trader

[1] Sbornik Dekretov, Postanovlenii, Rasporyazhenii i Prikazov po Narodnomu Khozyaistvu, No. 22 (43), July 1925, pp. 32-33.
[2] For a detailed study of private trade at this period see Chastnyi Kapital v Narodnom Khozyaistve SSSR, ed. A. M. Ginsburg (1927), pp. 111-154.
[3] VKP(B) v Rezolyutsiyakh (1941), i, 482.

in conditions of open competition ".[1]  At the same time the cooperatives were accorded certain exemptions from income-tax and industrial tax which gave them an advantage over the private trader.[2]  The increasing attention to problems of trade automatically brought back the consumer cooperatives into the centre of the picture.

This new reliance on the cooperatives led to a further move to restore their original status and prestige.  Since April 1918, membership of consumer cooperatives had in theory been universal and obligatory : all distinction between members and non-members had been lost.  No immediate change was made in this situation when the cooperatives regained their formal independence and their property after the introduction of NEP.[3]  But membership of producer cooperatives had, in the nature of things, remained voluntary ; and the spirit of NEP appeared to require that membership of the consumer cooperatives, as of the trade unions, should once more be placed on a voluntary basis.  The principle of voluntary individual membership seems to have been re-established in practice before the legal situation was altered.[4]  The first official move was a decree of December 28, 1923, " on the reorganization of consumer cooperatives on the principles of voluntary membership " ; [5] and a resolution of the party central committee in April 1924 [6] was followed by an extensive decree issued on May 20, 1924, on the eve of the thirteenth party congress, which was afterwards freely quoted as the charter of the cooperatives.  It placed consumer cooperatives on a voluntary basis, while limiting the entrance fee to 50 kopeks and the annual subscription to a maximum of 5 rubles.  All citizens possessing voting rights under the constitutions of the republics were eligible

[1] VKP(B) v Rezolyutsiyakh (1941), i, 550; for this resolution see The Interregnum, 1923–1924, pp. 113-116.
[2] Sobranie Uzakonenii, 1924, No. 16, art. 150.
[3] See The Bolshevik Revolution, 1917–1923, Vol. 2, pp. 121-122, 338-339.
[4] A speaker at the sixth trade union congress in November 1924 said that the cooperatives " had gone over to the principle of voluntary membership even before the special decree on the subject was issued " (Shestoi S"ezd Professional'nykh Soyuzov SSSR (1925), p. 302).  According to the same source, 62 per cent of industrial workers belonged at this time to consumer cooperatives ; the percentage reached 90 in some districts, but fell to 50 in Moscow, where there were plenty of shops and private traders (ibid. pp. 302-303).          [5] Sobranie Uzakonenii, 1924, No. 17, art. 173.
[6] Ekonomicheskaya Zhizn', April 24, 1924.

for membership.[1]  Non-members, however, might also be served. The consumer cooperatives were surprisingly allowed or encouraged to impinge on the preserves of the flagging industrial, agricultural and credit cooperatives.[2]  They could not only sell, but " collect and work up raw materials " ; they could act as agents for the sale of produce on behalf of their members, and supply their members with tools or raw material for their work ; and they could engage in credit operations.[3]

The encouragement accorded to the cooperatives in this decree was reinforced by the proceedings of the thirteenth party congress, the first since 1920 which had considered it necessary to devote a separate discussion and resolution to the cooperatives. Kamenev, in his report on internal trade, described the cooperatives as " not only a weapon in the struggle against the divergence between wholesale and retail prices, but at the same time a weapon of our social policy in the countryside, of the organization of the poor and middle strata of the peasantry round the proletariat and against the *kulaks* ".  Andreev, in his special report on the cooperatives, drew the familiar distinction between their rôle under capitalism and their rôle under socialism, though

---

[1] The formal distinction between " workers' cooperatives " and " all-citizen cooperatives ", which had been invoked at an earlier period in order to split the cooperative movement and break its independence (see *The Bolshevik Revolution, 1917–1923*, Vol. 2, pp. 239-238), was still maintained ; " workers' cooperatives " continued to exist, but had their own separate organization within the movement.

[2] A resolution of the party central committee of August 1922 had rather grudgingly sanctioned " mixed forms of cooperatives ", especially if these were formed " around consumer cooperatives " (*VKP(B) v Rezolyutsiyakh* (1941), i, 462) : later such composite cooperatives were banned.

[3] *Sobranie Uzakonenii, 1924*, No. 64, art. 645.  This was the first decree of the USSR on the cooperatives, being issued jointly by Sovnarkom and VTsIK ; most of the decrees on the cooperatives were issued by the republics concerned (see, for example, the decree of the RSFSR, *ibid.* No. 89, art. 894). The cooperatives were responsible, according to their various functions, to Narkomtorg and Narkomfin, which were unified commissariats, and to the Narkomzems of the republics.  The constitutional situation was ambiguous. L. Povolotsky, *Osnovnye Nachala Kooperativnogo Prava SSSR* (1925), pp. 7-19, argued at length, against claims by the Ukrainian cooperatives, that, while the cooperatives were not named in the constitution, the USSR had the constitutional right to lay down the " basic principles " of cooperative law ; this view seems to have prevailed in practice.  Tsentrosoyuz combined its old position as the central organ of the consumer cooperatives of the RSFSR with its new position as central organ for all cooperatives of the USSR (L. Povolotsky, *Kooperativnoe Zakonodatel'stvo* (3rd ed. 1926), p. 99).

he reproached them with being more interested in making profits than in supplying the needs of their members.[1]  The resolution began by defining the rôle of the cooperative as that of a middle-man between state industry (which accounted for 90 per cent of all industry) and the predominantly peasant consumer : it was anomalous in a proletarian state to leave this rôle to the private trader.  The cooperatives were instructed to apply firmly the rule of voluntary membership, and not to " disperse their efforts in order to serve the whole population " at the expense of the " organized consumer ".  Attempts at fusion of agricultural and consumer cooperatives were now declared to be " incorrect " ; the peasant as consumer and the peasant as producer should be organized separately.  At a time when wages were still not high enough " to cover fully all the needs of the worker " the co-operatives could help by reducing prices, offering discounts and granting credit.  State industry should extend maximum favours to the cooperatives, " both in offering them the best choice of goods, and in terms and conditions of credit ".  State trade should confine itself to the wholesale and wholesale-retail markets.[2]  In the summer of 1924 a further attempt was made to rectify an old grievance by instructing Vesenkha to return to agricultural, in-dustrial and credit cooperatives the buildings, warehouses and other property taken over by it.[3]

In spite, however, of the success of the cooperatives in securing a growing share in the rapidly expanding internal trade of the Soviet Union, the path was still beset with difficulties.

> Once again [said Kamenev at this time] we began to look at the cooperatives ; but, when we really took a look at them, it appeared that, as a weapon for carrying out our task of reducing prices, . . . they were very, very feeble, that the cooperatives were very, very little adapted for the real struggle against private capital.[4]

---

[1]  *Trinadtsatyi S"ezd Rossiiskoi Kommunisticheskoi Partii (Bol'shevikov)* (1924), pp. 416, 436-437, 442-443.    [2]  *VKP(B) v Rezolyutsiyakh* (1941), i, 584-588.
[3]  Decree of July 26, 1924 (*Izvestiya*, September 10, 1924) ; whether this order was any more effective than the previous decree of October 1921 (see *The Bolshevik Revolution, 1917-1923*, Vol. 2, pp. 338-339) is very uncertain. Another grievance, which was only slowly remedied, was that former property of the credit cooperatives had been handed over to the consumer cooperatives (*Chetyrnadtsataya Konferentsiya Rossiiskoi Kommunisticheskoi Partii (Bol'-shevikov)* (1925), p. 133).    [4]  L. Kamenev, *Stat'i i Rechi*, x (1927), 294.

Six months after the thirteenth party congress it was claimed that, as the result of " a desperate struggle between the cooperatives and retail traders ", cooperative prices were now fully competitive and, for certain standard commodities, lower than those of the private trader. But shortage of working capital was a constant handicap. On the one side, the cooperatives could only take goods from the manufacturers on credit : on the other side, they attempted to obtain advances from purchasers, and gave priority to purchasers who were able to pay in advance.[1] What was needed, declared the official economic journal, was " *systematic work to conquer the national market* " ; and this could not be achieved " *without increasing* the working capital of the cooperatives ".[2] This policy, however, met with keen opposition. The rivalries and jealousies which had marked earlier relations between the cooperatives and state economic organs [3] were still very much alive. At the sixth trade union congress in November 1924, Rykov depicted the credit-hunger of the cooperatives in unusually caustic terms.[4] The cooperatives expected " unlimited credit ", and came to town with Lenin's brochure on the cooperatives in their hands as sufficient proof of their creditworthiness. Backward in collecting subscriptions from their members, they already owed hundreds of millions of rubles to trusts and industrial concerns, which could no longer deplete their working capital for this purpose, and would be obliged to sell to private traders who could pay cash.[5] The spokesman of the cooperatives in his reply could make out a strong case :

Workers' cooperatives trade in conditions which preclude any accumulation of capital out of profits, and may for a long time have to trade without being able to accumulate capital

---

[1] *Shestoi S"ezd Professional'nykh Soyuzov SSSR* (1925), pp. 301-302, 308.

[2] Leading article in *Ekonomicheskaya Zhizn'*, October 23, 1924 ; a further article (*ibid.* January 25, 1925), pointed out that many cooperatives had increased their turnover threefold or fourfold, while their working capital had increased by only 10–12 per cent.

[3] See *The Bolshevik Revolution, 1917–1923*, Vol. 2, pp. 338-340.

[4] That the attack was something of a surprise is suggested by subsequent references to it at VTsIK, where Rykov was said to have " driven a stake through the cooperatives ", and his speech was described as having " fallen like a spark in a powder-barrel and almost blown up our cooperatives " (*SSSR : Tsentral'nyi Ispolnitel'nyi Komitet 2 Sozvya : 3 Sessiya* (1925), pp. 81, 121-122).

[5] *Shestoi S"ezd Professional'nykh Soyuzov SSSR* (1925), pp. 249-250, 283-284.

out of profits. We must say that a certain amount of credit
will have to be given to the cooperatives. This must be done
in view of the national utility of their economic work.[1]

But at a time when industry itself was hard pressed for credit
and its programmes of expansion were subject to curtailment,
little tenderness was likely to be shown towards the needs of the
cooperatives. These difficulties were symptomatic of the acute
shortage of capital throughout the Soviet economy which was a
continuing factor of the NEP period.

The battle continued throughout the winter of 1924–1925. As
a result of the general restriction of credit by the banks, the trusts
and syndicates, themselves hard pressed from above, began to
withhold credit from the cooperatives and to demand a larger
proportion of payment in cash for their consignments, with the
result of a large increase in the volume of protested bills and a
curtailment of retail trade owing to lack of funds.[2] But the cooperatives were not without powerful support. The Leningrad
provincial party conference of January 1925 denounced " the
attempt noticeable in recent months on the part of some economic
organizations to curtail their transactions with the cooperatives
by setting harder conditions of sale and purchase in dealings with
the cooperatives (restriction of credit, etc.) ".[3] In the end, the
forces which overcame the resistance to the expansion of heavy
industry also scored a victory for the consumer cooperatives. At
the session of VTsIK in March 1925, a resolution was passed to
vote 8 million rubles from the budget of the USSR " to increase
the working capital of the cooperatives in the form of a long-term
advance ".[4] Gosplan, campaigning hotly in favour of credits for
industry,[5] also took the consumer cooperatives under its protection.
Pointing to the imminent danger of a general contraction of trade,
it argued against the fallacy of supposing that private capital
could fill the gap in credits, and proposed that the state sub-

[1] *Shestoi S"ezd Professional'nykh Soyuzov SSSR* (1925), p. 315.
[2] *Ekonomicheskaya Zhizn'*, March 1, 1925 ; *Planovoe Khozyaistvo*, No. 3,
1925, pp. 40-51 ; No. 5, 1925, p. 111. The cooperatives at the end of 1924
had protested bills to the total of 20 million rubles (L. Kamenev, *Stat'i i Rechi*,
xi (1929), 283-284.    [3] *Leningradskaya Pravda*, January 31, 1925.
[4] For Kuibyshev's report on the subject see *SSSR : Tsentral'nyi Ispolnitel'-
nyi Komitet 2 Sozyva : 3 Sessiya* (1925), pp. 195-197. For the resolution
see *id. : Postanovleniya* (1925), p. 15 ; *Sobranie Zakonov, 1925*, No. 17, art. 125.
[5] See p. 346 above.

vention to trading organizations, including the consumer co-operatives, should be increased from 8 million rubles to 25 millions in the current financial year.[1]

The sequel to these proceedings was a long decree issued by STO on March 31, 1925, which reflected the general rise in prosperity and put the consumer cooperatives on their feet. It raised to 10 million rubles the sum to be advanced to the co-operatives by way of a long-term credit from the budget to re-inforce their fixed capital. Of this, 4 millions were to go to the consumer cooperatives, 4 millions to the agricultural cooperatives, and the remainder to industrial cooperatives, including those engaged in rural industries. The decree laid down not only that the credits granted by state industry to the cooperatives in the past should be continued, but that " the conditions of credit and settlement of accounts " for the cooperatives in their dealings with industry should be " made easier ". But these solid concessions were accompanied by some sharp words to the cooperatives. Henceforth they must rely for any further increase of their capital on the subscriptions collected from their members and on the profits earned by them in course of business : both profits and a reduction in prices should be achieved by cutting overheads and rationalizing their work.[2] At the fourteenth party conference in the following month the place of honour was given to the agri-cultural cooperatives.[3] But the consumer cooperatives also received some attention. Rykov sang the praises of the principle of voluntary membership, and claimed for the consumer co-operatives a membership of 8 millions — a total which, according to Khinchuk, the president of Tsentrosoyuz, included a large number of " dead souls ".[4] Complaints that all the leading posts in the cooperatives were occupied by party members, and com-plaints that the cooperative organs were full of " anti-Soviet people " and had inadequate links with the party, appeared to cancel one another out.[5] The conference resolution extended its

[1] *Planovoe Khozyaistvo*, No. 3, 1925, p. 276 ; No. 5, 1925, p. 300.
[2] The decree was published in *Pravda*, April 2, 1925.
[3] See pp. 280-282 above.
[4] *Chetyrnadtsataya Konferentsiya Rossiiskoi Kommunisticheskoi Partii (Bol'-shevikov)* (1925), pp. 90-91, 128.
[5] *Ibid.* pp. 98, 101, 127. On September 1, 1924, 11 per cent of the workers, and 19·4 per cent of the " responsible workers ", in the headquarters organization of Tsentrosoyuz were party members ; a year later the respective

blessing to " cooperative organizations of all types ", but insisted on " a strict delimitation of functions between them ", especially requiring that agricultural and consumer cooperatives should be kept apart. As regards consumer cooperatives, the resolution endorsed in vaguer and more guarded terms the main provisions of the decree of March 31. The tolerance shown to the continued participation of private capital in retail trade enhanced rather than diminished the rôle of the cooperatives, which were promised " more decisive support from the party and the state ", and " better conditions of credit and of settlement of accounts ", though they were also told " to overcome their inner short-comings ".[1] The final revision of the budget by the third Union Congress of Soviets in May 1925 brought a further token of official good-will, the sum assigned to credit for the cooperatives being once more increased to 12 million rubles.[2]

The decree of March 31, 1925, and the resolution of the fourteenth party conference consolidated the position of the consumer cooperatives, which was not again seriously challenged. In full enjoyment of the support of the party and of the government, they began step by step to oust the private trader.[3] The turnover of the consumer cooperatives rose from 2000 million rubles in 1923–1924 to 3500 millions in 1924–1925.[4] But friction continued for some time between them and the industrialists. As long ago as 1921 it had been laid down that state economic organs and enterprises, including the industrial trusts, should offer their pro-ducts in the first instance to the cooperatives and, only if these declined to handle them, to private traders.[5] This rule was

percentages were 13·6 and 28·2. Of members of boards of administration of cooperatives 71·5 per cent were party members on September 1, 1924, and 66·9 per cent a year later (*Partiinye, Professional'nye i Kooperativnye Organy i Gosapparat : k XIV S"ezdu RKP(B)* (1926), pp. 184, 188).
[1] *VKP(B) v Rezolyutsiyakh* (1941), ii, 15-16, 18, 21.
[2] *Sobranie Zakonov, 1925*, No. 40, art. 290.
[3] A source not biased in favour of the cooperatives gives an account of the transformation in a rural district in the province of Tver. When the con-sumer cooperatives resumed operations in 1923, there were two private shops which outbid the cooperatives by offering better and cheaper goods and having smaller overheads ; in the two following years the cooperatives progressed so rapidly that by 1926 all that was left of private trade was confined to " two wretched wooden booths conducting petty transactions at a few kopeks each " (A. M. Bolshakov, *Sovetskaya Derevnya, 1917–1927* (1927), pp. 121-126).
[4] *Ekonomicheskoe Obozrenie*, December 1925, p. 202.
[5] See *The Bolshevik Revolution, 1917–1923*, Vol. 2, p. 339.

specifically reiterated in the resolution of the fourteenth party conference of April 1925.[1] When, however, in the summer of that year the effort was made to enforce its observance by the trusts and syndicates, some of these pleaded the commercial independence conferred on them by the statute on trusts of April 10, 1923 [2] — a curious instance of an attempt to carry the principles of NEP to their logical conclusion ; and a further intervention by STO was required to overcome this resistance.[3] Even this did not end the difficulties.    In January 1926 Dzerzhinsky, as president of Vesenkha, and Khinchuk, as president of Tsentrosoyuz, issued a joint appeal to the organs of state industry and to the cooperatives for friendly collaboration, and agreed to set up a joint commission to which differences and disputes could be referred for settlement. This time the complaints do not appear to have been all on one side.    Khinchuk admitted that some cooperative organs had fixed their margins too high ; one of the purposes of the agreement was to limit the percentage which the selling organizations might add to the wholesale prices.[4]    An opposition critic in the journal of Gosplan protested against indirect measures of compulsion applied to industrial organizations to sell their products to the cooperatives on favourable terms.    This form of " favouring the cooperatives " and " protecting one economic organization at the expense of another " was resented by industry as " a weight that slows up its progress ".[5]

The control of prices had been introduced in the autumn of 1923 to overcome the scissors, and the adoption of this policy was the most important motive for the creation of Narkomvnutorg in the spring of 1924.    The new commissariat at once set out to strengthen the machinery of control established three months earlier by STO and Komvnutorg.[6]    Throughout the ensuing

---

[1]  *VKP(B) v Rezolyutsiyakh* (1941), ii, 21.
[2]  See *The Bolshevik Revolution, 1917–1923*, Vol. 2, p. 309.
[3]  The complaints of the cooperatives and the ruling of STO were recorded in *Ekonomicheskaya Zhizn'*, July 31, 1925.
[4]  *Ibid.* January 13, 19, 1926.
[5]  *Planovoe Khozyaistvo*, No. 2, 1926, p. 119.
[6]  *Sobranie Uzakonenii, 1924*, No. 62, art. 620 ; for the February decrees see *The Interregnum, 1923–1924*, p. 139.

period the battle for price control was a central point in all economic issues. It was recognized that price control was not at the present stage uniformly applicable to all commodities, though this was the ultimate goal of a planned socialist economy. In the summer of 1924, it was assumed that control of the prices of manufactured goods presented no great difficulty. This proposition was formally correct, since a preponderant proportion of manufactured goods of all categories was produced by state industry ; and it seemed to have been confirmed by the experience of 1923, when prices of industrial goods had been forced down in accordance with the dictates of public policy at a time when the instruments of control were far weaker than they were a year later. The control of agricultural prices was subject to different considerations, and required different methods, from the control of industrial prices. But this, too, in the autumn of 1924 seemed to be well within the scope of Narkomvnutorg, which was charged by VTsIK in October with the task of supervising the disposal of the harvest and of " bringing the prices of agricultural products on the one hand and of the products of industry on the other into a healthy relation and maintaining maximum stability of prices ".[1]

The fall in industrial prices which began in October 1923, and was responsible for closing one blade of the scissors,[2] continued without interruption, though at a decreasing rate, throughout the year 1924. With greater efficiency in organization, largely increased volume of production and the introduction of a stable currency, the policy of official control of wholesale prices worked smoothly and successfully. The scissors had opened to their widest extent, and industrial prices had reached their highest level, on October 1, 1923. By December 1, 1924, industrial prices had fallen, according to official figures, by an average of 27·4 per cent, though even these lower prices were, with few exceptions, from 50 to 100 per cent higher than the prices of 1913.[3] Dzerzhinsky, in his speech at Vesenkha of December 2, 1924, was enabled to pass somewhat delicately over the decision to slow down industrial expansion, by dwelling with justifiable pride on the suc-

[1] *Postanovleniya TsIK Soyuza SSR : 2 Sessiya* (1924), p. 13.
[2] See *The Interregnum, 1923–1924*, p. 118.
[3] Y. S. Rozenfeld, *Promyshlennaya Politika SSSR* (1926), pp. 438-439.

cesses of price policy.  He embarked on one of the rare excursions of this period into a theory of price control :

> The crisis operated with mathematical precision and forced us to consider the question of the interests of industry and of the national economy as mutually dependent. . . . We became convinced that the cost of production does not always determine the price, but that perhaps in our own country the price should determine the cost of production ; for the reduction which had to be effected in prices showed that the cost of production is not an absolute, but may be divided into elements dependent on the good will of the working class and on that of the economic authorities.

He claimed that between October 1, 1923, and October 1, 1924, wholesale prices of industrial products had fallen by 29 per cent ; in the metal and foodstuffs industries the fall had amounted to 35 per cent, in the leather industry to 33 per cent.  On October 1, 1923, the ratio between the agricultural price-index and the price-index for textiles had been 1 : 4·8 ; by October 1, 1924, it had fallen to 1 : 1·8.  Moreover, the decline in industrial prices, and the process of closing the scissors, were still continuing.[1]

One blot remained, however, on this record of a successful price policy.  Even when complete confidence was felt in the ability of the government to control wholesale prices, the control of retail prices was still a serious crux.  " If ", wrote Strumilin in one of his pleas for planning, " we can dictate not only wholesale, but retail, prices both to the producer and to the mass consumer, the task of overcoming NEP will be resolved." [2]  This dream was far from realization.  Throughout this period retail prices had fallen much less sharply, and had been less amenable to control, than wholesale prices.  In some standard commodities, such as paraffin, salt and sugar (the first commodities over which retail price control had been attempted),[3] the margin between wholesale and retail prices had been kept relatively stable.  In

---

[1] Dzerzhinsky's speech (see p. 337 above) was published in *Pravda*, December 4, 1924 ; Bukharin explained about the same time that, whereas the policy of the opposition in the autumn of 1923 had been to maintain high prices and high profits, the official policy now was " minimum profit per unit of merchandise ", combined with an increased volume of production (N. Bukharin, *Kritika Ekonomicheskoi Platformy Oppozitsii* (1926), p. 83).

[2] *Ekonomicheskaya Zhizn'*, November 7, 1924.

[3] See *The Interregnum, 1923-1924*, p. 112.

October 1924 Lezhava, the then People's Commissar for Internal Trade, assured TsIK that fixed uniform prices for the main staple articles of mass consumption — paraffin, sugar, matches and tobacco — had been established throughout the Soviet Union.[1] But these were only partial exceptions to the general rule. In spite of a fresh order requiring shopkeepers to exhibit prices of price-controlled goods,[2] the margin of retail over wholesale prices which, on April 1, 1924, stood at 33·5 per cent, rose by November 1924 to 45 per cent, this being from two to two-and-a-half times as great as in 1913.[3] The predominance of private capital in the retail sector largely neutralized official attempts at control. This was a recognized and admitted problem. Kamenev spoke of the divergence between wholesale and retail prices of manufactured goods as " the new scissors which cut in pieces both our industry and our good relations with the peasantry " ;[4] and the peasant began to ask why the state should control the price of grain but not the price of cotton cloth.[5] But, with growing success in the control of wholesale trade, the solution could, it was felt, not be far distant. Throughout the year 1924–1925, the consumer cooperatives worked hard to reduce their margins on consumer goods.[6] In the summer of 1925 the average margin between wholesale and retail prices was said to have been brought back to 33 per cent.[7]

The problem of agricultural prices, which had at first been regarded as secondary, now emerged as the real stumbling block. Down to August 1924, when the new harvest came in, the rising trend of agricultural prices had been welcomed as a further token of the closing of the scissors. When the harvest was not followed by the expected fall in prices, Narkomvnutorg attempted to hold them down by fixing maximum limits, well in excess of those of the previous year but below the level to which the market had

[1] SSSR : Tsentral'nyi Ispolnitel'nyi Komitet 2 Sozyva : 2 Sessiya (1924), p. 272.

[2] Sbornik Dekretov, Postanovlenii, Rasporyazhenii i Prikazov po Narodnomu Khozyaistvu, No. 13, October 1926, pp. 18-19 ; for the original order see The Interregnum, 1923–1924, p. 139.

[3] Y. S. Rozenfeld, Promyshlennaya Politika SSSR (1926), p. 453 ; Planovoe Khozyaistvo, No. 10, 1925, p. 68.

[4] L. Kamenev, Stat'i i Rechi, xi (1929), 226.

[5] Soveshchanie po Voprosam Sovetskogo Stroitel'stva 1925 g. : Yanvar' (1925), p. 143.    [6] Ekonomicheskoe Obozrenie, April 1926, pp. 120-124.

[7] Y. S. Rozenfeld, Promyshlennaya Politika SSSR (1926), p. 453.

carried them in August.  The attempt failed ;  and by November
and December 1924 the policy of grain collection at fixed prices
had completely broken down under competition from the private
market.[1]  But this mishap, however disconcerting, was attributed
to the accident of a bad harvest.  No immediate conclusions as to
the feasibility of a policy of price control were drawn from it.
Grain prices continued to rise in the ensuing months, while
industrial prices remained stable or fell, thus completing the
closing of the scissors of 1923.  At midsummer 1925, a pound of
rye would buy, on an average, approximately the same quantities
of manufactured goods of mass consumption as in 1913.[2]  This
was the period when party and governmental policy was most
firmly committed to support of the well-to-do peasant, so that
rising grain prices were not felt as a cause for serious alarm ;  and
even industry continued to expand under the stimulus of abundant
credit.  It was once more assumed — so great and so lasting was
the impression which the scissors crisis had left behind — that
the principal danger ahead was a collapse of grain prices after the
harvest.  Fixed prices for agricultural products were now discredited.  But " directive " prices were to be employed to keep
prices at reasonable levels.[3]

The sequel to the harvest shattered these expectations, and
agricultural prices rose steadily, and sometimes steeply, from
September 1925 into the following summer.  But the situation
was now radically different from that of the previous year.  Industrial prices, for the first time since the summer of 1923, also
began to rise even more rapidly than agricultural prices, so that
it was possible to speak of a partial " reopening " of the scissors.[4]
On a long view the whole process could now be seen as one of a
decline in the purchasing-power of the ruble due to the first stages
of a credit and currency inflation.[5]  But the immediate symptom

[1] See pp. 192-193 above.
[2] Ekonomicheskoe Obozrenie, February 1926, p. 31 ; a similar result is
obtained from statistics of sales by consumer cooperatives covering a larger
range of agricultural products (ibid. April 1926, pp. 124-127).
[3] See pp. 292-293 above.
[4] Kamenev used the phrase in his abortive report to the party central committee in October 1925, being careful, however, to point out that the phenomenon
now related especially to retail prices (L. Kamenev, Stat'i i Rechi, xii (1926), 359).
[5] For the beginning of this phenomenon and first attempts to diagnose it
see pp. 479-484 below.

was a renewed sharp increase in the margin between wholesale and retail prices, which, after reaching the low level of 33 per cent in the summer of 1925, began to rise again in August. In October 1925 the margin once more stood at 47 per cent,[1] and the situation seemed critical. An order of Vesenkha to all economic organs signed by Dzerzhinsky was published in the press demanding that an end should be put to " the bacchanalia of the rise of retail prices ", together with an instruction to the OGPU, also signed by Dzerzhinsky, to assist in enforcing the order by rigorous action against speculators.[2] But, with the growing signs of inflation, retail prices were soon completely out of hand. " Grain prices are high," wrote a commentator in January 1926, " but industrial prices in the villages are higher still. That is the crux." [3] *Pravda* set the shortage of industrial goods and the high prices of agricultural products side by side as the principal " economic difficulties " of the moment, and spoke openly of inflation.[4] A conference at Narkomtorg [5] in March 1926 decided on the establishment of a central price commission, with local commissions working under it, to deal with rising retail prices, though its powers and — still more important — its policy seem to have been undefined.[6] In the spring of 1926 retail prices were said to exceed wholesale prices by an average margin of 60 per cent, with resulting high profits for those engaged in trade.[7]

The party central committee at its session early in April 1926 made a perfunctory attempt to wrestle with the problem. It noted both " a rise in the general price level " and " a sharp divergence between wholesale and retail prices ". These were attributed,

[1] Y. S. Rozenfeld, *Promyshlennaya Politika SSSR* (1926), p. 453.

[2] Both were published on the front page of *Leningradskaya Pravda*, October 31, 1925 ; *Pravda* published a leader on the subject on October 30, 1925, but does not appear to have printed the documents.

[3] *Ekonomicheskaya Zhizn'*, January 14, 1926.

[4] *Pravda*, February 18, March 4, 1926.

[5] For the amalgamation of the People's Commissariats of Internal and Foreign Trade to form a single People's Commissariat of Trade (Narkomtorg) in November 1925 see p. 451 below.

[6] *Ekonomicheskaya Zhizn'*, March 18, 21, 23, 1926 ; Kamenev was appointed president of the central price commission (*ibid.* April 13, 1926).

[7] *Ekonomicheskoe Obozrenie*, May 1926, p. 52 ; the unpublished memorandum of April 2, 1926, on economic policy in the Trotsky archives (see p. 326 above) estimated profits from trade at 210 million rubles for 1923-1924, 435 millions for 1924-1925, and 800 millions for the current year.

partly to the rise in demand for manufactured goods in the countryside, following the good harvest, the reduction in the agricultural tax and the high prices of agricultural products, partly to the increased demand in the towns, due to the rise in wages and the increase in the number of industrial workers, especially through the development of heavy industry. The basic difficulty was thus the rate of investment in heavy industry, which produced no immediate counterpart in the form of consumer goods to balance the rising wages bill. But the diagnosis was, as usual, easier to find than the remedy. The resolution could do no more than insist on " the indispensability of achieving a decisive reduction in retail prices ", and urge that " the attention of trade unions, state industry, state trading organs and especially cooperatives " should be concentrated on this task.[1]  A few days later Rykov in a speech at VTsIK echoed the current opinion when he described " the excessive rise of retail prices " as " the chief danger menacing our economy ".[2]  But the notion of deliberately using prices as an instrument to promote accumulation and change the shape of the economy was still confined to a few visionaries or extremists ; and these cries of alarm merely pointed to the perennial problem of the Soviet economy without advancing a solution.

## (b) Foreign Trade

The first year in which any coherent plan or policy for foreign trade was achieved was the year 1922–1923, when the good harvest made modest exports of grain possible for the first time since the revolution, and the preparations for the stabilization of the currency kept Narkomfin keenly alive to the importance of a favourable trade balance. During the year a plan was drawn up providing for exports to the value of 210 million goods rubles ; and this was carried out, though with many deviations in particular items.[3] Recalculated in terms of pre-war prices, however, the total value of these exports reached only 133 million rubles or less than 10

---

[1] *VKP(B) v Rezolyutsiyakh* (1941), ii, 94, 97.

[2] *SSSR : Tsentral'nyi Ispolnitel'nyi Komitet 3 Sozyva : 2 Sessiya* (1926), p. 9.

[3] *Sotsialisticheskoe Khozyaistvo*, No. 2, 1924, pp. 184-186 ; No. 1, 1925, pp. 197-198.

per cent of the pre-war figure. Agricultural products accounted for 65 per cent of the total, timber and timber products for 16·8 per cent, and oil for 11·4 per cent ; other items were negligible.[1] Exports exceeded imports for 1922–1923 by 23 million rubles, though if the figures were recalculated in pre-war prices (when grain prices were lower and prices of manufactured goods higher) they showed a deficit of 14–15 millions.[2] Of imports in this year 53 per cent were absorbed by industry and 11·7 per cent by railways, posts and telegraphs.[3] Altogether 70 per cent of imports for 1922–1923 represented supplies required by industry, this being a great advance on the previous year when 60 per cent of imports had consisted of supplies for the consumer market, including foodstuffs.[4]

For the following year 1923–1924, an export and import plan was, for the first time, adopted before the beginning of the economic year, i.e. on September 7, 1923, and was twice revised in the course of the year in February and June 1924. In its final form it provided for exports to the value of 428 million rubles and imports to the value of 334 millions, the favourable balance being an important factor in the policy of currency stabilization. Both these figures were substantially exceeded, exports reaching 522 millions and imports 439 millions.[5] The maintenance for the second year running of an active trade balance was an important psychological factor in the stabilization of the currency. Of the exports 75 per cent were agricultural products.[6] These included substantial exports of grain, mainly rye ; not till 1930–1931 was the Soviet Union again to export so large a quantity of grain.

[1] *Sotsialisticheskoe Khozyaistvo*, No. 1, 1924, pp. 147, 160.
[2] L. B. Krasin, *Vneshnyaya Torgovlya SSSR* (1924), p. 10.
[3] *Sotsialisticheskoe Khozyaistvo*, No. 5, 1924, p. 352.
[4] L. B. Krasin, *Vneshnyaya Torgovlya SSSR* (1924), p. 10.
[5] *Ekonomicheskoe Obozrenie*, February 1926, pp. 66, 72 ; the final results are confirmed in *Kontrol'nye Tsifry Narodnogo Khozyaistva na 1926–1927 god* (1926), pp. 296-297. The Gosplan totals are expressed in three denominations " at pre-war prices ", " at current prices in goods rubles ", " at current prices in chervonets rubles " (the last being the denomination in common use). This sometimes confused Soviet statisticians as well as those who followed their calculations ; for example, *Sotsialisticheskoe Khozyaistvo*, No. 1, 1925, p. 198, quoted a figure of 370 millions as the total of exports in goods rubles, whereas the table in the control figures for 1926–1927 quoted above makes it clear that this is the total calculated at pre-war prices.
[6] *Kontrol'nye Tsifry Narodnogo Khozyaistva na 1926–1927 god* (1926), pp. 296-297.

Timber products and oil were once more the largest non-agricultural items. Of the imports nearly 75 per cent were absorbed by industry, mainly in the form of raw materials and semi-manufactured products.[1] The pattern of Soviet foreign trade was plainly set in these years. Imports went to supply the most urgent current needs of industry and of the consumer market ; re-equipment in the form of new machinery had scarcely yet begun. The exports required to finance these imports and to maintain the balance of payments consisted primarily of agricultural products and were therefore dependent on favourable harvests. This was the precarious situation which Krasin had vainly hoped to remedy by a policy of compromise with foreign capital. But, after the failure of Krasin's plea at the twelfth party congress in 1923,[2] the case for a foreign loan was no longer treated as practical politics. The policy of foreign concessions was pursued, but not on a scale sufficient to affect the balance of trade. Imports had to be paid for by exports.

Plans for the year 1924–1925 were made in a mood of rising optimism. With the currency stabilized, and production rapidly expanding, foreign trade would also expand : increased supplies of raw materials and equipment for industry would be paid for by increased agricultural exports. The partial failure of the harvest administered a shock to these expectations. Grain exports in 1924–1925 fell to less than half the value of the previous year ; and, though the deficiency was made up by increased exports of flax, timber, oil and manganese ore, the prices for flax and timber on the world market had slumped, and the total value of exports rose only from 522 million rubles in 1923–1924 to 558 millions in 1924–1925. Meanwhile an inflated import programme was further swollen by the necessity for emergency imports of grain and sugar in the first months of 1925 ; and total imports for the year rose to 720 million rubles, leaving a passive balance of 162 millions. Among the imports, machinery

---

[1] *Sotsialisticheskoe Khozyaistvo*, No. 5, 1924, p. 352 ; *Ekonomicheskoe Obozrenie*, December 1925, p. 228.

[2] See *The Interregnum, 1923–1924*, p. 19. Krasin was still pleading for long-term foreign loans in 1925 (extracts from a pamphlet of that year, *Why Foreign Loans are Necessary to Us*, are reprinted in L. B. Krasin, *Voprosy Vneshnei Torgovli* (1928), pp. 354-373), but was by this time virtually without support in party circles, even from Sokolnikov.

and tools for industry reached the modest figure of 48 millions,
and agricultural tools and machinery 40 millions ; both were more
than three times the figure for the previous year.[1]

The modest rise in the value of foreign trade even in the bad
harvest year 1924–1925 encouraged hopes of far more spectacular
progress after the expected good harvest of 1925 ; and the planners
who issued the first " control figures of the national economy "
in August of that year [2] gave full rein to those hopes. Restricting
themselves to trade with Europe and the west (figures for trade
across Asiatic frontiers were still incomplete and unreliable),[3] and
estimating actual exports for 1924–1925 at the unduly low figure
of 462 million rubles, they forecast for 1925–1926 an increase to
1100 millions, of which 950 millions would be agricultural exports,
or more than two-and-a-half times the corresponding figure for
1924–1925. This would allow for a rise in imports to 950 millions
and at the same time wipe out the passive balance incurred in the
previous year. These apparently extravagant figures were justified
by comparisons with the pre-war record. In 1924–1925, when
industrial production had reached 70 per cent, and agricultural
production 71 per cent, of pre-war values, foreign trade had
reached only 24 per cent. The cautious Krasin pointed out that
the discrepancy was due partly to a general decline in international
trade, and partly to the decreased proportion of Soviet grain which
came to the market, but believed that the gap could be narrowed.[4]

[1] The totals are in *Kontrol'nye Tsifry Narodnogo Khozyaistva na 1926–1927
god* (1926), pp. 296-297 ; more detailed but incomplete figures (for European
frontiers only) are in *Ekonomicheskoe Obozrenie*, December 1925, pp. 224,
228. Exports of rye were resumed in July and August 1925 when the prospects of a bumper harvest were apparent (*ibid.* March 1926, p. 43); but for
this last-minute recovery the deficiency in exports would have been still more
serious.

[2] See p. 500 below ; the importance attached in Gosplan to foreign trade is
shown by the regular appearance of a detailed bulletin on world trade in the
monthly journal *Planovoe Khozyaistvo*.

[3] Official figures for foreign trade from 1918 to September 1923 related
exclusively to trade over western or maritime frontiers (*Vneshnyaya Torgovlya
SSSR za 20 Let (1917–1937)* (1939), p. 6). The Soviet customs tariff of February
1922 (*Sobranie Uzakonenii, 1922*, No. 24, art. 259) was " a customs tariff for
European trade " : it was extended to the Far Eastern Region, but not to other
Asiatic frontiers, in September 1923 (*Sobranie Uzakonenii, 1923*, No. 83, art.
803). Trade with eastern countries will be dealt with in Part V in a subsequent
volume.

[4] L. B. Krasin, *Voprosy Vneshnei Torgovli* (1928), pp. 156-158.

The Gosplan figures for 1925–1926 still represented less than half the foreign trade figures for 1913, whereas both agricultural and industrial production were expected to reach about 90 per cent of the totals for that year.[1] They proved none the less fallacious. Unexpected difficulties in the collection of grain by state organs [2] limited the quantities available for export; and a fall in world prices in September and October 1925 made exports unprofitable.[3] Sokolnikov, as the custodian of financial orthodoxy, made himself the spokesman of those who insisted on reduced imports to match reduced exports, and urged, even at this unpropitious moment, the need for an active trade balance to replenish depleted gold reserves.[4] The argument seemed irresistible. Import programmes were drastically cut, and the plan scaled down to totals of 720 million rubles for exports and 685 millions for imports.[5] In the final result, exports failed to achieve this reduced target, reaching only 670 million rubles while imports rose to 735 millions.[6] The adverse balance of 65 million rubles, though less than half that of 1924–1925, remained substantial, and made its contribution to the weakness developed by the currency in the spring and summer of 1926.[7]

The organization of foreign trade was a constant theme of controversy in these years. Lenin's emphatic and successful defence, in the last months of his active life, of the monopoly of foreign trade secured that institution from any open renewal of the attack on it.[8] But acceptance of the monopoly of foreign trade did not preclude increasing inroads into the exclusive right

[1] *Kontrol'nye Tsifry Narodnogo Khozyaistva na 1925–1926 god* (1925), pp. 52-53 ; *Planovoe Khozyaistvo*, No. 2, 1926, p. 58. These calculations were accepted and repeated in the resolution of the party central committee of October 1925 on foreign trade, which spoke of reaching 60 per cent of the pre-war level of foreign trade (for this resolution see p. 450 below).

[2] See pp. 293-297 above.

[3] *Ekonomicheskoe Obozrenie*, No. 2, 1926, p. 123.

[4] G. Sokolnikov, *Finansovaya Politika Revolyutsii*, iii (1928), 19, 41-42 ; at the end of November 1925, however, Sokolnikov still counted on a total export of the value of 800 million rubles (*ibid.* iii, 231).

[5] *Na Agrarnom Fronte*, No. 4, 1926, p. 5.

[6] *Kontrol'nye Tsifry Narodnogo Khozyaistva na 1926–1927 god* (1926), pp. 296-297.           [7] See pp. 484-487 below.

[8] See *The Bolshevik Revolution, 1917–1923*, Vol. 3, pp. 464-466.

of the People's Commissariat of Foreign Trade (Vneshtorg) to conduct commercial business with foreign traders. As early as March
1922 Tsentrosoyuz had secured formal recognition of its right to
conduct foreign trade operations.[1] The claim of other economic
organs, within the general framework of Vneshtorg policy, to
conduct trading operations direct with foreign buyers and sellers
had been conceded in principle by the decree of October 16,
1922.[2] A month later conditions were laid down for the granting
of licences to the cooperatives, as well as to mixed companies
with foreign capital, to engage in foreign trade.[3] In the following
month, Trotsky, in the course of rebutting the general attack on
the monopoly of foreign trade, argued against a more limited
proposal to accord separate representation abroad to trusts and
syndicates.[4]

Since the early days of Vneshtorg, a clear distinction had been
drawn between its administrative functions, which were similar
to those of other People's Commissariats and presented no
particular problem, and its operational functions, which required
the employment of technical staff experienced in trade and
business. Vneshtorg, like other union commissariats, had its
plenipotentiary attached to the Sovnarkom of each of the union
republics ; and the functions of buying and selling on the home
market were entrusted to state trading establishments (gostorgi)
set up in each republic under the supervision of these representatives.[5] Vneshtorg was represented in every important foreign
capital by a trade delegation (torgovoe predstavitel'stvo, or
torgpred). A similar distinction was drawn between the adminis-

---

[1] *Sobranie Uzakonenii, 1922*, No. 24, art. 266.
[2] See *The Bolshevik Revolution, 1917-1923*, Vol. 3, pp. 463-464.
[3] *Sobranie Uzakonenii, 1922*, No. 76, art. 945. For the special rôle of the
cooperatives in early Soviet foreign trade see *The Bolshevik Revolution, 1917-
1923*, Vol. 3, pp. 155-156, 161 ; for the mixed companies see *ibid.* Vol. 3,
pp. 351-352, 367-368, 427.
[4] Note from Trotsky to Lenin of December 12, 1922, in the Trotsky
archives.
[5] The statute of Vneshtorg approved by VTsIK in November 1923 is in
*Sistematicheskoe Sobranie Deistvuyushchikh Zakonov SSSR*, i (1926), 88-96 ;
the gostorgs were officially described as " state organs immediately subordinated
to the People's Commissariat, but in their organization approximating in many
respects to share companies and working on the principle of *khozraschet*, with
separate capital balance-sheet and separate accounting " (L. B. Krasin, *Voprosy
Vneshnei Torgovli* (1928), p. 76).

trative and operational sections or departments of torgpreds ; and, though this distinction may not have been practically effective in the smaller torgpreds, the departments were kept strictly separate in the two most important of them — those of London and Berlin.[1]  No exception was taken to the administrative and policy-making aspects of Vneshtorg's work.  But its control over the business of buying and selling was widely and constantly challenged.  The main argument was that trusts and syndicates handling particular commodities were better qualified to negotiate the purchase or sale of their specialities than the general trading organs of Vneshtorg and the torgpreds, which carried the inevitable stigma of bureaucracy.  " Give us the possibility to travel abroad and buy what we need — we need no bureaucratic organs ", was the plausible plea of many " comrades who stand close to production ".[2]  If pressed to its logical conclusion, this argument ran counter to the whole principle of the monopoly.  But, couched in moderate terms and with lip-service to the over-riding authority of Vneshtorg, it carried conviction, and could be supported by innumerable stories of incompetence on the part of Vneshtorg and its agents.  The campaign was helped by Krasin's isolated position in the party, especially after Lenin's death, and by the prejudice easily excited against him as a man of bourgeois and western habits of life and an advocate of concessions to the capitalists.  The swollen staffs of Vneshtorg, the total number of whose employees rose from 18,900 on May 1, 1924, to 24,700 on October 1, 1925, came under attack.[3]  The torgpreds abroad became, in particular, a target for gossip and scandal.  Posts in them were currently believed to be coveted by the gilded youth of the party and by especially favoured nepmen, to be filled

[1] For a fairly full account of the organization see the report *ibid.* pp. 64-72. The operational section of the London torgpred was registered as a company under the name of Arcos ; the Berlin torgpred with its trading section had in 1925 a staff of 800, the London torgpred with Arcos slightly fewer (*ibid.* p. 116). Much general information about the work of Vneshtorg and its organs is also given in L. B. Krasin, *Vneshnyaya Torgovlya SSSR* (1924), which is said to be a revised and enlarged version of the same report.

[2] L. Kamenev, *Stat'i i Rechi*, xii (1926), 464 ; regional economic councils were given the right to appoint representatives to torgpreds (*Sobranie Uzakonenii, 1923*, No. 42, art. 453).

[3] *Partiinye, Professional'nye i Kooperativnye Organy i Gosapparat : k XIV S"ezdu RKP(B)* (1926), p. 134 ; on the first date 15·5 per cent, on the second 18·1 per cent, of those employed were party members.

largely through nepotism,[1] and to provide unrivalled opportunities for indulgence in bourgeois tastes, as well as for more direct forms of corruption.

Handicapped by its low rating in party circles and by the suspicion which easily attached to its activities, Vneshtorg was engaged throughout its career in a running battle to maintain the integrity of the monopoly and of the principle of planning in foreign trade against the encroachments of other trading agencies. Even outside the restricted circle of those authorized to transact business with foreign firms, independent action by powerful interests was difficult to prevent. In 1920, when the transport situation was still acute, a railway mission headed by a railway engineer, Lomonosov, visited western Europe, " placing orders ", as Krasin afterwards complained, " to the value of tens of millions of gold rubles without any previous coordination with the plan or with the resources of the state, grossly prejudicing the interests of our industry, and sometimes, as in the order for rails, on very unfavourable terms ". In the winter of 1923–1924, Nogin, the director of the textile trust and an energetic and influential member of the party, travelled to the United States (where Vneshtorg had as yet no representative), made extensive purchases of cotton, and established in New York a share company with Soviet capital called the All-Russian Textile Syndicate, without consulting either Vneshtorg or Narkomindel.[2] Krasin continued to deny that " even from the point of view of the technical execution of this or that operation there would be any sense in transferring it into the hands of particular state or economic organs ". He argued that, as individuals, representatives of these organs were no more inherently likely to be competent than representatives of Vneshtorg, and, finally, that the hatred which the monopoly of foreign trade inspired in capitalist circles was the best proof of

---

[1] According to *Sotsialisticheskii Vestnik* (Berlin), No. 17-18 (135-136), September 18, 1926, p. 12, important posts in the torgpred in Berlin had recently been held by a brother-in-law of Rykov, a brother-in-law of Joffe, and a brother of Lozovsky : such reports, whether true or not, were symptomatic. Kuibyshev referred, at the fourteenth party congress of December 1925, to an investigation, recently undertaken by the central control commission, of " Vneshtorg and all its foreign delegations, especially those in London and Berlin " (*XIV S"ezd Vsesoyuznoi Kommunisticheskoi Partii (B)* (1926), p. 546).

[2] L. B. Krasin, *Voprosy Vneshnei Torgovli* (1928), pp. 100-101.

the importance of maintaining it intact.[1]  By 1924 even mixed companies with foreign capital had lost their usefulness : " Vneshtorg itself and the cooperatives are more and more masters of the technique of foreign trade ".[2]

The increasing importance of foreign trade in the national economy, the prospects of a bumper harvest for 1925 giving every hope of its further expansion, the campaign for the development of heavy industry, and the new emphasis on the principle of planning, made it more and more difficult to tolerate the ambiguity of Vneshtorg's position.  The complaint was also heard that trading organizations were making profits from foreign transactions, thus further raising prices.[3]  During 1924–1925 an attempt was made to revise the list of organizations licensed by STO to engage in foreign trade : Donugol, Azneft, Grozneft and GUM were removed from it, and Lesoeksport (created on the analogy of Khlebeksport) and Maslotsentr added to it.[4]  It seems to have been Krasin who felt the necessity of bringing the issue to a head.  At the end of July 1925 he drew up a set of theses which constituted a summary exposition and defence of the principles and policies upheld by Vneshtorg.  The theses cunningly connected the attacks on Vneshtorg in " the white guard emigration and the foreign press " with similar attacks by " rising *kulak* elements " in the Soviet Union engaged in a struggle against the Soviet power, and associated Vneshtorg firmly with the supporters of planning :

> The monopoly of foreign trade presupposes a single state import-export plan of the union, worked out by the planning organs of Vneshtorg with the participation of all interested departments, confirmed by Gosplan and by Sovnarkom, and welded together on the basis of a calculation of the needs of the whole national economy, as well as of the resources of the union in export goods and in valuta.

For this reason " the partition of its functions among other departments would in fact mean a refusal to give effect to the

---

[1] *Ibid.* pp. 71, 90, 99.                    [2] *Ibid.* p. 61.
[3] *Vestnik Finansov*, No. 8, August 1925, p. 13.
[4] *God Raboty Pravitel'stva SSSR 1924–1925* (1926), p. 447 ; the decree of STO authorizing Maslotsentr to engage in foreign trade is in *Sobranie Zakonov, 1925*, No. 2, art. 26.

monopoly of foreign trade ". At the same time there was nothing inconsistent with this principle in allowing subordinate organs of Vneshtorg to carry out commercial operations on its behalf independently and on a commission basis, provided steps were taken to prevent competition between such organs as buyers or sellers on the foreign market. Greater discretion in the matter of prices and the conclusion of bargains must be left to the torgpreds and trading organs in foreign centres.[1]

This document was presented to the party central committee in September 1925 and provoked what was evidently a lengthy discussion in the committee on October 5. A draft prepared by Kuibyshev, the president of the party central commission, who was emerging at this time as a safe party man on controversial financial and economic questions, was adopted as the basis of a decision, and referred for further elaboration to a commission consisting of several members of the central committee, local party workers, and the heads of Vneshtorg. The final text of the resolution as drafted by this commission was approved by the Politburo, and published in *Pravda* on November 6, 1925.[2] The resolution confirmed the need to conduct foreign trade " through a specially created organ (Vneshtorg) ". It drew attention to the importance of foreign trade both for agriculture and for industry :

> A large number of important branches of agriculture at the present time (cultivation of wheat, barley, maize, flax ; dairy products, poultry, live-stock) can undergo a further substantial development on condition that they are assured of an expanding demand on the world market. On the other hand, the extension of the problem of basic capital for our industry is closely bound up with the further development of foreign trade.

Trade had, however, become more specialized, and more specialized treatment was required. In order to handle important items of export or import, the formation of special companies, associations or syndicates was desirable. The organization of such companies should be undertaken by Vneshtorg and confirmed by STO.

---

[1] L. B. Krasin, *Voprosy Vneshnei Torgovli* (1928), pp. 121-137.
[2] *VKP(B) v Rezolyutsiyakh* (1941), ii, 32-38.

This was the most important innovation in the resolution and gave some satisfaction to those who called both for the decentralization and for the commercialization of foreign trade.[1] But it made no radical alteration in existing practice, and, though couched in terms designed to conciliate all interests, adequately safeguarded the position of Vneshtorg.[2]

It did not, however, end Krasin's anxieties. These were expressed in a long article which appeared in the journal of STO on the day after the publication of the resolution, the coincidence being probably accidental. Krasin admitted that the principle of the monopoly of foreign trade was treated as sacrosanct. But " the uncontested recognition of the principle conceals within itself great dangers of various devious and masked movements and manœuvres ".[3] One of these manœuvres quickly came to a head. On November 18, 1925, a decree was issued amalgamating the two trade commissariats into a single People's Commissariat of Foreign and Internal Trade (Narkomtorg).[4] It was a logical reform which Gosplan had advocated ever since the creation of a People's Commissariat of Internal Trade.[5] The official explanation of the change dwelt on the growing importance of foreign trade in the economy and on the need to coordinate the requirements of foreign with those of internal trade. Lezhava, a former People's Commissar for Internal Trade, added that it was a safeguard against an " export deviation " — a determination to export at all costs regardless of the needs of the community.[6] If some commercial interests hoped that the amalgamation would lead to the extension of private enterprise to the sphere of foreign trade —

---

[1] According to G. Cleinow, *Neu-Siberien* (1928), p. 390, the result of the decision in the Siberian region was to place the gostorgs under the control of the regional authorities, which thus scored a victory over the central trading organs of the RSFSR.

[2] An obituary article on Krasin a year later in *Planovoe Khozyaistvo*, No. 12, 1926, pp. 14-24, treated the resolution of October 1925 as a triumph for Krasin's policy : but this was only in part true.

[3] L. B. Krasin, *Voprosy Vneshnei Torgovli* (1928), p. 42.

[4] *Sobranie Zakonov*, No. 78, art. 590.

[5] A resolution of Gosplan supporting the amalgamation in July 1923 is in G. Krzhizhanovsky, *Sochineniya*, ii (1934), 281 ; Krzhizhanovsky once more urged it in an article in *Ekonomicheskaya Zhizn'*, January 13, 1924.

[6] Statements by Tsyurupa appeared in *Pravda*, November 20, 1925 ; by Rykov, *ibid.* December 9, 1925 ; by Lezhava in *Ekonomicheskaya Zhizn'*, November 21, 23, 1925.

" a NEP in foreign trade " — such hopes were doomed to disappointment.[1] The most obvious consequence of the change was
to deprive Krasin of his post in Sovnarkom, Tsyurupa becoming
the new People's Commissar for Trade with the former commissars for foreign and internal trade, Krasin and Sheinman, as
his deputies.[2] Krasin, who had been polpred in France since
October 1924, and was now transferred to London, would spend
the remaining twelve months of his life abroad. But this was
no novelty. Since Lenin's death Krasin had been too little in
sympathy with the party leaders to play any rôle except in issues of
foreign trade ; and even here his authority was contested.

But if Krasin's enemies regarded the suppression of a separate
commissariat of foreign trade as the prelude to a radical change of
policy, they were disappointed. The foundations which Krasin
had laid in Vneshtorg between 1920 and 1925 proved too strong
to be disturbed. The loss of his governmental office, and his
death a year later in London, were followed not by a weakening,
but by a strengthening, of the monopoly which he had worked to
establish and administer. A month after the amalgamation of the
commissariats, the fourteenth party congress confirmed the drive
for industrial development and the decision to make the Soviet
Union, in Stalin's words, " an independent economic unit " ; [3]
and this in the long run implied support not only for the principle
of planning, but for more rigorous control of export and import
policies. A resolution of the Politburo of January 21, 1926,
ordered the accumulation of reserves of staple commodities of
foreign trade as a protection against price fluctuations and the
undue influence of market conditions.[4] In existing conditions of
scarcity the project was utopian, but indicated a new awareness
of the problems of foreign trade. Finally, in April 1926, the party
central committee adopted a resolution on economic policy which,
while introducing no innovations in the field of foreign trade,
contained an unequivocal recognition of its importance :

---

[1] An article in *Vlast' Sovetov*, No. 49, December 6, 1925, pp. 1-2, was
directed against this interpretation of the decree.
[2] *Ekonomicheskaya Zhizn'*, November 19, 1925.
[3] Stalin, *Sochineniya*, vii, 299.
[4] The resolution does not seem to have been published, but was referred to
and endorsed in the decision of the party central committee of April 1926
(*VKP(B) v Rezolyutsiyakh* (1941), ii, 97).

The tempo of the expansion of fixed capital and of the re-equipment of industry, as well as the technical improvement and intensification of agriculture, depends in the greatest degree on the successful development of our export operations and on the import from abroad of necessary equipment, raw material and semi-finished goods for our industry and of agricultural implements for the cultivation of the land.    Therefore the development of export is an indispensable condition for the industrialization of the country and for quickening the tempo of industrial development.[1]

The battle for the consolidation of the monopoly of foreign trade had been finally won when the fourteenth party congress voted for self-sufficiency and intensive industrialization.

The theoretical problems arising out of the expansion of Soviet foreign trade did not attract much attention in this period.   Whether or not the capitalist belief in the economic advantages of the international division of labour applied to relations between capitalist countries and a country aspiring to create a socialist economy, it was indisputable that the rapid building of socialism in the Soviet Union was dependent on extensive imports of capital equipment from more advanced industrial countries, and therefore also on finding lucrative markets in those countries for Soviet products.   It was Trotsky who, in his article *Towards Socialism or Capitalism?* in the autumn of 1925, first noted the unexpected community of interest between capitalist countries and the Soviet Union in maintaining a general level of prosperity :

A commercial and industrial depression in Europe, and still more a world depression, might lead to a wave of depression in our country.   Conversely, a commercial and industrial boom in Europe would at once be followed by a demand for essential raw materials for industrial purposes, such as timber and flax, and for grain, the consumption of which would increase with the increasing prosperity of the European peoples. . . . We thus reach a position where, as an economic state unit, it is to some extent, at any rate, in our interest to see improved conditions in capitalist countries.[2]

Trotsky refused to be embarrassed by this striking example of the "inconsistencies inherent in our so-called new economic policy ",

[1] *Ibid.* ii, 92.
*Pravda*, September 22, 1925 ; for this article see p. 505 below.

which at home and abroad involved a certain amount of collaboration between socialism and capitalism as well as an intensification of the struggle between them. He did not examine the potential influence of this collaboration on Soviet foreign policy. Nor was anyone else eager to pursue this issue. For the present it was enough that foreign trade was making an important contribution to the progress of industrialization.

Much attention was given in this period to the establishment of machinery for the granting of concessions to foreign capital. " A chief concessions committee " attached to STO had originally been set up in April 1922 in preparation for the Genoa conference ; [1] and neither the failure of negotiations with the western Powers nor the dramatic rejection of the Urquhart concession in the autumn of that year [2] ended hopes for the ultimate success of the policy. In 1923 the committee was put under the direct authority of Sovnarkom ; [3] and special concessions commissions were attached to the trade delegations in Berlin and London.[4] The year 1925 saw a recrudescence of interest in foreign concessions. Concessions commissions were attached to the torgpreds in Paris and in Rome ; [5] and a standing committee was attached to the chief concessions committee in Moscow " to verify the carrying out of concessions agreements ".[6] This was the year of the two most spectacular achievements of the concessions policy — the Lena Goldfields concession and the Harriman manganese concession in the Caucasus.[7]

Foreign concessions never fulfilled the extravagant hopes which had at first been placed on them, or played a significant rôle in the

[1] *Sobranie Uzakonenii, 1922*, No. 28, art. 320.

[2] See *The Bolshevik Revolution, 1917–1923*, Vol. 3, pp. 432-434.

[3] *Sobranie Uzakonenii, 1923*, No. 20, art. 246 ; for its statute approved in August 1923 see *Sistematicheskoe Sobranie Deistvuyushchikh Zakonov SSSR*, i (1926), 43-45.

[4] *Sobranie Uzakonenii, 1923*, No. 23, art. 259 ; No. 26, art. 307.

[5] *Sobranie Zakonov, 1925*, No. 21, art. 139 ; *Sbornik Dekretov, Postanovlenii, Rasporyazhenii i Prikazov po Narodnomu Khozyaistvu*, No. 24 (45), September 1925, pp. 24-25.

[6] *Sobranie Zakonov, 1925*, No. 52, art. 394.

[7] These concessions, as well as the general relation of the concessions to foreign policy, will be discussed in Part V in a subsequent volume.

Soviet economy.  A review of the situation in the spring of 1925 gave figures for the three preceding years :

|           | Proposals Received | Agreements Concluded |
|-----------|:------------------:|:--------------------:|
| 1921–1922 | 338                | 18                   |
| 1923      | 607                | 44                   |
| 1924      | 311                | 26                   |

As against this, only 30 proposals had been received, and only three agreements concluded, in the first four months of 1925. Germany held the first place on the list with 43 per cent of proposals and 24·2 per cent of agreements concluded, followed by Great Britain, the United States of America and France.  About 40 per cent of the concessions were classified as " industrial " (including mining and forestry), the remainder being devoted to agriculture, transport or trade.[1]  Wide discrepancies exist between different statements of the capital invested in concessions and the income derived from them.  According to one authoritative account, state revenue from the concessions amounted to 14 million, rubles in 1923–1924.  But Kamenev, in the latter part of 1925, estimated the income for the current year at no more than 4 millions of which 1,200,000 were to come from the Harriman concession.[2]  The endorsement by the fourteenth party congress of the policy of intensive industrialization on a basis of national self-sufficiency coincided with the growing recognition of the failure of the concessions policy to attract any significant volume of foreign capital on terms acceptable to the régime.

[1] *Bol'shevik*, No. 8, April 30, 1925, pp. 46-60 ;  G. Gerschuni, *Die Konzessionspolitik Sowjetrusslands* (1926), pp. 123-124, using Soviet press sources, gives the number of concessions in operation at the beginning of 1925 as 90 (including 22 German, 17 British and 8 American) : of these 26 were classified as trading, 17 manufacturing, 13 mining, 13 agricultural, 12 transport, 6 forestry and 3 " other ".

[2] *Bol'shevik*, No. 8, April 3c 1925, p. 57 ;  L. Kamenev, *Stat'i i Rechi*, xii (1926), 473.

# CHAPTER 9

# FINANCE AND CREDIT

THE currency reform marked the culmination of NEP in the financial sphere. It was the logical corollary of the return to freedom of trade and to a monetary economy, the advantage of which could not be fully enjoyed in the absence of a stable currency. It also crowned the policy of concessions to the peasant, who had borne the major share of the consequences, direct and indirect, of headlong currency depreciation. Its immediate effects, like those of NEP itself three years earlier, were almost wholly salutary, and quickly silenced those who had at first regarded it with mistrust or disapproval. It created fresh confidence all round, paved the way for a remarkable recovery both in agriculture and in industry, and for the first time made planning a serious possibility. At the same time it was significant that, at the moment when a stable currency based on the universal gold standard, and therefore immune from manipulation by the state, was being introduced, the state had been compelled to restore its control of prices of essential commodities. The forces which made it impossible to maintain a régime of " free " market prices would eventually prove fatal to the régime of free exchanges and a " free " currency on a fixed basis. But these difficulties still lay ahead. For the moment, immense pride was felt in the magnitude of the achievement and equally strong determination to take any measures, however irksome, which might be necessary to maintain it. During the first year of the reform, preoccupation about the stability of the currency was the dominant factor in financial policy.

The establishment of a stable currency had immediate repercussions on the state budget : indeed, the prospective difficulty of balancing the budget in the new conditions had been the chief argument used by the opponents of the reform. The budget for

1923–1924 was estimated to balance at approximately 1900 million rubles.  Of the receipts 17·8 per cent came from direct taxation ;  16·5 per cent from indirect taxation (excise, customs and other duties) ;  40·7 per cent came from non-tax revenues (more than three-quarters of this being, however, accounted for by transport, which was self-supporting and was included in balancing items on both sides of the budget) ; and 25 per cent came from extraordinary revenue, the largest single item of which consisted of profits from currency emission to the amount of 180 millions — a substantial figure, though only half that of the previous year.[1]  Since no corresponding resource would be available in 1924–1925, and any budget deficiency would have to be made good out of the unresponsive market for state loans, the utmost restraint was required in drawing up the budget.  When Sokolnikov addressed a financial conference in July 1924 he was in a cautious mood.

> The question of the dimensions of our budget [he said] depends in a large measure on the dimensions of the tax burden on the peasant, i.e. on the question of the proportional relation between state economy and peasant economy.  Following the prescriptions of comrade Lenin, following the general line of our party, our task, the task of our financial apparatus is to protect and maintain the possibility of developing the peasant economy.  It is on the basis of this growth of the peasant economy that the market for our industry can develop in the future, and on the basis of the growth of the peasant economy that our state budget can also develop.

Sokolnikov did not believe that the country, that is to say, the peasantry, could bear an increased burden of taxation, and fixed 2100 million rubles as the maximum budget total for the coming year.  This was a 10-per-cent increase on the previous year.[2]

Working within these limits, Narkomfin now produced so-called " control figures ", or preliminary estimates, for the 1924–

---

[1] *SSSR : Tsentral'nyi Ispolnitel'nyi Komitet 2 Sozyva : 2 Sessiya* (1924), p. 137.  Final figures quoted in R. W. Davies, *The Development of the Soviet Budgetary System* (1958), p. 82, show a substantially higher total of 2300 millions, a slightly higher percentage (12·6) for non-tax revenue other than transport, and a much lower percentage (5·4) for profits from currency emission.

[2] *Sotsialisticheskoe Khozyaistvo*, No. 5, 1924, pp. 12-13.

1925 budget.¹ These were submitted to Sovnarkom in the first
ten days of the new financial year, approved by it on October 14,
1924, and submitted to TsIK by Sokolnikov later in the same
month. Though these estimates did not yet constitute a formal
budget, they were an advance on anything yet attempted in the
unimpressive history of Soviet public finance. Sokolnikov could
boast that, for the first time since the revolution, " we have the
possibility of looking ahead at the beginning of the budgetary
year ". It was also the first budget covering the whole territory
of the USSR.² Of the budget total of 2100 million rubles, 843
millions were accounted for by balancing entries for transport
and communications, which were maintained only for formal
reasons on the state budget.³

The principal revenues from taxation, amounting to 46 per
cent of all revenues (or more than 70 per cent, if transport and
communications were excluded), were 250 millions from the
agricultural tax, 70 millions from the income-tax, 120 millions
(of which, however, only 66 millions were included in the union
budget, the remainder being taken by local budgets) from the
industrial tax, and 300 millions from excise.⁴ Direct taxation
still therefore provided the largest source of revenue. " Our

¹ The significance of the " control figures " was explained in *Sotsialisticheskoe Khozyaistvo*, No. 5, 1924, p. 46 : they were " a general outline which
provides directives for the People's Commissariats in drawing up their estimates ". They were indicative, but not imperative.

² *SSSR : Tsentral'nyi Ispolnitel'nyi Komitet 2 Sozyva : 2 Sessiya* (1924),
pp. 136-138 ; the budget for 1923-1924 had been the first budget of the USSR,
but in that year the Transcaucasian SFSR and the Far Eastern Territory (the
temporary successor of the Far Eastern Republic) had for technical reasons
retained their separate currencies and separate budgets.

³ It is difficult to say why the whole budgets of the People's Commissariats
of Communications and Posts (i.e. current receipts and expenditure) were
retained in the state budget, whereas all other industrial enterprises were independent for accounting purposes, only profits or losses being ultimately
transferred to the state budget. Sokolnikov defended the practice on the ground
that communications and posts were still working at a loss (G. Sokolnikov,
*Finansovaya Politika Revolyutsii*, ii (1926), 161-162) ; but the same was true of
other enterprises. The precedent of the pre-revolutionary period, when communications and posts had been the only " nationalized " enterprises and had
been carried on the budget for that reason, was probably the decisive factor.

⁴ The preliminary estimates as submitted to Sovnarkom and TsIK in
October 1924 are in *Sotsialisticheskoe Khozyaistvo*, No. 5, 1924, pp. 52-54 :
Sokolnikov's speech on their presentation to TsIK is in *SSSR : Tsentral'nyi
Ispolnitel'nyi Komitet 2 Sozyva : 2 Sessiya* (1924), pp. 135-196.

Soviet system ", said Sokolnikov, " has definitely taken its stand
on the path to the development of direct taxes " ; and he repeated
the axiom that " direct taxation is class taxation ".[1] The largest
items of expenditure were 400 million rubles for the administrative
organs of the union, central and local, and 378 millions for
defence.  Other noteworthy items were 59·6 millions for industry,
40 millions for agriculture, besides 46 millions for relief to
sufferers from the bad harvest, and 37·9 millions for " building
and electrification ".  Estimated expenditure exceeded estimated
revenue by 120 million rubles.  It was proposed to cover this deficit
by receipts from the issue of silver and copper coinage, estimated at 80 millions, and by state loans.  TsIK approved these
" control figures ", and instructed Narkomfin to prepare a formal
budget based on them for resubmission to Sovnarkom and to a
later session of TsIK.  It also issued a new decree reforming
the income-tax.  Peasants paying the agricultural tax and workers
earning less than 75 rubles a month (which included nearly all
manual workers at this time) were exempt from income-tax
altogether.  Other incomes were classified in 17 categories.  The
lowest, covering incomes of less than 500 rubles a year, paid
10 rubles in tax ; the highest, covering incomes of 8000 rubles a
year and upwards, paid 1500 rubles plus 300 rubles on every 1000
above 8000.[2]  The property tax, which had formed an integral
part of the original income-tax, was now abolished.  When the
tax was introduced in 1922, pains had been taken to bring
within its scope those members of the former ruling class who
had no income, but lived by selling their possessions.  This
class had now virtually disappeared, and receipts from the
property tax were too insignificant to be worth the cost of
collection.[3]

The decision approving the preliminary figures for the budget
of 1924–1925 was an occasion of some importance.  It was accompanied by the issue on October 29, 1924, of a statute on budgetary

    [1] *Ibid.* p. 139.
    [2] *Postanovleniya TsIK Soyuza SSR : 2 Sessiya* (1924), pp. 17, 32-42 ;
*Sobranie Zakonov, 1924*, No. 20, art. 196.  The tax had been attacked by Larin
at the second Union Congress of Soviets in January 1924 as not being steeply
enough graded (*Vtoroi S"ezd Sovetov SSSR* (1924), pp. 154-160).
    [3] *SSSR : Tsentral'nyi Ispolnitel'nyi Komitet 2 Sozyva : 2 Sessiya* (1924),
pp. 142-143.

rights which was designed to bring departmental expenditure under the strict control of Narkomfin and ensure that it was kept within the limits of revenue. The principle of the balanced budget was now firmly established in Soviet practice. It was the function of Narkomfin to examine the estimates submitted by the departments on the basis of the original " control figures ", to bring them into line with the estimates of revenue, and to combine them with the budgets of the republics (themselves the product of similar procedure at the republican level) into a single union budget, which, together with the observations of Gosplan on it, was submitted to Sovnarkom, and eventually to TsIK, for final approval.[1] The Soviet Union now had not only a stable currency, but a regular budgetary system. The irrepressible Larin described it as " the highest moment in the history of Narkomfin ", and suggested that the time had come when finance could lay down its " dictatorship ".[2] Like most of Larin's comments, it exhibited a certain prescience.

The buoyancy of revenue from excise, especially on sugar and vodka, and an improvement in receipts from transport — all indices of an increasing prosperity — soon justified an increase of the budget estimates. It was afterwards explained by a spokesman of Narkomfin that the original control figures had been prepared in the summer of 1924 before the beneficial effects of the currency reform had been fully felt.[3] Pressure from the spending departments, and the less pessimistic view now taken of the consequences of the poor harvest completed the process. The estimates were raised by 180 million rubles to 2280 millions.[4] But the mood was still cautious ; and it was only " after long debates " that the party central committee in January 1925 gave its sanction to " an expansion of the budget ".[5]

Another decision of some importance was taken at this time. It was a symptom of the shifting balance of power in the Soviet machine, and the first overt blow struck at the supremacy of

[1] *Sobranie Zakonov, 1924*, No. 19, art. 189 ; for the financial relations between the union and the republics as laid down in the statute see Note B : " The Budgets of the Republics " (pp. 530-534 below).

[2] *SSSR : Tsentral'nyi Ispolnitel'nyi Komitet 2 Sozyva : 2 Sessiya* (1924), pp. 294-295.

[3] *Id. : 3 Sessiya* (1925), p. 169.

[4] *Planovoe Khozyaistvo*, No. 1, 1925, pp. 57-58.

[5] *VKP(B) v Rezolyutsiyakh* (1941), i, 634.

Narkomfin. In 1923 the twelfth party congress had approved a reorganization of the People's Commissariat of Workers' and Peasants' Inspection (Rabkrin) in the form of an interlocking arrangement with the party central control commission ; [1] and Kuibyshev, the People's Commissar for Rabkrin, also became president of the control commission. Since both Rabkrin and Narkomfin exercised rights of supervision over the activities of other departments, it is not surprising that the reorganization of Rabkrin should have been followed by a dispute with Narkomfin on the respective control functions of the two organs.[2] At a moment when orthodox finance was still in the ascendant and the prestige of Narkomfin at its highest, the dispute was settled in favour of Narkomfin. Rabkrin received extended disciplinary powers in the way of checking the efficiency, regularity and honesty of the administration, but was relieved of all functions of financial control, which were vested exclusively in Narkomfin.[3] The powers of Narkomfin now seemed unchallenged. When the statute on budgetary rights was adopted on October 29, 1924, provision was made for the establishment in each of the union republics of a budget commission attached to the TsIK of the republic. It was noticeable that no such commission was created for the USSR, where the commission responsible for drafting the budget was a departmental body working within Narkomfin.

This situation at length gave Kuibyshev, who was a Stalin man and had the weight of the secretariat behind him, the opportunity to recoup himself for the exclusion of Rabkrin from financial affairs. In January 1925 the party central committee decided that a commission for the budget of the USSR should be set up and attached to TsIK in order to assist that august body to carry out its (hitherto formal) function of supervising the budget.[4] The decision was embodied in two decrees of the presidium of TsIK of March 7, 1925, which named the 56 original members of the

[1] See *The Bolshevik Revolution, 1917–1923*, Vol. 1, p. 228.
[2] *Vestnik Finansov*, No. 10, October 1925, pp. 35-44.
[3] See the statute of Rabkrin of November 1923 in *Sistematicheskoe Sobranie Deistvuyushchikh Zakonov SSSR*, i (1926), 189-193.
[4] The decision was not published, and rests on the authority of V. Dyachenko, *Sovetskie Finansy v Pervoi Faze Razvitiya Sovetskogo Gosudarstva*, i (1947), 426, which is, however, likely to be reliable.

commission. The statute was drawn up later in the same month. The functions of the commission were to review the draft budget approved by Sovnarkom, to make any other recommendations regarding the budget, and to prepare a report on it for eventual adoption by TsIK.[1] A decree formally appointing Kuibyshev president of the commission (he had occupied the position from the start) followed shortly afterwards.[2] A year later the membership of the commission had increased to 96 : of these 15 were representatives of the USSR, and the remainder representatives of the union republics in proportion to population, 47 being drawn from the RSFSR.[3]

It at once became apparent that a new power had been created, and the financial monopoly of Narkomfin broken, though this did not imply any immediate change in policy or outlook. The session of TsIK in March 1925 was held in Tiflis. Sokolnikov significantly did not make the journey, being represented by his deputy Bryukhanov ; and Kuibyshev automatically assumed the rôle of principal spokesman on the budget, reporting to TsIK at length on behalf of its new commission. The commission proposed further minor increases which carried the budget total to the new height of 2360 million rubles. But the most important points in Kuibyshev's speech related to the future. He asked that the budget should be approved not, as in previous years, as a " directive " budget, but as a " firm " budget, binding on those whose business it was to execute it ;[4] and he expressed the significant hope that future budgets would be able to rely to a larger extent on non-tax revenues. TsIK duly adopted a resolution amending

[1] *Sobranie Zakonov, 1925* No. 17, arts. 127, 128 ; No. 71, art. 520.
[2] *Ibid.* No. 38, art. 282.
[3] *SSSR : Tsentral'nyi Ispolnitel'nyi Komitet 3 Sozyva : 2 Sessiya : Postanovleniya* (1926), pp. 16-18.
[4] In December 1924, a conference of budget experts of Narkomfin and of the union republics had passed a resolution declaring that the budget of 1925–1926 was to be no longer a collection of estimates subject to modification from month to month, but a " firm budget for the year " (*Vestnik Finansov*, No. 1, January 1925, p. 96). One important purpose of the introduction of a " firm " budget was to do away with the system of block grants to departments, over the expenditure of which Narkomfin exercised no control. A " firm " budget implied " budgetary discipline ", sums being expendable only for the specific items for which they were allocated, and uniformity of administration being thus assured (*Ekonomicheskoe Obozrenie*, November 1925, pp. 13-15). How far these results were achieved in practice is another question.

and confirming the budget.[1]  Stalin at a party meeting commented enthusiastically on these successive increases, and drew a significant moral for the policies just beginning to shape themselves in his mind :

> You know that three times in the course of the past half year we had to change our state budget in view of the rapid growth of budget revenue items unforeseen in our original estimates.  In other words, our budget estimates and budget plans did not keep pace with the growth of state revenues, so that surpluses appeared in the state treasury.  That means that the springs of the economic life of our country are gushing forth with irresistible strength, upsetting all and sundry scientific plans of our financial specialists.  That means that we are experiencing a not less, perhaps even more, powerful economic and productive drive than took place, for example, in America after the civil war.[2]

Next, the budget came before the third Union Congress of Soviets in May 1925, and once more underwent a process of upward revision, being finally approved at a total of 2558 million rubles, with a specific authorization to the presidium of TsIK to proceed to a further revision if conditions justified it.  The congress also endorsed for the future the principle of " firm " annual budgets.[3]  Finally in June 1925 the presidium of TsIK, using the authority given to it, lifted the total of the budget to 2876 million rubles, the principal increases in revenue being from excise and from the industrial tax, and increased allocations being made to agricultural credit, to the cooperatives, and to the electrification and building programmes.[4]  Never was a budget of the USSR so long, so frequently or so exhaustively debated in public as that of 1924–1925, the first stable currency budget of the union. Rarely has any budget inspired so much enthusiasm and optimism : it was an important factor in the rising self-confidence of the summer of 1925.

When later the balance of the financial year 1924–1925 was

---

[1] For Kuibyshev's speech see *SSSR : Tsentral'nyi Ispolnitel'nyi Komitet 2 Sozvya : 3 Sessiya* (1925), pp. 189-203 ; for the resolution *id.: Postanovleniya*, pp. 13-17.          [2] Stalin, *Sochineniya*, vii, 128-129.

[3] *Tretii S"ezd Sovetov SSSR : Postanovleniya* (1925), pp. 30-31 ; *Sobranie Zakonov, 1926*, No. 35, art. 250.  The detailed figures of the budget are in *Vestnik Finansov*, No. 6, June 1925, pp. 170-175.

[4] *Sobranie Zakonov, 1925*, No. 48, art. 347.

struck, the results bore witness to the rapid recovery of the national economy and of state finances. Even after these repeated upward revisions the final estimates were exceeded. Revenue just topped 3000 million rubles, leaving a surplus of 32 millions over expenditure. While direct taxation contributed about the same proportion of revenue as in the previous year, the share of excise had risen from 10·4 per cent to 16·9 per cent : thanks mainly to this increase, taxation now yielded 44·2 per cent of revenue as against 33·9 per cent in the previous year. The self-balancing item of communications, which for the first time paid their way without subsidy, accounted for a slightly higher proportion at 35·8 per cent ; and non-tax revenues, principally profits from state industry and state forests, rose from 12·6 per cent to 13·4 per cent. On the other hand, revenue from currency issues had disappeared altogether from the budget, and revenue from loans and credit was negligible. The state was for the first time paying its way. On the expenditure side the major increases were in subsidies to agriculture due to the crop failure of 1924 (these eventually reached 171 million rubles), in subsidies to housing, in the cost of social services, including education, and in grants to local budgets. Subsidies to industry and costs of administration remained stationary ; and there was a 10-per-cent advance in defence costs.[1] The chief aim pursued and achieved in the budget of 1924–1925 was financial stability — to meet essential requirements out of current revenue. The budget was not yet being consciously and deliberately framed to promote economic or political ends. The budget of 1924–1925 was the last of which Narkomfin was the main or sole artificer.

Even before the budget of 1924–1925 had been finally approved, plans for the budget of 1925–1926 began to be laid. A mood of far-flung optimism had been engendered by the economic recovery of the past twelve months. The budget of 1924–1925 was balanced without severe strain at a level one-third above the original estimates. Further progress was confidently expected in the coming year. Heavy industry was clamouring for increased funds from the budget to finance the policy of expansion fostered

[1] See tables in R. W. Davies, *The Development of the Soviet Budgetary System* (1958), pp. 82-83.

by the party and by the government. At a conference of the People's Commissars for Finance of the union republics in April 1925, Sokolnikov sounded a note of moderation. At a time when the budget for 1924–1925 had still not reached 2500 million rubles, he set a figure of 3000 millions as the target for 1925–1926. A formal decree of July 3, 1925, instructed Narkomfin to submit the draft of a "firm" budget for 1925–1926 to Sovnarkom, and Gosplan to present its comments on the draft, not later than October 1, 1925.[1]

The most contentious issue was now the source of future revenue. A long-standing party tradition, going back to the second party congress in 1903, hymned the virtues of direct taxation, and had been endorsed by Lenin since the revolution.[2] Nevertheless substantial inroads had already been made on it. The need for new tax revenue after the introduction of NEP had quickly led to a restoration of excise on matches, candles, tobacco, wines, coffee, sugar and salt, the tax on salt being especially resented, since it had been abolished by the Tsarist régime as long ago as 1881.[3] The amount raised in excise increased from year to year.[4] In 1923 excise duties were imposed on textiles and on rubber galoshes,[5] and the vodka monopoly was reintroduced.[6] In 1924–1925 excise yielded over 500 million rubles (against an original estimate of 300 millions), out of which the vodka monopoly, throughout this period a bone of contention, accounted for 178 millions.[7] At the end of 1924 an article in the journal of

---

[1] *Sbornik Dekretov, Postanovlenii, Rasporyazhenii i Prikazov po Narodnomu Khozyaistvu*, No. 22 (43), July 1925, p. 35.

[2] See *The Bolshevik Revolution, 1917–1923*, Vol. 2, p. 141.

[3] For the salt tax see *Sobranie Uzakonenii*, *1922*, No. 19, art. 211; it was abolished in March 1927 (*Sobranie Zakonov, 1927*, No. 17, art. 186).

[4] An account of the restoration of excise duties is in *Planovoe Khozvaistvo*, No. 1, 1926, pp. 98–103.

[5] *Sobranie Uzakonenii, 1923*, No. 17, art. 214; No. 41, art. 436.

[6] See *The Interregnum, 1923–1924*, p. 35, note 2.

[7] G. Sokolnikov, etc., *Soviet Policy in Public Finance* (Stanford, 1931), p. 189. Stalin in a letter of 1927 referred to an alleged discussion in the party central committee in October 1924 when " certain members of the central committee objected to the introduction of vodka without, however, indicating any other sources from which it would have been possible to draw funds for industry ", and seven members of the committee, including himself, made a declaration reporting statements made by Lenin " on several occasions " in the summer and autumn of 1922 defending the introduction of a vodka monopoly as necessary " for the maintenance of the currency and the support of industry "

Narkomfin called for " a further tightening up of existing rates [of excise] especially on articles of mass peasant consumption ", explaining with unusual frankness that this was the only way of taxing the poor peasant :

> In practice it is infinitely difficult to reach strata of population which have minimum surpluses by direct taxation : here only sufficiently refined methods of more or less universal indirect taxation can help.[1]

In the spring of 1925 the policy of appeasing the well-to-do peasant and the pressure to reduce the agricultural tax could point to only one conclusion.  Zinoviev, still at this time the main champion of " Face to the countryside ", did not hesitate to draw it :

> We are approaching a time when, in one way or another, the peasantry must be freed from direct taxes. . . . It would be incorrect to repeat any longer the social-democratic catchwords and say that a progressive income-tax is justice and better than an indirect tax.[2]

Sokolnikov continued to fight a delaying action.  " We must ", he declared, " defend the system of direct taxation, as taxation which guarantees the possibility of a class approach, a class policy " ; anything else would be " a betrayal of the fundamental principles of socialism ".[3]  But at the third Union Congress of Soviets in May 1925 he bashfully admitted that revenue from vodka played " a fairly substantial rôle " in the budget, though he intended " in future years strictly to limit the production and consumption of alcohol ".[4]  Now that specific commitments had been taken by the fourteenth party conference in April, and confirmed by the congress of Soviets in the following month, to make a drastic reduction in the agricultural tax, no alternative seemed open.  It would have been quixotic in the present stringency to neglect so buoyant a source of revenue.  The rise in the price of vodka was to be rendered less unpalatable to the consumer by an improvement in quality.  The limitations on *rykovka* had been

(Stalin, *Sochineniya*, ix, 192).  This looks like a misdating of the debate on the original introduction of vodka in January 1923 (see preceding note) ; but there may have been some further discussion in October 1924.

[1] *Vestnik Finansov*, No. 11, November 1924, pp. 69, 77.
[2] *Leningradskaya Pravda*, March 10, 1925, reporting a speech of the previous day in the Leningrad Soviet.          [3] *Ekonomicheskaya Zhizn'*, April 9, 1925.
[4] *Tretii S"ezd Sovetov SSSR* (1925), p. 474.

quickly abandoned : the alcoholic strength of vodka had been raised from 20° to 30° in 1924, and was now raised to 40°.[1]

When, therefore, in June 1925, Sokolnikov offered to a professional audience of workers in Narkomfin a preview of the budget for 1925–1926, it was found to combine optimism with realism. A budget total of 3560 million rubles, representing an increase of nearly 1000 millions on the current estimates for the previous year, was now contemplated. This would provide, in addition to increased allocation for defence and costs of administration, 140 millions for agriculture, 85 millions for industry, 60 millions for electrification and housing and 25 millions for the cooperatives. On the revenue side direct taxation would account for 568 millions (this allowed for a reduction of receipts from the agricultural tax from 442 millions to 390 millions), indirect taxation for 870 millions (including an increase in excise from 500 millions to 750 millions), and customs for 130 millions ; nontax receipts were estimated at 1748 millions, of which transport and communications would account for 1250 millions. Even before such an audience the rise in estimated revenue from vodka, from 173 millions to 298 millions, was evidently the item which required most explanation. Nobody intended, said Sokolnikov, to resuscitate the " drunken budget " of Tsarist Russia ; but, whereas the production of spirit had in the current year been not more than 5 per cent of the pre-war figure, it might rise in the following year to a " firm limit " of 15 per cent. It was better, " since it is impossible to prohibit drunkenness, in any case to compel those who drink to pay something to the state budget ".[2] Nevertheless, it was, as Sokolnikov admitted a few weeks later, " a step backwards, a forced step backwards ".[3]

[1] G. Sokolnikov, etc., *Soviet Policy in Public Finance* (Stanford, 1931), pp. 189-190, 194-196 ; for *rykovka* see *The Interregnum, 1923–1924*, p. 35, note 2. Kamenev announced the impending decision to produce 40° spirit to the Moscow Soviet on April 10, 1925, defending it on the ground that it was the only way to stop illicit distilling (L. Kamenev, *Stat'i i Rechi*, xii (1926), 145-146).

[2] *Sotsialisticheskoe Khozyaistvo*, No. 4, 1925, p. 15 ; six months later Stalin defended the policy at the fourteenth party congress with the remark (twice repeated) that it was impossible to " build socialism in white gloves ", and that it was better to get revenue from vodka than to go cap in hand to foreign capitalists (*Sochineniya*, vii, 340-341). According to figures in *Planovoe Khozyaistvo*, No. 11, 1925, p. 150, budget receipts from vodka in 1924-1925 amounted, after adjustment to the change in values, to 16·2 per cent of the 1913 figure.

[3] G. Sokolnikov, *Finansovaya Politiko Revolyutsii*, iii (1928), 21.

In accordance with these prognostications the budget for 1925–1926 submitted to Sovnarkom by Narkomfin in the autumn of 1925 balanced at 3778 million rubles.[1] By this time the usual upward pressures were at work, especially in the form of urgent demands for the expansion of heavy industry; and, after a long and intensive examination, the budget was approved by Sovnarkom in January 1926 at a total of 4000 millions. The increase was covered on the revenue side by raising the cost of vodka to 1·5 rubles per vedro (equivalent to 2·70 gallons) and by higher railway and postal tariffs.[2] In the revenue eventually collected in 1925–1926, excise receipts rose to 840 million rubles, of which receipts from the alcohol monopoly accounted for 364 millions; consumption of vodka increased fourfold in this year.

In the crisis of the winter of 1925–1926, after the unexpected difficulties of the grain collection, the size of the budget inspired some anxieties; and it was in this cautious period that it came up for final examination at the session of TsIK in April 1926. After lengthy speeches by Bryukhanov, who had succeeded Sokolnikov in January as People's Commissar for Finance, and by Kuibyshev, the president of the budget commission of TsIK,[3] the budget was finally approved at a total of 3900 million rubles. Tax revenue was estimated at 1900 millions (including 1150 from indirect taxation), non-tax revenue at 2000 millions (including the item of transport and communications). Bryukhanov once more apologetically admitted that the decline in the proportion of direct taxation meant a retreat from " the symbol and principle of class ", but saw no alternative. On the expenditure side, industry was to get 155 millions (including 107 millions for heavy industry) as against 98 millions in the previous year, agriculture 157 millions as against 147 millions.[4] The implications of the budget were

[1] The full figures are in *Vestnik Finansov*, No. 11-12, November-December 1925, pp. 190-192; for an analysis of them see *Ekonomicheskoe Obozrenie*, November 1925, pp. 21-24. The control figures of Gosplan for 1925–1926 had contemplated a budget of from 3750 to 3850 million rubles.

[2] *Planovoe Khozyaistvo*, No. 2, 1926, p. 73.

[3] The commission had a large number of sittings to discuss the budget: these were summarily reported in *Ekonomicheskaya Zhizn'*, March 31, April 1, 4, 5, 7, 11, 1926.

[4] The speeches are in *SSSR : Tsentral'nyi Ispolnitel'nyi Komitet 3 Sozyva : 2 Sessiya* (1926), pp. 18-97, the resolution approving the budget in *id. : Postanovleniya* (1926), pp. 3-13.

clear, though it would have been inconvenient to avow them openly. A beginning had been made in financing the large-scale expansion of heavy industry in the only way in which, foreign loans being excluded, it could be financed, namely, by drawing upon the earnings of the peasantry. But, in view of the commanding position of the well-to-do peasant in the economy, and his resistance to any substantial increase in direct taxation, it was necessary to rely primarily, in defiance of socialist principles, on indirect taxation which bore equally on all groups of the peasantry. The budget of 1925–1926, which eventually balanced at 4000 million rubles,[1] marked the highest point of the influence of the *kulak* on fiscal policy.

The almost complete lack of liquid capital resources, and the consequent weakness of credit, was a serious handicap to public finance. State loans throughout this period were no more than an alternative method to direct taxation for drawing into the treasury as large a proportion as possible of the earnings of state or private enterprises. Attempts to attract the savings of individuals, and thus mop up surplus purchasing power, were unlikely to succeed on any significant scale or without some form of compulsion, which once again assimilated such levies to direct taxation. Borrowing in kind — the " grain loans " of 1922 and 1923 [2] — disappeared with the currency reform. The introduction of the stable currency had been followed in February 1924 by a new 8-per-cent gold loan, the bonds of which were placed exclusively in large denominations with government organs and institutions and were not negotiable, and in the following month by a 5-per-cent " peasant loan " in denominations from one ruble upwards, repayable by drawings between November 1924 and December 1926. In April 1924 a second 6-per-cent lottery loan repayable by annual drawings over five years was issued.[3] But this, too, enjoyed little spontaneous success, and soon became the

---

[1] R. W. Davies, *The Development of the Soviet Budgetary System* (1958), pp. 82-83.          [2] See *The Interregnum, 1923–1924*, pp. 35-36.

[3] *Sobranie Uzakonenii, 1924*, No. 34, art. 311 ; No. 45, art. 421 ; No. 55, art. 536 ; for a list of all these loans, with the amounts realized by them, see *Zadachi i Perspektivy Goskredita v SSSR*, ed. D. Loevetsky (1927), p. 12. For the original lottery gold loan see *The Interregnum, 1923–1924*, p. 100.

subject of forced placings : [1] some 60 million rubles had been
subscribed in this way before February 23, 1925, when a further
decree once more placed a veto on the practice of obligatory subscriptions.[2] It was a common practice for those who were compelled or persuaded to purchase bonds to deposit them at the
banks as security for advances, so that it was no exaggeration to
write at this time that government bonds " remained for the most
part sitting in the banks ".[3] At this time state loans changed
hands on the open market at not more than 40 per cent of their
nominal value.[4] The only loan which appears to have had some
popularity was the peasant loan, of which a second instalment was
issued in the spring of 1925. But peasants were attracted to it
entirely by the availability of the bonds for tax payments and did
not hold it beyond the limits of the season ; otherwise it seems to
have become a minor vehicle for small savings in the towns.[5] The
total amounts realized by state borrowing were still very small.
Net revenue from loans in the budget of 1924–1925 was only
64·3 million rubles, and in the budget of 1925–1926 28·4 millions.[6]
A further one-year 5-per-cent loan for 10 million rubles was issued
in April 1925, and a second peasant lottery loan in the autumn of
the same year.[7]

Persistent efforts to restore the habit of small savings were
successful on a minor scale. The State Workers' Savings Banks
(Gosudarstvennye Trudovye Sberegatel'nye Kassy), established in
1923, gradually expanded and attracted confidence, but never
became during this period a serious factor in financial policy.
Between October 1, 1924, and September 1, 1925, the number of
such banks or offices increased from 5000 to 9000 and their
deposits from 11 million to 29 million rubles ; but of the deposits
48 per cent came from institutions, 25 per cent from employers,

[1] *Sobranie Zakonov, 1924*, No. 15, art. 155.
[2] *Zadachi i Perspektivy Goskredita v SSSR*, ed. D. Loevstsky (1927), p. 62 ;
the decree is in *Sobranie Zakonov, 1925*, No. 13, art. 105.
[3] *Ekonomicheskoe Obozrenie*, December 1925, p. 129.
[4] *Ekonomicheskaya Zhizn'*, January 10, 1925.
[5] G. Sokolnikov, etc., *Soviet Policy in Public Finance* (Stanford, 1931),
p. 263.
[6] R. W. Davies, *The Development of the Soviet Budgetary System* (1958),
p. 126 ; for other calculations see the sources there quoted and *Planovoe
Khozyaistvo*, No. 11, 1925, p. 148.
[7] *Sobranie Zakonov, 1925*, No. 13, art. 100 ; No. 68, art. 505.

and only 7·5 and 1·8 per cent respectively from workers and peasants.[1] A decree of November 27, 1925, regulating their status was a symptom of their growing importance.[2] On December 1, 1925, there were 10,000 banks and offices (7700 in the RSFSR), 870,000 depositors (713,000 in the RSFSR) and deposits amounting to 42 million rubles (34 millions in the RSFSR). They had scarcely yet penetrated the countryside, and in the remoter regions they were still unknown.[3] One factor which is said to have discouraged savings in the country was that " the local authorities look on the depositor as a well-to-do element, an object of taxation ".[4]

The development of bank credit had begun slowly and painfully with the foundation of Gosbank in 1921 and of Prombank and a number of other specialized banks in the following year.[5] The creation of an agricultural bank was surprisingly delayed. It had been mooted at the ninth All-Russian Congress of Soviets in December 1921 ; but the decision then recorded was not carried out.[6] It was in February 1924 that the second Union Congress of Soviets finally decided to establish a Central Agricultural Bank (Tsentrosel'bank) to facilitate the provision of credit for agriculture. Its statutes were approved by TsIK in the following month.[7] Its foundation had actually been preceded by the foundation of an agricultural bank for the Ukraine in November 1923, and was followed by the foundation of similar banks for the other constituent republics, ending with an agricultural bank for the RSFSR in February 1925.[8] The banks of the republics were in effect branches of Tsentrosel'bank ; [9] and credit was channelled through them to local credit societies, including the credit cooperatives, which at first formed part of the agricultural

---

[1] *Planovoe Khozyaistvo*, No. 10, 1925, p. 108.
[2] *Sobranie Zakonov, 1925*, No. 81, art. 612.
[3] *Planovoe Khozyaistvo*, supplement to No. 12, 1925, p. 14 ; see *Vestnik Finansov*, No. 6, June 1925, pp. 126-135, for an informative account of the savings banks.
[4] *Ekonomicheskoe Obozrenie*, October 1925, p. 211.
[5] See *The Bolshevik Revolution, 1917–1923*, Vol. 2, pp. 356-357.
[6] *Sobranie Uzakonenii, 1922*, No. 4, art. 41 ; *S"ezdy Sovetov RSFSR* (1939), p. 209.
[7] *Sobranie Uzakonenii, 1924*, No. 29-30, art. 275 ; ii, No. 11, art. 31.
[8] *Na Agrarnom Fronte*, No. 11, 1926, p. 140.
[9] For the statutes of the Ukrainian Agricultural Bank see *Sobranie Zakonov, 1926*, ii, No. 4, art. 28.

cooperatives, but later became independent entities.[1]  The system
of agricultural credit depended mainly on the provision of funds
from the budget in the form of subventions or from Gosbank in the
form of advances :  the party central committee decided in April
1925 that 10 million rubles should be assigned to Tsentrosel'bank
from the budget in addition to an equal sum advanced by Gosbank.[2]  Only 15 per cent of the resources of Tsentrosel'bank and
its subordinate organs came from share-holdings and deposits.[3]

The organization of credit through the banks soon presented
fresh problems.  In 1923 Sokolnikov had stoutly maintained that
credit policy was the independent preserve of the banks, immune
from " the introduction of obligatory planning ", and the safeguard
of the market principles of NEP.[4]  At first sight this view might
seem to have been reinforced by the financial reform :  among the
prerequisites of a stable and independent currency was a sound
and independent credit policy.  In reality, however, the autonomy
of finance was no longer compatible with the more positive view
of the economic functions of the state which had emerged in the
aftermath of the scissors crisis.  Neither in agriculture nor in
industry could credit policy be guided any longer by purely
financial considerations.  In agriculture, the " sound " credit
policy which accorded a natural preference to the well-to-do
peasant could not withstand a change of the party line designed
to support the middle and poor peasant.[5]  In industry the practice
by which bank credit was virtually reserved to light industries
working at a profit, and the revival of heavy industry which could
not expect to earn quick profits was left to subventions from the
budget,[6] could hardly survive the new party directive to concentrate on the expansion of the metal industry.  The banks, as

---

[1] For the relations between agricultural and credit cooperatives see pp. 280-
281 above ; a detailed description of the growth and organization of agricultural
credit is given in an article in *Entsiklopediya Gosudarstva i Prava*, iii (1925–1927),
829-836.                    [2] *VKP(B) v Rezolyutsiyakh* (1941), i, 646.
[3] *Bol'shevik*, No. 9-10, May 30, 1926, p. 64.
[4] See *The Interregnum, 1923–1924*, pp. 107-108.
[5] In November 1925 an article in the Ukrainian party journal *Kommunist*
(quoted in *Leningradskaya Pravda*, December 2, 1925) argued that support for
the poor peasant was incompatible with earlier party directives for a sound
credit policy, objected to " the regularization or popularization of special
privileges for the poor peasant in obtaining the services of agricultural credit ",
and demanded the elimination of any " social service conception " from credit
policy.                    [6] See *The Interregnum, 1923–1924*, p. 8.

public institutions, would be called on to play their part in furthering the new policy. ·

The first corollary of this new conception of the rôle of the
banks as instruments of policy was to bring order into the banking
system. A halt had to be called to the haphazard creation of a
multiplicity of banks of varying and often ill-defined functions,
sometimes pursuing different policies and competing with one
another, and sometimes uniting to contest the supremacy of Gosbank. At the end of April 1924 the party central committee, in
the course of a resolution mainly concerned with the regulation of
internal trade, touched on the question of the control of credit :

> It is indispensable to organize a committee of banks, whose
> task should be the organization of bank credit and the avoidance
> of duplication, the preliminary examination of directive plans
> of credit, the fixing of coordinated discount rates, and the
> appropriate distribution of banking facilities among different
> regions and branches of industry.

A leading article in *Ekonomicheskaya Zhizn'* pointed the moral.
The decision meant " a deepening of the principle of planning ",
and put an end to the controversy whether " planned credit " was
possible or necessary.[1] The committee was constituted by a
decree of Sovnarkom of June 24, 1924 : the banks belonging to it,
apart from Gosbank, were Prombank, Vsekobank, Mosgorbank
(Moscow Municipal Bank), Tsentrosel'bank, Vneshtorgbank (the
Bank of Foreign Trade) and the Association for Mutual Credit.[2]
In 1925 the number of banks engaged in financing the development programme was increased by the creation of a Joint Stock
Bank for Electrification (Aktsionernyi Bank po Elektrifikatsii or

---

[1] *Ekonomicheskaya Zhizn'*, April 24, 25, 1924.

[2] See A. Z. Arnold, *Banks, Credit and Money in Soviet Russia* (N.Y., 1937),
p. 266 ; the list of banks is taken from an announcement in *Ekonomicheskaya
Zhizn'*, August 22, 1924. For the origin of Vneshtorgbank see A. Z. Arnold,
*op. cit.*, pp. 313-316 ; it was at this time still commonly known by its former
name of Roskombank (Russian Commercial Bank) ; for the other banks see
*The Bolshevik Revolution, 1917–1923*, Vol. 2, p. 357. The Association for
Mutual Credit was the central organ of a number of mutual credit associations
throughout the country, which catered for the nepman and the private trader,
and relied for their funds on private deposits. These grew steadily : deposits
are said to have risen in 1925 by 14 million rubles (*Ekonomicheskoe Obozrenie*,
January 1926, p. 7). On October 1, 1925, there were 167 such associations, of
which 91 were in the RSFSR and 65 in the Ukraine, with 57,000 members and
balances amounting to 67 million rubles ; the rate of interest on advances
varied from 4 to 10 per cent per month (*ibid.* March 1926, pp. 146-153).

Elektrobank), and of a Central Bank for Communal Economy and Housing (Tsentral'nyi Bank Kommunal'nogo Khozyaistva i Zhilishchnogo Stroitel'stva or Tsekombank).[1]

This expansion of banking facilities provided the organizational framework for an extraordinarily rapid expansion of credit to meet the needs of expanding production. " Loans and discounts " in the accounts of the State Bank rose from 312 million rubles on October 1, 1923, to 598 millions on October 1, 1924, and 1425 millions a year later.[2] Advances from other banks and credit institutions swelled in about the same proportions. The stream of agricultural credit flowed from Gosbank through Tsentrosel'- bank to the agricultural banks of the republics, and thence to credit societies and local agricultural or credit cooperatives. Rates of interest on advances to peasants are said to have fallen from 8 per cent for long-term and 12 per cent for short-term credits in 1924 to 6 and 10 per cent respectively in 1925.[3] Of advances to peasants, 27 per cent were for purchase of implements, 23 per cent for purchase of working animals, and 9 per cent for purchase of seeds : credits were also given for land improvement, electrification, development of specialized forms of agriculture and purchase of equipment for processing agricultural products.[4]   Prombank

[1] Details of the increase in the number of banks and of their branches in 1924 and 1925 are in A. Z. Arnold, *Banks, Credit and Money in Soviet Russia* (N.Y. 1937), pp. 284-285.

[2] *The State Bank of the USSR* (Moscow, 1927), pp. 31-32.  A. Z. Arnold, *Banks, Credit and Money in Soviet Russia* (N.Y. 1937), pp. 252-253, on the basis of material published in various issues of *Vestnik Finansov*, attempts to break down the " loans and discounts " into different categories of borrowers, with " state enterprises " heavily predominating, followed by " cooperatives " and " credit institutions " : advances to private firms and to agriculture (which was, however, heavily represented in advances to cooperatives and to other credit institutions) were negligible.  The State Bank of the USSR was not at this period, in the technical sense, a central bank, i.e. a bank primarily concerned in financing other banks.  It conducted all forms of banking business on its own account : in 1925, 75 per cent of its business was transacted not by its central office, but by its branches, numbering 400 or 500, throughout the country (*Planovoe Khozyaistvo*, No. 5, 1925, p. 289).  A list of industrial and commercial trusts and enterprises, *ibid.* No. 11, 1925, p. 27, shows that many of them obtained credits simultaneously from three or four banks, of which Gosbank was one ; Gosbank was " the chief credit institution competing with other banks " (*ibid.* No. 11, 1925, pp. 31-32).

[3] *Na Agrarnom Fronte*, No. 1, 1926, pp. 145-146 : slightly higher rates are quoted in *Planovoe Khozyaistvo*, No. 11, 1925, p. 79.

[4] *Na Agrarnom Fronte*, No. 3, 1926, p. 54 ; *Planovoe Khozyaistvo*, No. 11, 1925, p. 80.

and Elektrobank distinguished in their accounts between discounts, loans against commodities, and " earmarked " (i.e. long-term) loans. This last category became important only after the autumn of 1925 ; up to that time nearly all bank lending was on a short-term basis.[1] Vsekobank, Mosgorbank and the Association for Mutual Credit were alone dependent largely on deposits : they also received some credits from the State Bank. Tsentrosel'bank and Tsekombank, both founded for specific purposes of state policy, were financed mainly from the state budget.[2] All contributed to the rapid increase of finance and credit. In the summer of 1924 even Sokolnikov was a convert to a constructive view of the function of credit, and announced it as his aim, " starting from real financial plans, to proceed to real economic plans ".[3] In January 1925 Kamenev described " centralized credit " as " *this new ' commanding height ' which we have created practically out of nothing* ", and as " *the decisive factor in the regulation of the economy, the factor which introduces decisive correctives, and is capable both of causing and of preventing crises* " ; and Krzhizhanov-sky, who still mistrusted " *the hieroglyphs of banking accountancy and the back-stage secrets of banking concerns* ", was none the less ready to welcome " credit and the plan " as " blood brothers in a single system of socialization ".[4]

The expansion of credit called for a corresponding expansion of the note issue, and this occurred in full measure. The total

[1] Between October 1, 1924, and October 1, 1925, 90 per cent of all credits to industry came from Gosbank and Prombank, the former predominating (*Ekonomicheskoe Obozrenie,* December 1925, p. 132) ; Gosbank made only short-term advances.

[2] Figures for all these banks are collected in A. Z. Arnold, *Banks, Credit and Money in Soviet Russia* (N.Y., 1937), pp. 289, 294, 298, 304, 309, 311, 314-315 ; a brief history of Prombank is in *Ekonomicheskoe Obozrenie,* November 1925, pp. 139-149, of Mosgorbank in *Vestnik Finansov,* No. 3, March 1925, pp. 145-150.

[3] *Sotsialisticheskoe Khozyaistvo,* No. 5, 1924, p. 23. It appears from an exchange of pencilled notes between Trotsky, Pyatakov, Krasin and Sokolnikov at a meeting of STO on July 2, 1924, preserved in the Trotsky archives, that the three first were mistrustful of the credit policy of Gosbank and of its then president Sheinman, and wanted a " clean-up " (*sanirovanie*) of Gosbank's portfolio of bills ; Sokolnikov promised to undertake this, but wished to avoid a formal intervention by STO.

[4] *Planovoe Khozyaistvo,* No. 1, 1925, pp. 19, 30-31.

value of notes in circulation on January 1, 1924, was 237 million
rubles.  After a period of comparative restraint following the
financial reform,[1] the bank note issue rose rapidly throughout the
autumn to 346 million rubles on October 1, 1924, and to 410
millions on January 1, 1925.  After a slight reduction in the early
months of 1925 [2] the total issue soared rapidly in the summer and
autumn, touching 651 millions on October 1, 1925 (or nearly
double the figure of the previous year), and 719 millions on
November 1.  The rise was still more substantial when the total
of small denomination treasury notes and silver and copper coinage
(490 million rubles on October 1, 1925, compared with 280
millions a year earlier) was added to the account.[3]  At the time
of the introduction of the financial reform, the note and currency
issue had been covered to the extent of more than 50 per cent by
gold and foreign currency.  On October 1, 1924, the cover was
38 per cent and on October 1, 1925, 23 per cent.[4]  Since the
active foreign trade balance of 1923–1924, which had made possible the building up of strong reserves of gold and foreign currency, was followed by the passive balance of 1924–1925, there

[1] The mood of caution engendered by the financial reform brought about
a temporary cessation of the expansion of credit in the spring of 1924 and a
slackening of trade : this phenomenon, which lasted about three months, is
discussed at length in *Sotsialisticheskoe Khozyaistvo*, No. 5, 1924, pp. 94-103.

[2] The contraction of the note issue in the winter months after the realization of the harvest, and its expansion in the late summer and autumn, was a
long-standing and familiar feature of Russian finance.

[3] The following table (in millions of rubles) is taken from *Nashe Denezhnoe
Obrashchenie*, ed. L. Yurovsky (1926), pp. 154-155, and *Kontrol'nye Tsifry
Narodnogo Khozyaistva na 1926-1927 god* (1926), pp. 382-383 (where there is
a misprint in the figures of Treasury notes for October 1, 1925) :

|                  | Bank Notes | Treasury Notes and Coinage | Total  |
|------------------|------------|----------------------------|--------|
| October 1, 1924  | 346·5      | 280·7                      | 627·2  |
| January 1, 1925  | 410·8      | 331·9                      | 742·7  |
| April 1, 1925    | 402·4      | 363·3                      | 765·7  |
| July 1, 1925     | 460·1      | 386·0                      | 846·1  |
| October 1, 1925  | 652·0      | 490·9                      | 1142·9 |
| January 1, 1926  | 726·6      | 542·7                      | 1269·3 |
| April 1, 1926    | 693·4      | 510·8                      | 1204·2 |

[4] *Nashe Denezhnoe Obrashchenie*, ed. L. Yurovsky (1926), pp. 132-133 ;
the legal cover of 25 per cent related only to bank notes and was not at this time
in danger.

was no immediate prospect of adding fresh reserves to cover the increased currency emission.

Towards the end of 1924 the rising rate of currency emission attracted the attention of professional observers, some of whom were not unwilling to invoke the bogy of inflation.[1] But the situation was sufficiently unlike that of the earlier inflation to make it easy to dismiss these fears as misguided or exaggerated. In the long run it may have mattered little whether an excess of expenditure over revenue in the national economy was expressed financially in the form of a budget deficit or of an expansion of credit through the banks. These were alternative means of attaining the same end, and both had the same inflationary consequences. But for contemporary observers, faced with unfamiliar contingencies, the superficial differences obscured the fundamental identity. Before the reform of 1924, paper currency had been issued in order to enable the state to pay its way in a period of chaos and acute economic crisis. Now the budget was balanced, the machinery of public finance was in good order, and the economy was expanding at a rapid rate : it was precisely this expansion of real values which demanded the expansion of credit and of the currency. If the amount of currency in circulation had enormously increased, so also had its uses. The single agricultural tax was being paid for the first time in cash ; the payment of wages in kind had virtually ceased ; everywhere the last vestiges of a " natural " economy had given place to a money economy. The argument was not groundless, though it was not realised that, once this adjustment had been made, the rapid and painless absorption of an expanding currency would not continue indefinitely. At a discussion of the question among workers in

---

[1] An article in *Ekonomicheskaya Zhizn'*, November 23, 1924, opened with the remark that " a definition of the limits of the bank note emission becomes ever more topical ", and attacked the " involuntary inflationists " of Gosplan. In the same month *Vestnik Finansov*, No. 11, November 1924, pp. 79-86, detected symptoms of a " quasi-inflation " in the unwillingness of the peasant to market his grain, the shortage of industrial goods and the discrepancy between official and " free " prices, and argued that an increase in the note issue would merely aggravate the crisis : prices should be left to find their own level. A more moderate article in *Ekonomicheskaya Zhizn'*, January 14, 1925, ended with the recommendation of " an energetic policy of development of credit relations not involving the issue of new monetary tokens ". But this stated the dilemma without resolving it. A brief comment on the controversy from the side of Gosplan appeared in *Planovoe Khozyaistvo*, No. 1, 1925, pp. 289-291.

Gosplan in January 1925 Strumilin argued that the present rate of currency emission gave no cause for anxiety. What mattered was the proportion of currency in circulation to trade turn-over. Strumilin purported to show that, if currency emission were to increase as rapidly as trade turn-over had done since October 1, 1924, the total would reach 884 million rubles by May 1, 1925 (on which date the actual figure was only 780 millions) : if trade turn-over continued to increase at the same rate, the total amount of currency in circulation could rise without danger to 1254 millions by January 1, 1926.[1]

These conclusions seem to have been generally accepted. Pressure for credit expansion, which must sooner or later carry with it an expansion of the currency issue, became irresistible. In a report to STO on March 30, 1925, even Sokolnikov conceded that, while the principle of credit " rations " would not be formally abandoned in the coming quarter, " the possibility of discounting bills in excess of the established credit ration " need not be excluded, and suggested that " our stable currency can to a far greater degree than hitherto be put to the service of an expansion of our economy ".[2] A few days later the president of Gosbank announced " a certain relaxation " in credit policy.[3] Sokolnikov now seemed reconciled to the principle of planning, coining the phrase, " to plan is to dispose of reserves ", and declaring that " the utilization of credit reserves " could now proceed without fear of inflation.[4] In the spring of 1925 all the symptoms appeared to justify an almost reckless optimism about the future of the chervonets. In spite of the large increase in the currency issue, the slight general rise in prices did not seriously invalidate Sokolnikov's boast at the third Union Congress of Soviets in May 1925 that " the purchasing-power of our money . . . in the domestic market has been completely stable throughout the year ".[5] Down

---

[1] *Planovoe Khozyaistvo*, No. 5, 1925, pp. 115-135, gives the substance of Strumilin's argument at the Gosplan meeting in January 1925 brought up to date in May 1925.

[2] A signed article by Sokolnikov in *Ekonomicheskaya Zhizn'*, April 4, 1925, was a slightly revised version of this report.          [3] *Ibid.* April 16, 1925.

[4] G. Sokolnikov, *Finansovaya Politika Revolyutsii*, iii (1928), 248.

[5] *Tretii S"ezd Sovetov SSSR* (1925), p. 422. In a later passage of his speech Sokolnikov uttered a warning on the limits of credit expansion, but in the mildest terms : " Here we must maintain a firm line, and shall try to maintain it, though it is sometimes hard going, since we are pressed on all sides "(*ibid.* p. 445).

to March or April 1925 both foreign currency and hoarded gold (mainly gold coins of the Tsarist period) were being freely offered in Moscow in exchange for chervontsy — a striking tribute to the general confidence in the currency.[1]

The first break came in or about May 1925, when an unusually persistent demand for foreign currency in exchange for chervontsy began to be experienced. This was attributed, partly to state enterprises making purchases abroad (the concentration of the execution of such transactions in the hands of Vneshtorg was at this period still not complete), partly to illicit imports,[2] and partly to speculators on what was currently known as the " American market ", or the black bourse.[3] But the dimensions of the crisis were not at first recognized. At the financial conference in June 1925 Sokolnikov, tacitly endorsing Strumilin's estimate of the previous January, forecast a further 50 per cent increase in the note issue (from 800 million to 1200 million rubles) by January 1, 1926 ; [4] and the financial authorities were content throughout the summer to gamble on the belief that " the fact of a favourable harvest and the continuously progressive development of industry are creating in this respect also favourable conditions for us ".[5] In July a leading article in the official financial journal cautiously deprecated the view that the struggle against inflation was " a struggle with imaginary opponents ", and pointed out that conditions in the Soviet Union were too different from those in capitalist countries for capitalist precedents to have any value in fixing the limits of currency emission. But no specific conclusion

[1] *Planovoe Khozyaistvo*, No. 5, 1926, pp. 97-98.

[2] Illicit imports of textiles, clothing and luxury goods were smuggled into the Soviet Union at this time on a fairly extensive scale, mainly over Asiatic frontiers. For particulars of this trade, including lists of contraband goods confiscated in 1924-1925, see *ibid.* No. 5, 1926, pp. 92-94 ; it was considered impossible to stop it entirely. Trotsky wrote of a contraband trade in small articles " which is at present draining the country of millions of rubles of gold currency " (*Pravda*, September 22, 1925).

[3] One explanation given at this time of the revival of currency speculation was that " the private trader, being driven by us on the one hand out of the grain trade and on the other out of trade in industrial goods, has forced his way into the black bourse and into valuta operations, undermining the exchange rate of our currency " (*Planovoe Khozyaistvo*, No. 2, 1926, p. 29). For the " American market " see *ibid.* No. 2, 1926, p. 90 ; *Ekonomicheskoe Obozrenie*, January 1926, pp. 7-9.

[4] G. Sokolnikov, *Finansovaya Politika Revolyutsii*, iii (1928), 207.

[5] *Planovoe Khozyaistvo*, No. 9, 1925, p. 39.

was recorded.[1] A decree of July 1925 prohibited payments in foreign currency except for foreign trade transactions or other purposes provided for by law.[2]

A new stage was marked by the publication in August 1925 of the Gosplan control figures of 1925–1926. Boldly pursuing the line of rapid industrial development, Gosplan argued that in the current period " the volume of money should grow more rapidly than trade turnover, and credit more rapidly than the volume of money ". On these principles it forecast that the volume of currency in circulation would rise from 1157 million rubles on October 1, 1925, to 1973 millions on October 1, 1926 (an increase of 78 per cent as against an increase of 97 per cent in the preceding year), that deposits and current accounts in the banks would rise in the same period from 1067 millions to 2400 millions, and loans and advances from 1900 millions to 3800 millions.[3] These figures provoked sharp opposition from spokesmen of orthodox finance. One critic described incursions of the planners into the sphere of monetary policy as " inadmissible in principle " ; [4] another made a frontal attack on the view implicit in the control figures that credit was " something that can be created by the state ", and prolonged his argument into a general critique of planning :

> The single fact that the market for money capital is closely bound up with a stable currency, and that it has a definite price which is dependent both on the internal market for capital and on the price of capital on the world money market, shows that the limits and possibilities of planning are fairly restricted. For, as is well known, the foundation of our credit — our stable currency — depends both on a balanced budget and on an active

[1] *Vestnik Finansov*, No. 7, July 1925, pp. 3-12. In the following issue an article was published in the form of a discussion article which argued that " hoarding of foreign valuta " was irrational, and that the need for such a reserve would decline (*ibid.* No. 8, August 1925, pp. 9-12) ; in the next issue another discussion article by a " Narkomfin professor " reviewed at length foreign theories of currency and inflation, and reached the sound conclusion that, " since under the monopoly of foreign trade the exchange rate does not play the same rôle as under a free trade system, a divergence between the exchange rate and the price level can in these conditions be maintained for an indefinite period of time " (*ibid.* No. 9, September 1925, pp. 30-66). The confusion of mind on these issues in official circles was evidently great.
[2] *Sobranie Zakonov, 1925*, No. 45, art. 530.
[3] *Kontrol'nye Tsifry Narodnogo Khozyaistva na 1925–1926 god* (1925), pp. 33-35 ; for the preparation and fate of the control figures see pp. 500-505 below. [4] *Ekonomicheskoe Obozrenie*, October 1925, pp. 28-38.

balance of foreign payments. . . . Thus one of the most powerful sources of " spontaneity " in our economy is money capital.[1]

Even more moderate critics, who refrained from raising the issue
of principle, still saw in the Gosplan figures the " danger of
inflation ".[2]  Sokolnikov, reverting to an earlier attitude, accused
Gosplan of holding that " the policy of monetary circulation
should be subordinated to the policy of the development of
credit ", and called this " a formula of inflation ".[3]

The optimism of Gosplan could have been justified only by a
rapid and progressive growth of economic activity following the
harvest.  In a situation where the expansion of the currency had
already outstripped the expansion of trade, the refusal of the
peasant to bring his crops to the market hastened the inevitable
crisis.  In November 1925 Gosbank had for the first time to
throw significant quantities of gold and foreign currency on to the
market in order to keep the exchange stable,[4] though the fact that
the chervonets was still not quoted on any of the world's major
exchanges [5] made this operation easier than might have been
expected.  Thanks to these efforts, the official price of gold
which had risen in September was kept stable at the higher figure
for the remainder of the year.[6]  But transactions on the " American
market " more than doubled in volume between October and
December 1925 ; and the demand for gold remained persistent.[7]

The reappearance, eighteen months after the consummation
of the currency reform, of unmistakable symptoms of inflation
provoked no drastic reaction in party circles.  This was due, in
part, to the preoccupation of the leaders at this moment with the
internal party struggle, and, in part, to the obscure and esoteric
character of the problem presented : questions of currency and
credit, in the words of a report of Gosplan to STO in February

[1] *Sotsialisticheskoe Khozyaistvo*, No. 5, 1925, pp. 15-16.
[2] *Ekonomicheskoe Obozrenie*, October 1925, p. 40.
[3] *Ekonomicheskaya Zhizn'*, September 24, 1925 ; the attack occurred in the
course of Sokolnikov's comments on Gosplan's control figures for 1925-1926
(see p. 504 below).
[4] G. Sokolnikov, *Finansovaya Politika Revolyutsii*, iii (1928), 235 ; *Planovoe
Khozyaistvo*, No. 5, 1926, pp. 98-99.
[5] Since April 1925 it had been quoted in Rome (*Ekonomicheskaya Zhizn'*,
April 23, 1925), but on no other exchange in central or western Europe.
[6] *Vestnik Finansov*, No. 11-12, November-December 1925, pp. 175-178.
[7] *Ekonomischeskoe Obozrenie*, January 1926, p. 5.

1926, " belong to the most complex and least studied sphere of the national economy ".[1]  Sokolnikov, in a pamphlet apparently written in October 1925 under the title *Autumn Hesitations and Problems of Economic Expansion*, started from the consoling platitude that " goods hunger has arisen in the USSR only as a result of lack of correspondence between the mass of money in circulation and the degree of development of the circulation of goods ", and concluded that it was still possible to overcome " the elements of disorganization " in the market, and to " bring the monetary circulation into complete order ", though he cryptically added that " the maintenance of a stable currency is a problem which passes out of the sphere of economics into that of politics ".[2] At this time it could still be pretended that what was wrong was not " an excess of money " but " an insufficiency of goods ",[3] with the implication that continued currency expansion would promote continued expansion of production.  But this light-hearted attitude was soon overtaken by events.  In a speech to a party meeting at the end of November 1925 Sokolnikov, admitting for the first time that the reserves of Gosbank were being drained to support the currency, reminded his audience that the currency was guaranteed by a gold cover (" about this we have somehow forgotten "), and proposed an import of gold to build up depleted reserves.[4]  But the speech was not published ; and in a public speech at Gosbank a few days later Sokolnikov asserted " with complete confidence " that " *our present seasonal economic difficulties do not, and cannot, inspire any alarm for the fate of our stable currency* ".[5]  It was left to Bukharin, at the Moscow party conference in December 1925, to speak openly of " *the danger of fluctuations in our valuta* ", and to add that " this danger even now hangs over us ".[6]  In the heat of the fourteenth party congress, which

---

[1] *Planovoe Khozyaistvo*, No. 2, 1926, p. 64.

[2] The pamphlet was reprinted in G. Sokolnikov, *Finansovaya Politika Revolyutsii*, iii (1928), 31-47 (for the quotations in the text see pp. 38, 45), where it bears the date November 1925.  Internal evidence suggests that it was written not later than October : it is full of derogatory references to planning evidently provoked by the September discussion of the Gosplan control figures (see pp. 503-505 below).

[3] *Ekonomicheskoe Obozrenie*, November 1925, p. 114.

[4] G. Sokolnikov, *Finansovaya Politika Revolyutsii*, iii (1928), 232-233.

[5] *Ibid.* iii, 257 ; the speech was published in *Ekonomicheskaya Zhizn'*, December 1, 1925.                    [6] *Pravda*, December 10, 1925.

occupied the second half of the month, the obscure and distasteful problem of the currency and of the danger of inflation was not mentioned by any of the leaders.

The issue was, however, like everything else in the Soviet economy, decisively affected by the basic decision of the congress to press forward with the policy of industrialization ; for financial policy hinged on the demand for expanding credits for industry. The outstanding novelty of the past twelve months had been the flow of credit into heavy industry.  This credit, while it created consumer demand by expanding salary and wages bills, produced no corresponding immediate output of consumers' goods, thus leading directly to the classic inflationary situation of too much currency chasing too few goods.  The solution of the problem was, however, hard to find.  The financial reform had taken away the resource through which the recovery of industry had at first been financed — the use of the printing press — and had not provided an alternative.  In November 1925 Gosbank had reacted to the currency crisis in what might be called the orthodox way by restricting credit to industry.  During the next three or four months long-term credit to industry remained stationary, while short-term credits were actually reduced.  A halt was called to the rapid expansion of industry ; [1] and these measures, which helped to promote the contraction of the currency normal in the winter months,[2] gave a momentary impression that the inflationary tendencies were under control.  But these appearances were deceptive.  The situation had changed radically since the last occasion when, in August 1923, Gosbank had called industry sharply to order by a restriction of credit.[3]  The power of industry and its rôle in the economy had increased enormously ;  and the moment when the fourteenth party congress had just proclaimed intensive industrialization as the prime goal of party policy was not a propitious one for attempting to curb or curtail it.  Starved of credit by the banks, the big industrial trusts proved strong enough, by dint of enforcing cash payments on their customers and by a large increase in their holding of bills, to provide for their own financial needs.  During the period of restriction of

[1] See p. 349-351 above for this set-back.
[2] The total of bank notes in circulation declined from 1269 million rubles, on January 1, 1926, to 1204 millions on April 1, 1926 (*Ekonomicheskaya Zhizn'*, April 16, 1926).          [3] See *The Interregnum, 1923-1924*, pp. 96-99.

bank credit, their cash holdings increased by 55 per cent, their portfolios of undiscounted bills by 53 per cent, and their current accounts at the bank by 24 per cent ; the restriction of currency was met in part by a largely increased circulation of bills. In these circumstances " the contraction of credit proved in large measure fictitious ".[1] The inflationary tendencies of industrial expansion not balanced by savings in any other sector of the economy were still at work. What had hitherto been thought of as a temporary rise in prices now began to be seen as a fall in the purchasing power of the ruble ; and this fall continued at an accelerated rate throughout the winter of 1925–1926.[2]

These developments provoked a lively controversy among economists. Gosplan, being above all concerned not to interrupt the process of industrialization, was on the defensive and dis-inclined to attach too much importance to fears for the currency. Preobrazhensky openly attacked the Narkomfin policy of support-ing the chervonets, and accused those responsible for it of monetary fetichism :

> In a country which has no gold circulation and is obliged in the sphere of economic control to replace the spontaneous reason of gold, as a regulating instrument under the law of value, by a planned policy of allocating resources between means of production and means of consumption through the medium of a paper currency, they systematically appeal to the golden reason of the black bourse and, in the event of a divergence between the paper chervonets and its gold equivalent, fall into panic alarm, and engage in unnecessary " gold interventions " at a loss to the state, permitting nepmen to exchange their paper chervontsy into gold.[3]

Smilga, speaking at the Communist Academy on February 2, 1926, admitted that " knots of inflation " existed, but attributed them principally to speculators, and thought that " a consistent policy of the industrialization of the country " was the only remedy. He was prepared to maintain that " the régime of a

---

[1] *Planovoe Khozyaistvo*, No. 6, 1926, pp. 110-111.
[2] The value of the chervonets fell in terms of the Gosplan price-index from 5·36 pre-war rubles on September 1, 1925, to 4·56 pre-war rubles on February 1, 1926 (*Zadachi i Perspektivy Goskredita v SSSR*, ed. D. Loevetsky (1927), p. 54).
[3] E. Preobrazhensky, *Novaya Ekonomika* (1926), pp. 201-202 : the passage is taken from a paper read to the Communist Academy in January 1926.

stable currency has not been shaken ", and that the currency emission could safely continue at, or near, its present rate, and attacked both those who wanted deflation by credit restriction and the " inflationists " who proposed either to alter the gold parity of the chervonets or to detach it from gold and restore the goods ruble.[1]  A memorandum addressed about the same time by Gosplan to STO issued a strong warning against " a deflationàry policy, a contraction of the volume of money in circulation ". Such a policy would " strike at industry, the sector which at the present time needs the greatest support ".[2]

It is not surprising that representatives of Narkomfin were more acutely conscious of the embarrassments of the *status quo*, as well as of the cost of intervention to sustain the sagging exchange value of the chervonets.  Bronsky, now an official of Narkomfin, put the blame squarely on the unwarranted expansion of credit :

> Credit inflation is the cause of the rise in prices both of agricultural and of industrial products.  It brings about a depreciation of the currency, increases the effective demand of both town and country, . . . makes export difficult, and provokes the critical symptoms of a goods famine.[3]

Bryukhanov, the newly appointed People's Commissar for Finance, admitted that " our miscalculations in the autumn " had led to a danger of inflation, but expressed determination to resist the inflationary blandishments of "a few of our comrade-industrialists ".[4] Bukharin cryptically informed a Komsomol congress that current financial policy was engaged in two contradictory operations — the curtailment of credit to trading organs in order to restrict currency emission, and an expansion of the flow of goods to the market.[5]  Milyutin touched on an aspect of the question which it was fashionable to ignore, when he attacked the inflationary expansion of agricultural credit through subsidies from the budget.[6] The banks reported unanimously against any further expansion of the currency in existing conditions, and advised a contraction

[1] *Planovoe Khozyaistvo*, No. 2, 1926, pp. 29, 35, 38.
[2] *Ibid.* No. 2, 1926, p. 77.
[3] *Sotsialisticheskoe Khozyaistvo*, No. 1, 1926, p. 24.
[4] *Pravda*, February 5, 1926.
[5] *VII S"ezd Vsesoyuznogo Leninskogo Kommunisticheskogo Soyuza Molodezhi* (1926), pp. 252-253.
[6] *Na Agrarnom Fronte*, No. 3, 1926, pp. 100-104.

of anything from 25 to 100 million rubles in the current quarter.[1]
In this predicament, some " Narkomfin professors ", apparently
supported by some workers in Vesenkha and Gosplan, wished to
devalue the chervonets and stabilize at a lower level. The more
orthodox view in financial circles was that such a step would provide only the most fleeting of remedies for the disease. This was
true enough. But, beyond a vague injunction to " renounce inflation and correct the inflation which was allowed to occur last
autumn ", these stalwarts had no practical course to recommend.[2]

Out of this confusion of policies and ideas, in which harassed
party leaders showed no inclination to intervene,[3] a solution
gradually imposed itself. It had been pointed out in the course
of the controversy that the conditions which affected the purchasing-power of the currency on the domestic market were
different and separate from those which threatened its exchange
value in relation to foreign currencies, and that its convertibility
in terms of gold was the only factor which linked these two elements
together. It was unthinkable that the provision of the credit and
currency necessary to further industrial development should be
suspended or hampered. It was impossible to continue indefinitely
to squander gold or foreign currency in order to protect the exchange value of the currency against speculators. The part played
by the chervonets in foreign trade transactions was negligible :
these were nearly all conducted in foreign currency. From this it
was a short step to the conclusion that " we are not in the least
interested in the quotation of the chervonets on the ' American
market'"", and that "there is no connexion between the purchasing

---

[1] *Ekonomicheskoe Obozrenie*, February 1926, p. 44.

[2] The orthodox financial view was set out at length in two articles by
Shanin in *Planovoe Khozyaistvo*, No. 2, 1926, pp. 92-103 ; No. 5, 1926, pp. 91-
106. The second article, though not published till May, seems to have been
written shortly after the first, which was published in February. Among those
who suggested a devaluation of the chervonets was Stetsky, a disciple of Bukharin
(see his article in *Pravda*, February 6, 1926). An anonymous article of April 2,
1926, in the Trotsky archives (see p. 326 above), puts the point more bluntly
than any published document of the period : " Perhaps accumulation within our
whole national economy was so insignificant that we could effect the expansion
that was necessary only by the artificial method of lowering the rate of the ruble,
i.e. by something in the nature of a tax on all holders of money ".

[3] The bewilderment in party circles was reflected in a leading article in
*Pravda*, February 21, 1926, which displayed anxiety for the fate of the currency,
but had no positive recommendation to make.

power of our money and the gold exchange rate of the chervonets ".[1]  Once this point had been reached, the rest followed.
In March 1926 the Treasury, apparently without any formal
decision or announcement, ceased to offer gold and foreign currency at gold parity for the chervonets, which thereupon gradually
began to decline in value on the black bourse.  At the same time
retail prices rose.  Rumours of a serious curtailment of industrial
production began to circulate, and created, according to one
observer, " a certain panic in the capitals ".[2]  In fact, no such
curtailment was, or could have been, contemplated.  The resolution of the fourteenth party congress in favour of intensive
industrialization stood as the corner-stone of policy.  Financial
orthodoxy melted away in face of its requirements.

Simultaneously with the tacit abandonment of the chervonets
to its fate, a settlement was announced in the lengthy controversy
on the means of providing long-term credit for industry.  The
Vesenkha scheme for an " industrial fund " independent of the
banks was not revived.[3]  But the decision was taken to create
within Prombank a special department of long-term credit for
industry with separate accounts of its own, thus separating the
organization of long-term credit from that of short-term credit.
The decision indicated the intention of resuming a more generous
credit policy.[4]  In April 1926 the party central committee,
momentarily turning its attention for the first time for two years
to the question of credit and currency, pronounced on " the
necessity over the coming months of achieving a balance between
the volume of money in circulation in the country and the turnover of commodities, and of allowing an expansion of the currency
emission only in so far as successes are achieved in raising the
purchasing power of the ruble ".[5]  But this was a counsel of per-
fection and an empty token of unwillingness to recognize officially

[1] *Planovoe Khozyaistvo*, No. 2, 1926, p. 91.
[2] *Ibid.* No. 6, 1926, p. 36.
[3] It was, however, being discussed in *Ekonomicheskaya Zhizn'* as late as
March 5, 1926.
[4] A. Z. Arnold, *Banks, Credit and Money in Soviet Russia* (N.Y., 1937),
p. 292.
[5] *VKP(B) v Rezolyutsiyakh* (1941), ii, 97 ; it may have been a coincidence
that, on the day when the resolution was published, *Pravda* carried prominently
an article on the need to increase the production of gold (*Pravda*, April 13,
1926).

that the basis of the currency reform of 1924 was being abandoned. Exchange transactions now became criminal and counter-revolutionary. Much publicity was given in the press of May 6, 1926, to an announcement that three officials of Narkomfin had been shot, and some others condemned to imprisonment, for " speculating in gold, currency and state securities ", thus raising the demand for gold and foreign currency and adversely affecting the exchange. The process was completed by a decree of July 9, 1926, prohibiting the export of chervontsy : thereafter all chervontsy offered abroad were treated as contraband, and no further obligation accepted to redeem them.[1] This act formally sealed the abandonment of the short-lived attempt to maintain a Soviet currency based on gold and linked by its gold parity to the international monetary system.

Several morals could be drawn from the failure of this enterprise. The first was that the national finances were now strong enough, as they had not been in 1923 and 1924, to sustain the weight of a managed currency, and no longer required a gold backing to create confidence in its stability. The tradition of the flight from the ruble, endemic in the years of the great inflation, had been overcome, though it was significant that no official admission was ever forthcoming that the gold basis of the currency had been abandoned.[2] The second moral was that a fully planned economy, on which the Soviet régime was just about to embark, was incompatible with the submission of so vital an element in the economy as its currency and credit policy to the laws of the market : what had, in fact, led to the abandonment of the gold basis of the currency was the inability, so long as that basis was maintained, to drive forward the process of industrialization at a rate which seemed practicable and desirable to the directors of Soviet policy. Finance was no longer to play the rôle of a quasi-autonomous regulating factor in the economy, but was to become an instrument of policy in the hands of the planners. The third moral was the tenuous nature of the link between the Soviet economy and that of the capitalist world. Foreign trade played

---

[1] *Sobranie Zakonov, 1926*, No. 48, art. 348; A. Z. Arnold, *Banks, Credit and Money in Soviet Russia* (N.Y., 1937), p. 263.

[2] As late as October 1927 Sokolnikov asserted that " the system of gold circulation has been replaced by a system of gold guarantee " (G. Sokolnikov, *Finansovaya Politika Revolyutsii*, iii (1928), 290) — a gross travesty of the situation.

no great rôle in the Soviet economy ; and the rôle of trade with the Soviet Union in the world economy was altogether insignificant. The chervonets, for all its gold backing, had never really won for itself a place on the world money market : it had always been more convenient to conduct Soviet foreign trade in other currencies. Nor in 1926 was there any prospect of a change in this respect. The Dawes plan and its sequel had merely emphasized the financial isolation of the Soviet from the integrated capitalist world. The abandonment of the gold basis of the chervonets, coming at this moment, might serve as an unconscious reflexion or symbol of the movement towards socialism in one country.

# CHAPTER 10

# PLANNING

THE spring of 1924 was marked by a cautious advance in the party towards acceptance of the principle and practice of planning. The prospects of planning after Lenin's final collapse in the spring of 1923 seemed by no means reassuring. The loudly advertised association of Trotsky and of the opposition group of the 46 in the autumn of 1923 with the demand for more planning made it impossible for the triumvirate to espouse the same cause, especially in view of Lenin's well-known opposition to the more extreme ambitions of the planners.[1] But, as so often happened throughout this period, underlying economic forces continued to operate, and to impose on the leaders courses of action which they had condemned when originally proposed by others, so that the defeat of an opposition by no means always implied the ultimate rejection of its policy. The condemnation of Trotsky and of the opposition at the thirteenth party conference of January 1924 was unexpectedly followed by a series of measures which for the first time gave planning a central place in economic policy : the appointment of Tsyurupa as president of Gosplan with the concurrent appointment of deputy president of Sovnarkom ; the instruction of the central control commission and Rabkrin to Gosplan " to establish a general perspective plan of the economic activity of the USSR for a number of years (five or ten) " ; and above all the new attention now being given by the party to the revival of the metal industry.[2] All these steps betokened a conscious or unconscious change of attitude to planning.

Much, however, remained to be done before the conception of centralized planning was established and accepted. Down to this time, in spite of official pronouncements in favour of the " single

---

[1] See *The Bolshevik Revolution, 1917–1923*, Vol. 2, p. 376.
[2] For these measures see *The Interregnum, 1923–1924*, pp. 143-144.

economic plan ", the practice of planning consisted mainly in the drawing up of plans for particular industries by Vesenkha with the general advice and assistance of Gosplan. A first five-year plan for the metal industry, the prognostications of which were to be far exceeded, was drawn up in 1922–1923.[1] Such projects often reflected a sudden crisis in a particular industry or a crucial demand for its products, and recalled the appeals for *udarnichestvo* or " shock work " in the days of war communism : [2] in more orderly times, they were found to create confusion and one-sided development. The first general industrial plan prepared by Vesenkha in 1923 was no more than an attempt to amalgamate a group of such individual plans.[3] In the autumn of the same year Narkomzem, with encouragement from Gosplan, began to work on a five-year plan for agriculture, and a pamphlet under the title *Foundations of a Perspective Plan of the Development of Agriculture and Forestry* appeared at the end of 1924. But this was still little more than a series of specific plans — a plan for land survey and reclamation, a veterinary plan, a forestry plan and so forth — representing in statistical terms the desiderata which it might be hoped to achieve by 1928.[4] Krzhizhanovsky's complaint in the summer of 1924 that " after three years of work of Gosplan . . . we still lack the ' single economic plan ' " [5] was well justified ; and an instruction of the central control commission and Rabkrin that the planning functions of all other organs should be transferred to Gosplan, and that " the People's Commissariats, not merely their planning commissions, should be accountable to Gosplan " [6] was a dead letter.

Nevertheless several causes contributed to the rapid practical advances in planning made at this time. In the first place, the scissors crisis, by revealing the untoward consequences of exclusive reliance on the spontaneous working of the market, had made many unwilling, and sometimes unconscious, converts to

[1] *Sotsialisticheskoe Khozyaistvo*, No. 1, 1925, p. 82.
[2] See *The Bolshevik Revolution, 1917–1923*, Vol. 2, p. 217.
[3] *Planovoe Khozyaistvo*, No. 3, 1926, p. 91 ; see also an article by Krzhizhanovsky in *Ekonomicheskaya Zhizn'*, April 19, 1923, on *Gosplan After Two Years' Work*, and an interview with him in the same issue stressing the importance of the single economic plan.
[4] *Osnovy Perspektivnogo Plana Razvitiya Sel'skogo i Lesnogo Khozyaistva* (1924).    [5] G. Krzhizhanovsky, *Sochineniya*, ii (1934), 155.
[6] For this instruction see *The Interregnum, 1923–1924*, p. 143.

planning. The control of prices, wholesale and retail, was accepted as an empirical necessity, though its implications were not recognized, and even explicitly denied. Yet much could be said for the thesis that the adoption of price control represented an " ending of the economic retreat " and a " revision " of NEP.[1] Once the state intervened to alter the terms of trade by controlling prices, no sector of the economy could ultimately escape from its influence ; and, consciously or unconsciously, intervention must be governed and shaped by some broad view of ends to be pursued, in other words, by a general plan for the economy. Specific plans directed to particular objectives were no longer adequate or relevant. The interdependence of all sectors of the economy was once more demonstrated. It was a significant move when Gosplan decided in December 1923 to set up a " *Konjunktur* council ", with a trade section attached to it, to study the fluctuations and working of the market.[2] This now seemed the necessary starting-point of any serious attempt to establish a comprehensive planned control over the economy. Just as NEP, originally conceived as a licence to exchange products in market conditions, had gradually extended its sway, by a logical chain of development, over every branch of the economy, reshaping each in turn to the pattern of a free market, so now an apparently limited and purely empirical decision to re-establish the balance between town and country by controlling the prices of certain basic commodities led by a gradual and inevitable process to the extension of control to other sectors of the economy, and finally to the adoption of an all-

[1] See *The Interregnum, 1923–1924*, pp. 112-113.

[2] The decision was taken by the presidium of Gosplan on December 13, 1923 (G. Krzhizhanovsky, *Sochineniya*, ii (1934), 191). A report of Gosplan on the new institution throws light on the economic thought of the period : " Gosplan has recently created, by way of addition to its already functioning sections, special organs in the form of a *Konjunktur* Council with a trade section attached to it, and side by side with this has laid the foundation for an active control of the operations of our banks. We believe that the apparatus of Gosplan is now strong enough for us to regard the first stages of its work, which were naturally directed towards what is the basis of every economy, towards production, as already passed. Now we can undertake a far more active intervention in the distributive mechanism, in the processes of the circulation of goods. Here we approach the secrets of the whole monetary-capitalist system " (quoted in *Planovoe Khozyaistvo*, No. 4, 1926, pp. 14-15) ; the " *Konjunktur* institutions " of Gosplan were established " under the influence of the great sales crisis " of the autumn of 1923. At the end of 1924 regional *Konjunktur* sections had also been created (*ibid.* No. 4, 1926, pp. 59-60).

embracing plan. The five years that passed between the scissors crisis and the inception of the first Five-Year Plan covered the history of this process.

Secondly, it was no accident that renewed emphasis on planning should have coincided with the decision, also registered by the fourteenth party conference, to promote the metal industry " to the front rank " for attention and support, and should have been followed by the mandate to Dzerzhinsky to give effect to the decision.[1] Recognition that the expansion of heavy industry, of the production of the means of production, was a condition of the advance to socialism was a commonplace in all party discussions. The only alternative to this proposition was offered by those who believed that the Soviet Union should revert to the status of Tsarist Russia as a large exporter of grain and agricultural products and import its major requirements of industrial goods ; and this willingness to allow the country to become a " colonial depend- ency " of the capitalist world was never shared by any influential sector of party opinion. Yet the conditions of NEP did nothing to foster the expansion of heavy industry, and were indeed inimical to it. In the primitive and predominantly agricultural Soviet economy it was true, and long remained true, that the development of heavy industry meant planning, and that planning meant, first and foremost, the development of heavy industry. Heavy industry was the first, as agriculture was the last, section of the economy to become amenable to the procedures of planning.

A third factor which emerged in the middle nineteen-twenties to add fresh urgency to planning was the persistence of mass un- employment and the recognition of a rapid population increase as one of its determining causes. An unemployment crisis had swept over much of the capitalist world in 1920 and 1921 ; but a recovery had set in with the so-called " stabilization " of capitalism in 1923 and the succeeding years. In Soviet Russia, unemployment had been endemic since the second year of NEP and was still steadily rising. About 1924 it began to be realized that this was a different phenomenon from that of industrial unemployment in developed capitalist countries. It was in that year that Preobrazhensky drew attention to " the colossal concealed unemployment in the

[1] See *The Interregnum, 1923-1924*, pp. 115, 143-144.

countryside ", and Rykov attributed the unemployment crisis to
" the influx from the village into the town ".[1]   So long as the natural
rate of population increase stood at nearly 2 per cent per annum,[2]
nothing but the traditional antidotes of war, famine and migration
could prevent the unlimited pressure of surplus labour power on
the industrial labour market.   When the country had turned its
back on war and civil war, and had recovered from the famine of
1921–1922, unemployment quickly became an acute problem.
Foreign outlets for emigration were closed.   Greater efficiency in
agriculture would on the whole mean the employment of less,
not more, man-power on the land.[3]   Migration from the land-
hungry provinces of European Russia to the relatively uncrowded
and uncultivated steppes of Asiatic Russia was tried, but required
too much capital to make it a practicable solution on any significant
scale.[4]   In 1924 and 1925 the drive for the expansion of industry,
now beginning to achieve spectacular results, presented the last
hopes of absorbing at any rate some part of the surplus ;  and this
laudable purpose seemed one of those which planning could be
invoked to serve.

In the winter of 1924–1925, as the machinery of planning was
being brought slowly into action, a new controversy arose, not a
controversy about the relative interests of agriculture and industry,
but a controversy among the planners themselves about fundamental methods.   Bolshevik theorists who had concerned themselves with future economic organization had always assumed that
the economic laws which governed capitalist society would have
no application to the new social order.   In his *Economics of the
Transition Period*, published in 1920, Bukharin proclaimed that
" the end of capitalist commodity society will be at the same time
the end of political economy ", since political economy was " the
science of a social economy based on the production of *commodities*,
i.e. the science of an *unorganized* national economy ".[5]   Trotsky
in the same year declared that " as time goes on, political economy

---

[1] See *The Interregnum, 1923–1924*, pp. 49, 144.

[2] See p. 269, note 1 above.

[3] Groman, in drawing attention to this point, pessimistically observed that
" ten Americas would be necessary to absorb this rapid accumulation of unused
labour forces " (*Planovoe Khozyaistvo*, No. 8, 1925, p. 129).

[4] See pp. 523-529 below.

[5] N. Bukharin, *Ekonomika Perekhodnogo Perioda* (1920), pp. 7-8.

will more and more have only historical significance " ; [1] and Preobrazhensky later spoke of the conflict between " the law of value " and " the element of planning ", and predicted that the law of value would die away with the transition to socialism.[2] But how, in the light of these preconceptions, was the task of the planners to be interpreted ? Were they, in computing the " control figures " which were to serve as a beacon light for socialist construction, to take as their starting-point the calculations and methods of the capitalist past ? This hypothesis appeared to admit that the laws of political economy had validity and significance for the building of socialism. Or were the planners to be guided solely by some inner vision of the potentialities of a socialist future ? This hypothesis appeared to treat planning as a matter of intuition rather than of science. Neither conclusion was free from embarrassment.

The controversy was brought into the open by two articles in the first two issues of the regular journal of Gosplan in January and February 1925 under the title *On Certain Regularities Empirically Discoverable in our National Economy*. They were the work of Groman, one of the most distinguished economists of Gosplan, who, by way of refuting the scepticism and indeterminism of the Rykov school, endeavoured to establish certain economic laws or " regularities " in the national economy which justified prediction about future trends. To understand these " regularities " and to extrapolate the data derived from them into the future development of the economy was the essence of planning :

> Even if [wrote Groman] we proceed at the very beginning to a conscious transformation of society, the methods and forms of such a transformation are dictated by the objective tendencies of development inherent in it.[3]

[1] Trotsky, *Sochineniya*, xii, 141.
[2] E. Preobrazhensky, *Novaya Ekonomika* (1926), pp. 28-29, 36-37. This continued to be orthodox doctrine for twenty years ; the standard text-book of the nineteen-twenties, Lapidus and Ostrovityanov, *Politicheskaya Ekonomika* (1928), started from the assumption that political economy and its laws related to the spontaneous workings of a capitalist economy and not to a planned economy, and that the " law of value " was in process of dying away.
[3] *Planovoe Khozyaistvo*, No. 1, 1925, pp. 88-101 ; No. 2, 1925, pp. 125-141. Groman spoiled a sensible argument by an eccentric attempt to demonstrate that agricultural and industrial output for the market had reverted in the

The challenge was quickly takeń up by those who were more ready to dwell on the " conscious " element in planning.  The economic journal associated with STO published an article inquiring what economic categories were held to justify the extrapolation of data from the pre-revolutionary economy into the economic curves of the revolutionary period.  Were the economic laws and categories of capitalism still to be accepted as valid ?  Did not the period of transition to socialism call for " a special system of economics " of its own ? [1]  In the journal of Gosplan another economist attacked Groman's diagnosis of " objective tendencies inherent in society " as reflecting " the views of the historical school ", and boldly declared that any attempt at planning had " to some extent the character of inner intuition ".[2]

The dispute between what came to be known respectively as the " genetic " and " teleological " conceptions of planning was not free from an element of unreality.  The resolution of the twelfth party congress drafted by Trotsky had recommended an approach to planning which combined " economic prediction and the instruction of the appropriate economic organs in regard to these or those phenomena which will inevitably or in all probability arise in a given economic situation " with " the maximum concretization of such prediction for separate branches of industry or regions, with model dated directives in regard to the measures necessary to turn the expected situation to advantage ".[3]  It was not denied by the " teleologists " that scientific prediction on the basis of ascertained facts was an essential part of planning : nobody insisted more vehemently than the " super-industrialist " Preobrazhensky on the economic laws whose consequences " are dictated to us with an externally compulsive force ", and in turn " dictate to us *inter alia* definite propositions for the alienation of the surplus product of the countryside for the purposes of expanded socialist reproduction ".[4]  It was not denied by the " geneticists " that conscious and purposeful direction of the economy was possible and necessary.  The discovery of a

recovery period to a pre-war ratio of 63 : 37 ; this exposed him later to the unjustified charge that he had treated this ratio as unalterable.

[1] *Ekonomicheskoe Obozrenie*, March 1925, pp. 63-71.
[2] *Planovoe Khozyaistvo*, No. 7, 1925, pp. 151-166.
[3] *VKP(B) v Rezolyutsiyakh* (1941), i, 479.
[4] *Bol'shevik*, No. 15-16, August 31, 1926, p. 73.

profound political antipathy between the two standpoints, the identification of the " genetic " approach with a Menshevik attitude to the revolution [1] and of the " teleological " approach with a Bolshevik attitude — all this still lay in the future. The issue partly turned on the character of the period. So long as policy was directed primarily to the task of recovery, to a restoration of a level of production and efficiency already attained in the past, the " genetic " approach satisfied practical requirements. But so soon as the period of recovery gave place to a period of fresh advance, the need for a " teleological " conception of planning became difficult to refute. It would be unfair and misleading to confuse the " geneticists " among the planners with the sceptics who questioned the practicability of planning. Even if they sometimes found themselves in apparent agreement with the sceptics in preaching caution to their bolder colleagues, they were committed more deeply than anyone to a belief in the validity of economic prediction as the basis of planning, which was precisely what the sceptics denied. It was only when the sceptics had been routed, and the main battle of planning won, that the rift between " geneticists " and " teleologists " became really acute.

A certain difference of emphasis between the two schools on concrete issues of policy did indeed make itself felt quite early. It was natural that the " geneticists " should be those who stressed most strongly the problems presented to the planners by the predominant influence of a backward agriculture on the economy, and the " teleologists " those who most loudly demanded priority for the expansion of industry. The report of Narkomzem on planning put the issue clearly and fairly :

> In a plan of development for agriculture, where a mass of decentralized, dispersed, isolated households exists, where the elemental factors of development have a predominant influence, the rôle of teleological constructions obviously diminishes and has a subordinate importance, and the rôle of genetic elements correspondingly increases.[2]

---

[1] Groman's Menshevik past was constantly brought up against him : in a speech of October 1924, Kamenev referred to " a little corner " of Mensheviks in Gosplan (L. Kamenev, *Stat'i i Rechi*, xi (1929), 202).

[2] *Osnovy Perspektivnogo Plana Razvitiya Sel'skogo i Lesnogo Khozyaistva* (1924), p. 6 ; for this report see p. 491 above.

And the spokesman of Narkomzem who later presented the report
to Gosplań explained :

> We considered that it was necessary to establish first of all
> the peculiarities of agriculture and the direction in which it is
> evolving in order to make clear to ourselves how it is practically
> possible to reconstruct it.  Therefore, without refusing to set
> definite goals for the reconstruction of agriculture, we kept
> these goals linked with the actual tendencies of agricultural
> evolution.  We do not aspire, and do not think it possible in
> five years, to bring about a complete revolution in agriculture
> or to realize within that time all the tasks that confront us.  In
> taking note of them as a goal, we at the same time investigate
> the evolutionary tendencies of agriculture and its concrete con-
> ditions and thus try to make clear the limits of that revolutionary
> action which the state power can successfully apply to it within
> the given period.  Therefore our method was basically genetic,
> though it also contained teleological elements.[1]

Dzerzhinsky, speaking for heavy industry at the fourteenth party
conference in April 1925, tactfully tried to bridge the gap in non-
theoretical terms :

> If we think that we can introduce communism by sitting in
> an office, surrounding ourselves with books and drawing up an
> ideal plan, we know for certain that with such a plan we shall
> fail.  Our plan is a process of bringing into the open the inter-
> connexions of our state industry and its parts with one another,
> and of each of these parts with our market, with those for whom
> we work, i.e. the peasantry. . . . This process is not yet per-
> fected, not yet completed ;  and for us, for heavy industry,
> which unites a few million workers as against the hundred
> millions of the peasantry, the task of bringing it into the open,
> in order to work out this plan which leads to communism — it
> is for that we must live, for that we must struggle, for that we
> must solve a whole series of problems.[2]

The question of the rate and extent to which pressure could
be applied to the economy as a whole, and to its predominant

---

[1] *Planovoe Khozyaistvo*, No. 8, 1925, pp. 100-101.  The speaker added,
however, at a later stage of the discussion : " Groman recommends us to study
the process of recovery and to frame our plan on the basis of the regularities of
this process.  But we cannot wait till these regularities finally become clear,
since we are bound in practice here and now to assist the process of recovery
and the process of development " (*ibid.* No. 8, 1925, p. 140).

[2] *Chetyrnadtsataya Konferentsiya Rossiiskoi Kommunisticheskoi Partii (Bol'-
shevikov)* (1925), p. 212.

agricultural sector, in the interest of investment in heavy industry underlay every political issue, and every theoretical controversy, about planning.

The summer of 1925 witnessed a great outburst of planning activity, reflecting the mood of optimism inspired by the rapid development of the past year in every branch of the economy. In February 1925, Narkomvnutorg established a trade planning commission (Vnutorgplan) to draw up annual and perspective plans for the internal trade of the Soviet Union.[1] In March 1925 the central statistical administration produced an abstract of a projected balance-sheet for the whole economy in the economic year 1923–1924.[2] In July 1925 the Narkomzem of the RSFSR presented its five-year plan to the presidium of Gosplan, and provoked a substantial debate on agricultural planning ; [3] later in the year, the Narkomzem of the Ukrainian SSR, not to be outdone, produced a seven-year plan for agriculture.[4] In July 1925, Vesenkha, advancing from its previous practice of framing plans for particular industries or regions, issued a comprehensive report on *Prospects of Industry for the Economic Year 1925–1926*.[5] On August 14, 1925, STO issued two decrees which gave the impression (though this turned out afterwards to be somewhat

[1] *Sobranie Zakonov, 1925*, No. 13, art. 106 ; " the drawing up of a general perspective plan of development of the trade turnover of the USSR and its coordination through Gosplan with the general plan of the national economy of the USSR " was one of the functions assigned to Narkomvnutorg in the decree establishing it nine months earlier (*Sobranie Uzakonenii, 1924*, No. 50, art. 473).

[2] *Ekonomicheskaya Zhizn'*, March 29, 1925 : this abstract was reviewed, belatedly and rather grudgingly, in *Planovoe Khozyaistvo*, No. 2, 1926, pp. 254–256. The instruction to prepare the balance-sheet by October 1, 1924, was given by STO on July 21, 1924 (S. G. Strumilin, *Ocherki Sovetskoi Ekonomiki* (1928), p. 311) ; the delay in carrying it out was a source of constant complaints by Gosplan. The full balance-sheet was eventually published as *Trudi Tsentral'-nogo Statisticheskogo Upravleniya*, xxix, in 1926, by which time interest in it had evaporated.

[3] For the plan see p. 491 above ; for the discussion reported in *Planovoe Khozyaistvo*, No. 8, 1925, pp. 100–140, see pp. 497–498 above.

[4] *Ibid.* No. 3, 1926, p. 23 ; G. Krzhizhanovsky, *Sochineniya*, ii (1934), 191. The plan was published under the title *Perspektivny Plan po Sel'skomu Khozyaistvu Lesostepi i Poles'ya Ukrainy* (Kharkov, 1925).

[5] The report was reviewed and summarized in *Planovoe Khozyaistvo*, No. 10, 1925, pp. 309–311 ; the document itself has not been available.

premature) of a whole-hearted conversion to the principle of planning. The first instructed Vesenkha and the other commissariats concerned with the development of industry to prepare and submit to Gosplan not later than September 5, 1925, " production and financial plans " on the basis of the general Gosplan figures for 1925–1926 ; these sectional and regional plans were then to be collated by Gosplan into a " general industrial plan ". The second decree looked forward to the next series of " control figures " for 1926–1927, which were to be drawn up by Gosplan not later than August 1, 1926.[1] Finally, on August 20, 1925, Gosplan submitted to STO, and published, its *Control Figures of the National Economy for the Year 1925–1926*, on which it had been at work since the beginning of the year.[2] The term " control figures " was borrowed by Gosplan from Narkomfin, which had used it for the preliminary estimates issued in advance of the formal budget.[3] The control figures of Gosplan bore the same relation to the plan of campaign of the national economy for the coming year as the control figures of Narkomfin had borne to the " firm " budget. As Smilga cautiously explained in submitting the figures to STO, they had no binding character. What Gosplan said to the departments concerned was : " Make your plans by taking our control figures into account ".[4] The control figures were in the main the work of the three leading economists of Gosplan, Groman, Strumilin and Bazarov. Since Groman was the leading exponent of the " genetic ", and Strumilin of the " teleological ", approach, it is clear that the difference between the two views was not at this time a barrier to agreement.[5] The opposition,

[1] *Sobranie Zakonov, 1925*, No. 56, arts. 422, 423.

[2] *Planovoe Khozyaistvo*, No. 5, 1926, p. 59 ; Krzhizhanovsky gave a forecast of the general form of the " control figures " to a session of the presidium of Gosplan on June 23, 1925, comparing them with the Goelro plan of electrification in 1920 (*ibid.* No. 7, 1925, pp. 9-28). The essence of the new plan was, however, as Krzhizhanovsky later explained, that it " combined productive technique, economic analysis and a financial programme " (G. Krzhizhanovsky, *Sochineniya*, ii (1934), 335-336).                         [3] See pp. 457-458 above.

[4] *Planovoe Khozyaistvo*, No. 8, 1925, p. 13.

[5] *Ibid.* No. 7, 1925, p. 105. Groman himself wrote of the control figures : " We united both elements — the elements of the objective tendency of development, and the teleological element, the goals which the state sets for itself. We said that the control figures were an organic synthesis of prediction of objective development and of consciousness of those goals which the state sets for itself ; we said that the statistical expression of economic processes is organically connected with a definite system of economic policy " (*ibid.* No. 5, 1926, p. 60).

which all planners, whatever their school, were fighting to overcome, came from the sceptics who more or less openly doubted
the practicability of planning.

The introduction to the slim volume of 96 pages in small
format, which contained the whole of the *Control Figures of the
National Economy for the Year 1925–1926*, explained that three
methods had been used in arriving at the figures. The first,
defined as the " method of dynamic coefficients ", was based on
the extrapolation into the future of certain statistical trends of the
years of recovery since 1921 ; it was a reasonable assumption that,
other things being equal, the forces making for recovery during
these years would continue to operate in the same proportions or in
proportions governed by the same laws of development. The
second method, that of " expert estimates ", was the utilization
of the calculations of officials and managers concerned in different
branches of production ;  this was a rough-and-ready empirical
method, the value of which depended entirely on the skill, knowledge and acumen of those consulted. The third method,
described as the " method of control confrontations with pre-war
data ", consisted in checking against corresponding pre-war figures
the results arrived at by the two first methods ;  this was a natural
expedient to adopt at a time when the attainment of pre-war
standards was still an ideal and a convenient measuring-rod,
though the authors of the plan apologetically explained that it was
to be treated " not as a standard, not as a model for perspective
calculations ".[1]  These introductory remarks were followed by a
section containing the totals of the current year and estimated
totals of 1925–1926 for agricultural and industrial production, for
the volume of trade, for price movements, for export and import,
for wages, for building and transport, for capital investment, for
monetary circulation and credit, and for the state credit. Then
came a section of policy recommendations of a somewhat general
character necessary for the realization of the plan. These were
intended as the foundation for an " operational " as opposed to a
" perspective " plan. They included " the maximum forcing of
export ", the lowering of prices, the raising of wages, the increase
of the number of horses and, especially, of tractors employed in

[1] *Kontrol'nye Tsifry Narodnogo Khozyaistva na 1925–1926 god* (1925),
pp. 9-15.

agriculture, and an expansion of money and credit.[1]   Finally, the second half of the volume was occupied by detailed tables of the figures on which these calculations rested.   All the basic figures of production were expressed in three denominations :   in pre-war ruble prices, in price-index rubles at contemporary prices, and in chervonets rubles at contemporary prices.

Judged in retrospect, the Gosplan control figures for 1925–1926 were a remarkable achievement, and were to a substantial extent borne out by results.   The figures of industrial production recorded at the end of the year [2] showed a slight excess over the Gosplan estimates in pre-war prices and a somewhat larger excess in chervonets rubles, reflecting the unexpected rise in prices. Agricultural production was also estimated with considerable accuracy ;   the prospects of the harvest could be gauged when the figures were compiled.   The three major errors, all on the side of undue optimism, admitted a year later by Gosplan were an under-estimate of " the difficulties connected with the realization " of the harvest ;   an " exaggerated figure of proposed currency emission " ; and an " exaggerated figure of exports and corresponding im-ports ".[3]   At the moment, however, so striking and important a novelty as the *Control Figures of the National Economy for the Year 1925–1926* was unlikely to escape criticism.   It had been produced by " the unaided resources of the workers of Gosplan " ;[4] the cooperation of other departments had not been secured — by whose fault, is not clear.   The time of its presentation turned out to be unfortunate.   When the figures were compiled in June and July, and even when they were submitted to STO on August 20, 1925, optimism and self-confidence still reigned supreme.[5]   Within the next few weeks, rain damaged the last stages of the harvest ; first reports of difficulties in the grain collection began to come in ;

---

[1] *Kontrol'nye Tsifry Narodnogo Khozyaistva na 1925–1926 god* (1925), pp. 15-46.

[2] *Kontrol'nye Tsifry Narodnogo Khozyaistva na 1926–1927 god* (1926), pp. 288-289.                                                            [3] *Ibid.* pp. 4-5.

[4] G. Krzhizhanovsky, *Sochineniya*, ii (1934), 336.

[5] J. M. Keynes, who was on a visit to the USSR when the control figures appeared, said in an interview : " If you have a good harvest two years running, the economic situation of Soviet Russia will change incomparably — so much so that the increase of wealth in Soviet Russia will be greater than the largest credit which you will obtain or can obtain abroad " (*Leningradskaya Pravda*, September 8, 1925).

and there was a sharp reaction against the generous flow of credit
which industry had enjoyed since the spring.[1]

These changes of fortune, and the controversies and recriminations provoked by them, created an atmosphere inimical to the
ambitious projects of Gosplan. The core of the opposition was to
be found in Narkomfin. Sokolnikov had apparently repented his
momentary conversion to planning in the spring,[2] and at a Narkomfin conference on September 10, 1925, delivered a massive attack
on the control figures.

> Gosplan has not succeeded [were his opening words] in its
> first attempt to furnish control figures in the strict sense of the
> term for the national economy of the USSR. The figures of
> Gosplan have only an *auxiliary* perspective significance, and
> cannot, of course, be accepted as a practical directive.

He was particularly indignant at the claim of Gosplan to trespass
on the preserves of Narkomfin. The size of the budget could not
be made to depend " on the volume of production or of its marketable part ". The currency issue might rise in the coming year to
1600 or 1650 million rubles, but not to 1900 millions, as Gosplan
predicted. Sokolnikov repeated his favourite argument that the
development of industry could be assured only through the
development of agriculture. Since foreign credits for the import
of machinery and equipment were unobtainable, the only practicable course was to aim at " the rapid development of agricultural
exports ".[3] A rather more technical article in the official journal
of Narkomfin attacked all three methods used by Gosplan in
computing its figures, and, as regards the third method (comparison with pre-war figures), remarked severely that " figures
calculated in this way cannot be satisfactory, especially in relation
to the budget and, *a fortiori*, to monetary circulation and credit ".[4]

The first set discussion of the control figures took place at a
meeting of STO under the presidency of Kamenev on September

[1] See pp. 291-295, 483 above.
[2] See p. 478 above ; the phrase which he had coined on that occasion (" to
plan is to dispose of reserves ") was twice repeated by him in the autumn of
1925 (G. Sokolnikov, *Finansovaya Politika Revolyutsii*, iii (1928), 33, 217), but
now with the implication that this made planning impracticable.
[3] *Ibid.* iii, 63-66. The statement *ibid.* iii, 347, that this speech was printed
in *Ekonomicheskaya Zhizn'*, September 24, 1925, is incorrect: it was the speech
of September 18 (see below) which appeared there.
[4] *Vestnik Finansov*, No. 9, September 1925, pp. 116-118.

18, 1925. Pyatakov, on behalf of Vesenkha, thought that the potential rate of industrial development had been understated and the amount of grain which could be brought to the market over-estimated. Svidersky, speaking for Narkomzem, and Sokolnikov, speaking for Narkomfin, both complained that too much attention had been given to industry and the interests of agriculture neglected. Sokolnikov was once more particularly hostile, describing the practical usefulness of the control figures as " minimal " and decrying the possibility of " an organized and planned utilization of all resources ", since " an enormous number of elements lie outside our planning will " — the usual argument of the sceptics. The Gosplan figures of currency emission were, he declared, " a formula of inflation ". Kamenev summed up cautiously, but not very favourably. He agreed with most other speakers in thinking the Gosplan control figures too optimistic. He also — a more puzzling charge — thought them too un-systematic : they offered " rows of figures " instead of " a system of figures ".[1] The meeting ended with the appointment of a commission under Kamenev's presidency to study the importance of the figures and draft an appropriate resolution. But this proved to be merely a polite way of shelving the discussion ; the commission never met.[2] One of the " Narkomfin professors ", referring a little sarcastically to this " new *Tableau Économique* ", expressed satisfaction that, as the result of the decision of STO,

[1] Accounts of this meeting appeared in *Pravda* and *Ekonomicheskaya Zhizn'*, September 24, 1925, and *Planovoe Khozyaistvo*, No. 2, 1926, pp. 31, 44 ; Kamenev's speech is in L. Kamenev, *Stat'i i Rechi*, xii (1926), 344-346. *Leningradskaya Pravda*, September 18, 1925 (the day of the meeting), published Kamenev's earlier speech of September 4 on the economic situation (see pp. 292, 299 above), with a note by Kamenev that he had quoted Gosplan figures as they were the only ones available, though they " probably contain mistakes " ; the campaign against the control figures had evidently been worked up in the intervening fortnight.

[2] Members of Gosplan continued to complain of this treatment (*Planovoe Khozyaistvo*, No. 2, 1926, pp. 31, 44, 84). Constant references occur in the literature of the period to the unwillingness of other departments to cooperate with Gosplan ; *Kontrol'nye Tsifry Narodnogo Khozyaistva na 1925-1926 god* (1925) was circulated with a note inviting economic departments and organs to send their corrections of any figures in the report, but none of them responded (*Planovoe Khozyaistvo*, No. 1, 1926, p. 40). Relations between Gosplan and the central statistical administration continued to be " abnormal " (*ibid.* No. 2, 1926, p. 57). Jealousy among the older commissariats of the pretensions of an upstart department certainly played a negative, though minor, rôle in the early history of planning.

the Gosplan figures had " lost their quality of ' control ' ", and
had been " recognized as merely a working hypothesis, the
statistical expression of which cannot bind departments in their
planning work ".[1]

A few days after the meeting of STO *Pravda* published in two
instalments a long article written by Trotsky, then on vacation in
the Caucasus, under the immediate impression of the publication
of the control figures. It bore the title *Towards Socialism or
Capitalism ?*, and opened with a dithyrambic eulogy of the " dry
columns of figures " drawn up by Gosplan, in which Trotsky
discerned " the glorious music of the rise of socialism ". Their
publication was an event which should be celebrated in the
Soviet calendar. The statisticians of Gosplan were not in the
position of astronomers who " try to grasp the dynamics of
processes completely outside their control ". They were the active
leaders of economic policy, for whom every figure was " not
merely a photograph, but a command ". The figures represented
a " dialectical coupling of theoretical precision with practical cir-
cumspection, i.e. of a calculation of objective conditions and
trends with a subjective definition of the tasks of the workers'
and peasants' state ".[2] In the autumn of 1925 the support of
Trotsky's lone voice was an embarrassment rather than an asset.
Praise from this quarter did nothing to commend the innovation
of the control figures to the party leadership.

External factors soon arose to encourage a revival of the
campaign against planning. Disillusionment over the results of
the harvest weakened a not very solid faith in the value of economic
forecasting. The crisis over the grain collection was used to discredit not merely particular estimates which had been based on
the assumption of falling grain prices after the good harvest, but
the whole principle of planning for a peasant economy.[3] Rykov
explained once more that, so long as the country was subject to

---

[1] *Ekonomicheskoe Obozrenie*, October 1925, p. 28.

[2] *Pravda*, September 20, 22, 1925 ; a passage in the first article states
that it was being written on August 28, just a week after the publication of the
figures. The articles were republished as a pamphlet, and an English translation appeared early in 1926 with a special preface, dated Kislovodsk, November
7, 1925. Stalin, a year later (*Sochineniya*, viii, 275-276), poked fun at Trotsky's
grandiloquent phrase about " the music of the rise of socialism ".

[3] This argument emerged from a leading article in *Ekonomicheskaya Zhizn'*,
October 1, 1925.

such crises as harvest failures " which bring with them the ruin
of millions of peasant households ", so long would deliberate
planning remain ineffective.   The disappointment over the
harvest of 1925 came as a reminder of " the lack of correspondence
between plans and the real processes of life ".[1]   Kamenev at a
party meeting pointed the same moral :

> Our plans for the development of industry, built on the
> calculation of a good harvest, have already undergone modifica-
> tions.   The peasant element has introduced a series of modifica-
> tions into our plans.

Kamenev concluded with a franker avowal than had hitherto been
made of the nature of the dilemma : " It turned out that our plan
and the peasant's understanding of his interests did not coincide ".[2]
A few days later Rykov developed the same theme.   " The plan of
Gosplan, Rykov and Kamenev " had been upset by " the plan of
the peasant household ".   Hitherto the state had extracted grain
from the peasant, first by requisitions, later by the agricultural
tax : now for the first time the peasant had " entered into economic
relations with the town and the factory as a kind of ' equal '
power ".   The result had been disconcerting :

> *We are facing the first test of a free economic exchange of*
> *goods between town and country, and at the moment we are not*
> *standing up to it very well.*[3]

Sokolnikov followed the same line :

> When it came to carrying out the plan of grain collection,
> the peasant plan spontaneously took the field against Gosplan.
> And, when the two plans clashed, Gosplan had to retreat before
> the peasant plan.[4]

As the credit crisis grew, the opposition of Narkomfin to Gosplan
and its control figures became more outspoken.   In November
1925, Sokolnikov made a speech at the Business Club which was

---

[1] *Izvestiya*, October 4, 1925 ;  *Ekonomicheskoe Obozrenie*, October 1925, p. 8.
[2] Speech of October 16, 1925, in *Pravda*, October 20, 1925.
[3] The speech was published (with a misprint in the date, which should be
October 22), in *Na Agrarnom Fronte*, No. 10, 1925, pp. 3-16.
[4] G. Sokolnikov, *Finansovaya Politika Revolyutsii*, iii (1928), 47 ; for this
article see p. 482, note 2 above.

described as " not merely against the control figures, but against planned economy in general ", and reverted to the old view that salvation could come only through " the greatest possible forcing of agriculture " in order to make large surpluses of agricultural products available for export.[1] All that could be expected of Gosplan, wrote Bronsky about the same time, was " an approximate explanation of the direction of development of the most important branches of the national economy " : this would facilitate " *an adaptation of our economic activity to the spontaneous forces of the national economy* ".[2] The antipathy of the financial authorities towards planning seemed to some to reflect bureaucratic exclusiveness ; according to Krzhizhanovsky, Sheinman, at this time president of Gosbank, looked on planning as " an encroachment on the independence of the financial organs ".[3] The People's Commissar for Agriculture of the RSFSR declared that the miscalculations of the grain collection had been due not to the supposed ill-will of the *kulak*, but to " paper plans out of touch with reality, too casually and hastily drawn up ".[4]

Throughout the autumn of 1925 the climate remained unpropitious to the cause of planning, and the control figures received little or no attention from the party leaders. When the party central committee met at the beginning of October 1925, Kamenev's report on the economic situation was shelved by the device of referring it to the Politburo,[5] and the controversial issues raised by Gosplan were avoided. A leading article in *Pravda* made it clear that the question of the control figures remained intact :

> The final assumptions of our calculations for the forthcoming year will probably become apparent in December, at the time of the party congress. For this reason the plenum of the central committee decided not to bind the party by any final appraisal of the economic situation, preferring to reserve this for the party congress and leaving in force for the present the general orientation expressed in the control figures.[6]

[1] *Planovoe Khozyaistvo*, No. 1, 1926, pp. 33-34.
[2] *Sotsialisticheskoe Khozyaistvo*, No. 5, 1925, pp. 25-26.
[3] G. Krzhizhanovsky, *Sochineniya*, ii (1934), 289, note 1.
[4] A. P. Smirnov in *Pravda*, December 22, 1925.
[5] See p. 306 above.          [6] *Pravda*, October 15, 1925.

But this could scarcely be regarded as a victory for planning. No " operational " plan based on the control figures had been adopted, or even considered. At the end of the calendar year neither Vesenkha's industrial plan, nor the budget, nor the credit plan, for 1925–1926 had yet been approved.[1] At the all-important fourteenth party congress in December 1925 Stalin's report contained no more than the conventional references to planning. The subject was once more kept in the background, and was not mentioned in the resolutions of the congress.

This silence notwithstanding, the fourteenth party congress was a decisive landmark in the progress of Soviet planning. The issue between the supporters and the antagonists of planning, between the enthusiasts and the sceptics, was the issue which underlay every economic, and almost every political, problem of the Soviet régime : the relation of industry to agriculture, of the state to the peasant. If the Soviet Union was to seek to advance along the line of peasant agriculture, developing agricultural exports and importing industrial products in the ultimate hope of a gradual and painless development of Soviet industry, then planning would remain an insignificant and ineffective factor : the " commanding heights " of the Soviet economy would be subject to the vagaries of the climate, of the powerful and unorganized individual peasant, and of the world market.[2] On this hypothesis, Sokolnikov and Rykov — to name only the most prominent and outspoken of the sceptics — were perfectly right. If, on the other hand, the Soviet Union was to seek to advance, through the intensive development of heavy industry, towards a position of self-sufficiency, both economic and military, and to regard the rate of this development as limited only by the extent of the strain which the proletariat, and, above all, the peasantry could be compelled to bear, then comprehensive planning became the fulcrum of all Soviet economic policy ; for, as had been shown over and over again, it was only in a planned economy that heavy industry — the production of the means of production

[1] *Planovoe Khozyaistvo*, No. 2, 1926, pp. 7-8.

[2] In a western country, as a Soviet commentator remarked, public control of heavy industry, transport and foreign trade, such as the Soviet Government exercised, would amount in practice to control of the whole economy ; in the USSR this did not hold good (*Vestnik Finansov*, No. 7, July 1925, p. 7).

— could be developed and fostered. The victory for planning was really won when the thirteenth party congress of May 1924 pronounced, through the mouth of Zinoviev, that the expansion of the metal industry was now a principal party objective. But the meaning of the decision was still not realized, and the desultory battle over planning continued to be fought with varying results over the next eighteen months. The victory was consummated — though once more without full understanding of what was involved — when the fourteenth party congress of December 1925, through the mouth of Stalin, proclaimed its intention " to make our country an economically independent country ", and " to preserve our country from economic dependence on the system of world capitalism ", and declared that it was " the fate of the countryside to march behind the town, behind heavy industry ".[1] It was no accident that Sokolnikov, whom Stalin attacked as the principal promoter of the policy of keeping Soviet Russia an agrarian country dependent on imports of industrial goods from abroad,[2] should also have been the principal enemy of planning. The fourteenth congress of December 1925 came to be known in party history as the " congress of industrialization ", while the fifteenth congress just two years later was the " congress of the five-year plan ". In fact, they marked successive stages of the same road. The decision to industrialize was the foundation on which the five-year plans rested.

It was some time before these consequences of the fourteenth party congress could make themselves felt. The first three months of 1926 were a period of continuing anxiety about the currency, of further attempts to restrict credit for the expansion of industry, and of muted interest in the advance of planning. It was no doubt by an unconscious stroke of irony that Sokolnikov, deposed in January 1926 from his office as People's Commissar for Finance, was appointed a deputy president of Gosplan.[3] But it showed that Gosplan was not yet thought of as one of the key positions in the economy. In February 1926, Gosplan submitted revised figures to take account of the less favourable economic prospect since the control figures had been drawn up six months earlier.

[1] Stalin, *Sochineniya*, vii, 299-300, 311.        [2] *Ibid.* vii, 354-356.
[3] The reshuffle of offices after the fourteenth party congress will be discussed in Part III in the following volume.

Opposition still centred round Narkomfin ; and Smilga, while deprecating the extreme view that the Soviet economy could already afford to defy economic laws, and while willing to admit that " Left " enthusiasts exaggerated the capacities of planning and ignored objective conditions, complained that " some elements of the state apparatus (and the apparatus is also in large measure a heritage of the old order) are against the plan in general ".[1] At the beginning of March 1926, Sokolnikov seized the occasion of an all-Union congress of planners for a further expression of scepticism about planning and about the potentialities of industrial expansion.[2]

What was officially described as " the first all-Union congress of planning organs ", central and regional, met in Moscow on March 10, 1926 ; its importance was marked by a leading article in *Pravda* on that day.[3] Rykov, more supple than Sokolnikov, appeared at the congress with a speech of welcome to the delegates which was in effect a somewhat naïve recantation of his earlier views. The country was entering " the so-called period of reconstruction " ; and " to realize the work of this period without a plan is, of course, impossible ".[4] Krzhizhanovsky, after giving some details of the organization of Gosplan, presented to the congress a simplified sketch of the process of planning as conceived in Gosplan at this time :

> The control figures of Gosplan begin . . . to play the rôle of a core round which planning work is organized. In future, of course, the procedure ought to be that the control figures of Gosplan anticipate the working out of annual operational plans by the departments in charge of economic business.
> The material thus worked over will return to Gosplan to be combined into a single annual operational economic plan for the country. The control figures in their turn will also

---

[1] *Planovoe Khozyaistvo*, No. 2, 1926, pp. 30, 41-42.

[2] G. Sokolnikov, *Finansovaya Politika Revolyutsii*, iii (1928), 69-81. The speech, delivered on March 12, 1926, is incorrectly described *ibid.* iii, 347 as a speech to the presidium of Gosplan.

[3] The stenographic record of the congress under the title *Problemy Planirovaniya : Itogi i Perspektivy* (1926) has not been available. But the congress was fairly fully reported in *Ekonomicheskaya Zhizn'*, March 11, 14, 16, 17, 18, 20, 1926, and in *Planovoe Khozyaistvo*, No. 4, 1926 ; Krzhizhanovsky's speech is reprinted in his collected works (*Sochineniya*, ii (1934), 286-301).

[4] *Planovoe Khozyaistvo*, No. 4, 1926, p. 7.

be subjected to revision, thus making a second appearance, no
longer as a preface, but as a conclusion. . . . The control
figures should finally be converted into the legal balance-sheet
of the annual operational plan, i.e. of the executive working
blue-print of economic activity.

This annual plan must, however, inevitably appear merely
as a section of the perspective plan which looks forward to a
cycle of work for, say, a five-year period. . . . The perspective
five-year plan will, in its turn, fall into its right place only if it is
accompanied by a properly drawn up general [operational] plan
for the national economy.

Krzhizhanovsky went on to divide the work of Gosplan into three
branches — " *a general plan, a perspective five-year plan, and
annual operational plans with a corresponding system of* control
figures ". Gosplan was at the moment " in the period of preparation of control figures for an economic perspective plan of the
country for the coming quinquennium ".[1]

More detailed reports followed. Strumilin brought " the five-
year perspective plan ", which was his special concern, into line
with new party and governmental policy and with the teleological
conception of planning :

We set ourselves as our fundamental task the industrialization of the country on the basis of electrification and the
expansion of the whole economy, so far as possible without
crises, and with an annual reinforcement of its socialist outposts
at the expense of a corresponding contraction of the elements
of a private economy.

It goes without saying that all these general regulative
principles and the planning directives which correspond to them
must find some kind of *statistical* expression in our five-year
plan, which represents in this way a systematic conspectus not
only of our forecasts, but of our prescriptions.

Strumilin went on to present detailed estimates, in rubles and in
percentages, of the development of industry, agriculture, transport
and construction for the five years from 1925-1926 to 1929-1930
inclusive — the first concrete five-year plan for the whole Soviet
economy.[2] This was followed by some discussion " whether such

[1] G. Krzhizhanovsky, *Sochineniya*, ii (1934), 286-301.
[2] *Planovoe Khozyaistvo*, No. 4, 1926, pp. 31-58.

five-year plans were necessary, whether it was not better to stick
simply to general plans ".[1]   In such a gathering the more ambitious
view was likely to prevail.   Groman presented a report on the work
of the *Konjunktur* department and complained of the inadequacy
of the statistical material.   In a further speech he drew attention
to what was to become a significant change in the history of
planning.   The control figures for 1925–1926 had been based on
the conception of " 100 per cent utilization of all existing productive forces " ;   the control figures for 1926–1927 would have
to take account of " the necessity of intensifying capital investments ".[2]   Hitherto it had been assumed that the task was to
rebuild up to the level already attained before 1914.   Now that
it was possible for the first time to think in terms of an advance
beyond that level, new questions of policy arose and planning
took on a new significance.   The issue of capital investment
emerged more clearly than ever as the crux of planning.[3]

The resolution of the party central committee in April 1926,
which sought to give effect to the decision of the fourteenth
party congress on the advance of industrialization, made a number
of pronouncements on planning.   It demanded " the reinforcement of the planning principle, and the introduction of planning
discipline into the activity of all state organs ", as well as " a
struggle for the suppression of separatism in planning and slovenli-
ness in the composition and execution of plans ".   Industrialization and planning were firmly marked as aspects of the same
process :

> In an increase of accumulation, in the practical utilization of
> accumulated resources, and in a realization, far more rigid than
> hitherto, of the planning principle, must be found the current
> tasks of the coming period of economic development.[4]

A few days later, at the session of TsIK, Rykov, in his new rôle
as the champion of planning, described it as " a mighty weapon

[1]  G. Krzhizhanovsky, *Sochineniya*, ii (1934), 337.
[2]  The report is in *Planovoe Khozyaistvo*, No. 4, 1925, pp. 59–79 ; the speech
in *Ekonomicheskaya Zhizn'*, March 17, 1926.
[3]  The unpublished memorandum of April 2, 1926, preserved in the Trotsky
archives (see p. 326 above), remarked that the chief defect of the 1925 control
figures was the absence of any estimate of actual or potential capital appreciation :  this would have to be remedied in the 1926 figures.
[4]  *VKP(B) v Rezolyutsiyakh* (1941), ii, 93.

for the practical and most effective utilization of our resources ".
He spoke of " the indispensability of planning discipline ", protested that sometimes one province set up a new saw-mill without
inquiring whether the neighbouring province did not already
possess a suitable one, and explained that " now, when our
economy has come up against an inadequacy of resources for
investment in industry and agriculture, every kopek spent un-
necessarily is a crime ".[1] This change of front indicated the
defeat of the long rearguard action fought by Narkomfin to
maintain the ascendancy of sound finance as the controlling lever
of the national economy. The factors which brought about a
sharp change in the official attitude towards planning were the
same factors which were soon to lead to the abandonment of the
uphill struggle to maintain the effective gold parity of the chervonets, and, by implication, to the relaxation of the restrictions on
credit for industry.[2] In pursuance of the decision of the fourteenth
congress, the expansion of industry, beginning with the production
of the means of production, and leading to the development of an
independent and self-sufficient national economy on a socialist
foundation, was henceforth the central focus of economic policy ;
and this carried with it a recognition of the supremacy of planning
over the forces of a " free " market and an " international "
currency. From the spring of 1926 Soviet economic policy was
firmly set on this path. The questions which still provoked
controversy were how rapidly, and by what means, the advance
could proceed along it.

By the middle nineteen-twenties " planning " had come to
connote the preparation of annual or quinquennial projects by
Gosplan covering the whole national economy. But the specific
" plans " out of which this comprehensive planning had arisen
were not forgotten. Though progress had at first been slow, by
the spring of 1926 almost the whole of the original Goelro plan
of electrification had been carried out. Five new generating
stations had been built in various parts of the Soviet Union,
and the Moscow station had been enlarged, accounting for a total

---

[1] *SSSR : Tsentral'nyi Ispolnitel'nyi Komitet 3 Sozyva : 2 Sessiya* (1926),
p. 10.                                        [2] See p. 487 above.

new production of 100,000 kw. of electric power.[1]  Of the original
Goelro projects, only the largest, the construction of a hydro-
electric station at Volkhovstroi in the neighbourhood of Leningrad,
with a capacity of 54,000 kw., was still outstanding.  The first
plans for this project had been drawn as long ago as 1913.  It had
been approved by the Provisional Government, and once again in
1918, before it was embodied in the Goelro plan of 1921.  Even
then shortage of funds postponed the beginning of work till 1924
when it benefited from the great industrial drive of that year.  By
September 1925, when it was officially inspected by a government
commission, the constructional work was virtually complete, the
electrical equipment had been received mainly from abroad (the
turbines and generators from Sweden), and the production of the
first current was scheduled for May or June 1926.[2]  This estimate
proved optimistic, and the Volkhovstroi station was eventually put
into service on December 19, 1926.[3]  As the first important
project in the Soviet Union for generating electricity from water-
power, it obtained wide publicity, and was hailed as a symbol of
the great industrial development of the future.  But, throughout
the nineteen-twenties, the Soviet Union continued to produce the
greater part of its electricity from coal (including lignite), wood
and peat.

Meanwhile other large-scale projects were being actively canvassed, the two most ambitious being the construction of a canal
to link the Volga with the Don and the building of a giant hydro-
electric station on the Dnieper (Dnieprostroi).  The Volga-Don

[1] *Planovoe Khozyaistvo*, No. 9, 1925, p. 18 ; G. Krzhizhanovsky, *Sochineniya*, ii (1934), 277-278.  According to later statistics, generating capacity grew
very slowly in the nineteen-twenties, though the amount of energy actually
generated increased somewhat faster :

|  | Generating Capacity (in millions of kilowatts) | Energy Generated (in thousands of millions of kilowatt hours) |
|---|---|---|
| 1921 | 1·2 | 0·5 |
| 1924 | 1·3 | 1·6 |
| 1925 | 1·4 | 2·9 |
| 1926 | 1·6 | 3·5 |

(*Sovetskoe Stroitel'stvo na 1935 g.* (1936), p. 97).
[2] *Leningradskaya Pravda*, September 24, 25, 1925.
[3] *Pravda*, December 19, 1926 ; *Malaya Sovetskaya Entsiklopediya*, ii (1934),
646-647.

canal had a long history behind it, having first been mooted in
the optimistic days of 1918.  An estimate by Gosplan in the spring
of 1925 put the cost at 140 million rubles ; and it was decided to
postpone the canal in favour of the more urgent Dnieprostroi
project.[1]  Interest in this was naturally strongest in the Ukraine.
The ninth All-Ukrainian Congress of Soviets in May 1925, which
was attended by Kamenev, made mention in its general resolution of " the task of posing the question of a rapid beginning of
work on the construction of the Dnieper hydro-electric station
and on the damming of the Dnieper ", and reverted to the project
in its special resolution on industrial development.[2]  In the late
summer of 1925 a technical commission visited the site ; and a
representative of Gosplan argued against further postponement of
the scheme.[3]  Trotsky wrote enthusiastically of " a combine uniting a powerful electrical station with a range of industries and
transport concerns needing cheap power ".[4]  It may well have
been Trotsky's advocacy which inspired Stalin to refer to the
project as late as April 1926 with mistrust and contempt.[5]  But
the real obstacle was lack of resources.  The scheme was too
ambitious to be undertaken at this stage without foreign technical
aid and foreign financial support.  It was not till 1927 that these
were eventually forthcoming, and the work begun.

The conception of planning in Russia had always been mixed
up with the conception of regionalism.  In the Witte period it
had been assumed that a reorganization of the Russian economy
would have, as one of its essential features, a new subdivision of
Russia into economic regions and a development of the special
economic resources and potentialities of different regions.  In the
Soviet period the advocates of planning were the foremost ex-
ponents of new schemes of regionalization.[6]  Though the theorists
of planning had from the first spoken in terms of " a single economic plan " for the whole country, planning in practice began
rather in the form of partial plans — not only plans for particular

[1] *Planovoe Khozyaistvo*, No. 4, 1925, pp. 315-316.
[2] For these resolutions see p. 270 note 3 above.
[3] *Izvestiya*, September 8, 1925 ; *Leningradskaya Pravda*, September 9, 1925.
[4] *Pravda*, September 22, 1925.                    [5] See p. 355 above.
[6] These will be discussed in Part IV in the following volume.

industries, but plans for particular regions.  This tendency was
facilitated by the constitutional division of the country into
national units, and further encouraged by the new process of
regionalization.  Between 1923 and 1925, the Ukrainian, Transcaucasian and White Russian SSRs, and finally the RSFSR,
established their own Gosplans, with varying degrees of autonomy
or subordination to the central Gosplan of the USSR.[1]  At the
beginning of 1926, the Gosplan of the RSFSR alone employed
950 workers and had an annual budget of over 2 million rubles.
Subordinate to it were three Gosplans of autonomous republics,
and 12 regional, 42 provincial and 43 departmental planning commissions.[2]  Of the newly created regions, the Ural region was not
only the first in the field, but the most advanced in its development of planning.  By the end of 1923 it had no less than three
planning organs at work — a central plan for the region, and
plans for the Perm and Tyumen provinces — and the problem
had arisen of fusing them into one.[3]  The Ukrainian, Transcaucasian and Ural regional Gosplans all published independent
plans for 1924–1925.[4]  A " Siberian section of Gosplan " was
planning the economic development of the " west Siberian
region " on the basis of a link between the coal of the Kuznets
basin and the industry of the Urals.[5]  Elsewhere the impetus to
regional planning had to come entirely from the centre.  In 1925
an expedition was sent from Moscow to investigate the natural

[1] As was to be expected, the Ukrainian Gosplan was the most active of these
organs ; its programme of work for 1923-1924 was confirmed by the central
Gosplan in December 1923 (*Ekonomicheskaya Zhizn'*, December 8, 1923).  An
account of the beginnings of planning in the Ukraine was given by the president
of the Ukrainian Vesenkha in December 1924 : " Under the influence of the
trade crisis of the autumn of 1923, the Vesenkha of the Ukrainian SSR set
itself the first task of seeking methods of planned forecasting for the prevention
of similar crises.  About the middle of 1924 this task was to some extent accomplished : we *created a general industrial plan*, which gave us the possibility to
put into operation a planned leadership of Ukrainian industry " (*ibid*. December
5, 1924).  A " plan " for Donugol, the Ukrainian coal trust, for 1923–1924,
which was over-fulfilled and led to an excessive accumulation of stocks, is also
mentioned (*Planovoe Khozyaistvo*, No. 4, 1925, p. 315).  The decree creating
the Gosplan of the RSFSR was dated February 13, 1925 (*Sobranie Uzakonenii*,
*1925*, No. 20, art. 140).
[2] G. Krzhizhanovsky, *Sochineniya*, ii (1934), 286.
[3] *Ekonomicheskaya Zhizn'*, December 11, 1923 ; the establishment of the
Ural region will be described in Part IV in the following volume.
[4] *Ekonomicheskoe Obozrenie*, March 1926, pp. 189-192.
[5] *Planovoe Khozyaistvo*, No. 10, 1925, p. 259.

resources and economic potentialities of the Karachaevo-Cher-
kassian autonomous region in the North Caucasus.[1]

In spite of the emphasis at this time on the regional aspects of
planning, Gosplan's first control figures for 1925–1926 were
entirely inspired by the conception of the " single economic plan "
for the whole USSR, contained no regional figures, and made no
attempt to break down totals into their regional components.
Even if such an analysis had been desired, adequate regional
statistics were not available, and those that were available, having
been compiled by different regional authorities on their own
initiative, can scarcely have been uniform or comparable. At the
conference of planning organs at Gosplan in March 1926, it was
decided to remedy this defect and include in the control figures
for 1926–1927 figures for union and autonomous republics and
for regions. But again the material was not available ; and, when
the control figures appeared in September 1926, they contained
regional figures only for two republics — the Ukrainian and White
Russian SSRs — and for two regions — the North-western region
(from which the Karelian autonomous region and the Murmansk
province were omitted for lack of figures) and the Ural region (for
which only summary figures were forthcoming).[2] The experiment was, however, not repeated in the following year. The
control figures for 1927–1928, while the introductory text contained a chapter on *The Economy of the Regions*, produced no
regional figures. Nor was this failure accidental. While regional
initiative played a notable part in the early history of Soviet
planning, and while the development of the specific resources of
different regions (especially those hitherto accounted backward)
continued to be an important aim of the planners, it became
obvious as time went on that planning was essentially a centralizing
factor in the direction of the economy. Planning meant, in the
last resort, the taking of major decisions of economic policy by a
single authority : local planning organs must submit their statistics
and their estimates in accordance with a uniform prescribed
pattern, and must be subordinate to the central organ in the
execution of policy. Since, moreover, the essence of Soviet

[1] Its report is in *Planovoe Khozyaistvo*, No. 1, 1926, pp. 288-309.
[2] *Kontrol'nye Tsifry Narodnogo Khozyaistva na 1926–1927 god* (1926),
pp. 223-224, 238-243 (Ukraine), 254-257 (White Russia), 258, 268-272 (North-
western region), 282 (Ural region).

planning lay in the expansion of industry, and provision of the capital necessary for such expansion was entirely dependent on decisions of policy taken in Moscow, the power of decision in the hands of regional planning authorities was restricted to matters of detail.  Regionalism in planning involved a measure of administrative devolution, and a particular attention to the development of the resources of backward regions.  But, even before the period of the first five-year plan was reached, the authority of Gosplan over local planning organs was absolute.  No substantial difference of status could be discerned between the Gosplans of the union and autonomous republics, or between the planning organs of the autonomous regions and regions: all were, in effect, local agents of the central Gosplan.

# NOTE A

## MIGRATION AND COLONIZATION

A BY-PRODUCT of the agrarian policy of these years was a revival of the process of internal migration and colonization which had been a striking feature of the last phase of Tsarist Russia. For two centuries before the emancipation of the serfs the expansion of Russia had followed the military pattern proper to a feudal order of society — occupation by military garrisons and the settlement of quasi-military Cossack colonies. The emancipation shattered the old order and, in giving fresh impetus to the traditional land hunger of the Russian peasant, opened the possibility of new outlets. For the first time the peasant was free to wander over the face of the vast Russian Empire in search of virgin soil to till, and for the first time the authorities were inclined to encourage him in the search. Here, as in the process of industrialization, the results of the emancipation did not develop fully till the last decade of the century. The years from 1861 to 1890 witnessed a movement of peasants, on a small scale and mainly unorganized, from the overcrowded central provinces, first to the Volga region and the northern Caucasus, later to Siberia. In the eighteen-eighties peasants were crossing the Urals into Siberia at the rate of 27,000 a year.[1] This was the beginning of a new movement. Hitherto, the European population of Asiatic Russia had consisted mainly of political or criminal deportees (of whom more than a million were said to have entered Siberia in the hundred years before 1914), agents, military and civilian, of the Russian Government, and merchants seeking to exploit the natural riches of the country and to profit by trade in primitive conditions with the native population. Now for the first time the foundations were laid of a considered policy designed, on the one hand, to relieve the population pressures of European Russia and, on the other, to open up for cultivation new and fertile tracts of land in Asia. This was the specifically Russian version of the expansion of Europe.

The policy came to a head when the construction of the Trans-Siberian Railway began in 1891 — a measure, like the whole policy of industrialization, dictated primarily by strategic motives and dependent for its inception on European example and European capital. In 1892

[1] Statistics quoted in G. von Mende, *Studien zur Kolonisation in der Sovetunion* (Breslau, 1933), p. 11 ; this is the best available study of the subject.

Witte formed a " committee for the Siberian railway ", which had
among its other functions that of promoting settlement and colonization in the territory to be served by the railway ; and in 1896 this
function had been far enough developed to be handed over to a special
department of the Ministry of the Interior.   From this time migration
to Siberia began to be organized as a large-scale operation receiving
substantial subsidies from public funds.   In the 20 years from 1885
to 1905 about 1,885,000 peasants migrated across the Urals.   In the
years from 1906 to 1913 the total number recorded was 3,274,000.[1]
The census of 1897 returned the population of Russia in Asia at
13,506,000 ; an official estimate of 1915 put it at 21,632,000.   Annual
expenditure, which started in the eighteen-nineties at 2·5 or 3 million
rubles, had risen by 1912 by 26 millions ; most of the funds were spent
on reclaiming land for settlement and on subsidies to the settlers.
Stolypin was an enthusiastic supporter of the scheme, which fitted in
well with his policy of basing the economy of rural Russia on prosperous
individual peasant agriculture.   In 1910, shortly before his assassina-
tion, he made an official visit to Siberia together with Krivoshein, the
Minister for Agriculture ;  and the account of the visit published on
their return is an important official *apologia* for the policy.   The conclusion had by this time been reached that " the emigration does not
cover half the natural increase in population ", and that, " however
seductive the idea may be of using migration to solve agrarian questions
in European Russia, one must completely renounce this idea ".   The
emphasis was now laid on the constructive purpose of developing
Asiatic Russia.[2]

The process of migration and colonization was primarily directed
to Siberia and inspired by the building of the railway.   Of the total
number of migrants between 1896 and 1914, two-and-a-quarter
millions settled in Siberia west of Lake Baikal and 350,000 in the Far
East, these two contingents amounting to over 70 per cent of the whole.
The remaining million migrants[3] established themselves in the fertile

[1] A convenient summary of this migration with tables of statistics is in
*Entsiklopedicheskii Slovar' Russkogo Bibliograficheskogo Instituta Granat*, xxxi
(2nd ed. 1933), 531-548.   During the period 1906-1914 more than 3,000,000
Russian subjects emigrated to North and South America :  a large majority of
these, unlike the migrants to Asiatic Russia, belonged to national minorities,
including Jews.

[2] P. A. Stolypin and A. V. Krivoshein, *Die Kolonisation Siberiens* (German
transl. 1912), pp. 99, 101.

[3] The figures are in G. von Mende, *Studien zur Kolonisation in der Sovetunion* (Breslau, 1933), p. 60, note 6.   Of the small number of migrants officially
recorded as settling in Turkestan, a high proportion went to Semirechia which,
though politically part of Turkestan down to 1924, belonged geographically to
the steppe region (the later Kazakhstan) ;  migration to Turkestan proper

regions of the Asian steppe hitherto occupied almost exclusively by nomad Kazakhs (or Kirgiz, as they were still at this time officially called).[1] These attempts were resisted, and presented an issue between immigrant Russian settlers and native nomads in possession which was defined in the report of Stolypin and Krivoshein with exemplary clarity :

> The essential is to organize not the Kirgiz themselves, but the Kirgiz steppe, and to think not of the future of individual nomads, but of the future of the whole steppe.[2]

The Kazakh revolt of 1916, which was officially attributed to resistance to the imposition of military conscription on the Kazakhs, was certainly not unconnected with the peasant invasion. Projects were also made to develop other areas of Asiatic Russia by migration from European Russia. Visits by Krivoshein to Turkestan and Transcaucasia for this purpose took place in 1912 and 1913 respectively.

The movements of population caused by the upheaval of the war left few lasting effects, though a certain number of prisoners of war transported to Asiatic Russia remained there and were ultimately absorbed into the population. The advent of the Soviet régime with its insistence on the self-determination of the lesser nationalities of the Russian Empire, and the unpopularity of everything that had been done under the Tsars, brought a strong reaction against colonizing policies in Asia ; and recent Russian settlers in Turkestan, Kazakhstan and the northern Caucasus were apparently driven out or in some cases murdered.[3] Belief that the land hunger of the peasant could be satisfied by distribution of the landlords' estates momentarily removed the impulse to further migration, even if the civil war had not intervened to make it impossible. The end of the civil war and the famine of 1921 once more set in motion the process of migration from Russia in Europe to Russia in Asia.[4] But these were fugitives

where pressure of existing population was unfavourable to settlement by Russian peasants, was negligible.

[1] For the nomenclature see *The Bolshevik Revolution, 1917–1923*, Vol. 1, p. 316, note 2.

[2] P. A. Stolypin and A. V. Krivoshein, *Die Kolonisation Sibiriens* (German transl. 1912), p. 112 ; a government circular of 1912 quoted in *Entsiklopedicheskii Slovar' Russkogo Bibliograficheskogo Instituta Granat*, xxxi (2nd ed. 1933), 532, describes the motive of migration to Kazakhstan as being " in order that the orthodox may have predominance . . . over the natives ".

[3] Sources quoted in G. von Mende, *Studien zur Kolonisation in der Sovetunion* (Breslau, 1933), pp. 32, 35-37.

[4] According to official figures quoted *ibid.* p. 30, the total of migrants for 1920 was 85,000 (including 59,000 to Siberia and 25,000 to Kazakhstan), and for 1921, 72,000 (including 52,000 to Siberia and 17,000 to Kazakhstan).

rather than settlers; and many of them later returned to their homes.[1]

While these chaotic movements flowed and ebbed, it was some time before organized migration could be undertaken. As early as 1922 an " Institute for Scientific Enquiry into State Colonization " (Goskolonit) was established.[2]  The agrarian code of the RSFSR of December 1922 contained several articles (arts. 222-226) designed to regulate migration from the over-populated to the unoccupied regions of the RSFSR, power being reserved to TsIK to declare particular areas open or closed to settlement.[3]  But migration was treated as a " free and voluntary " process undertaken at the cost of the migrant : only in exceptional cases of " forced " migration were state funds to be drawn on.  The first attempt at official organization of migration was outlined in Narkomzem's five-year plan for agriculture, which provided for the migration of 630,500 persons to unoccupied territories in the Volga and Ural regions and in Siberia in the five years beginning with 1923-1924.  The plan recognized for the first time that large funds would be necessary to finance the operation, and estimated for an expenditure of 26 million rubles in the five years.  But it proposed to derive only one-third of this sum from the state budget, the rest being raised by contributions from various sources, including levies on the population of the region from which the migrants came.[4]  These hopes proved fallacious.  In 1923-1924 the state advanced the trivial sum of 500,000 rubles to promote migration, and in 1924-1925 only 1,500,000 rubles ;[5] and no funds were available from other sources.  In these circumstances little could be done.  Out of 107,000 would-be migrants who registered in 1923-1924, 15,000 actually migrated, and virtually all of these were " voluntary " migrants travelling on their own initiative and their own resources.[6]

It was some time before this confusion of policies was cleared up. The well-to-do peasant who could help himself rarely desired to migrate, especially as he could claim no compensation for the land

[1] The " decolonization " of western Siberia resulting from the outflow of these returning refugees engaged the attention of the authorities, and a commission was sent to investigate it in 1923 (G. von Mende, *Studien zur Kolonisation in der Sovetunion* (Breslau, 1933), p. 30).

[2] For an account of its early work see *Trudy Gosudarstvennogo Kolonizatsionnogo Nauchno-Issledovatel'skogo Instituta*, i (1924), 299-341.

[3] *Sobranie Uzakonenii, 1922*, No. 68, art. 901.

[4] *Osnovy Perspektivnogo Plana Razvitiya Sel'skogo i Lesnogo Khozyaistva* (1924), pp. 59-61 (for this plan see p. 491 above) ; *Ekonomicheskoe Obozrenie*, No. 3, 1929, pp. 146-148.

[5] Unpublished report quoted in G. von Mende, *Studien zur Kolonisation in der Sovetunion* (Breslau, 1933), p. 35.

[6] *Ekonomicheskoe Obozrenie*, No. 3, 1929, pp. 146, 152-153.

abandoned by him.[1]  At the other end of the scale, the *batrak* usually
lacked the initiative to move at all ;  and the potential migrant was
generally the middle or relatively poor peasant[2] who could not move
without provision for his journey and resettlement being made by the
government.  In the summer of 1924 Goskolonit pessimistically recommended that " any extension of *the agricultural colonization of Siberia
should be abandoned as a current task of the immediate future* ", since the
accommodation of further migrants must depend on the development
of local resources.[3]  A decree of the RSFSR of August 7, 1924, warned
voluntary migrants that they would be exposed to great hardships on
reaching their destination and could not count on government help.
Two days later a further decree encouraged the formation of " migration societies " of would-be migrants on a basis of voluntary self-help.[4]
In the summer of 1925 the Siberian provinces of Omsk, Novo-
nikolaevsk, Tomsk, Irkutsk and Yeniseisk were declared open for free
migrants.[5]  But such measures touched only the fringes of the problem.
In 1924–1925 most of the small number of migrants were still
" voluntary ".[6]  Moreover, up to this time, the number of eastward
migrants was almost balanced by those who returned to the old home-
lands (the proportion of returners to migrants in 1924–1925 was 80
per cent) ;[7] " decolonization " was still proceeding nearly as fast as
colonization.  It was only from 1925–1926, when state funds were
available on a substantial scale, that the planned settlement of migrants
in the eastern borderlands and in Asia really became effective.

The first step towards a coherent migration policy seems to have
been a decree of STO of October 17, 1924, which also represented the
first intervention of the organs of the USSR in the question.  Hitherto
migration had been treated as an expedient to relieve population pressure in the hungry and overcrowded regions of European Russia.  The
new decree defined the purpose of migration as being to bring uncultivated land into production and so increase the agricultural and

---

[1] *Na Agrarnom Fronte*, No. 10, 1926, p. 73.

[2] Lenin, in *The Development of Capitalism in Russia*, published in 1899, noted
that " those who move out from the emigration regions are mainly peasants
*of the middle category*, and those who remain at home are mainly the extreme
groups of the peasants " (Lenin, *Sochineniya*, iii, 133).

[3] *Trudy Gosudarstvennogo Kolonizatsionnogo Nauchno-Issledovatel'skogo Instituta*, i (1924), 353-355.

[4] *Sobranie Uzakonenii, 1924*, No. 68. arts. 679, 681.

[5] *Sobranie Uzakonenii, 1925*, No. 49, art. 371.

[6] *Ekonomicheskoe Obozrenie*, No. 3, 1929, p. 153, gives a total of 111,000 for
this year, of whom 80 per cent were voluntary ; but the total is almost certainly
exaggerated.  A contemporary source counted 12,500 " planned " and 57,000
" voluntary " migrants (*Planovoe Khozyaistvo*, No. 12, 1925, pp. 232-233).

[7] *Ekonomicheskoe Obozrenie*, No. 3, 1929, p. 152.

industrial output of the country. Having this end in view, it set up a colonization committee attached to TsIK with the ambitious mandate to carry out the settlement of nomadic peoples, the settlement of migrants who had moved into unoccupied lands on their own responsibility, and finally the colonization by organized migration of lands still unoccupied.[1] Six months later an executive organ was set up in the form of an All-Union Migration Committee (Vsesoyuznyi Pereselencheskii Komitet or VPK) to draw up annual and perspective migration plans, and to superintend the movement and settlement of migrants. It consisted of 23 members, including representatives of the union republics, nominated by TsIK; and its working organ was a presidium of five.[2] This creation of institutions was accompanied by ambitious estimates of what could be immediately achieved. Early in 1925 Gosplan made plans for a migration of 130,000 peasants in the current year to the Volga region (50,000), to Siberia (50,000), and to the Far East (30,000); [3] and this target was raised to 165,000 after a discussion in the third Union Congress of Soviets in May of that year.[4] The seven-year agricultural plan of the Narkomzem of the Ukraine contained proposals for the resettlement in the southern Ukraine of 350,000 peasants from the northern Ukraine; and the Narkomzem of the RSFSR had a three-year plan to settle 800,000 persons in the Volga region, the North Caucasus, the Urals, Siberia and the Far East.[5] Later in the year a still more utopian plan looked forward to the settlement of 1,200,000 migrants in Siberia and an eventual settlement of four millions.[6] These were symptoms of the growing popularity of planning in administrative quarters and of the wave of optimism which swept over the country in the summer and autumn of 1925.

Under the impetus of such projects, and with the help of the newly created organs, the year 1925–1926 saw the first practical results of the

---

[1] Quoted by G. Cleinow, *Neu-Siberien* (1928), p. 261, from a special collection of decrees which has not been available; it is also quoted as a decisive pronouncement in *Ekonomicheskoe Obozrenie*, No. 3, 1929, p. 146. Its omission from the general collection of decrees was probably accidental.

[2] *Sobranie Zakonov, 1925*, No. 30, arts. 193, 194.

[3] *Planovoe Khozyaistvo*, No. 3, 1925, pp. 274.

[4] *Na Agrarnom Fronte*, No. 9, 1925, pp. 143-144; on the other hand, the official decree of July 6, 1925 (*Sobranie Uzakonenii, 1925*, No. 49, art. 371), reverted to the figure of 130,000.

[5] *Na Agrarnom Fronte*, No. 5-6, 1925, p. 92; for the report of the Narkomzem of the RSFSR submitting its plan to Gosplan see p. 497 above. For the Ukrainian plan see p. 499 above; plans for settlement in the steppe region of the southern Ukraine apparently foundered on the opposition of the local *kulaks*, who succeeded in renting, on short leases, the land originally designed for settlement of migrants (*Na Agrarnom Fronte*, No. 9, 1925, p. 18).

[6] *Ibid.* No. 12, 1924, p. 233.

policy inaugurated by the decree of October 1924. But the institution of managed migration and settlement raised many problems not susceptible of rapid solution. A decree of August 1925 recognized for the first time the need for organization, not so much at the centre as in the regions where the migrants were to be settled. District (*raion*) migration administrations were established at Rostov for the North Caucasian region, in Sverdlovsk for the Ural region, in Novosibirsk for the Siberian region, and in Khabarovsk for the Far Eastern region.[1] These administrations were formally responsible to the Narkomzem of the RSFSR. But, since more than 70 per cent of the costs of migration and settlement were borne on the budget of USSR,[2] the authority of VPK doubtless remained paramount. More realistic estimates also prevailed of the number of migrants who could be successfully settled. The decree of August 1925 fixed the number for the Siberian region at 35,000; and this was modified by a further decree of March 1926 which left the fixing of the total to the discretion of the Narkomzem of the union republic concerned (in virtually all cases, the RSFSR).[3] In fact, the number of migrants in 1925–1926 was said to have reached 120,000, of whom more than half were planned and aided ; the number of returners sank to 22,000.

The two motives which inspired official encouragement of migration — to relieve population pressures in central Russia and to develop productive resources in the eastern territories — were closely intertwined. By 1924 the effects of war and famine had been overcome, and population was again increasing throughout the country. No less than 1,600,000 persons were said to have moved from the country to the towns during that year ; [4] and the growth of unemployment made it impossible for the towns to absorb the surplus. A circular of the

[1] *Sobranie Zakonov, 1925*, No. 57, art. 453.
[2] *Ekonomicheskoe Obozrenie*, No. 3, 1929, p. 148.
[3] *Sobranie Zakonov, 1926*, No. 20, art. 153.
[4] *Na Agrarnom Fronte*, No. 5-6, 1925, p. 86. A table in *Osnovy Perspektivnogo Plana Razvitiya Sel'skogo i Lesnogo Khozyaistva* (1924), p. 24, showed a rapid increase of density of population between 1916 and 1923 in the western regions (i.e. White Russia), in the Ukraine, and in the north-eastern, north-western and central regions of the RSFSR ; population had been stationary in the Volga regions, doubtless owing to the famine of 1921–1922. A report of Goskolonit treated the rural population as liable to indefinite expansion, since it was limited only by " the physiological level of existence ". Using alternative calculations of four or five desyatins of land to every full agricultural worker, Goskolonit reached figures of 14 and 19 millions respectively of " surplus " population in the four central and western regions of European Russia in 1923 (*Trudy Gosudartsvennogo Kolonizatsionnogo Nauchno-Issledovatel'skogo Instituta*, iii (1926), 535-536, 549) ; the calculations have an artificial element about them (especially as non-agricultural earnings were apparently excluded), but indicate something of the magnitude of the problem.

Narkomzem of the RSFSR of March 1925 referred to pressure from
provincial authorities to drain off surplus population as a motive for
organized migration, though it added that more rational agricultural
methods were the best remedy against rural over-population.[1] About this
time the problem of surplus population began to figure in party and
Soviet resolutions.[2] The third Union Congress of Soviets in May
1925 decided, " in order to give the peasantry of regions with little
land the possibility of migrating to free lands ", to press on with the
preparation of such lands for settlement and to supply migrants with
the necessary inventory, " increasing the grant of state funds for these
purposes ".[3] In 1924–1925, 35 per cent of the migrants came from
the central region of the RSFSR and 22·6 per cent from the Middle
Volga region.[4] These proportions fell to 17·8 per cent and 14·5 per
cent in 1925–1926 and lower still in the following years : the decline
probably reflected the increasing absorption of the surplus population
of these regions in industrial development. The proportion of migrants
from the Ukraine was 17 per cent for each of these years and afterwards
rose steeply. The Ukraine, more than any other part of the USSR,
was a chronic sufferer from rural over-population, and constantly complained of neglect of Ukrainian needs in schemes of migration and
settlement.[5] In 1924–1925 the western region of the RSFSR and the
White Russian SSR accounted respectively for only 4·9 per cent and
5·7 per cent of the migrants ; in 1925–1926 for 14·2 per cent and 17 per
cent, these higher proportions being maintained or exceeded in subsequent years : the low figures for 1924–1925 are probably explained
by the late development of the organization in these regions, which
were over-populated and thinly industrialized.[6] No other region

[1] Quoted in G. von Mende, *Studien zur Kolonisation in der Sovetunion*
(Breslau, 1933), p. 37. [2] See p. 269, note 1 above.
[3] *Tretii S"ezd Sovetov SSSR : Postanovleniya* (1925), p. 26.
[4] These and the following percentages in this paragraph have been calculated
in G. von Mende, *Studien zur Kolonisation in der Sovetunion* (Breslau, 1933),
p. 38, from figures in *Statisticheskii Spravochnik SSSR za 1928 g.* (1929),
pp. 66-67.
[5] Grinko, the president of the Ukrainian Gosplan, asserted in March 1926
that " inter-republic migration is still almost completely unorganized " (*Ekonomicheskaya Zhizn'*, March 14, 1926) ; since virtually all lands available for
settlement were in the territory of the RSFSR, this implied that Ukrainians
were at a disadvantage. He added the not particularly relevant comparison
that, whereas the average number of inhabitants per 100 desyatins in the
USSR was 19, in the Ukraine it was 67. The same complaint was repeated at
the session of TsIK in the following month (*SSSR : Tsentral'nyi Ispolnitel'-
nyi Komitet 3 Sozyva : 2 Sessiya* (1926), pp. 468-469).
[6] A resolution of the TsIK of the White Russian SSR of October 31, 1925,
described " agrarian over-population " as the major evil from which the republic
was suffering (*Zbor Zakonau i Zahadau BSSR, 1925*, No. 48, art. 381) ; there

contributed significantly to the eastward flow of migrants.

While the pressure of surplus rural population was, in the middle nineteen-twenties, the main driving force which made migration policies urgent and popular, Marxist orthodoxy shunned the conception of " over-population ".[1]   Most official pronouncements on migration dwelt on the need to promote increased agricultural production ; and this theme was more intensively stressed as planning became more effective.   At the outset intention in this respect far outran performance. The initial decree of October 17, 1924, laid down the principle that settlement should be effected in regions where it might be expected to achieve the greatest productivity in the shortest time and at the lowest cost ;   this was mainly the quest for a quick and cheap solution of a pressing problem.   The dilemma was clearly put in the report of the Narkomzem of the RSFSR in the summer of 1925.   Since " colossal " means would be required for settlement of " the empty areas ", it was indispensable to settle migrants in the immediate future in already inhabited regions ;   but, on the other hand, these regions could not easily accommodate any large number of new settlers.[2]   Migrants who travelled on their own initiative created a particular problem :

> While the wave of migrants increases from year to year [wrote Gosplan in 1927], *the provision of a land fund for the migrants is relatively behindhand,* and the newly arrived households, being to a considerable extent obliged to settle in already populated villages, fall into the position of semi-proletarians, are frequently exploited by the local *kulaks,* and *in the end furnish substantial reserves of unemployed* to the Siberian towns.[3]

A visitor to Siberia in 1926 found many of the towns surrounded by colonies of new immigrants, who lived in caves or in wooden huts or barracks constructed by themselves, and worked for well-to-do local peasants.[4]   In certain areas new immigrants are said to have obtained land at the expense of earlier settlers, " particularly Old Believers ", who had been driven from their homes and compelled to settle further

---

was a constant flow of surplus population from White Russia to the Donbass " in search of employment " (*SSSR : Tsentral'nyi Ispolnitel'nyi Komitet 3 Sozyva : 3 Sessiya* (1927), 69).

[1] An article in *Bol'shevik,* No. 9-10, June 1, 1925, pp. 81-94, took Lubny-Gertsyk and the other " professors " in Goskolonit to task for exaggerating the problem of rural over-population, but admitted that it existed, and could be solved only by industrialization.

[2] *Planovoe Khozyaistvo,* No. 8, 1925, p. 113 (for this report see p. 498 above).

[3] *Kontrol'nye Tsifry Narodnogo Khozyaistva SSSR na 1927–1928 god* (1928), p. 432.          [4] G. Cleinow, *Neu-Siberien* (1928), pp. 267-268.

north in the frozen steppe (*taiga*).[1]  Siberia occupied pride of place among the regions receiving migrants : it accounted for 62·6 per cent of migrants in 1925 and 55·2 per cent in 1926.[2]  Of the 1925 migrants, only 4·9 per cent settled in the Far Eastern region, and 10·3 per cent in 1926.  These proportions rose in subsequent years when more funds were available.  No less than 21·8 per cent of the migrants in 1924–1925 entered Kazakhstan, apparently without official support. But their reception by the local population was hostile ; and in the three following years, when official control over migration had become partially effective, the net entry of migrants into Kazakhstan fell to insignificant dimensions.[3]  Kazakhstan remained virtually closed to immigration till 1929.  The Central Asian republics were, for the most part, closed territory ; extensive irrigation was required before more land could be brought into cultivation, and the fertile areas were already thickly settled.  Next to Siberia and the Far Eastern region the Volga region offered most land for settlement ; in 1925, 3·3 per cent of the migrants, and in 1926, 7·9 per cent, were settled there.  The North Caucasian region, which attracted some of the early migrants,[4] soon fell off as a receiving area, whether because the available land was used up, or because it had been occupied by *kulaks*.  It cannot be said that up to this time migration had made any great contribution either to the relief of over-population or to the expansion of agricultural production.  But by 1926 the chaotic and spontaneous movements of the early nineteen-twenties had been brought under control.  The machinery of organized migration and settlement had been established.  Thereafter the number of migrants gradually increased from year to year.

A by-product of these general migration schemes was a project for the settlement of Jews on the land.  A committee for this purpose was set up in 1923, but apparently achieved no results till, in the following year, an American Jewish organization established an American Jewish Joint Agricultural Corporation (" Agro-Joint ") to promote " the mass transfer to productive occupations " of as many as possible of the 2,700,000 Jews living in the Soviet Union.  An agreement was reached by which funds for this enterprise would be provided in equal proportions by Agro-Joint and by the Soviet authorities.  In the autumn of 1924 a " committee for the settlement on the land of Jewish toilers " (Komzet) was established by the presidium of the Soviet of Nationalities, and drew up a programme for the settlement of 100,000

---

[1] *Planovoe Khozyaistvo*, No. 12, 1925, pp. 232-233.

[2] For the analysis of the regions to which migrants went see *Statisticheskoe Obozrenie*, No. 5, 1930, p. 87.

[3] The figures of migrants and returners are in *Statisticheskii Spravochnik SSSR za 1928 g.* (1929), pp. 66-67.

[4] *Planovoe Khozyaistvo*, No. 10, 1925, p. 35.

Jewish families. Land was put at the disposal of the committee in the southern Ukraine and in the Crimea, with the promise of further allocations in the Volga region and in the North Caucasus.[1] In 1925, in spite of some local resistance, 100,000 Jews were in fact settled, and the number had risen to 250,000 by 1928, mainly in the Ukraine and in the Crimea. Settlement was almost exclusively in the form of kolkhozy; individual Jewish settlers were rare. The scheme had no political implications, though Petrovsky, the president of the Ukrainian Sovnarkom, went so far as to suggest to the ninth Ukrainian Congress of Soviets in May 1925 the creation of " separate Jewish districts or even a Jewish region ", and hopes were expressed elsewhere that the project might one day lead to the foundation of a Jewish Soviet republic.[2]

[1] The White Russian SSR seems to have been first in the field with decrees of July and October 1924 (*Zbor Zakonau i Zahadau BSSR, 1924*, No. 20, arts. 183, 184); but this haste indicated the extent of the Jewish problem in the republic rather than the availability of land or funds.

[2] The authorities for this episode are articles in *Na Agrarnom Fronte*, No. 5-6, 1925, pp. 112-122; *Vlast' Sovetov*, No. 14, April 15, 1925, p. 10; *American Jewish Year Book*, xxvii (1925), 58-62; xxviii (1926), 59, 77-81; *Universal Jewish Encyclopedia*, i (1939), 253-256; iii (1941), 291. For a decree of September 25, 1925 defining the powers of Komzet and authorizing it to establish subordinate committees attached to the TsIKs of union or autonomous republics, see *Sobranie Zakonov, 1925*, No. 69, art. 509 (amended by resolution of March 21, 1928—see *Sobranie Zakonov, 1928*, No. 21, art. 188.

## NOTE B

## THE BUDGETS OF THE REPUBLICS

A MINOR problem of budgetary policy was provided by the subordinate budgets of the constituent republics of the union.  The
financial relations between the USSR and the republics were determined with unusual precision by article 1 of the constitution of July 6,
1923, which included the following items among those falling within
the competence of the supreme organs of the union :

> The approval of a single state budget of the USSR, in which
> are incorporated the budgets of the union republics ;  the deter-
> mination of the general union taxes and revenues, and also of the
> deductions therefrom and additions thereto which are included in
> the budgets of the union republics ;  the authorization of supple-
> mentary taxes and levies forming part of the budgets of the union
> republics.
> The establishment of a single system of money and credit.

The result of these provisions was to confer unlimited financial and
fiscal powers on the organs of the USSR.  Narkomfin was a unified
commissariat, so that each constituent republic had its own subordinate
Narkomfin ;  but the republican Narkomfins would, in effect, be no
more than agencies of the central organ.  Since the Ukrainian and
Transcausian SSRs [1] had had their independent budgets before the
Soviet Union came into existence, this represented a formal curtailment of the previous powers of the republics.  But, since the budgets
of the period before 1923 had been largely fictitious, and the other
republics had in fact been financially dependent on the RSFSR, the
curtailment was more nominal than real.  The immediate sequel of the
new arrangements was a measure of almost complete financial centralization.  In the budget of 1923–1924 96 per cent of all revenues were
collected by the union, and 87 per cent of all expenditure was expenditure by the union.  The RSFSR collected 2·8 per cent and spent
10 per cent ;  the Ukrainian SSR collected 1 per cent and spent 2·4
per cent.[2]

---

[1] Presumably also the White Russian SSR, though evidence has not been
found on this point.

[2] *Sotsialisticheskoe Khozyaistvo*, No. 2, 1924, p. 6.

When Sokolnikov made his budget speech to TsIK in October 1924, the inconvenience of excessive centralization had been recognized ; and he was able, not only to present an optimistic and confident review of the finances of the union as a whole,[1] but to appear with the air of one making concessions to the ambitions of the republics for financial autonomy.  While it was " completely indispensable " to preserve the financial unity of the union, the example of *glavkizm* had shown that " not all centralization is useful ".  The purpose of the draft statute on budgetary rights which he presented to TsIK was to allow to each republic " a real deployment of those resources of which it disposes at a republican level for the better construction of its financial economy ".[2] Apart, however, from a convenient measure of decentralization, the concession to the republics was more apparent than real.  The republics handled less than 20 per cent of the total revenues of the union, and 22·5 per cent of the expenditure ; and none of them had a budget that balanced.  Even if this situation were remedied — as it was, in part, in later years — by the allocation of larger revenues to the republics and the increase of the share of the republics in the union budget, the principle of " the unity of the budget ", on which Sokolnikov insisted, was firmly anchored in the constitution.  The draft decree approved by Sovnarkom and now submitted to TsIK was designed to give effect to it.  In the debates in both chambers, only Skrypnik, speaking in the Council of Nationalities, made a serious protest, asking for a division of both direct and indirect taxes between the union and the republics in fixed proportions, and complaining that the republics were placed in the position of poor relations who would always have a deficit.[3]  The other spokesmen of the republics were for the most part content to plead for minor concessions and adjustments.

At the conclusion of the debates the statute on budgetary rights was adopted with a few verbal amendments.  Each union republic was to have its separate budget, prepared by its own Narkomfin and approved by its own central executive committee.  But these budgets were subject

[1] For this part of his speech see p. 458 above.
[2] *SSSR : Tsentral'nyi Ispolnitel'nyi Komitet 2 Sozyva : 2 Sessiya* (1924), p. 164 ; for the statute see pp. 459-460 above.
[3] *Ibid.* pp. 327-328.  When the draft decree was discussed a few weeks earlier in the TsIK of the RSFSR, it was pointed out that out of 600 million rubles collected on the territory of the RSFSR in direct taxation, only 130 millions would go to the budget of the RSFSR ; of non-tax revenue 90 millions would go to the RSFSR as against 150 millions to the union (*Vserossiiskii Tsentral'nyi Ispolnitel'nyi Komitet XI Sozyva : Vtoraya Sessiya* (1924), pp. 174-175).  The draft was also stated to have been discussed in the TsIK of the Ukrainian SSR (*SSSR : Tsentral'nyi Ispolnitel'nyi Komitet 2 Sozyva : 2 Sessiya* (1924), p. 163) ; but records of this have not been available.

to revision by the Sovnarkom [1] of the USSR, and were then incorporated in the budget of the USSR, of which they formed an integral part. On the revenue side, all indirect taxes entered the union budget. Direct taxes were equally fixed by the union, but a part of them could be handed over to the republics in one of two ways : either a percentage of the tax received was deducted and paid over to the republic, or a supplement was added to the tax and collected for the benefit of the republic. Non-tax revenues, mainly profits from public enterprises, were divided between the USSR and the republics according to the character of the enterprise concerned ; and the republics were authorized to raise certain dues and taxes on their own account. On the expenditure side, not only the cost of the republican administrations, but the cost of educational, cultural, health, labour and agricultural services fell mainly on the republics, as well as the cost of financing the autonomous republics and regions (which had budgets of their own) and the local Soviets and their executive committees (which at this time had no budgets). In theory the purpose was proclaimed of balancing each republican budget within the union budget. But, while decentralization in expenditure was convenient, the central authority was unlikely to relax its control of taxes which were, as one commentator explained, " not only sources of revenue but an instrument of the economic and social policy of the union ".[2] Though progress was made towards equilibrium in the republican budgets, the principle of " the unity of the budget ", exhibited primarily in the central control of revenue, was never seriously relaxed. Since currency and credit policy were also in the hands of the central organs, the USSR was and remained, in all essential financial relations, a highly centralized unitary state.

The budgetary system established by the decree of October 29, 1924, remained unchanged for two-and-a-half years. The subsidizing of the republican budgets from the budget of the USSR continued to arouse criticism from those who wished to increase the independence of the republics. The party central committee in January 1925 recommended " the closer participation of representatives of the union republics and regions " in the drawing up of the budget.[3] At the third Union Congress of Soviets in May 1925 Sokolnikov declared his intention in the next budget " to assure to the union republics firm sources

[1] Under art. 12 of the statute, the Sovnarkom of the republic concerned was allowed two weeks in which to consider the proposed revisions, after which they would presumably be carried over its head ; a complaint was recorded that " the budget of each union republic has to pass nine authorities before it is finally confirmed " (*Ekonomicheskoe Obozrenie*, November 1925, p. 15).

[2] *Ibid.* September 1925, p. 12.

[3] *VKP(B) v Rezolyutsiyakh* (1941), i, 634.

of revenue " in order that subsidies might be dispensed with.[1]  A
resolution of the congress confirmed this intention and demanded a
speedy delimitation of " property and enterprises " between the union
and the republics in order to " increase the volume of property and
enterprises reserved for the union republics ".[2]  Some attempt was
made by Narkomfin to carry this ruling into effect.  The budget
estimates for 1925–1926, submitted by Sokolnikov to Narkomfin in
November 1925, provided for expenditure to be balanced by receipts
in the budgets of the RSFSR and the Ukrainian SSR, while the poorer
White Russian Transcaucasian, Turkmen and Uzbek republics continued to cover their deficits by subsidies, though on a reduced scale ;
out of a total revenue of 3620 million rubles, 648 millions, or 17·5 per
cent, were allocated to the republics.[3]  These estimates proved, in the
sequel, too optimistic.  When the total budget was increased to 3900
million rubles, the proportion of receipts allocated to the republics was
also raised ; but of the republican budgets only the budget of the
RSFSR was made to balance without subsidy.  The final totals of
revenue and expenditure for the republics in the financial years 1924–
1925 and 1925–1926 were as follows (in millions of rubles) : [4]

|  | Revenue | | Expenditure | | Deficit | |
|---|---|---|---|---|---|---|
|  | 1924–5 | 1925–6 | 1924–5 | 1925–6 | 1924–5 | 1925–6 |
| RSFSR | 407·2 | 733·7 | 464·8 | 681·4 | 57·6 | — |
| Ukrainian SSR | 77·3 | 178·3 | 96·5 | 186·2 | 19·2 | 7·9 |
| White Russian SSR | 15·5 | 35·7 | 18·7 | 43·2 | 3·2 | 7·5 |
| Transcaucasian SFSR | 23·3 | 36·4 | 45·3 | 74·3 | 22·0 | 37·9 |
| Turkmen SSR | 4·5 | 4·7 | 7·7 | 15·3 | 3·2 | 10·6 |
| Uzbek SSR | 13·8 | 23·3 | 26·0 | 41·4 | 12·2 | 18·1 |
| Totals | 541·6 | 1012·1 | 659·0 | 1041·8 | 117·4 | 82·0 |

The significant features of this picture were the rapid increase in the
budgets of all the republics, and the increase in the deficits of all except

    [1] *Tretii S"ezd Sovetov SSSR* (1925), pp. 429–430 ; in the course of the
debate a Ukrainian delegate complained that " we scarcely have a budget ",
and that the Ukrainian Narkomfin " makes up the budget in the course of its
work " (*ibid.* p. 460).                    [2] *Id. : Postanovleniya* (1925), p. 31.
    [3] *Ekonomicheskaya Zhizn'*, November 14, 1925 ; it was announced that
Sovnarkom devoted two meetings on November 21 and 28, 1925, to an examination of the republican budgets (*ibid.* December 3, 1925).
    [4] *Kontrol'nye Tsifry Narodnogo Khozyaistva SSSR na 1927–1928 god*
(1928), pp. 554–555, 558–559.  Since the Turkmen and Uzbek SSRs were only
created in the course of the year 1924–1925, the figures of their budgets for
that year must be partly conjectural ; the surplus of the RSFSR for 1925–1926
was apparently credited to the joint budget under another heading and not used
to balance the deficits of the other republics.

the RSFSR and the Ukrainian SSR.  This spoke well for the material development of the union, and especially of its more backward territories.  But it also revealed the extent to which material development intensified the dependence of the smaller and weaker republics on the central authority.  The progress towards the regularization of the fiscal systems of the republics, which continued in subsequent years, was indicative of more efficient measures of devolution rather than of any relaxation of central control of fiscal policy.  Taxation was too burning a social issue, and too important an instrument of social policy, to be left to local initiative, except on the most limited scale ; and in this respect, as in others, the advent of planning proved a weighty factor on the side of central control.  On January 12, 1926, Sovnarkom instructed Narkomfin to prepare a draft of amendments to the statute of October 29, 1924, to take fuller account of the requirements of the union republics.[1]  At the session of TsIK in April 1926, Kuibyshev once more admitted in principle that " the budgetary rights and possibilities of the republics should be widened ".  But the question required further working out, and was postponed to the next session for " final examination and confirmation ".[2]

[1] *Vestnik Finansov*, No. 5-6, May-June 1926, pp. 221-223.

[2] *SSSR : Tsentral'nyi Ispolnitel'nyi Komitet 3 Sozyva : 2 Sessiya* (1926), pp. 96-97.  The budgets of autonomous republics had much the same relation to the budget of the union republic to which they belonged as had the budgets of the union republics to the budget of the USSR ; provincial and local finances, which were not included in the union or republican budgets (except in so far as they depended on subsidies), will be discussed in Part IV in the following volume.

# ADDENDA

P. 326, note 3

Dzerzhinsky in his speech at the committee openly coupled Trotsky and Kamenev as aiming at "the creation of a new platform which would have approximated to the replacement of the recent slogan 'Face to the countryside' by the slogan 'Fist to the countryside'" (F. Dzerzhinsky, *Izbrannye Proizvedeniya*, ii (1957), 259).

P. 336, note 5

Dzerzhinsky later revealed that this decision was taken against his vote and by a majority of one (F. Dzerzhinsky, *Izbrannye Proizvedeniya*, ii (1957), 266-267).

P. 355, note 2

Stalin in his speech at the Leningrad party meeting after the session again used the simile of a peasant who bought "an outsize gramophone" instead of repairing his plough, but without specific mention either of Dnieprostroi or of Trotsky (*Sochineniya*, viii, 130).

P. 399, note 3

For the whole passage see F. Dzerzhinsky, *Izbrannye Proizvedeniya*, ii (1957), 208-210 ; the speech was delivered on December 11, 1925, at the Moscow provincial party conference.

P. 440, note 2

Both are reprinted in F. Dzerzhinsky, *Izbrannye Proizvedeniya*, ii (1957), 169-172.

P. 441, note 1

Dzerzhinsky in his speech at the session fiercely attacked the inefficiency of the state and cooperative trading apparatus (F. Dzerzhinsky, *Izbrannye Proizvedeniya*, ii (1957), 263-264).

# LIST OF ABBREVIATIONS

Comintern = Kommunisticheskii Internatsional (Communist International).

Dobrokhim = Obshchestvo Druzei Khimicheskoi Oborony (Society of Friends of Chemical Defence).

Donbass = Donetskii Bassein (Donets Basin).

Elektrobank = Aktsionernyi Bank po Elektrifikatsii (Joint Stock Bank for Electrification).

Glavelektro = Glavnoe Upravlenie Elektricheskoi Promyshlennosti (Chief Administration of the Electrical Industry).

Glavmetal = Glavnoe Upravlenie Metallicheskoi Promyshlennosti (Chief Administration of the Metal Industry).

Goelro = Gosudarstvennaya Komissiya po Elektrifikatsii Rossii (State Commission for the Electrification of Russia).

Gosbank = Gosudarstvennyi Bank (State Bank).

Gosizdat = Gosudartsvennoe Izdatel'stvo (State Publishing House).

Goskolonit = Gosudarstvennyi Kolonizatsionnyi Nauchno - Issledovatel'skii Institut (State Institute for the Scientific Study of Colonization).

Gosplan = Gosudarstvennaya Obshcheplanovaya Komissiya (State General Planning Commission).

Gossel'sindikat = Gosudarstvennyi Sel'skokhozyaistvennyi Sindikat (State Agricultural Syndicate).

Kolkhoz = Kollektivnoe Khozyaistvo (Collective Farm).

Kombedy = Komitety Bednoty (Committees of Poor Peasants).

Komnezamozhi (KNS) = Komiteti Nezamozhikh Selyan (Ukrainian Committees of Poor Peasants).

Komsomol = Kommunisticheskii Soyuz Molodezhi (Communist League of Youth).

Komzet = Komitet po Zemel'nomu Ustroistvu Trudyashchikhsya Evreev (Committee for the Settlement on the Land of Jewish Toilers).

KPD = Kommunistische Partei Deutschlands (German Communist Party).

MOPR = Mezhdunarodnaya Organizatsiya Pomoshchi Bortsam Revolyutsii (International Association for Aid to Revolutionaries).

| | |
|---|---|
| Mosgorbank | = Moskovskii Gorodnyi Bank (Moscow Municipal Bank). |
| Narkomfin | = Narodnyi Komissariat Finansov (People's Commissariat of Finance). |
| Narkomindel | = Narodnyi Komissariat Inostrannykh Del (People's Commissariat of Foreign Affairs). |
| Narkompros | = Narodnyi Komissariat Prosveshcheniya (People's Commissariat of Education). |
| Narkomtorg | = Narodnyi Komissariat Torgovli (People's Commissariat of Trade). |
| Narkomtrud | = Narodnyi Komissariat Truda (People's Commissariat of Labour). |
| Narkomvnudel (NKVD) | = Narodnyi Komissariat Vnutrennykh Del (People's Commissariat of Internal Affairs). |
| Narkomvnutorg | = Narodnyi Komissariat Vnutrennei Torgovli (People's Commissariat of Internal Trade). |
| Narkomzem | = Narodnyi Komissariat Zemledeliya (People's Commissariat of Agriculture). |
| NEP | = Novaya Ekonomicheskaya Politika (New Economic Policy). |
| NOT | = Nauchnaya Organizatsiya Truda (Scientific Organization of Labour). |
| ODVF | = Obshchestvo Druzei Vozdushnogo Flota (Society of Friends of the Air Fleet). |
| OGPU | = Ob"edinennoe Gosudarstvennoe Politicheskoe Upravlenie (Unified State Political Administration). |
| OSO | = Obshchestvo Sodeistviya Oborone (Society for the Promotion of Defence). |
| Osvok | = Osoboe Soveshchanie po Vosstanovleniyu Osnovnogo Kapitala (Special Conference for the Restoration of Fixed Capital). |
| Polpred | = Polnomochnyi Predstavitel' (Plenipotentiary Representative). |
| Proletkult | = Organizatsiya Predstavitelei Proletarskogo Iskusstva (Organization of Representatives of Proletarian Art). |
| Prombank | = Torgovo-Promyshlennyi Bank (Bank of Industry and Trade). |
| Rabkor | = Rabochii Korrespondent (Worker Correspondent). |
| Rabkrin (RKI) | = Narodnyi Komissariat Rabochei i Krest'yanskoi Inspektsii (People's Commissariat of Workers' and Peasants' Inspection). |
| RKK | = Rastsenochno-Konfliktnye Komissii (Assessment and Conflict Commissions). |
| RKP(B) | = Rossiiskaya Kommunisticheskaya Partiya (Bol'shevikov) (Russian Communist Party (Bolsheviks)). |

| | |
|---|---|
| RSFSR | = Rossiiskaya Sotsialisticheskaya Federativnaya Sovet-skaya Respublika (Russian Socialist Federal Soviet Republic). |
| Sel'kor | = Sel'skii Korrespondent (Village Correspondent). |
| Sovkhoz | = Sovetskoe Khozyaistvo (Soviet Farm). |
| Sovnarkom | = Sovet Narodnykh Komissarov (Council of People's Commissars). |
| SR | = Sotsial-Revolyutsioner (Social-Revolutionary). |
| STO | = Sovet Truda i Oborony (Council of Labour and Defence). |
| Torgpred | = Torgovoe Predstavitel'stvo (Trade Delegation). |
| TOZ | = Tovarishchestvo dlya Obshchego Zemlepol'zovaniya (Association for Common Cultivation of Land). |
| Tsekombank | = Tsentral'nyi Bank dlya Kommunal'nogo Khozyaistva i Zhilishchnogo Stroitel'stva (Central Bank for Communal Services and Housing Construction). |
| Tsentrosel'bank | = Vsesoyuznyi Tsentral'nyi Sel'skokhozyaistvennyi Bank (All-Union Central Agricultural Bank). |
| Tsentrosoyuz | = Vserossiiskii Tsentral'nyi Soyuz Potrebitel'skikh Obshchestv (All-Russian Central Union of Consumers' Societies). |
| TsIK | = Tsentral'nyi Ispolnitel'nyi Komitet (Central Executive Committee). |
| VAPP | = Vserossiiskaya Assotsiatsiya Proletarskikh Pisatelei (All-Russian Association of Proletarian Writers). |
| Vesenkha | = Vysshii Sovet Narodnogo Khozyaistva (Supreme Council of National Economy). |
| VKP(B) | = Vsesoyuznaya Kommunisticheskaya Partiya (Bol'-shevikov) (All-Union Communist Party (Bolsheviks)). |
| Vneshtorg | = Narodnyi Komissariat Vneshnei Torgovli (People's Commissariat of Foreign Trade). |
| VPK | = Vsesoyuznyi Pereselencheskii Komitet (All-Union Migration Committee). |
| Vsekobank | = Vsesoyuznyi (Vserossiiskii) Kooperativnyi Bank (All-Union (All-Russian) Cooperative Bank). |
| Vserabotzemles | = Vserossiiskii Professional'nyi Soyuz Rabotnikov Zemli i Lesa (All-Russian Trade Union of Agricultural and Forestry Workers). |

## TABLE OF APPROXIMATE EQUIVALENTS

| | |
|---|---|
| 1 arshin | = 2 ft. 4 ins. |
| 1 chervonets (gold) | = 1 £ sterling (gold) |
| 1 desyatin | = 2·7 acres |
| 1 pud | = 36 lbs. |
| 1 verst | = ·66 mile |

# INDEX

Abortion, 29, 33
Academy of Sciences: All-Union [*formerly* Russian, *previously* Imperial], 122; Communist [*formerly* Socialist], 202, 206, 231 n.
Accumulation, Socialist, 202-208. *See also* Agrarian policy; Industrial policy; Labour policy
Adler, A., 145
Adler, V., 145 n.
Administrators, *see under* Intelligentsia
Agrarian policy: and thirteenth party congress, 189, 233-235, 240-241, 275-278, 282; and 1924 harvest failure, 189-201, 334, 336; and grain exports, 190, 193, 200 n., 210, 239, 291, 295, 313, 316, 349, 352; and credit, 190-191, 222, 268, 275-276, 279, 281, 294, 307, 328, 471-472, 474; and taxation, 191-192, 239, 247, 249-256, 261, 265, 268-271, 293-294, 308, 318-320, 326-327, 354-355; and prices, 191-195, 199, 244-245, 268, 272, 292-295, 306, 317; and scissors crisis, 191, 193, 206 n., 209, 223, 275, 292, 317; and fourteenth party congress, 194 n., 201, 253 n., 284 n., 286 n., 287 n., 301 n., 311-315, 319, 323; and agrarian discontent, 196-198; and *sel'kors*, 196-198, 242, 247, 259 n.; conflicting views on, 198-213, 240-247, 276-277, 283-290, 297-304, 309-311; and NEP, 202, 209, 213-214, 216, 219-220, 222, 241, 257 n., 263, 277, 280, 282-283, 287, 289, 295, 297-298, 321, 329; and socialist accumulation, 202-208, 245, 259-260, 315-317; and industry, 209, 218, 223, 241, 245-246, 261, 271, 295, 297, 315-317, 320; and land tenure, 209-225, 239, 247-249, 257, 268, pre-revolutionary, 210-212; and collective cultivation, 210-222, 239-242, 267, 272, 280-281, 312, 321-323; and size of holdings, 210,

212, 215, 226, 228-229, 239; and socialization of land, 212-214; five-year plan for, 213; and growth of rural capitalism, 222-236, 240, 243, 260-262, 265, 282, 329; and twelfth party congress, 234, 249 n., 251 n., 252; and redistribution of land, 239, 268, 274; dilemma of, 239-240, 315-316; and German Social-Democratic Party, 241; and individual peasant enterprise, 241, 282, 322; and rural over-population, 266-270, 272-273, 324, 361, 365-366; and rural industries, 272, 323; and *smenovekhovtsy*, 284 n., 300; and 1925 harvest, 290-293, 299, 306; and grain purchasing organizations, 293-297, 327-328; and marketing, 295-297, 315-316; and mechanization, 320-325; and electrification, 323. *See also* Industrial policy; Migration; Peasantry
Agro-Joint, *see* American Jewish Joint Agricultural Corporation
Aikhenvald, Yu., 65
Aksakov, I., 59 n.
All-Russian Association of Proletarian Writers (VAPP), 49, 51, 64
All-Russian Congress of Engineers, December 1924, 121
All-Russian Congress of Proletarian Writers, October 1920, 49
All-Russian Congress of Scientific Workers: first, November 1923, 121
All-Russian Textile Syndicate, 448
All-Russian Union of Industrial Co-operatives, 360
All-Union Communist Party (Bolsheviks) [*formerly* Russian Communist Party (Bolsheviks), *previously* Russian Social-Democratic Workers' Party]: name of, 16; second congress, 1903, 16; Bolshevik-Menshevik split, 16; westerners and easterners in, 16-21; tenth

541

201 ; and twelfth party congress, 155, 333-334 ; and "Leninism", 155, 303-305 ; and "Trotskyism", 155 ; as leader, 155-157 ; criticism of, 156-157 ; and Molotov, 156 ; and fourteenth party congress, 156-157 ; and opposition, 157, 298, 352 ; and Bukharin, 157, 170, 173, 209-210, 263 n., 301-304 ; as orator, 158 ; and German Independent Social-Democratic Party, 158 ; downfall, 158 ; and NEP,

195, 287, 300-301, 303 ; and *rabkors* and *sel'kors*, 196 n., 198 ; and Georgia, 199 ; and agrarian policy, 262, 274, 283, 285-287, 289 n., 290 n., 298, 300-304, 307, 334 ; and Krupskaya, 303 ; and socialism in one country, 304-305 ; and industrial policy, 333-334, 342, 345, 349, 351-352, 509 ; and labour policy, 366, 384, 386, 389, 412 ; and financial policy, 466 ; and planning, 509

Zoshchenko, M., 52

THE END